Edited by
C. N. R. Rao and A. K. Sood

Graphene

Related Titles

Martin, N., Nierengarten, J.-F. (eds.)

Supramolecular Chemistry of Fullerenes and Carbon Nanotubes

2012

ISBN: 978-3-527-32789-8

Jorio, A., Dresselhaus, M. S., Saito, R., Dresselhaus, G.

Raman Spectroscopy in Graphene Related Systems

2011

ISBN: 978-3-527-40811-5

Kumar, C. S. S. R. (ed.)

Carbon Nanomaterials

Series: Nanomaterials for the Life Sciences (Volume 9)

2011

ISBN: 978-3-527-32169-8

Saito, Y. (ed.)

Carbon Nanotube and Related Field Emitters

Fundamentals and Applications

2010

ISBN: 978-3-527-32734-8

Krüger, A.

Carbon Materials and Nanotechnology

2010

ISBN: 978-3-527-31803-2

Guldi, D. M., Martín, N. (eds.)

Carbon Nanotubes and Related Structures

Synthesis, Characterization, Functionalization, and Applications

2010

ISBN: 978-3-527-32406-4

Edited by C. N. R. Rao and A. K. Sood

Graphene

Synthesis, Properties, and Phenomena

WILEY-VCH Verlag GmbH & Co. KGaA

The Editors

Prof. Dr. C. N. R. Rao
Chemistry of Materials Unit
Jawaharlal Nehru Centre
Jakkur P.O.
Bangalore 560 064
India

Prof. Dr. A. K. Sood
Indian Institute of Science
Department of Physics
Bangalore 560 012
India

Library of Congress Card No.: applied for

British Library Cataloguing-in-Publication Data
A catalogue record for this book is available from the British Library.

Bibliographic information published by the Deutsche Nationalbibliothek
The Deutsche Nationalbibliothek lists this publication in the Deutsche Nationalbibliografie; detailed bibliographic data are available on the Internet at <http://dnb.d-nb.de>.

Print ISBN: 978-3-527-33258-8
ePDF ISBN: 978-3-527-65115-3
ePub ISBN: 978-3-527-65114-6
mobi ISBN: 978-3-527-65113-9
oBook ISBN: 978-3-527-65112-2

Cover Design Formgeber, Eppelheim
Typesetting Laserwords Private Limited, Chennai, India
Printing and Binding Markono Print Media Pte Ltd, Singapore

Contents

Preface

Graphene is a fascinating subject of recent origin, its first isolation being made possible through micromechanical cleavage of a graphite crystal. Since its discovery, graphene has caused great sensation because of its unusual electronic properties, and scientists from all over the world have been working on the varied facets of graphene. Thus, there has been much effort to synthesize both single-layer and few-layer graphenes by a number of methods. A variety of properties and phenomena have been investigated, and many of the studies have been directed toward understanding the physical and chemical properties of graphene. Raman spectroscopy has been particularly useful in unraveling various aspects of graphene. A graphene field-effect transistor, a basic building block of nanodevices, is a single-element laboratory to study electron–phonon interactions using Raman scattering. The low-frequency electrical noise or the flicker noise in graphene devices defines the figure of merit of a device and has contrasting behavior for single- and bilayer-graphene devices. Magnetic properties have been of equal interest with the indication that graphene may be ferromagnetic at room temperature, exhibiting magnetoresistance. Graphene nanoribbons have attracted attention because of their unique electronic structure and properties. Graphene also provides a playground for exploring many quantum field related phenomena such as Klein tunneling, antilocalization, zitterbewegung, vacuum collapse by Lorenz boost and so on. Suspended graphene devices have been used to study nanoscale electromechanics and quantum Hall effect.

A variety of applications of graphene have come to the fore. Its use in supercapacitors and batteries has been explored. Other properties of graphene, which are noteworthy, are those that enable its use in nanoelectronics, field emission and catalysis. Biological aspects of graphene have been investigated by a number of workers, with emphasis on its toxicity and its possible use for drug delivery.

In this book, we have tried to cover many of the salient aspects of graphene, which are of current interest. Although the book mostly deals with graphene, we have included some material on graphene-like inorganic layered materials. It is possible, however, that some topics have been left out owing to constraints on the size of the book and possible errors in judgement. We trust that the book will be useful to students, teachers, and practitioners, and serves as an introduction to those who want to take part in the exciting developments of this subject.

June 2012

C.N.R. Rao
A.K. Sood

List of Contributors

Ganapathy Baskaran
The Institute of Mathematical Sciences
C.I.T. Campus
Chennai 600 113
Tamil Nadu
India

and

Perimeter Institute for Theoretical Physics
31 Caroline Street North
Waterloo
Ontario
Canada N2L 2Y5

Santanu Bhattacharya
Indian Institute of Science
Department of Organic Chemistry
Bangalore 560 012
Tala Marg
Karnataka
India

and

The Chemical Biology Unit
Jawaharlal Nehru Centre for Advanced Scientific Research
Bangalore 560 064
Jakkur
Karnataka
India

Biswanath Chakraborty
Indian Institute of Science
Department of Physics
Bangalore 560012
Karnataka
India

Mandar M. Deshmukh
Tata Institute of Fundamental Research
Department of Condensed Matter Physics and Materials Science
Homi Bhabha Road
Mumbai 400005
India

M. Samy El-Shall
Virginia Commonwealth University
Department of Chemistry
Richmond
VA 23284-2006
USA

and

King Abdualziz University
Department of Chemistry
Jeddah 21589
Kingdom of Saudi Arabia

Toshiaki Enoki
Tokyo Institute of Technology
Department of Chemistry
Ookayama
Meguro-ku
Tokyo 152-8551
Japan

Arindam Ghosh
Indian Institute of Science
Department of Physics
C.V. Raman Avenue
Bangalore 560012
Karnataka
India

Srijit Goswami
Indian Institute of Science
Department of Physics
C.V. Raman Avenue
Bangalore 560012
Karnataka
India

Vidya Kochat
Indian Institute of Science
Department of Physics
C.V. Raman Avenue
Bangalore 560012
Karnataka
India

Manzoor Koyakutty
Amrita Vishwa Vidyapeetham
University
Amrita Centre for Nanosciences
and Molecular Medicine
Elamakkara
Kochi 682041
Kerala
India

Urmimala Maitra
Jawaharlal Nehru Centre for
Advanced Scientific Research
International Centre for Materials
Science
Chemistry and Physics of
Materials Unit, and CSIR Centre
of Excellence in Chemistry
Jakkur P.O.
Bangalore 560 064
Karnataka
India

H. S. S. Ramakrishna Matte
Jawaharlal Nehru Centre for
Advanced Scientific Research
International Centre for Materials
Science
Chemistry and Physics of
Materials Unit, and CSIR Centre
of Excellence in Chemistry
Jakkur P.O.
Bangalore 560 064
Karnataka
India

Shantikumar Nair
Amrita Vishwa Vidyapeetham
University
Amrita Centre for Nanosciences
and Molecular Medicine
Elamakkara
Kochi 682041
Kerala
India

Atindra Nath Pal
Indian Institute of Science
Department of Physics
C.V. Raman Avenue
Bangalore 560012
Karnataka
India

Ganganahalli Kotturappa Ramesha
Indian Institute of Science
Department of Inorganic and Physical Chemistry
Bangalore 560 012
Karnataka
India

C. N. R. Rao
Jawaharlal Nehru Centre for Advanced Scientific Research
International Centre for Materials Science
Chemistry and Physics of Materials Unit, and CSIR Centre of Excellence in Chemistry
Jakkur P.O.
Bangalore 560 064
Karnataka
India

Suman K. Samanta
Indian Institute of Science
Department of Organic Chemistry
Bangalore 560 012
Tala Marg
Karnataka
India

Srinivasan Sampath
Indian Institute of Science
Department of Inorganic and Physical Chemistry
Bangalore 560 012
Karnataka
India

Biplab Sanyal
Uppsala University
Department of Physics and Astronomy
Box-516
75120 Uppsala
Sweden

Sharmila N. Shirodkar
Theoretical Sciences Unit
Jawaharlal Nehru Centre for Advanced Scientific Research
Srirampura Cross
Jakkur P.O.
Bangalore 560 064
Karnataka
India

Vibhor Singh
Tata Institute of Fundamental Research
Department of Condensed Matter Physics and Materials Science
Homi Bhabha Road
Mumbai 400005
India

A. K. Sood
Indian Institute of Science
Department of Physics
Bangalore 560012
Karnataka
India

Abhilash Sasidharan
Amrita Vishwa Vidyapeetham
University
Amrita Centre for Nanosciences
and Molecular Medicine
Elamakkara
Kochi 682041
Kerala
India

Umesh V. Waghmare
Theoretical Sciences Unit
Jawaharlal Nehru Centre for
Advanced Scientific Research
Srirampura Cross
Jakkur P.O.
Bangalore 560 064
Karnataka
India

1 Synthesis, Characterization, and Selected Properties of Graphene

C. N. R. Rao, Urmimala Maitra, and H. S. S. Ramakrishna Matte

1.1 Introduction

Carbon nanotubes (CNTs) and graphene are two of the most studied materials today. Two-dimensional graphene has specially attracted a lot of attention because of its unique electrical properties such as very high carrier mobility [1–4], the quantum Hall effect at room temperature [2, 5], and ambipolar electric field effect along with ballistic conduction of charge carriers [1]. Some other properties of graphene that are equally interesting include its unexpectedly high absorption of white light [6], high elasticity [7], unusual magnetic properties [8, 9], high surface area [10], gas adsorption [11], and charge-transfer interactions with molecules [12, 13]. We discuss some of these aspects in this chapter. While graphene normally refers to a single layer of sp^2 bonded carbon atoms, there are important investigations on bi- and few-layered graphenes (FGs) as well. In the very first experimental study on graphene by Novoselov *et al.* [1, 2] in 2004, graphene was prepared by micromechanical cleavage from graphite flakes. Since then, there has been much progress in the synthesis of graphene and a number of methods have been devised to prepare high-quality single-layer graphenes (SLGs) and FGs, some of which are described in this chapter.

Characterization of graphene forms an important part of graphene research and involves measurements based on various microscopic and spectroscopic techniques. Characterization involves determination of the number of layers and the purity of sample in terms of absence or presence of defects. Optical contrast of graphene layers on different substrates is the most simple and effective method for the identification of the number of layers. This method is based on the contrast arising from the interference of the reflected light beams at the air-to-graphene, graphene-to-dielectric, and (in the case of thin dielectric films) dielectric-to-substrate interfaces [14]. SLG, bilayer-, and multiple-layer graphenes (<10 layers) on Si substrate with a 285 nm SiO_2 are differentiated using contrast spectra, generated from the reflection light of a white-light source (Figure 1.1a) [15]. A total color difference (TCD) method, based on a combination of the reflection spectrum calculation and the International Commission on Illumination (CIE) color space

Graphene: Synthesis, Properties, and Phenomena, First Edition. Edited by C. N. R. Rao and A. K. Sood.

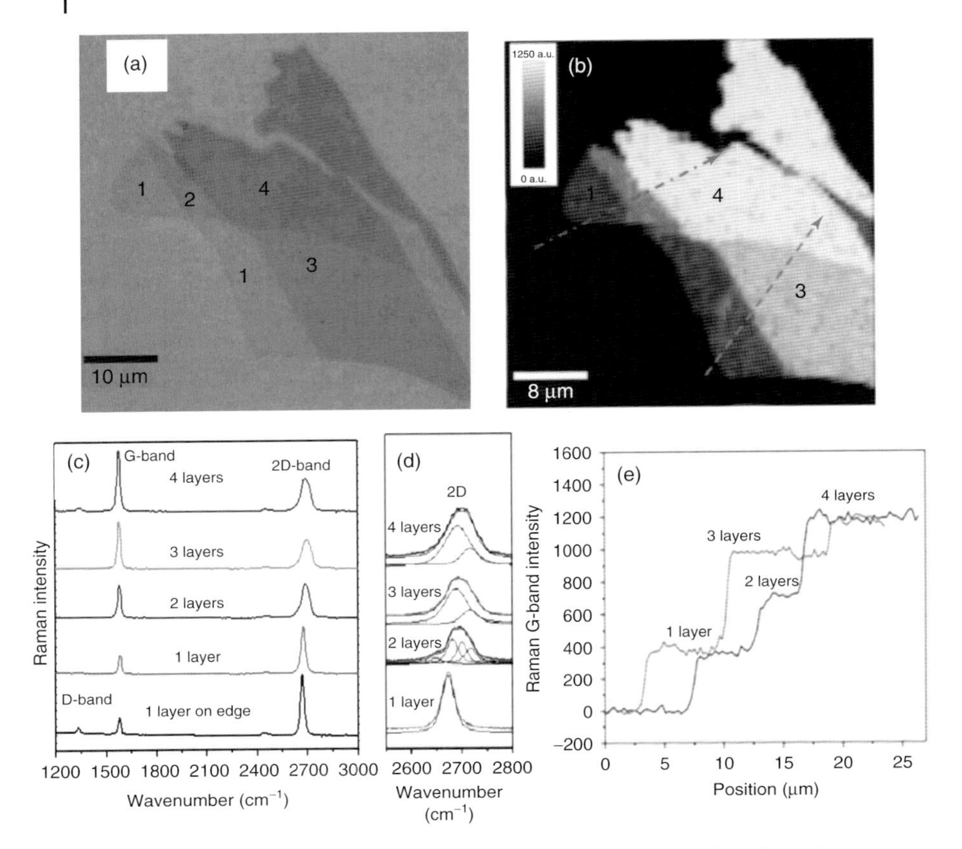

Figure 1.1 (a) Optical image of graphene with one, two, three, and four layers; (b) Raman image plotted by the intensity of G-band; (c) Raman spectra as a function of the number of layers; (d) zoom-in view of the Raman 2D-band; and (e) the cross section of the Raman image, which corresponds to the dashed lines in (b). (Source: Reprinted with permission from Ref. [15].)

is also used to quantitatively investigate the effect of light source and substrate on the optical imaging of graphene for determining the thickness of the flakes. It is found that 72 nm thick Al_2O_3 film is much better at characterizing graphene than SiO_2 and Si_3N_4 films [16].

Contrast in scanning electron microscopic (SEM) images is another way to determine the number of layers. The secondary electron intensity from the sample operating at low electron acceleration voltage has a linear relationship with the number of graphene layers (Figure 1.2a) [17]. A quantitative estimation of the layer thicknesses is obtained using attenuated secondary electrons emitted from the substrate with an in-column low-energy electron detector [18]. Transmission electron microscopy (TEM) can be directly used to observe the number of layers on viewing the edges of the sample, each layers corresponding to a dark line. Gass *et al.* [19] observed individual atoms in graphene by high-angle annular

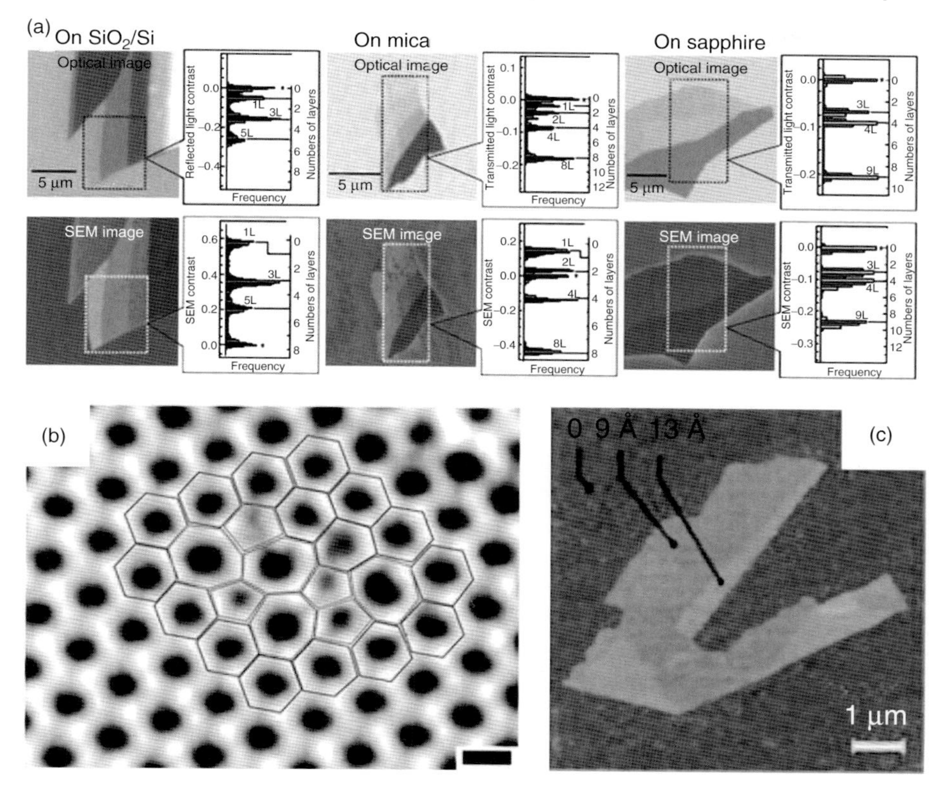

Figure 1.2 (a) Comparison of the counting of layers by optical microscopy and SEM for graphene on SiO_2/Si, mica, and sapphire. For each figure is shown a histogram of the distribution of graphene layers within the rectangular area indicated by a dotted line. (Source: Reprinted with permission from Ref. [17].) (b) High-resolution transmission electron microscopic image showing the Stone–Wales defects in graphene. (Source: Reprinted with permission from Ref. [20].) (c) Atomic force microscopic image of single-layered graphene. Folded edge shows a height increase of 4 Å indicating single-layer graphene. (Source: Reprinted with permission from Ref. [3].)

dark-field (HAADF) scanning transmission electron microscopy (STEM) in the aberration-corrected mode at an operation voltage of 100 kV. Direct visualization of defects in the graphene lattice, such as the Stone–Wales defect, has been possible by aberration-corrected TEM with monochromator (Figure 1.2b) [20]. Electron diffraction can be used for differentiating the single layer from multiple layers of graphene. In SLG, there is only the zero-order Laue zone in the reciprocal space, and the intensities of diffraction peaks do not therefore, change much with the incidence angle. In contrast, bilayer graphene exhibits changes in total intensity with different incidence angles. Thus, the weak monotonic variation in diffraction intensities with tilt angle is a reliable way to identify monolayer graphene [21]. The relative intensities of the electron diffraction pattern from the {2110} and {1100} planes can be used to determine the number of layers. If $I_{\{1100\}}/I_{\{2110\}}$ is

>1, it is reported as SLG, and if the ratio is <1, it is multilayer graphene [22]. Thickness of graphene layers can be directly probed by atomic force microscopy (AFM) in tapping mode. On the basis of the interlayer distance in graphite of 3.5 Å [3], the thickness of a graphene flake or the number of layers is determined as shown in Figure 1.2c [3]. Scanning tunneling microscopy (STM) also provides high-resolution images of graphene.

Raman spectroscopy has been extensively used as a nondestructive tool to probe the structural and electronic characteristics of graphene [3]. Figure 1.1c shows typical Raman spectra of one-, two-, three-, and four-layered graphene prepared using micromechanical cleavage technique and placed on SiO_2/Si substrate. The Raman spectrum of graphene has three major bands. The D-band located around 1300 cm^{-1} is a defect-induced band. The G-band located around 1580 cm^{-1} is due to in-plane vibrations of the sp^2 carbon atoms. The 2D-band around 2700 cm^{-1} results from a second-order process. The appearance of the D- and 2D-bands is related to the double resonance Raman scattering process [23], and with the increasing the number of layers, the 2D-band gets broadened and blue shifted. A sharp and symmetric 2D-band is found in the case of SLG as shown in Figure 1.1d. The Raman image obtained from the intensity of the G-band is shown in Figure 1.1b. A linear increase in the intensity profile of the G-band with increase in the number of layers along the dashed line is shown in Figure 1.1e [15]. Surface area, which also forms an important characteristic of graphene, is discussed later in the chapter.

1.2 Synthesis of Single-Layer and Few-Layered Graphenes

SLG and FG have been synthesized by several methods. In Table 1.1, we have listed some of these methods. The synthesis procedure can be broadly classified into exfoliation, chemical vapor deposition (CVD), arc discharge, and reduction of graphene oxide.

Table 1.1 Synthesis of single- and few-layered graphene.

Graphene synthesis	
Single layer	**Few layers**
Micromechanical cleavage of HOPG	Chemical reduction of exfoliated graphene oxide (2–6 layers)
CVD on metal surfaces	
Epitaxial growth on an insulator (SiC)	Thermal exfoliation of graphite oxide (2–7 layers)
Intercalation of graphite	Aerosol pyrolysis (2–40 layers)
Dispersion of graphite in water, NMP	
Reduction of single-layer graphene oxide	Arc discharge in presence of H_2 (2–4 layers)

1.2.1
Mechanical Exfoliation

Stacking of sheets in graphite is the result of overlap of partially filled p_z or π orbital perpendicular to the plane of the sheet (involving van der Waals forces). Exfoliation is the reverse of stacking; owing to the weak bonding and large lattice spacing in the perpendicular direction compared to the small lattice spacing and stronger bonding in the hexagonal lattice plane, it has been tempting to generate graphene sheets through exfoliation of graphite (EG). Graphene sheets of different thickness can indeed be obtained through mechanical exfoliation or by peeling off layers from graphitic materials such as highly ordered pyrolytic graphite (HOPG), single-crystal graphite, or natural graphite. Peeling and manipulation of graphene sheets have been achieved through AFM and STM tips [24–29]. Greater control over folding and unfolding could be achieved by modulating the distance or bias voltage between the tip and the sample [29]. Zhang [30] obtained 10–100 nm thick graphene sheets using graphite island attached to tip of micromachined Si cantilever to scan over SiO_2/Si surface. Folding and tearing of the sheets arise due to the formation of sp^3-like line defects in the sp^2 graphitic network, occurring preferentially along the symmetry axes of graphite.

Novoselov *et al.* [1] pressed patterned HOPG square meshes on a photo resist spun over a glass substrate followed by repeated peeling using scotch tape and then released the flakes so obtained in acetone. Some flakes got deposited on the SiO_2/Si wafer when dipped in the acetone dispersion. Using this method, atomically thin graphene sheets were obtained. This method was simplified to just peeling off of one or a few sheets of graphene using scotch tape and depositing them on SiO_2 (300 nm)/Si substrates. Although mechanical exfoliation produces graphene of the highest quality (with least defects), the method is limited due to low productivity. Chemical exfoliation, on the other hand, possesses the advantages of bulk-scale production.

1.2.2
Chemical Exfoliation

Chemical exfoliation is a two-step process. The first step is to increase the interlayer spacing, thereby reducing the interlayer van der Waals forces. This is achieved by intercalating graphene to prepare graphene-intercalated compounds (GICs) [21]. The GICs are then exfoliated into graphene with single to few layers by rapid heating or sonication. A classic example of chemical exfoliation is the generation of single-layer graphene oxide (SGO) prepared from graphite oxide by ultrasonication [31–36]. Graphene oxide (GO) is readily prepared by the Hummers method involving the oxidation of graphite with strong oxidizing agents such as $KMnO_4$ and $NaNO_3$ in H_2SO_4/H_3PO_4 [31, 33]. On oxidation, the interlayer spacing increases from 3.7 to 9.5 Å, and exfoliation resulting in SLG is achieved by simple ultrasonication in a DMF/water (9 : 1) (dimethyl formamide) mixture. The SGO so prepared has a high density of functional groups, and reduction needs to be carried out to obtain graphene-like properties. Chemical reduction has been achieved with hydrazine monohydrate to give well-dispersed SLG sheets [32, 35]. Thermal exfoliation and reduction of graphite oxide also produce good-quality graphene, generally referred to as reduced graphene oxide (RGO).

Rapid heating ($>200\,^{\circ}C\,min^{-1}$) to 1050 °C also breaks up functionalized GO into individual sheets through evolution of CO_2 [37, 38]. A statistical analysis by AFM has shown that 80% of the observed flakes are single sheets [38]. Exfoliation of commercial expandable graphite has also been carried out by heating at 1000 °C in forming gas for 60 s [39]. The resultant exfoliated graphite was reintercalated with oleum and tetrabutylammonium hydroxide (TBA). On sonication in a DMF solution of 1,2-distearoyl-*sn*-glycero-3-phosphoethanolamine-*N*-[methoxy-(polyethylene glycol)-5000] (DSPE-mPEG) for 60 min, the graphite-containing oleum and TBA get exfoliated to give a homogeneous suspension of SLG. These sheets can be made into large, transparent, conducting assembly in a layer-by-layer manner in organic solvents. On rapid heating, decomposition rate of the epoxy and hydroxyl groups of GO exceeds the diffusion rate of the evolved gases resulting in pressures that exceed the van der Waals forces holding the graphene sheets together and then exfoliation occurs. Exfoliated graphene sheets are highly wrinkled and have defects. As a result, these sheets do not collapse back to graphite but remain as highly agglomerated graphene sheets. Guoqing *et al.* [40] used microwaves to give thermal shock to acid-intercalated graphite oxide in order to carry out exfoliation. When irradiated in microwave oven, eddy currents are generated because of the stratified structure of GO, yielding high temperatures by Joule's heating. Decomposition and gasification of the intercalated acids in graphite leads to a sudden increase in interlayer spacing and thereby reduces van der Waals interaction. Further sonication yields SLG and FG sheets. Liang *et al.* [41] patterned FG on SiO_2/Si substrates using the electrostatic force of attraction between HOPG and the Si substrate. Laser exfoliation of HOPG has also been used to prepare FG, using a pulsed neodymium-doped yttrium aluminum garnet (Nd:YAG) laser [42]. The product depends on laser fluence, a fluence of $\sim 5.0\,J\,cm^{-2}$, yielding high-quality graphene with ultrathin morphology.

GICs can be prepared by the intercalation of alkali metal ions. Viculis *et al.* [43] prepared K-, Cs-, and NaK_2-intercalated graphite by reacting alkali metals with acid-intercalated exfoliated graphite in Pyrex sealed tubes. GICs were treated with ethanol causing a vigorous reaction to yield exfoliated FG. A schematic representation of the reaction is presented in Figure 1.3a. Potassium-intercalated GICs are also prepared using the ternary potassium salt $K(THF)_xC_{24}$, and they get readily exfoliated in *N*-methylpyrrolidone (NMP), yielding a dispersion of negatively charged SLG that can then be deposited onto any substrate [44].

Solution-phase EG in an organic solvent such as NMP results in high SLG yields [22]. In this case, the energy required to exfoliate graphene is balanced by the solvent–graphene interaction. Such solvent–graphene interactions are

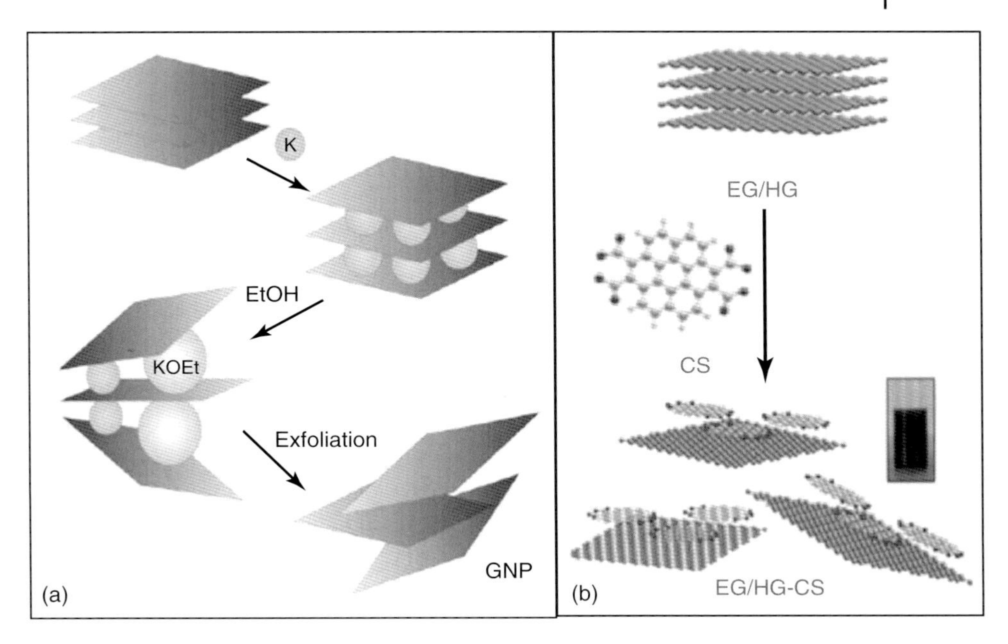

Figure 1.3 (a) Schematic diagram showing the intercalation of potassium between layers followed by violent reaction with alcohol to produce exfoliated ~30 layers of thin slabs of graphite. (Source: Reprinted with permission from Ref. [43].) (b) A schematic illustration of the exfoliation of few-layer graphene with coronene tetracarboxylate (CS) to yield monolayer graphene–CS composites. (Source: Reprinted with permission from Ref. [45].)

also used to disperse graphene in perfluorinated aromatic solvents [46], orthodichloro benzene [47], and even in low-boiling solvents such as chloroform and isopropanol [48]. Hernandez *et al.* [49] carried out a detailed study on dispersibility of graphene in 40 different solvents and proposed that good solvents for graphene are characterized by the Hildebrand and Hansen solubility parameters. Greater than 63% of observed flakes had less than five layers in most solvents. Direct exfoliation and noncovalent functionalization and solubilization of graphene in water are achieved using the potassium salt of coronene tetracarboxylic acid (CS) to yield monolayer graphene–CS composites (Figure 1.3b) [45]. Stable high-concentration suspensions of FG were obtained by direct sonication in ionic liquids [50]. Exfoliation, reintercalation, and expansion of graphite yields highly conducting graphene sheets suspended in organic solvents [39]. Gram quantities of SLG have been produced from ethanol and sodium [51]. Under solvothermal conditions, alcoholic solutions of the metal get saturated with the metal alkoxide, and at autogenerated pressures of around 10^{-2} bar, the free alcohol gets encapsulated into the metal alkoxide in a clathrate-like structure. This is then pyrolized to yield a fused array of graphene sheets, and sonicated to yield SLG.

1.2.3 Chemical Vapor Deposition

The most promising, inexpensive, and readily accessible approach for the deposition of reasonably high quality graphene is CVD onto transition-metal substrates such as Ni [52], Pd [53], Ru [54], Ir [55], and Cu [56]. The process is based on the carbon saturation of a transition metal on exposure to a hydrocarbon gas at high temperature. While cooling the substrate, the solubility of carbon in the transition metal decreases and a thin film of carbon is thought to precipitate from the surface [57]. Different hydrocarbons such as methane, ethylene, acetylene, and benzene were decomposed on various transition-metal substrates such as Ni, Cu, Co, Au, and Ru [57].

A radio frequency plasma-enhanced chemical vapor deposition (PECVD) system has been used to synthesize graphene on a variety of substrates such as Si, W, Mo, Zr, Ti, Hf, Nb, Ta, Cr, 304 stainless steel, SiO_2, and Al_2O_3. This method reduces energy consumption and prevents the formation of amorphous carbon or other types of unwanted products [58–60]. Graphene layers have been deposited on different transition-metal substrates by decomposing hydrocarbons such as methane, ethylene, acetylene, and benzene. The number of layers varies with the hydrocarbon and reaction parameters. Nickel and cobalt foils that measure $5 \times 5\ mm^2$ in area and 0.5 and 2 mm in thickness, respectively, have been used to carry out the CVD process at around 800–1000 °C; with nickel foil, CVD is carried out by passing methane (60–70 sccm) or ethylene (4–8 sccm) along with a high flow of hydrogen (around 500 sccm) at 1000 °C for 5–10 min. With benzene as the hydrocarbon source, benzene vapor diluted with argon and hydrogen was decomposed at 1000 °C for 5 min. On a cobalt foil, acetylene (4 sccm) and methane (65 sccm) were decomposed at 800 and 1000 °C, respectively. In all these experiments, the metal foils were cooled gradually after the decomposition. Figure 1.4 shows high-resolution TEM images of graphene sheets obtained by CVD on a nickel foil. Figure 1.4a shows graphenes obtained by the thermal decomposition of methane on the nickel foil, whereas Figure 1.4b shows graphene obtained by thermal decomposition of benzene. The insets in Figure 1.4a,b show selected area electron diffraction (SAED) patterns [61, 62]. All these graphene samples show G-band at 1580 cm^{-1} and 2D band around 2670 cm^{-1}, with a narrow line width of 30–40 cm^{-1}. Figure 1.4c (i,ii) shows the Raman spectra of the graphene samples in Figure 1.4a,b, respectively. The narrow line width and relatively high intensity of the 2D-band confirm that these Raman spectra correspond to graphenes having one to two layers [57]. Graphene obtained by CVD process can be transferred to other substrates by etching the underlying transition metal and can be transformed into any arbitrary substrate.

1.2.4 Arc Discharge

Synthesis of graphene by the arc evaporation of graphite in the presence of hydrogen has been reported [61, 63]. This procedure yields graphene arc discharge graphene in H_2 atmosphere (HG) sheets with two to three layers having flake

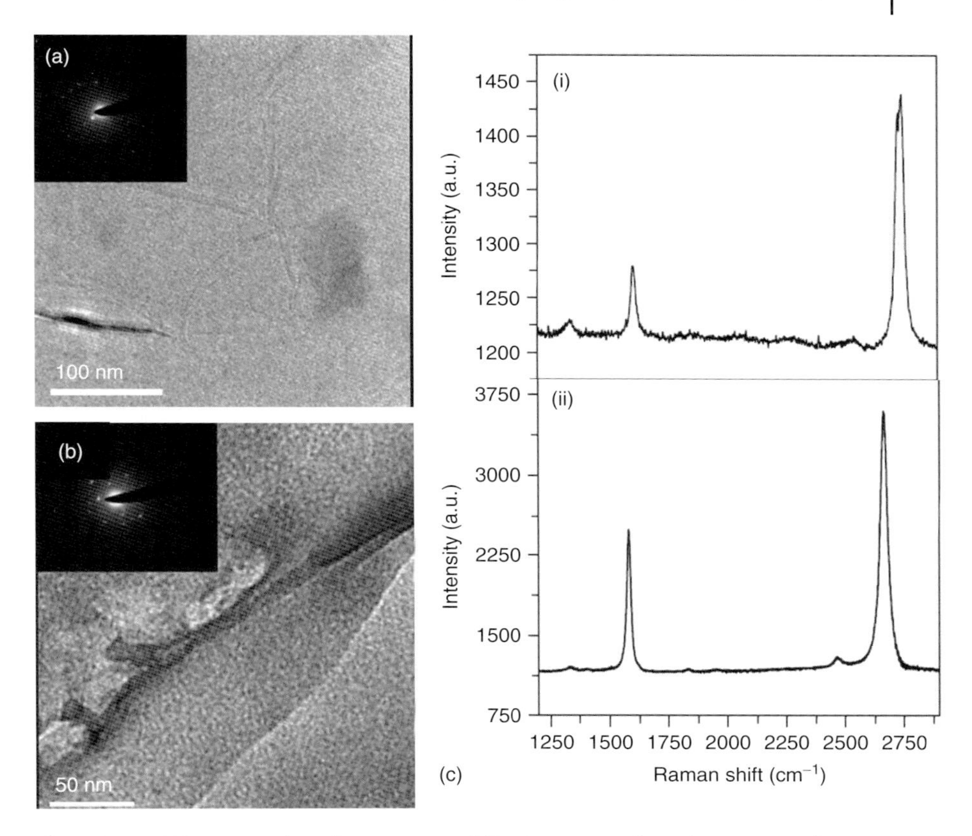

Figure 1.4 TEM images of graphene prepared by the thermal decomposition of (a) methane (70 sccm) at 1000 °C and (b) benzene (Ar passed through benzene with flow rate of 200 sccm) at 1000 °C on a nickel sheet. Insets show electron diffraction pattern from the corresponding graphene sheets, and (c) the Raman spectra of graphene prepared from the thermal decomposition of (i) methane and (ii) benzene. (Source: Reprinted with permission from Ref. [61].)

size of 100–200 nm. This makes use of the knowledge that the presence of H_2 during arc discharge process terminates the dangling carbon bonds with hydrogen and prevents the formation of closed structures. The conditions that are favorable for obtaining graphene in the inner walls are high current (above 100 A), high voltage (>50 V), and high pressure of hydrogen (above 200 Torr). In Figure 1.5a,b, TEM and AFM images of HG sample are shown, respectively. This method has been conveniently used to dope graphene with boron and nitrogen [64]. To prepare boron-doped graphene (B-HG) and nitrogen-doped graphene (N-HG), the discharge is carried out in the presence of H_2 + diborane and H_2 + pyridine or ammonia, respectively. Later, based on these observations, some modifications in the synthetic conditions also yielded FG in bulk scale. Cheng *et al.* [65] used hydrogen arc discharge process as a rapid heating method to prepare graphene from GO. Arc discharge in an air atmosphere resulted in graphene nanosheets that

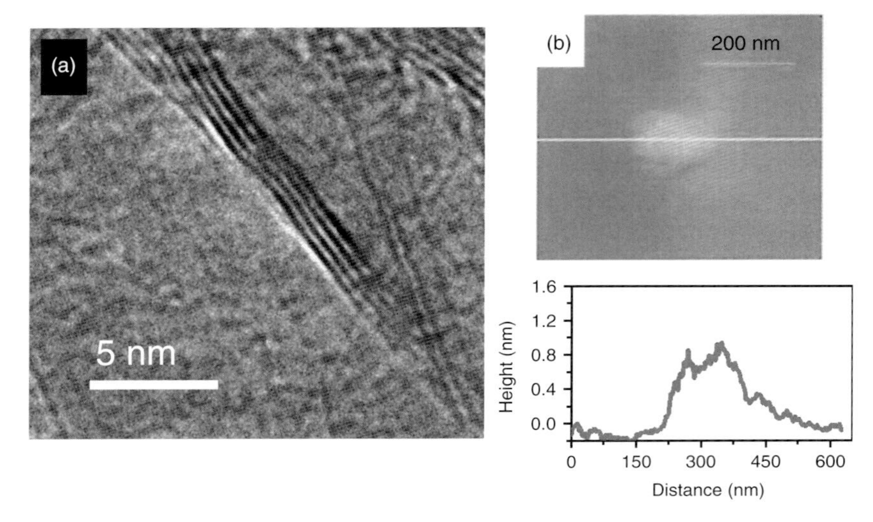

Figure 1.5 (a) TEM and (b) AFM image of HG prepared by arc discharge of graphite in hydrogen. Below is the height profile for the same. (Source: Reprinted with permission from Ref. [63].)

are ~100–200 nm wide predominantly with two layers. The yield depends strongly on the initial air pressure [66]. Li *et al.* [67] have synthesized N-doped multilayered graphene in He and NH_3 atmosphere using the arc discharge method. Arc discharge carried out in a helium atmosphere has been explored to obtain graphene sheets with different number of layers by regulating gas pressures and currents [68].

1.2.5 Reduction of Graphite Oxide

Chemical reduction of graphite oxide is one of the established procedures to prepare graphene in large quantities [33]. Graphite oxide when ultrasonicated in water forms a homogeneous colloidal dispersion of predominantly SGO in water. RGO with properties similar to that of graphene is prepared through chemical, thermal, or electrochemical reduction pathways [69]. While most strong reductants have slight to strong reactivity with water, hydrazine monohydrate does not, making it an attractive option for reducing aqueous dispersions of graphene oxide [70]. Syn addition of H_2 occurs across the alkenes, coupled with the extrusion of nitrogen gas. Large excess of $NaBH_4$ has also been used as a reducing agent [71]. Other reducing agents used include phenyl hydrazine [72], hydroxylamine [73], glucose [74], ascorbic acid [75], hydroquinone [76], alkaline solutions [77], and pyrrole [78]. Electrochemical reduction is another means to synthesize graphene in large scale [79–81]. The reduction initiates at −0.8 V and is completed by −1.5 V, with the formation of black precipitate onto the bare graphite electrode. Zhou *et al.* [82] coupled electrochemical reduction with a spray coating technique to prepare large-area and patterned RGO films with thicknesses ranging from a single monolayer to several micrometers on various conductive and insulating substrates. Organic dispersions of graphene oxide can be thermally reduced in polar organic solvents under reflux conditions to afford electrically conductive, chemically active reduced graphene oxide (CARGO) with tunable C/O ratios, dependent on the boiling point of the solvent. The dispersing medium must have a boiling point above 150 °C (the initiation point of the mass loss feature in the thermogravimetric analysis (TGA) profile of graphene oxide) and be able to disperse both graphene oxide and CARGO, for example, DMF, dimethyl sulfoxide, and NMP have been used for the purpose [81].

Photothermal and photochemical reduction of GO is a rapid, clean, and versatile way to form RGO. Ding *et al.* [83] reduced GO using UV irradiation to obtain single- to few-layered graphene sheets without the use of any photocatalyst. Cote *et al.* [84] prepared RGO by photothermal reduction of GO using xenon flash at ambient conditions and patterned GO or GO/polymer films using photomask. Nanosecond laser pulses of KrF eximer laser or 335 and 532 nm were shown to effectively reduce dispersions of GO to thermally and chemically stable graphene [85]. High-quality RGO has been prepared by irradiating GO with sunlight, ultraviolet light, and KrF excimer laser [61]. The reduction of GO to graphene by excimer laser irradiation results in the change of color of the solid GO film from brownish yellow to black. (Figure 1.6 shows darkening on reduction). Carbonyl and other oxygen functionalities on the surface of the GO film nearly disappear after irradiation, as can be seen from the infrared spectra shown in Figure 1.6a,b. The electrical conductivity increases by 2 orders of magnitude after laser irradiation of the GO film as shown in Figure 1.6c. Photochemical reduction of GO and SGO to graphene has also been exploited for patterning. For this purpose, GO films deposited on Si substrates were subjected to excimer laser radiation (Lambda Physik KrF excimer laser, 248 nm wavelength, 30 ns lifetime, 300 mJ laser energy, 5 Hz repetition rate, 200 shots), after inserting a TEM grid as the mask and covering them with a quartz plate [86, 87]. Figure 1.7a shows a schematic representation of the process of laser patterning using TEM grid as mask, and Figure 1.7b shows the optical microscopic image of the pattern achieved after excimer laser reduction of graphene oxide. Electron-beam-induced reduction of GO has been reported [88]. Electron-beam patterning of GO films has been used to obtain patterns of RGO as thin as 240 nm, as shown in Figure 1.7c [87].

Graphene oxide can be reversibly reduced and oxidized using electrical stimulus. Controlled reduction and oxidation in two-terminal devices containing multilayer graphene oxide films was demonstrated by Ekiz *et al.* [89] and by Yao *et al.* [90] Microwave irradiation (MWI)-induced heating has been used as a rapid way to synthesize graphene sheets. Owing to the difference in the solvent and reactant dielectric constants, selective dielectric heating can provide significant enhancement in the transfer of energy directly to the reactants, which causes an instantaneous internal temperature rise and thereby reduction of GO [91]. Dry GO absorbs MWI strongly with a sudden increase in surface temperature of the GO, up to ~400 °C, within just 2 s, leading to an ultrafast reduction of GO to RGO [92].

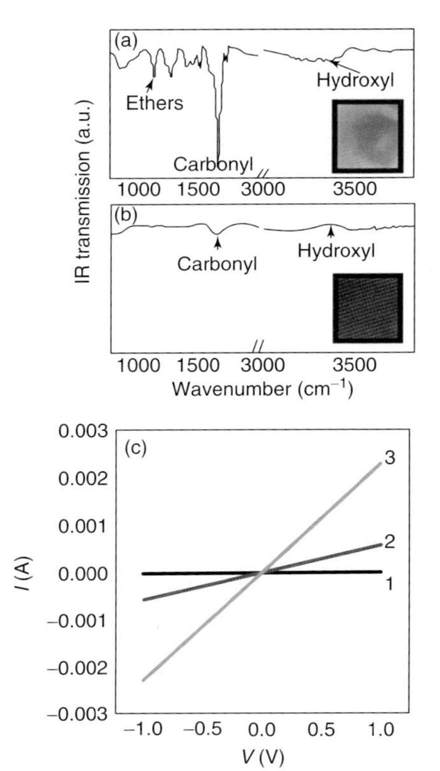

Figure 1.6 FTIR spectra of GO (a) before and (b) after laser reduction (laser reduced graphene oxide, LRGO). Insets show photographs of GO before and after reduction, respectively. (c) Current–voltage characteristics of 1, GO; 2, LRGO; 3, LRGO-Pt. (Source: Reprinted with permission from Ref. [87].)

1.3
Synthesis of Graphene Nanoribbons

Graphene nanoribbons (GNRs) can be thought of as thin strips of graphene or unrolled CNTs. GNRs have attracted attention because of their novel electronic and spin transport properties [5, 93–95]. GNRs of width 10–100 nm and 1–2 μm were prepared for the first time by oxygen plasma etching of graphene sheets [96]. A negative tone e-beam resist, hydrogen silsesquioxane (HSQ), is used to protect the underlying graphene layer, while the unprotected layer gets etched away by the oxygen plasma [96]. Tapaszto *et al.* [97] etched geometrically and crystallographically oriented GNRs from graphene sheets by applying constant bias potential (higher than that used for imaging) and simultaneously moving the STM tip with constant velocity over the surface. These methods did not produce GNRs of widths less than 20 nm and had edge roughness of ~5 nm. Li *et al.* [93] chemically prepared sub-10 nm width GNRs of varying lengths from thermally exfoliated graphite by dispersing it in a 1,2-dichloroethane (DCE) solution of poly(*m*-phenylenevinylene-*co*-2,5-dioctoxy-*p*-phenylenevinylene) (PmPV) by sonication and removing the larger pieces by centrifugation. Cano-Marquez *et al.* [98] prepared 20–300 nm few-layered GNRs in bulk scale by CVD of ethanol, with ferrocene and thiophene acting as catalysts.

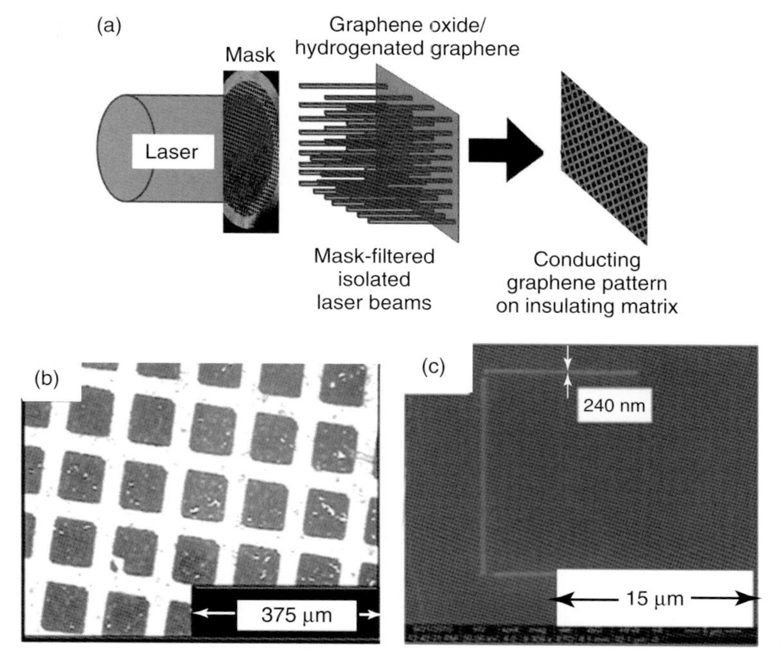

Figure 1.7 (a) Schematic diagram illustrating masked laser patterning. (Source: Reprinted with permission from Ref. [86].) (b) Optical microscopic image of the pattern achieved after excimer laser reduction of graphene oxide, and (c) electron-beam pattern with 240 nm wide lines of RGO on GO films. (Source: Reprinted with permission from Ref. [87].)

Longitudinal unzipping of CNTs has been used for synthesis of GNRs. While Kosynkin *et al.* [99] carried out oxidative unzipping using $KMnO_4/H_2SO_4$ mixture, Higginbotham *et al.* [100] used a second acidlike trifluoroacetic acid (TFA) or H_3PO_4 to get more controlled oxidation (protection of the vicinal diols formed on the basal plane of graphene during the oxidation, thereby preventing their overoxidation to diones and the subsequent hole generation), yielding high-quality GNRs with lesser holes. Figure 1.8a shows a schematic of potassium intercalation and sequential longitudinal splitting of the CNT walls to yield a nanoribbon stack. Jiao *et al.* [101] carried out mild gas-phase oxidation to create defects on CNTs that were then dispersed in DCE solution of PmPV by sonication and obtained high-quality unzipped nanoribbons. Cano-Marquez *et al.* [98] and Kosynkin *et al.* [102] unzipped CNTs by alkali metal intercalation and exfoliation either by protonation or with acid treatment and abrupt heating. Jiao *et al.* [103] carried out controlled unzipping of partially embedded CNTs in poly(methyl methacrylate) (PMMA) by Ar plasma etching. GNRs have also been obtained by sonochemical cutting of graphene sheets involving oxygen-induced unzipping of graphene sheets [104]. Laser irradiation of undoped and doped multiwalled CNTs by an excimer laser (energy ~200–350 mJ) also yielded GNRs (Figure 1.8b) [105].

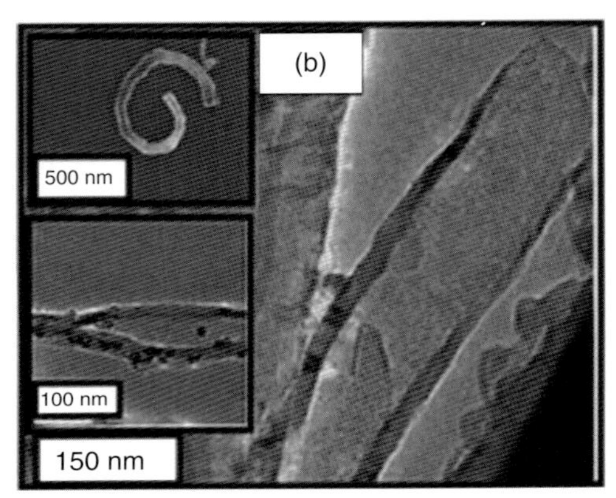

Figure 1.8 (a) Schematic of the splitting process produced by potassium intercalation between the nanotube walls and sequential longitudinal splitting of the walls followed by unraveling to a nanoribbon stack. (Source: Reprinted with permission from Ref. [44].) (b) Boron-doped CNT irradiated with laser energy of 250 mJ. Top inset shows the corresponding FESEM (field emission scanning electron microscopic) image. Bottom inset shows a TEM image of partially opened boron-doped CNT irradiated at 200 mJ. (Source: Reprinted with permission from Ref. [105].)

GNRs have also been obtained by PECVD on Pd nanowire templates. On removing the Pd nanowires, tubular graphene layer collapsed to yield edge-closed nanoribbons, while the graphene layers on the top part of the metal nanowire were selectively etched by O_2 plasma to yield edge-opened GNRs [106]. Wang and Dai [107] prepared 20–30 nm wide GNR arrays lithographically and used gas-phase etching chemistry to narrow the ribbons down to <10 nm, thereby achieving a high on/off ratio up to $\sim 10^4$. Bottom-up fabrication provides precise control over topologies and widths of GNRs. Surface-assisted coupling of molecular precursors into linear polyphenylenes and their subsequent cyclodehydrogenation have been used to prepare GNRs with predefined edge structure and morphology [108]. Yang *et al.* [109] carried out the Suzuki–Miyaura polymerization of the bis-boronic esters with diiodobenzenes to prepare polyphenylenes resembling GNRs.

1.4 Selected Properties

1.4.1 Magnetic Properties

Occurrence of high-temperature ferromagnetism (FM) in graphite-related materials is a topic of considerable interest. Yazyev *et al.* [110] showed that magnetism in graphene can be induced by vacancy defects or by hydrogen chemisorption. Some workers suggest that the zig-zag edges are responsible for the magnetic properties of graphene [111]. Inhomogeneous distribution of FM structures of nanographene sheets has been observed below 20 K [112]. Microporous carbon exhibits high-temperature FM originating from topological disorder associated with curved graphene [113]. Nanosized diamond particles implanted with nitrogen and carbon show FM hysteresis at room temperature [114]. Wang *et al.* [8] reported room temperature FM in a graphene sample prepared by the partial reduction of graphene oxide with hydrazine followed by annealing the samples at different temperatures in an argon atmosphere. Magnetic properties of graphene samples prepared by EG, conversion of nanodiamond (DG), and arc evaporation of graphite in hydrogen (HG) have been studied. The number of graphene layers in EG, DG, and HG was estimated to be six to seven, four to five, and two to three, respectively [3, 9, 61]. All these samples show divergence between the field-cooled (FC) and zero-field-cooled (ZFC) data, starting around 300 K. In Figure 1.9a, we show the temperature dependence of magnetization of EG and HG samples measured at 500 Oe. The divergence nearly disappears on application of 1 T as can be seen from the insets in Figure 1.9a. The divergence between the FC and ZFC data in the graphene samples is comparable to that in magnetically frustrated systems such as spin glasses and superparamagnetic materials. The Curie–Weiss temperatures obtained from the high-temperature inverse susceptibility data were negative in all these samples, indicating the presence of antiferromagnetic interactions. Interestingly, we observe well-defined maxima in the magnetization at low temperatures, the maxima becoming prominent in the data recorded at 1 T (see insets in Figure 1.9a). Such magnetic anomalies are found when antiferromagnetic correlations compete with FM order. Application of high fields can align the FM clusters and decrease the divergence between FC and ZFC data, as indeed observed. It is possible that the data correspond to percolation type of situation, wherein different types of magnetic states coexist. The FM clusters in such a case would not be associated with a well-defined global FM transition temperature. This behavior is similar to that of microporous carbon and some phase-separated members of the rare earth manganite family, $Ln_{1-x}A_xMnO_3$ (Ln = rare earth, A = alkaline earth) [115, 116]. Theoretical calculations predict the presence of antiferromagnetic states in the sheets and FM states at the edges of graphene [117].

The graphene samples show magnetic hysteresis at room temperature (Figure 1.9b) and the M_S increases with increase in temperature. Of the three samples, HG shows the best hysteretic features with saturation. While DG shows

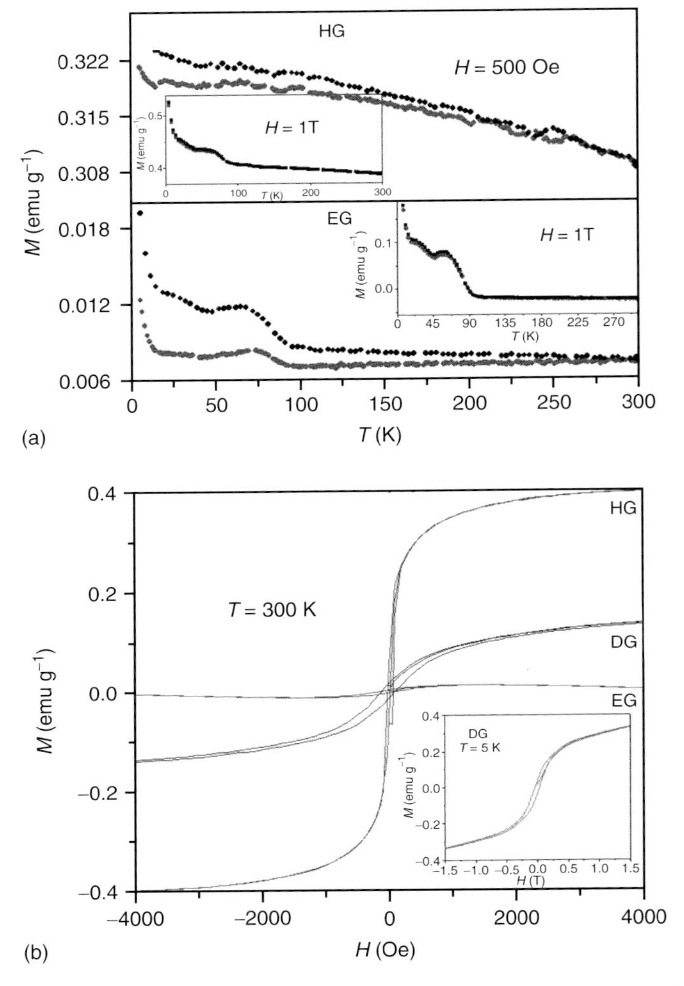

Figure 1.9 (a) Temperature variation of magnetization of few-layer graphenes EG and HG at 500 Oe, showing the ZFC and FC data. The insets show the magnetization data at 1 T. (b) Magnetic hysteresis in EG, DG, and HG at 300 K. Inset shows magnetic hysteresis in DG at 5 K. (Source: Reprinted with permission from Ref. [9].)

saturation magnetization, M_S, it is low when compared to HG. We see that θ_p, M_R, and M_S are the highest in case of HG, which also shows a higher value of magnetization than the other samples at all temperatures. The values of the various magnetic properties of the samples (M_S at 300 K) are plotted in Figure 1.9b to demonstrate how the properties vary as HG > DG > EG. It is noteworthy that both the area and the number of layers vary in the order of EG > DG > HG. It is likely that edge effects would be greater in samples with a smaller number of layers as well as small areas. In the case of HG, hydrogenation occurred to some extent, thereby favoring FM. Magnetic properties of DG samples prepared at different temperatures show a systematic decrease in magnetization with increase in the temperature of preparation. Electron paramagnetic resonance (EPR) measurements in the 2.5–300 K range on EG, DG, and HG show a signal with a line width of 0.7–2.9 mT and a g-value in the 2.006–2.013 range. The small value of the line width and the small deviation in the g-value from the free-electron value suggest that the spins do not originate from transition-metal impurities but from the spin species in the graphene sheets.

Adsorption of different guest molecules on graphene gives rise to a reversible low-spin/high-spin magnetic switching phenomenon, which depends on the nature of the guest species. Adsorption of H_2O [118], interaction with acids [119], and intercalation with potassium clusters reduce the magnetization of nanographite [120]. The reduction in magnetization has been interpreted as due to the interaction with lone pair orbitals as well as charge transfer with graphene sheets. The edge sites participating in host–guest interactions can give rise to magnetic phenomenon. Guest molecules accommodated through physisorption mechanically compress the flexible nanographite domains, leading to a significant reduction in the internanographene-sheet distance. Such a reduction in the intersheet distance could align the magnetic moments antiparallely and reduce the net magnetic moment [121]. Adsorption of benzene solutions of tetrathiafulvalene (TTF) and tetracyanoethylene (TCNE) is found to profoundly affect the magnetic properties of FG. In Figure 1.10a, we show typical results on the effect of adsorbing 0.05 M solution of TTF on the magnetic properties of HG. The value of the magnetization drastically decreases on adsorption of TTF and TCNE, although the basic trend in the temperature variation of magnetization remains the same. Thus, the graphene sample continues to show room-temperature hysteresis. On increasing the concentration of TTF or TCNE, the magnetization value decreases progressively. Interestingly, TTF has a greater effect than TCNE, even though the magnitude of adsorption of TCNE on HG is greater. The value of M_S at 300 K decreases on adsorption of TTF and TCNE, the decrease being larger in the case of former. Clearly, charge-transfer interaction between graphene and TTF (TCNE) [13] is responsible for affecting the magnetic properties. The reversible concentration-dependent effects of TTF and TCNE on the magnetic properties of graphene support the idea that the magnetic properties of the graphene samples are intrinsic.

Hydrogenation of graphene can induce magnetism since the formation of tetrahedral carbons can reduce the connectivity of the π-sheets and the $\pi-\pi$ energy gap of the localized double bonds and hence the ring current diamagnetism. Such changes in structure can therefore cause an increase in magnetic susceptibility [122]. Hydrogenated graphene samples with varying hydrogen contents have been prepared using the Birch reduction [123]. The samples of 2, 3, and 5 wt%, designated HGH_1, HGH_2, and HGH_3 respectively, have been examined for their magnetic properties. We observe a gradual increase in the magnetic moment, with an increase in the hydrogen content. An anomaly is also observed in magnetism from 50 to 80 K in the case of ZFC of HGH_2 when compared to HG probably due to percolation type of situation arising from different types of magnetic states. It appears that there is a change in the magnetic structure in HGH_2 compared to HG. In

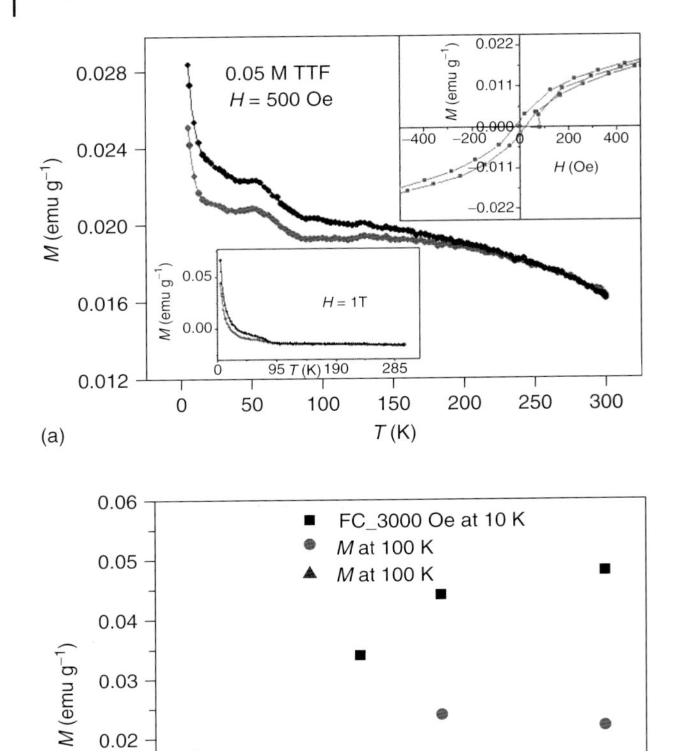

Figure 1.10 (a) Temperature variation of the magnetization of the few-layer graphene HG (500 Oe) after adsorption of 0.05 M solution of TTF. The magnetization data given in the figure are corrected for the weight of adsorbed TTF. Magnetic hysteresis data at 300 K and magnetization data at 1 T are shown in the insets. Magnetization data of HG with adsorbed TCNE are similar to those with TTF, except that the decrease in magnetization relative to pure HG is much smaller. (Source: Reprinted with permission from Ref. [9].) (b) Comparison of the magnetic properties of the hydrogenated few-layer graphene HG: HGH_1, HGH_2, and HGH_3. (Source: Reprinted with permission from Ramakrishna Matte *et al. Chem. Sci*, doi: 10.1039/c1sc00726b.)

Figure 1.10b, the remanent magnetization (M_r), saturation magnetization (M_S), and magnetization at 3000 Oe (FC at 10 K) of HG with different weight percentages of hydrogen are shown. The values of these properties increase with hydrogen content. On dehydrogenation at 500 °C for 4 h, the samples exhibit a decrease in the magnetic moment. This observation confirms that the increase in the magnetic properties is due to hydrogenation. On dehydrogenation, the hydrogenated samples revert to the initial graphene samples.

1.4.2
Electrical Properties

Intrinsic graphene is a semimetal or a zero-gap semiconductor. FGs show semiconducting nature, with the resistivity showing little change in the 100–300 K range. Conductivity, on the other hand, shows a sharp increase from 35 to 85 K, and the slope of temperature versus conductivity curve reduces thereafter. Resistance of FG decreases markedly if it is heated to high temperatures. Resistivity decreases markedly with increase in the number of layers as demonstrated by EG, HG, and RGO (ρ of RGO $<$ HG $<$ EG) samples with three to four layers, two to three layers, and single layer, respectively. Room-temperature thermal conductivity of graphene has been measured using a noncontact optical technique. The conductivity of graphene goes up to $(5.30 \pm 0.48) \times 10^3$ W mK^{-1} outperforming CNTs [4]. Experiments with field-effect transistors (FETs) on micromechanically cleaved graphene by Novoselov *et al.* [1] have revealed that the sheet resistivity (ρ depends on the gate voltage (V_g)) exhibits a sharp peak to a value of several kiloohms and decays back to 100 Ω on increasing V_g (Figure 1.11a). At the same V_g where ρ had its peak, the Hall coefficient showed a sharp reversal of its sign, thus revealing ambipolar character (Figure 1.11b). FET characteristics of EG, DG, HG, N-HG, and B-HG have been investigated by us in comparison with RGO, and RGO showed ambipolar transfer characteristics on sweeping the V_{gs} between −20 and +20 V and V_{ds} at 1 V (Figure 1.12a), while all the FGs showed n-type behavior. The highest mobility was found with HG possessing two to three layers and with the least defects. FETs based on B-HG and N-HG show n-type and p-type behavior, respectively (Figure 1.12b,c) [124]. Novoselov's micromechanically cleaved graphene showed extremely high mobilities of ~15000 cm^2 (V s)$^{-1}$ at room temperature with electron and hole concentrations of 10^{13} cm^{-2} with ballistic transport up to submicrometer distances [1]. Two- to three-layered HG samples have shown mobilities of 10 428 cm^2 (V s)$^{-1}$, while all other few-layered samples showed much lower mobilities [125]. Different factors such as the average number of layers, surface functionality, and concentration of defects are found to be responsible for observing different characteristics in different samples. HG with the smallest number of layers exhibits the highest mobility. It is remarkable that transistor characteristics are found even in few-layer samples with defects [61]. The linear dispersion relation of graphene predicts that the resistivity of graphene due to isotropic scatterers is independent of the carrier density. Hwang and Das Sarma [126] theoretically calculated the phonon-scattering-limited extrinsic electron mobility in graphene to be a function of temperature and carrier density, with the room-temperature intrinsic mobility reaching the values of above 10^5 cm^2 (V·s)$^{-1}$. Chen *et al.* [4] have shown that electroacoustic phonon scattering in graphene is independent of the carrier density and contributes just 30 Ω to room temperature resistivity of graphene, with the intrinsic mobility of graphene being 200 000 cm^2 (V·s)$^{-1}$. The

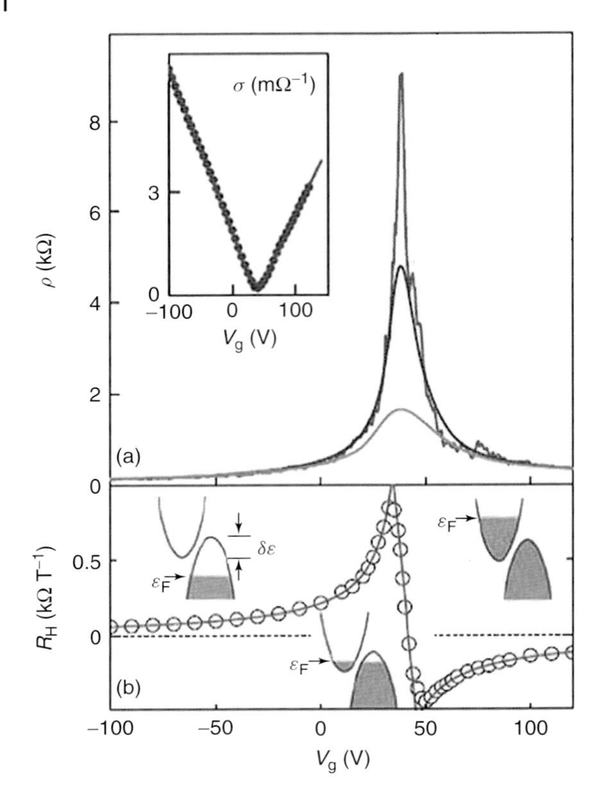

Figure 1.11 (a) Resistivity ρ of few-layer graphene on gate voltage (V_g) for different temperatures ($T = 5$, 70, and 300 K from top to bottom), with the inset showing the change in conductivity $\sigma = 1/\rho$ (at 70 K). (b) The Hall coefficient R_H versus V_g for the same at 5 K. (Source: Reprinted with permission from Ref. [1].)

actual mobility is, however, dependent on scattering by various extrinsic factors such as surface phonons [4, 125–128], charged impurities on top of graphene or in the underlying substrate [128, 129], and ripples and corrugation in the graphene sheet [130]. Dramatically reduced carrier scattering was reported in suspended graphene devices by Du *et al.* [131] allowing the observation of a very high mobility of 120 000 cm^2 $(V{\cdot}s)^{-1}$ near room temperature ($T \sim 240$ K).

Since the band gap of graphene is 0, devices with channels made of large-area graphenes cannot be switched off and therefore are not suitable for logic applications. However, the band structure of graphene can be modified to open a band gap by constraining large-area graphene in one dimension to form GNRs. Han *et al.* [5] first investigated electronic transport in lithographically patterned graphene ribbons and demonstrated band gap opening due to lateral confinement of charge carriers in case of narrower nanoribbons; band gap increases with decrease in nanoribbons' width. The sizes of these energy gaps were investigated by measuring the conductance in the nonlinear response regime at low temperatures [132]. Chen *et al.* [96]

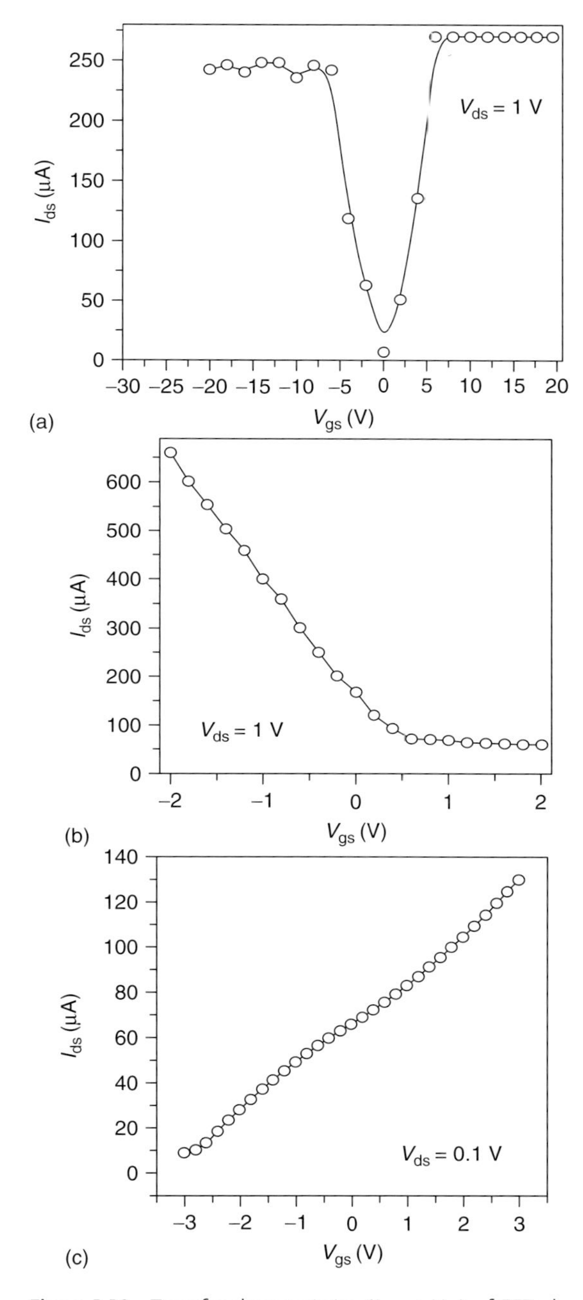

Figure 1.12 Transfer characteristics (I_{ds} vs V_{gs}) of FETs based on (a) RGO, (b) B-HG, and (c) N-HG. Here, I_{ds}, V_{ds}, and V_{gs} stand for source to drain current, source to drain voltage, and gate to source voltage, respectively. (Source: Reprinted with permission from Ref. [124].)

have fabricated GNR FETs with 20–40 nm width and measured FET characteristics and ON/OFF ratios. FETs with sub-10 nm nanoribbons prepared by Wang *et al.* [94] showed much greater I_{ON}/I_{OFF} of 10^5 at room temperature. The device had 20 times higher current density and 100 times higher transconductance per micrometer due to the larger band gaps and high GNR quality with better edge smoothness.

Graphene is considered to be the next-generation electrode material due to its extraordinary thermal, chemical, and mechanical stability. Transparent conducting films made of RGO have been fabricated [133]. These films are similar to HOPG in electronic and structural properties. Graphene films prepared by direct CVD on Ni substrates have been transferred onto polyethylene terephthalate (PET) substrate (thickness, 100 μm) coated with a thin polydimethylsiloxane (PDMS) layer (thickness, 200 μm) to prepare flexible, stretchable, foldable, transparent (80% optical transparency), conducting (sheet resistance of only ~280 Ω per square) films [52]. Conductivity as a function of bending radii has been studied and results hold promise for the application of these films as highly conducting, macroscopic, flexible transparent conducting electrodes. Mak *et al.* [134] studied optical reflectivity and transmission properties of graphene over photon energies of 0.2 and 1.2 eV and explained the properties based on the noninteracting massless Dirac fermions.

Band structure of graphene can be tuned by confining it to one dimension, as in the case of GNRs, to generate highly spin-polarized currents [135]. Bai *et al.* [136] obtained almost 100% magnetoresistance at low temperatures, with almost 50% remaining at room temperature for GNRs fabricated using the nanowire etch mask technique [137]. Such high magnetoresistance devices can find use in spin valve devices. Graphene-based superconducting transistors were reported by Heersche *et al.* [138] Although graphene is not superconducting by itself, it shows supercurrents over short distances when placed between superconducting electrodes because of the Josephson effect. Using the nonequilibrium Green's function method, transmission of superconductor-graphene-superconductor junctions has been examined theoretically and the possibility of superconducting switch has been predicted [139]. Palladium sheets sandwiched between graphene sheets give rise to a superconducting transition around 3.6 K [140]. Superconductivity here occurs in the Pd sheets.

1.4.2.1 **Supercapacitors**

Electrochemical properties of a few graphenes prepared by different methods have been investigated using the redox reactions with potassium ferrocyanide [141]. Among EG, DG (see Section 4.1 for description), and a graphene prepared by CVD over Ni and Co foils (CG), EG shows a behavior similar to the basal plane in graphite, whereas DG and CG show slightly better kinetics. Vivekchand *et al.* [142] prepared electrochemical supercapacitors with different graphene samples as electrode materials in aqueous H_2SO_4 as well as in an ionic liquid (*N*-butyl-*N*-methylpyrrolidinium bis(trifluoromethanesulfonyl)imide, PYR14TFSI) that were used as electrolytes. EG and DG exhibit high specific capacitance in aqueous H_2SO_4, the value reaching up to 117 and 35 F g^{-1}, respectively. Voltammetric characteristics of a capacitor built from graphene electrodes (5 mg each), at a scan rate of 100 mV s^{-1} using aqueous H_2SO_4 (1 M) (Figure 1.13a,b), show specific capacitance as a function of scan rate for different graphene samples. Using an ionic liquid, the operating voltage has been extended to 3.5 V (instead of 1 V in the case of aqueous H_2SO_4), the specific capacitance values being 75 and 40 F g^{-1} for EG and DG, respectively. High-surface-area graphite prepared by ball milling showed a large specific capacitance of 33 μF cm^{-2} in aqueous medium, which might be due to high open surface area, lattice defects, and oxygen functional groups in the sample [143].

Chemically modified graphene sheets obtained by the reduction of graphene oxide with hydrazine when used as electrode material in supercapacitors gave specific capacitances of 135 and 99 F g^{-1} in aqueous and organic electrolytes, respectively [144]. 3D CNT/graphene sandwich structures with CNT pillars grown in between the graphene layers have been used as high-performance electrode materials for supercapacitors, and a maximum specific capacitance of 385 F g^{-1} could be obtained

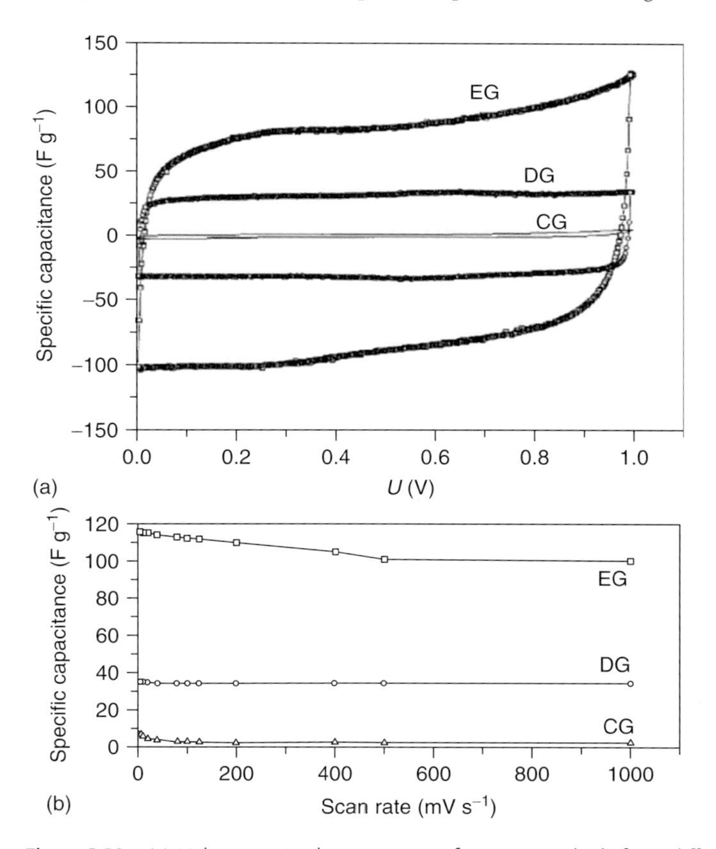

Figure 1.13 (a) Voltammetric characteristics of a capacitor built from different graphene electrodes (5 mg each) at a scan rate of 100 mV s^{-1} in aqueous H_2SO_4 (1 M), and (b) specific capacitance as a function of scan rate. (Source: Reprinted with permission from Ref. [142].)

at a scan rate of 10 mV s^{-1} in 6 M KOH aqueous solution [145]. Some novel strategies to synthesize graphene-based nanocomposites containing polyaniline [146] and $Co(OH)_2$ [147], and so on, for enhancing the electrochemical capacitance aqueous solution have been explored. Graphene/polyaniline composites with an appropriate weight ratio prepared using *in situ* polymerization exhibited a higher specific capacitance of 1046 F g^{-1} at a scan rate of 1 mV s^{-1} due to the synergistic effect between graphene and polyaniline. Graphene/$Co(OH)_2$ nanocomposite shows a capacitance as high as 972.5 F g^{-1}, leading to a significant improvement. Graphene nanosheets show high lithium storage capacity for lithium secondary batteries, the value reaching 540 mAh g^{-1}. This storage capacity can be further improved to 730 and 784 mAh g^{-1}, respectively, by incorporating CNTs and C60 [147].

1.4.2.2 Photovoltaics and Photodetectors

Photovoltaic devices fabricated with a bulk heterojunction (BHJ) architecture using solution-processible graphene as electron-acceptor material are reported. A power conversion efficiency of 14% is obtained using simulated 100 mW cm^{-2} AM 1.5G illumination [148]. The optical transparency and conductivity of graphene can be exploited for many photonic devices. For example, liquid-crystal devices with electrodes made of graphene show excellent performance with a high contrast ratio [149]. Conducting films of graphene for solar cell applications can also be prepared by a bottom-up approach [150]. Polymer photovoltaic cells based on solution-processible graphene are reported [151]. Because of its unique electronic structure, graphene shows useful photonic properties such as absorption of significant fraction of incident white light [6] and strong, tunable interband transitions [152]. Above ~0.5 eV, absorbance of graphene is additive resulting in strong graphene–light interactions. This has made possible fabrication of FETs for ultrafast photodetection [153]. Solution-processed thin films prepared using GO enable easy material processing and mechanical flexibility, making them useful candidates for use in large-area devices.

GNRs with substantial gaps have been used as phototransistors [154], specially for far-infrared detection [155]. It has been possible to prepare highly selective, sensitive, and high-speed nanoscale photodetectors and photoelectronic switches by drop-casting RGO and GNR on two-terminal 15 μm gap Cr (5 nm)/Au (300 nm) electrodes [156]. Electrical conductivities of RGO and GNR increase with IR laser irradiation. An RGO detector can sense the IR radiation emitted from a human body. The *detector current responsivity* (R_λ), defined as the photocurrent generated per unit power of the incident light on the effective area of a photoconductor, and the external quantum efficiency (EQE), defined as the number of electrons detected per incident photon for RGO photoconductors, are 4 mA W^{-1} and 0.3%, respectively, whereas for GNR, these values are higher, being 1 A W^{-1} and 80%, respectively, for an incident wavelength of 1550 nm at 2 V. RGO and GNRs. On absorbing light from an IR source, electron–hole (e-h) pairs are generated [157] because of a barrier like the Schattky barrier at the metal/graphene contact. The e-h pairs generated in graphene would normally recombine on a timescale of tens of picoseconds, depending on the quality and carrier concentration of the graphene

[125, 157, 158]. On application of an external field, the pairs get separated and a photocurrent is generated. A similar phenomenon can occur in the presence of an internal field formed by photoexcitation [159–161]. Graphene is also a very good UV absorber [162]. It has been possible to prepare UV detectors using RGO. The photodetecting responsivity is found to be 0.12 A W^{-1} with an EQE of 40% [163].

1.4.2.3 Field Emission and Blue Light Emission

Recently, there have been several attempts to investigate field emission properties of graphene films [164–166]. To take advantage of the high field enhancement, graphene sheets would have to stand on their edges and not lay laterally flat on the substrate. A spin-coated graphene–polystyrene composite film was reported to exhibit a threshold field of 4 V μm^{-1} (at 10^{-8} A cm^{-2}) with a field enhancement factor of 1200 [164]. Malesevic *et al.* [165] grew vertically aligned FG films by CVD and found these films to exhibit favorable turn-on field but decays after five cycles. Besides geometrical factors, spatial distribution can also tailor the work function and provide another means to improve electron field emission. Field emission properties of undoped arc-discharge-prepared graphene (HG), as well as B-HG and N-HG, have been studied. Electrophoretic deposition was used for depositing vertically oriented graphene sheets [167]. N-HG showed the lowest turn-on field of 0.6 V μm^{-1} with an emission current density of 10 μA cm^{-2} (Figure 1.14a). Emission current was generally stable for almost 3 h or more.

Aqueous solutions of acid-treated graphene or RGO show blue emission centered at 440 nm on being excited by UV of 325 nm [169]. On mixing the blue-light-emitting graphene samples with the yellow-light-emitting zinc oxide nanoparticles, it is possible to get a bluish-white light as can be seen from the PL spectra in Figure 1.14b. A plausible cause of the blue photoluminescence in RGO is the radiative recombination of e-h pairs generated within localized states. The energy gap between the π- and π^*-states generally depends on the size [170] of the sp^2 clusters or the conjugation length [171]. Interaction between nanometer-sized sp^2 clusters and finite-sized molecular sp^2 domains could play a role in optimizing the blue emission in RGO. The presence of isolated sp^2 clusters in a carbon-oxygen sp^3 matrix can lead to the localization of e-h pairs, facilitating radiative recombination of small clusters.

1.4.3 Molecular Charge Transfer

Interaction of carbon nanostructures with electron-donor and electron-acceptor molecules causes marked changes in their electronic structure and properties [13]. C_{60} is known to exhibit charge-transfer interaction with electron-donating molecules, such as organic amines, both in the ground and excited states [13]. Here, we discuss charge-transfer interaction of graphene with organic molecules, a property with potential utility in device applications. Raman spectroscopy is eminently effective in probing molecular charge-transfer interactions.

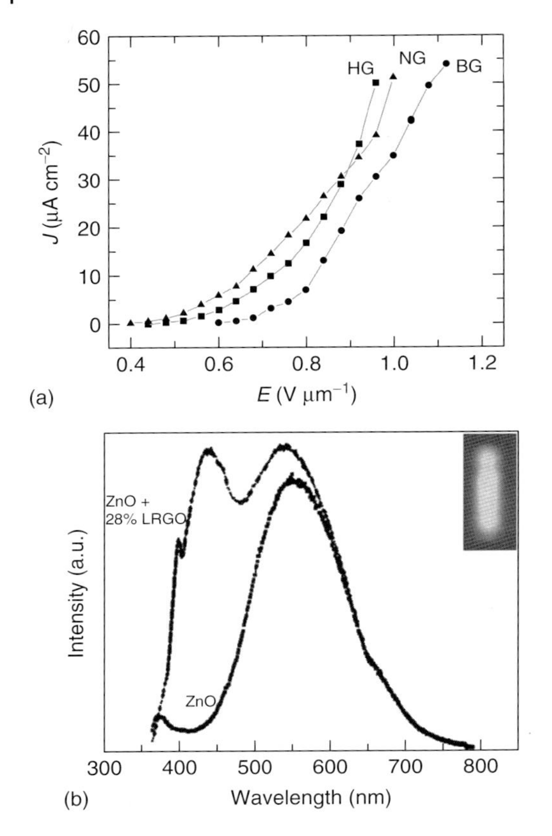

Figure 1.14 (a) Current density (J) of undoped HG, boron-doped BG, and nitrogen-doped NG graphenes as a function of electric field. (Source: Reprinted with permission from Ref. [167].) (b) White-light emission from ZnO-LRGO nanocomposite. (Source: Reprinted with permission from Ref. [168].)

Electron donors such as aniline and TTF soften (i.e., shift to lower frequency) the G-band of FG progressively with the increasing concentration, while electron acceptors such as nitrobenzene and TCNE stiffen (i.e., shift to higher frequency) the G-band, as can be seen in Figure 1.15a,b [12, 13, 172]. Both electron donors and electron acceptors broaden the G-band. The full-width at half maximum (FWHM) of the G-band increases on interaction with these molecules. The intensity of the 2D-band decreases markedly with the concentration of the either donor or acceptor molecule. The ratio of intensities of the 2D- and G-bands, $I_{(2D)}/I_{(G)}$, is a sensitive probe to examine doping of graphene by electron-donor and electron-acceptor molecules. SLGs have also shown similar results. Dong *et al.* [173] have studied the adsorption of various aromatic molecules on SLG films, which cause stiffening or softening of the G-band frequencies because of electronic effects.

Investigations of charge-transfer doping of FG (one layer (1 L) to four layer (4 L)) with Br_2 and I_2 vapors have shown that charge-transfer effects are greater

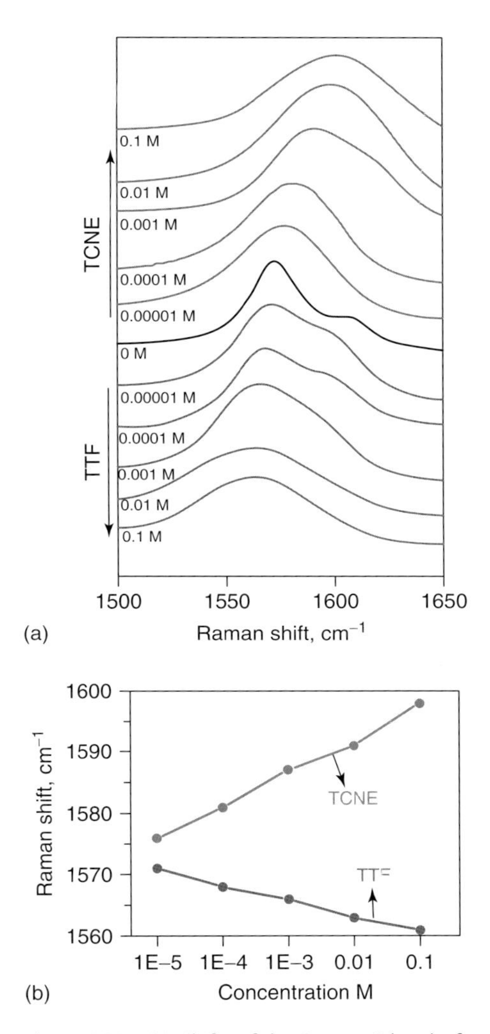

Figure 1.15 (a) Shifts of the Raman G-band of graphene caused by interaction with varying concentrations of TTF and TCNE, and (b) variation in the Raman G-band position of graphene on interaction with varying concentrations of electron-donor (TTF) and electron-acceptor (TCNE) molecules. (Source: Reprinted with permission from Ref. [13].)

on SLGs and bilayer graphenes compared to three- and four-layer graphenes [174]. Detailed studies of the interaction of halogen molecules with graphene have been carried out [175]. Both stiffening of the Raman G-bands on treating with the different halogen molecules and the emergence of new bands in the electronic absorption spectra point to the fact that the halogen molecules are involved in molecular charge transfer with the nanocarbons. The magnitude of molecular charge transfer between the halogens and the nanocarbons generally varies in the order ICl $>$ Br_2 $>$ IBr $>$ I_2 (Figure 1.16), which is consistent with the expected

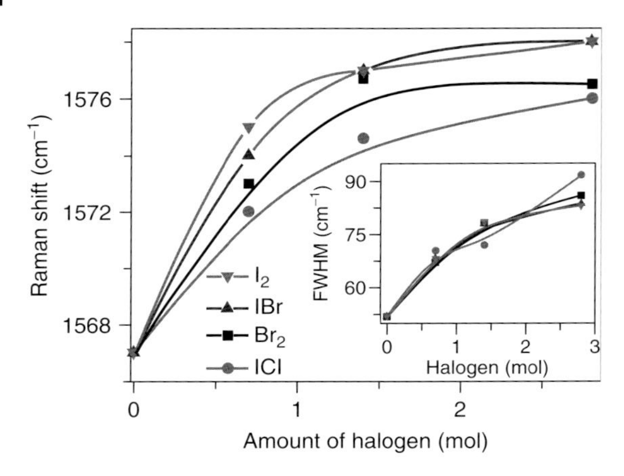

Figure 1.16 G-band characteristics for graphene (inset shows variation in FWHM) for various amounts of different halogens for a fixed molar amount of graphene (1 mol). (Source: Reprinted with permission from Ref. [175].)

order of electron affinities [175]. The occurrence of charge-transfer doping in FG covered with electron-acceptor (TCNE) and electron-donor (TTF) molecules is also evidenced in the electronic absorption spectra and X-ray photoelectron spectroscopy (XPS) [176]. Quantitative estimates of the extent of charge transfer in these complexes have been obtained through XPS. Electrical resistivity of graphene films with and without charge-transfer interactions shows the expected changes [12]. For example, the resistance is lowest in the presence of nitrobenzene and highest in the presence of aniline. There is a systematic dependence of resistance with the electron-donating and electron-withdrawing power of the substituents. The magnitude of interaction between graphene and donor/acceptor molecules seems to depend on the surface area of the graphene sample. Molecular charge transfer affects the magnetic properties of graphene [9]. Magnetization of graphene decreases on adsorption of TTF and TCNE, the interaction with TTF having a greater effect than with TCNE.

Density functional theory (DFT) calculations confirm the occurrence of charge-transfer-induced changes in graphene giving rise to midgap molecular levels with tuning of band gap region near the Dirac point and show how they are different from the effects of electrochemical doping [177, 178]. It has been shown that n-type and p-type graphenes result from charge-transfer interaction of graphene with donor and acceptor molecules, respectively. It is also predicted that the extent of doping depends on the coverage of organic molecules.

1.4.4
Decoration with Metal and Oxide Nanoparticles

Nanocarbons have been used as support materials for the dispersion and stabilization of metal nanoparticles because of their large chemically active surface and stability at high temperatures [177]. Decoration with metal nanoparticles results in changing the electronic structure of nanocarbons through Coulombic charge transfer [178]. Combinations of these two materials may lead to a successful integration of their properties in hybrid materials, with possible use in catalysis, nanoelectronics, optics, and nanobiotechnology [179, 180].

Graphene has been decorated with metal nanoparticles such as Au, Ag, Pt, Pd, and Co by different chemical methods [181]. Decoration of graphene with metal nanoparticles can be followed by absorption spectroscopy and electron microscopy [182]. The influence of metal nanoparticles on the electronic structure of graphene has been examined by Raman spectroscopy and first-principles calculations [183]. There is stiffening in the position of G- and D-bands, and the intensity of the 2D-band relative to that of G-band decreases, whereas the intensity of the D-band relative to that of G-band increases (Figure 1.17a). In comparison with pristine graphene, the FWHM of the G-band shows a significant broadening in graphene–metal composites. The shifts in the G- and D-bands show meaningful trends with the ionization energies of the metals as well as the charge-transfer energies. In Figure 1.17b, we have plotted the frequency shifts of the G-band of EG against the ionization energy (IE) of the metal. Note that the IE varies as Ag < Pd < Pt < Au. Interestingly, the magnitude of the band shifts generally decreases with increase in IE of the metal.

Decoration of graphene with Pt nanoparticles leads to a drastic increase in capacitance value, which is due to high surface area of the composite arresting the aggregation of graphene sheets [184]. Palladium nanoparticle–graphene hybrids are used as efficient catalysts for the Suzuki reaction [185]. Three-dimensional Pt-on-Pd bimetallic nanodendrites supported on graphene nanosheets are used as advanced nanoelectrocatalysts for methanol oxidation [186]. Composites of positively charged gold nanoparticles (GNPs) and pyrene-functionalized graphene (PFG) showed strong electrocatalytic activity and high electrochemical stability [187]. Au films deposited on SLG are used in surface-enhanced Raman scattering (SERS) substrates for the characterization of rhodamine R6G molecules [188]. Silver-decorated graphene oxide (Ag-GO) can be used as an antibacterial material, with a superior antibacterial activity toward *Escherichia coli* [189].

Graphene oxide (GO) is known to interact with nanoparticles of semiconducting oxides such as ZnO and TiO_2 through excited-state electron transfer [181, 190, 191]. The magnetic properties of graphene composites with nanoparticles of ZnO, TiO_2, Fe_3O_4, $CoFe_2O_4$, and Ni have been studied [192]. Raman studies of the composites of ZnO and TiO_2 with graphene reveal significant shifts in the G-band, with ZnO acting as an electron donor and TiO_2 as an acceptor. These composites also yield higher values of saturation magnetization compared to those of the individual particles or their mechanical mixtures with graphene. Composites of Fe_3O_4 and $CoFe_2O_4$ with graphene show softening of the G-band revealing a similar charge-transfer interaction, while the saturation magnetization remained without charge. First-principles DFT calculations reveal that the weak charge-transfer interaction and the magnetic coupling are directly linked to the IE and electron affinity of the deposited nanoparticles.

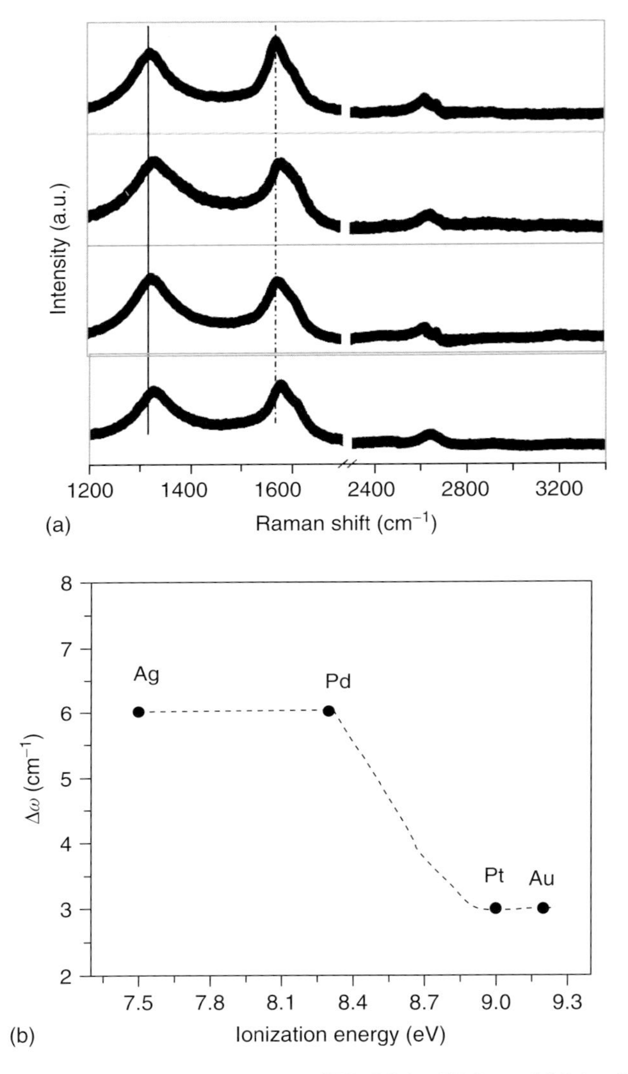

Figure 1.17 (a) Raman spectra of EG, EG-Ag, EG-Pt, and EG-Au. (b) Variation in the position of the G-band with the ionization energy of the metal. The broken curve is given as a guide to the eye. (Source: Reprinted with permission from Ref. [183].)

1.4.5
Surface Area and Gas Adsorption

SLG is theoretically predicted to have a large surface area of 2600 m^2 g^{-1} [10], while the surface area of FG is 270–1550 m^2 g^{-1} [11]. Patchkovskii *et al.* [193] carried out computations considering the contribution of quantum effects to the free energy and the equilibrium constant and suggested that H_2 adsorption capacities

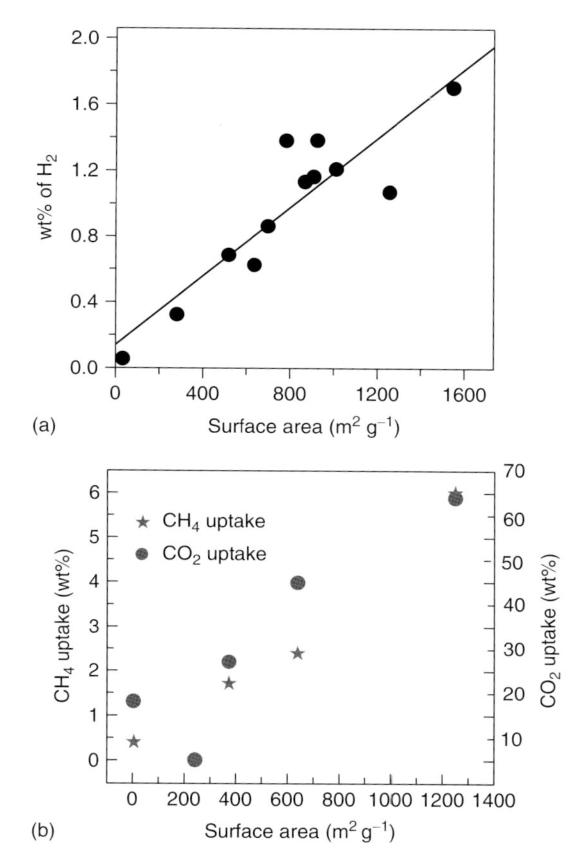

Figure 1.18 (a) Linear relationship between the BET (Brunauer-Emmett-Teller) surface area and weight percentage of hydrogen uptake at 1 atm pressure and 77 K for various graphene samples. (Source: Reprinted with permission from Ref. [194].) (b) Plot of weight percentage of CO_2 uptake (at 195 K and 1 atm) and methane uptake (at 298 K and 5 MPa) versus surface area for different graphene samples. (Source: Reprinted with permission from Ref. [195].)

on graphene can approach values set by the US Department of Energy (DOE) (6.5 wt% and 62 kg of H_2 per cubic meter). H_2 adsorption studies by Ghosh *et al.* [194] on FG samples prepared by the EG and transformation from nanodiamond (ND, DG) have revealed a H_2 uptake value of 1.7 wt% at atmospheric pressure and 77 K. Adsorption of H_2 was found to be directly proportional to the surface area of the samples (Figure 1.18a). A maximum adsorption of 3 wt% was achieved at 298 K and 100 atm for EG.

Uptake of CO_2 and CH_4 by graphenes (EG, HG, RGO, and SGO) was compared with that of activated charcoal [195], and adsorption was found to be dependent on surface areas of the studied samples, with EG showing the highest surface area (640 $m^2\ g^{-1}$) and SGO showing the lowest (5 $m^2\ g^{-1}$), while activated charcoal had a surface area of 1250 m^2 g^{-1}. Activated charcoal showed 64 wt% uptake of CO_2 at 195 K and 1 atm, while uptake of CO_2 by EG at 298 K and 50 bar was 51%. The uptake values varied between 5 and 45 wt% in the case of graphene samples at 195 K and 0.1 MPa, with EG exhibiting the highest uptake. EG and RGO samples with relatively high CO_2 uptake capacity contain oxygen functionalities on the surface, while HG with relatively clean surface did not show considerable uptake. However, all the graphenes exhibit smaller uptake capacity for CO_2 compared to activated charcoal, which also has a huge number of surface functional groups. Adsorption of methane on the graphenes and activated charcoal was measured at 273 and 298 K, respectively. The weight uptake of methane by activated charcoal is 7 and 6 wt% at 273 and 298 K and 5 MPa, respectively. The CH_4 uptake of the graphene samples varies between 0 and 3 wt% at 273 K and 5 MPa. Figure 1.18b shows the CO_2 and methane uptake of graphene samples as well as activated charcoal against their surface areas.

1.4.6
Mechanical Properties

Lee *et al.* [7] measured the elastic properties and intrinsic breaking strength of free-standing monolayer graphene membranes by nanoindentation in an AFM. They showed that defect-free monolayer graphene sheets possess excellent mechanical properties such as an elastic modulus of ~1 TPa, a strength of ~130 GPa, and a breaking strength of 42 N m^{-1}. This has led to the exploration of graphene-reinforced polymer matrix composites [196]. Ramanathan *et al.* [197] reported that just ~1 wt% addition of graphene to PMMA leads to increases of 80% in elastic modulus and 20% in ultimate tensile strength. A comparative study by these researchers shows that among all the nanofiller materials considered, single-layer functionalized graphene gives the best results (Figure 1.19a). They proposed that nanoscale surface roughness results in an enhanced mechanical interlocking with the polymer chains. Functionalized graphene sheets containing pendant hydroxyl groups across the surfaces may form hydrogen bonds with the carbonyl groups of PMMA and, consequently, stronger interfacial interactions with PMMA. A combined effect of these two enhanced interactions with the polymer matrix is better load transfer between matrix and the fiber resulting in enhancement of mechanical properties. A significant increase of 35 and 45% in the elastic modulus and hardness, respectively, was observed on addition of just 0.6 wt% of graphene to PVA (poly(vinyl alcohol)) [198]. Rafiee *et al.* [199] compared the mechanical properties of epoxy composites of 0.1 wt% of graphene with those of CNTs and found that graphene composites showed much greater increase in Young's modulus (by 31%), tensile strength (by 40%), and fracture toughness (by 53%) than in nanotube–epoxy composites. The fatigue suppression response of nanotube/epoxy composites degrades dramatically as the stress intensity factor amplitude is increased; the reverse effect is seen for graphene-based nanocomposites. Planar geometry of graphene and better matrix adhesion and interlocking arising

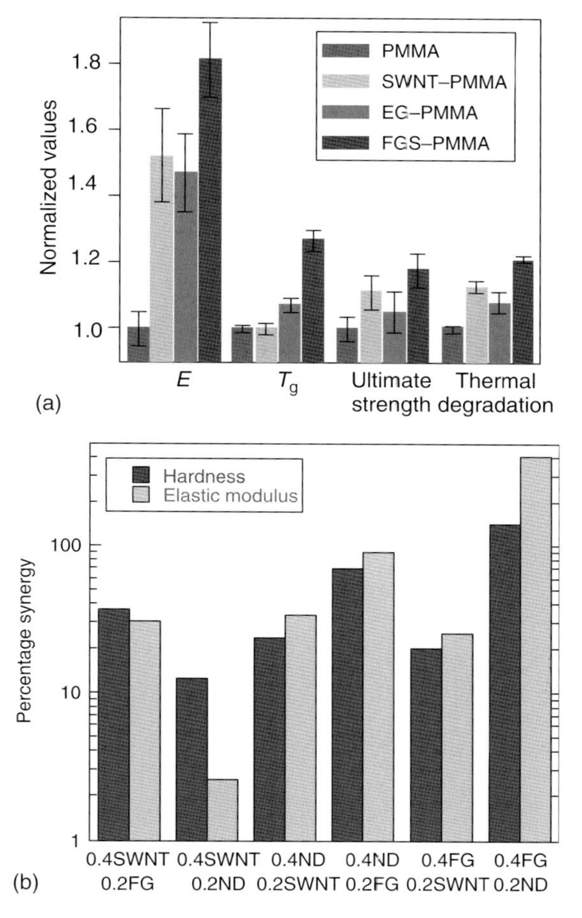

Figure 1.19 (a) Summary of thermomechanical property improvements for 1 wt% single-layer functionalized graphene–PMMA compared to SWNT–PMMA and EG–PMMA composites, with all property values normalized to the values for neat PMMA. (Source: Reprinted with permission from Ref. [197].) (b) Percentage synergy in mechanical properties of different binary mixtures of nanodiamond (ND), SWNTs, and few-layered graphene (FG). (Source: Reprinted with permission from Ref. [200].)

from their wrinkled surface is presumed to be the cause of better mechanical properties of graphene composites [199].

Detailed studies of the mechanical properties of binary combinations of ND, FG, and single-walled nanotubes (SWNTs) in PVA matrices have been carried out [200]. The mechanical properties of the resulting composites, evaluated by the nanoindentation technique, showed extraordinary synergy with improvement by as much as 400% in stiffness and hardness compared to those obtained with single nanocarbon reinforcements. The synergistic effect was dramatic in the ND plus FG composites (PVA: 0.4FG-0.2ND) with 4- and 1.5-fold increases in elastic modulus and hardness, respectively (Figure 1.19b). Variation in the percentage crystallinity (%) of the polymer matrix composite (PMC)s with the two nanocarbons is around 2%, suggesting that increase in crystallinity was not the cause of the observed synergy.

1.4.7
Quenching of Fluorescence of Aromatics

An aspect of graphene chemistry that needs understanding relates to its function in donor–acceptor hybrids with semiconducing organic molecules and polymers. Interaction of electron-donor and electron-acceptor molecules with graphenes has been exploited recently to modify to the electronic properties of graphene through ground-state molecular charge-transfer interactions [12, 13, 201–203]. Fluorescence quenching properties of graphene have been exploited for use in sensitive and selective detection of biomolecules [204]. More recently, the quenching phenomenon has been used for high-contrast visualization of graphene oxide [205] and also in resonance Raman spectroscopy as a substrate to suppress fluorescence [206]. Quenching of the fluorescence of porphyrin by graphene and photophysical properties of porphyrin–graphene complexes has been reported [207, 208]. Theoretical studies show that long-range energy transfer is operative in the fluorescence quenching of a dye molecule in the presence of graphene. The quenching of the green emission of ZnO nanoparticles accompanying the photoreduction of graphene oxide is, however, caused by electron transfer from ZnO. Electron transfer has been similarly invoked in the case of TiO_2/graphene oxide [181].

The interaction of graphene with pyrene-butanaoic acid succinimidyl ester, (**PyBS**), **I**, and oligo(*p*-phenylenevinylene) methyl ester (**OPV ester**), **II**, was investigated using a graphene derivative, acid treated thermally exfoliated graphene (EGA), soluble in chloroform and DMF [209]. Absorption spectra of **PyBS**, **I**, in DMF and **OPV ester**, **II**, in chloroform solution (10^{-5} M) are shown in Figure 1.20a,b, respectively, in the presence of varying concentrations of the graphene, EGA. These spectra show characteristic absorption bands of **I** and **II**. The increase in intensities of these bands with the graphene concentration is entirely accounted for the increasing intensity of the graphene absorption band around 270 nm. Thus, electronic absorption spectra of **I** + EGA and **II** + EGA show no evidence of interaction between the two molecules in the ground state. No new absorption bands attributable to charge transfer are also seen. Unlike the absorption spectra, fluorescence spectra of **I** and **II** show remarkable changes on the addition of EGA. The intensity of the fluorescence bands decreases markedly with the increase in EGA concentration, as illustrated in the Figure 1.20c,d. Fluorescence decay measurements on **I** monitored at 395 nm could be fitted to a three-exponential decay [210] with lifetimes of 1.8, 5.7, and 38.7 ns. EGA addition causes a significant decrease in all the three lifetimes, with the values being 1.2, 4.6, and 29.1 ns, respectively, for the addition of 0.3 mg of EGA.

In Figure 1.21a, we compare the transient absorption spectrum of the pure **I** with that of **I** after the addition of 0.3 mg of graphene. The spectrum of **PyBS**

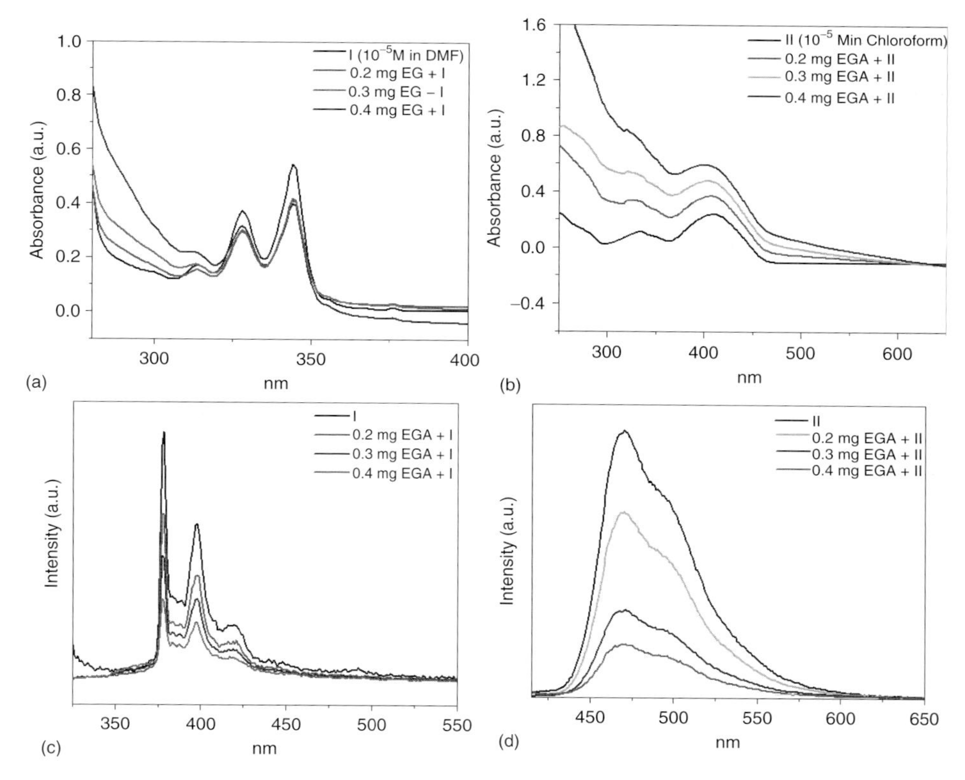

Figure 1.20 Electronic absorption spectra of (a) **PyBS**, **I**, (10^{-5} M in DMF) and (b) **OPV ester**, **II**, (10^{-5} M in chloroform), and fluorescence spectra of (c) **PyBS**, **I**, (10^{-5} M in DMF) and (d) **OPV ester**, **II**, (10^{-5} M in chloroform) with increasing concentration of graphene (EGA). (Source: Reprinted with permission from Ref. [209].)

shows an absorption maximum around 430 nm together with a broad band in the 450–530 nm range due to the triplet state [211]. On addition of EGA, new bands emerge around 470 and 520 nm in the transient absorption spectrum at 500 ns. The 470 nm band can be assigned to the pyrenyl radical cation as reported in the literature [210], suggesting the occurrence of photo-induced electron transfer from the **PyBS** to the graphene. Accordingly, we observe the transient absorption around 520 nm, which we assign to the graphene radical anion. The decay of the radical cation formed in the presence of graphene was fast, as evidenced from the appearance of a short-lived component (900 ns) in the decay profile (Figure 1.21b). However, the decay of the transient absorption of pure **PyBS** monitored at 470 nm (see inset in Figure 1.21b) shows a long-lived triplet with a lifetime of 6.17 μs. The transient absorption at 520 nm decays simultaneously with that of the pyrene radical cation indicating that it is due to the graphene radical anion.

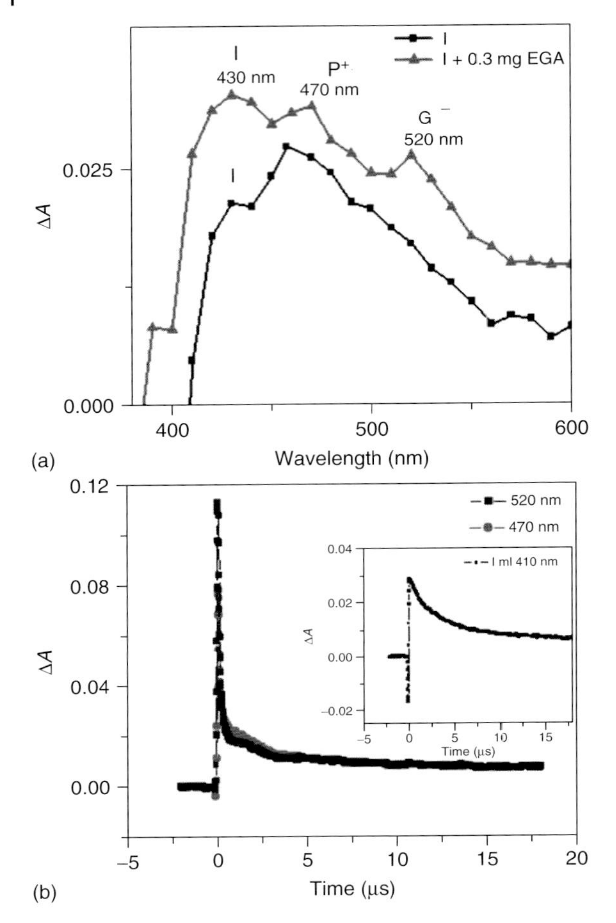

Figure 1.21 (a) Effect of addition of EGA on the transient absorption spectrum of **PyBS**, **I**, ($\lambda_{exc} = 355$ nm) after 500 ns. (b) Lifetime decay of transient species of **PyBS** + EGA recorded at 470 and 520 nm. Inset shows the decay of pure **PyBS** at 470 nm. (Source: Reprinted with permission from Ref. [209].)

1.4.8 Chemical Storage of Hydrogen and Halogens

Hydrogenation of graphene has been carried out with hydrogen plasma [212–215] and also from molecular H_2 by catalytic hydrogenation [216]. Elias *et al.* [212] reported reversible hydrogenation of graphene films prepared by micromechanical cleavage of graphite. Hydrogenation was obtained with cold hydrogen plasma containing hydrogen–argon mixture (10% H_2). The hydrogenated sample showed electronic behavior quite different from that of graphene, with evidence of metallic to insulator transition. The original properties of graphene were regained on heating the hydrogenated sample at 450 °C for 24 h. Catalytic hydrogenation of graphene using the radio frequency catalytic chemical vapor deposition (rf-cCVD) method was adopted by Zheng *et al.* [216] Ni (8 wt%) in Al_2O_3 was used as catalyst. Hydrogenation was confirmed by the appearance of peaks at 2920 and 2853 cm^{-1} in (C–H stretching modes) in IR spectra and increase in intensity of D-band in the Raman spectrum of hydrogenated samples. They also observed that on hydrogenation, samples turned hydrophobic, while the original graphene sample was hydrophilic. We have carried out plasma hydrogenation of graphene samples prepared by arc discharge in H_2 (HG). Elemental analysis of samples for two different plasma conditions (100 W, 20 min), one at room temperature and other at 170 °C, showed the presence of ~1.25 and 1.78 wt% of hydrogen, respectively. On plasma hydrogenation, the intensity of the D-band in the Raman spectrum increased with respect to the G-band, while that of the 2D-band decreased. Increased defect in graphene lattice caused origin of new band at 2909 cm^{-1}(D + G). Chemical hydrogenation of various graphene samples (EG and HG) by the Birch reduction has been performed [123]. Evidence of hydrogenation were obtained from appearance of C–H stretching modes in the IR spectra of hydrogenated samples, as can be seen from Figure 1.22a. An increase in the intensity of D-band relative to G-band in the Raman spectrum of hydrogenated exfoliation of graphite (EGH) and hydrogenated HG (HGH) samples reflects an increase in the sp^3 character. The UV spectrum of graphene is also affected on reduction, wherein the intensity of the 260 nm band decreases progressively, with a new band appearing around 235 nm. Magnetization of the graphene samples increases on hydrogenation. Elemental analysis of reduced graphene samples showed the hydrogen content to be around 5 wt%. Thermal analysis of EGH and HGH showed that these samples were stable at room temperature for prolonged period. Heating initiates the H_2 loss, as can be seen from Figure 1.22b. Almost all the H_2 was lost by 500 °C and the sample regained its original properties. Dehydrogenation could also be obtained by irradiation of the hydrogenated samples with UV rays or with a KrF excimer laser. On dehydrogenation, the sample regains all its properties and becomes graphene-like. The Birch reduction of GNRs prepared by oxidative unzipping of CNTs showed H_2 uptake of 3 wt%. Thermal analysis of the sample showed H_2 loss initiating at 300 °C and completing by 600 °C, with the release of 3.05 wt% H_2 during this period. Maximum H_2 loss occurs at 400 °C. The sample starts degrading at temperatures beyond 550 °C, which might be due to the presence of some functional groups from the initial oxidation treatment used to unzip the nanotubes.

We have examined the UV irradiation of graphene (HG and EG) in liquid chlorine medium resulting in graphene chlorination of 56 wt% [217]. The core level X-ray photoelectron spectrum of the product showed three features centered at 284.6, 285.8, and 287.6 eV corresponding to sp^2- and sp^3-hybridized carbons and C–Cl, respectively, as shown in Figure 1.22. The composition of the sample as determined by the ratio of the intensity of Cl 2p to that of the C 1s peak (taking into account the atomic sensitivity factors of Cl 2p and C 1s) was 30 at% of chlorine (~56 wt%). Interestingly, chlorination too was reversible. A temperature-dependent stability study of the chlorinated sample showed that the sample was stable at room

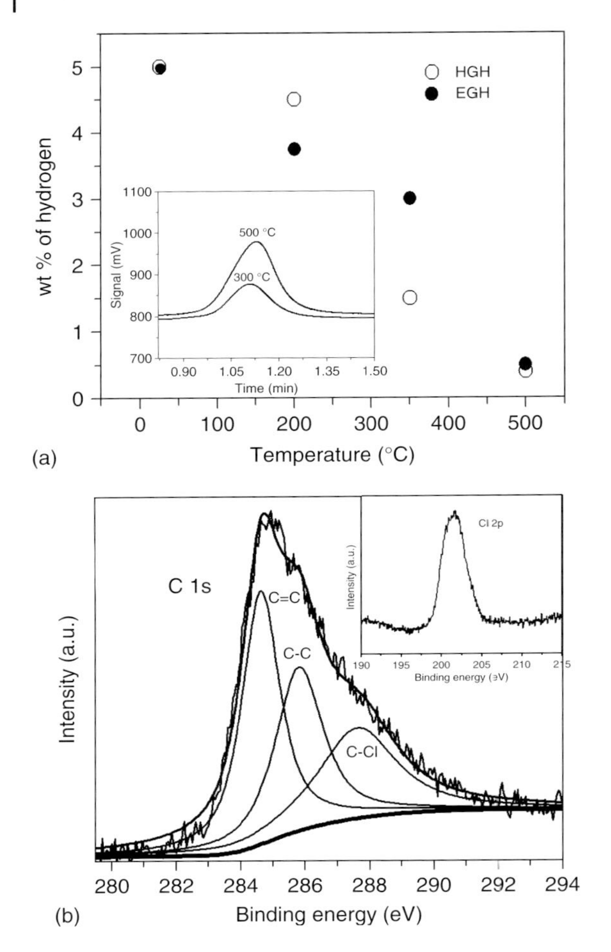

Figure 1.22 (a) Change in weight percentage of hydrogen released from EGH and HGH with temperature. In inset is shown the evolution of hydrogen as recorded by gas chromatography. (Source: Reprinted with permission from Ref. [123].) (b) C 1s core level XP spectrum of photo-chlorinated graphene. Inset shows the Cl 2p signal in XPS. (Source: Reprinted with permission from Ref. [217].)

temperature for long periods, but slowly lost Cl_2 on progressive heating, with complete loss of Cl_2 by 500 °C. IR spectra of samples taken at various stages of heating showed progressive decrease in the intensity of the C-Cl band at 790 cm^{-1}. We could also eliminate all the chlorine on irradiation with a laser (Lambda Physik KrF excimer laser ($\lambda = 248$ nm, $\tau = 30$ ns, rep. rate $= 5$ Hz, laser energy $= 370$ mJ)). Dechlorination appears to be associated with a small barrier just as the decomposition of hydrogenated graphene. The strain in the chlorinated sample appears to drive the dechlorination to form the more stable

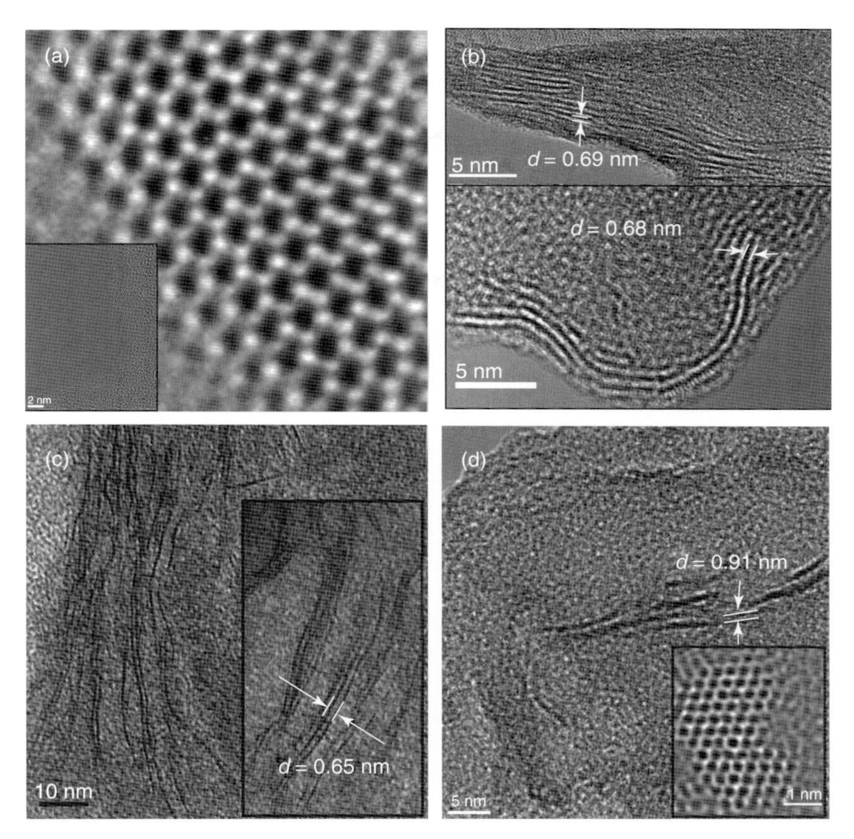

Figure 1.23 TEM images of MoS_2 layers obtained by (a) heating molybdic acid with an excess of thiourea at 773 K, (b) hydrothermal reaction between MoO_3 and KSCN, (c) high-resolution TEM image of layered MoS_2 from hydrothermal conditions, and (d) images of WS_2 layers obtained from Li intercalation and exfoliation of bulk WS_2 and heating molybdic acid with an excess of thiourea at 773 K, respectively. The bends in the layers may arise from defects. (Source: Reprinted with permission from Ref. [218].)

graphene. Bromination of graphene up to 25 wt% is achieved, and the bromine is fully eliminated by 500 °C. The study demonstrates that FG can be used to store chlorine and bromine.

1.5 Inorganic Graphene Analogs

There are many inorganic compounds with layered structures, the most well known being MoS_2, WS_2, and BN. Fullerene-type structures of these materials were made some years ago, soon followed by nanotube structures of these materials. It is, therefore, not surprising that one should be able to make graphene analogs of such layered materials. During the past year, graphene-like structures of MoS_2, WS_2, $MoSe_2$, WSe_2, BN, and other materials have indeed been prepared and characterized. The layered structures of MoS_2, WS_2, $MoSe_2$, and WSe_2 with different numbers of layers have been made by chemical methods and also by sonication in polar solvents [218–223]. The chemical methods include Li intercalation and exfoliation, hydrothermal synthesis, and reaction of molybdic and tungstic acids with excess of thiourea or selenourea in a N_2 atmosphere at 773 K [218, 223]. Figure 1.23a,b show graphene-like MoS_2 layers obtained by reaction of molybdic acid with thiourea and hydrothermal synthesis, respectively, with a layer separation in the range of 0.65–0.7 nm. The high-resolution image in Figure 1.23c shows the hexagonal structure formed by Mo and S atoms with a Mo–S distance of 2.30 Å. BN with different numbers of layers has also been made by chemical methods involving reaction of boric acid and urea at high temperatures [224]. WS_2 obtained by both hydrothermal and intercalation methods mostly consist of bilayers and single layers, as can be seen in the TEM images in Figure 1.23d,e. The spacing between the WS_2 layers in the bilayer sample is in the range of 0.65–0.70 nm. WS_2 layers obtained by the thiourea method show an interlayer spacing of 0.9 nm. BCN is another graphene analog obtained by the reaction of high-surface-area activated charcoal with a mixture of boric acid and urea or by vapor-phase synthesis from a mixture of BBr_3, ethylene, and ammonia [225, 226]. Besides their structural features, some of the properties of inorganic graphene analogs have been studied. For example, transistors have been made out of one- or few-layer MoS_2 and $MoSe_2$ [227]. Mechanical properties of polymer composites containing different number of layers of BN have been studied [228]. Greater improvement in mechanical properties is found to occur when BN with a fewer number of layers is imported into the composites. Clearly, several new graphene-like inorganic materials will be prepared in the next few years, and many of them may indeed possess interesting and useful properties.

References

1. Novoselov, K.S., Geim, A.K., Morozov, S.V., Jiang, D., Zhang, Y., Dubonos, S.V., Grigorieva, I.V., and Firsov, A.A. (2004) *Science*, **306**, 666.
2. Novoselov, K.S., Geim, A.K., Morozov, S.V., Jiang, D., Katsnelson, M.I., Grigorieva, I.V., Dubonos, S.V., and Firsov, A.A. (2005) *Nature*, **438**, 197.
3. Rao, C.N.R., Sood, A.K., Subrahmanyam, K.S., and Govindaraj, A. (2009) *Angew. Chem. Int. Ed.*, **48**, 7752.
4. Chen, J.-H., Jang, C., Xiao, S., Ishigami, M., and Fuhrer, M.S. (2008) *Nat. Nanotechnol.*, **3**, 206.
5. Han, M.Y., Oezyilmaz, B., Zhang, Y., and Kim, P. (2007) *Phys. Rev. Lett.*, **98**, 206805.
6. Nair, R.R., Blake, P., Grigorenko, A.N., Novoselov, K.S., Booth, T.J., Stauber, T., Peres, N.M.R., and Geim, A.K. (2008) *Science*, **320**, 1308.
7. Lee, C., Wei, X., Kysar, J.W., and Hone, J. (2008) *Science*, **321**, 385.
8. Wang, Y., Huang, Y., Song, Y., Zhang, X.Y., Ma, Y.F., Liang, J.J., and Chen, Y.S. (2009) *Nano Lett.*, **9**, 220.

9. Matte, H.S.S.R., Subrahmanyam, K.S., and Rao, C.N.R. (2009) *J. Phys. Chem. C*, **113**, 9982.
10. Peigney, A., Laurent, C., Flahaut, E., Bacsa, R.R., and Rousset, A. (2001) *Carbon*, **39**, 507.
11. Rao, C.N.R., Sood, A.K., Voggu, R., and Subrahmanyam, K.S. (2010) *J. Phys. Chem. Lett.*, **1**, 572.
12. Das, B., Voggu, R., Rout, C.S., and Rao, C.N.R. (2008) *Chem. Commun.*, 5155.
13. Rao, C.N.R. and Voggu, R. (2010) *Mater. Today*, **13**, 34.
14. Abergel, D.S.L., Russell, A., and Falko, V.I. (2007) *Appl. Phys. Lett.*, **91**, 063125.
15. Ni, Z.H., Wang, H.M., Kasim, J., Fan, H.M., Yu, T., Wu, Y.H., Feng, Y.P., and Shen, Z.X. (2007) *Nano Lett.*, **7**, 2758.
16. Gao, L., Ren, W., Li, F., and Cheng, H.-M. (2008) *ACS Nano*, **2**, 1625.
17. Hiura, H., Miyazaki, H., and Tsukagoshi, K. (2010) *Appl. Phys. Exp.*, **3**, 095101.
18. Kochat, V., Nath Pal, A., Sneha, E.S., Sampathkumar, A., Gairola, A., Shivashankar, S.A., Raghavan, S., and Ghosh, A. (2011) *J. Appl. Phys.*, **110**, 014315.
19. Gass, M.H., Bangert, U., Bleloch, A.L., Wang, P., Nair, R.R., and Geim, A.K. (2008) *Nat. Nanotechnol.*, **3**, 676.
20. Meyer, J.C., Kisielowski, C., Erni, R., Rossell, M.D., Crommie, M.F., and Zettl, A. (2008) *Nano Lett.*, **8**, 3582.
21. Wu, Y.H., Yu, T., and Shen, Z.X. (2010) *J. Appl. Phys.*, **108**, 071301.
22. Hernandez, Y., Nicolosi, V., Lotya, M., Blighe, F.M., Sun, Z., De, S., McGovern, I.T., Holland, B., Byrne, M., Gun'Ko, Y.K., Boland, J.J., Niraj, P., Duesberg, G., Krishnamurthy, S., Goodhue, R., Hutchison, J., Scardaci, V., Ferrari, A.C., and Coleman, J.N. (2008) *Nat. Nanotechnol.*, **3**, 563.
23. Geim, A.K. and Novoselov, K.S. (2007) *Nat. Mater.*, **6**, 183.
24. Hiura, H., Ebbesen, T.W., Fujita, J., Tanigaki, K., and Takada, T. (1994) *Nature*, **367**, 148.
25. Ebbesen, T.W. and Hiura, H. (1995) *Adv. Mater.*, **7**, 582.
26. Bernhardt, T.M., Kaiser, B., and Rademann, K. (1998) *Surf. Sci.*, **408**, 86.
27. Roy, H. (1998) *J. Appl. Phys.*, **83**, 4695.
28. Lu, X., Yu, M., Huang, H., and Ruoff, R.S. (1999) *Nanotechnology*, **10**, 269.
29. Roy, H.V., Kallinger, C., and Sattler, K. (1998) *Surf. Sci.*, **407**, 1.
30. Zhang, Y. (2005) *Appl. Phys. Lett.*, **86**, 073104.
31. Marcano, D.C., Kosynkin, D.V., Berlin, J.M., Sinitskii, A., Sun, Z., Slesarev, A., Alemany, L.B., Lu, W., and Tour, J.M. (2010) *ACS Nano*, **4**, 4806.
32. Park, S., An, J., Jung, I., Piner, R.D., An, S.J., Li, X., Velamakanni, A., and Ruoff, R.S. (2009) *Nano Lett.*, **9**, 1593.
33. Hummers, W.S. and Offeman, R.E. (1958) *J. Am. Chem. Soc.*, **80**, 1339.
34. Allen, M.J., Tung, V.C., and Kaner, R.B. (2009) *Chem. Rev.*, **110**, 132.
35. Tung, V.C., Allen, M.J., Yang, Y., and Kaner, R.B. (2009) *Nat. Nanotechnol.*, **4**, 25.
36. Paredes, J.I., Villar-Rodil, S., Marti'nez-Alonso, A., and Tasco'n, J.M.D. (2008) *Langmuir*, **24**, 10560.
37. Schniepp, H.C., Li, J.-L., McAllister, M.J., Sai, H., Herrera-Alonso, M., Adamson, D.H., Prud'homme, R.K., Car, R., Saville, D.A., and Aksay, I.A. (2006) *J. Phys. Chem. B*, **110**, 8535.
38. McAllister, M.J., Li, J.-L., Adamson, D.H., Schniepp, H.C., Abdala, A.A., Liu, J., Herrera-Alonso, M., Milius, D.L., Car, R., Prud'homme, R.K., and Aksay, I.A. (2007) *Chem. Mater.*, **19**, 4396.
39. Li, X., Zhang, G., Bai, X., Sun, X., Wang, X., Wang, E., and Dai, H. (2008) *Nat. Nanotechnol.*, **3**, 538.
40. Guoqing, X., Wontae, H., Namhun, K., Sung, M.C., and Heeyeop, C. (2010) *Nanotechnology*, **21**, 405201.
41. Liang, X., Chang, A.S.P., Zhang, Y., Harteneck, B.D., Choo, H., Olynick, D.L., and Cabrini, S. (2008) *Nano Lett.*, **9**, 467.
42. Qian, M. (2011) *Appl. Phys. Lett.*, **98**, 173108.

43. Viculis, L.M., Mack, J.J., Mayer, O.M., Hahn, H.T., and Kaner, R.B. (2005) *J. Mater. Chem.*, **15**, 974.
44. Valles, C., Drummond, C., Saadaoui, H., Furtado, C.A., He, M., Roubeau, O., Ortolani, L., Monthioux, M., and Pénicaud, A. (2008) *J. Am. Chem. Soc.*, **130**, 15802.
45. Ghosh, A., Rao, K.V., George, S.J., and Rao, C.N.R. (2010) *Chem. Eur. J.*, **16**, 2700.
46. Bourlinos, A.B., Georgakilas, V., Zboril, R., Steriotis, T.A., and Stubos, A.K. (2009) *Small*, **5**, 1841.
47. Hamilton, C.E., Lomeda, J.R., Sun, Z., Tour, J.M., and Barron, A.R. (2009) *Nano Lett.*, **9**, 3460.
48. ONeill, A., Khan, U., Nirmalraj, P.N., Boland, J., and Coleman, J.N. (2011) *J. Phys. Chem. C*, **115**, 5422.
49. Hernandez, Y., Lotya, M., Rickard, D., Bergin, S.D., and Coleman, J.N. (2009) *Langmuir*, **26**, 3208.
50. Wang, X., Fulvio, P.F., Baker, G.A., Veith, G.M., Unocic, R.R., Mahurin, S.M., Chi, M., and Dai, S. (2010) *Chem. Commun.*, **46**, 4487.
51. Choucair, M., Thordarson, P., and Stride, J.A. (2009) *Nat. Nanotechnol.*, **4**, 30.
52. Kim, K.S., Zhao, Y., Jang, H., Lee, S.Y., Kim, J.M., Kim, K.S., Ahn, J.-H., Kim, P., Choi, J.-Y., and Hong, B.H. (2009) *Nature*, **457**, 706.
53. Kwon, S.-Y., Ciobanu, C.V., Petrova, V., Shenoy, V.B., Bareno, J., Gambin, V., Petrov, I., and Kodambaka, S. (2009) *Nano Lett.*, **9**, 3985.
54. Sutter, P.W., Flege, J.-I., and Sutter, E.A. (2008) *Nat. Mater.*, **7**, 406.
55. Coraux, J., N'Diaye, A.T., Busse, C., and Michely, T. (2008) *Nano Lett.*, **8**, 565.
56. Li, X., Cai, W., An, J., Kim, S., Nah, J., Yang, D., Piner, R., Velamakanni, A., Jung, I., Tutuc, E., Banerjee, S.K., Colombo, L., and Ruoff, R.S. (2009) *Science*, **324**, 1312.
57. Reina, A., Jia, X., Ho, J., Nezich, D., Son, H., Bulovic, V., Dresselhaus, M.S., and Kong, J. (2008) *Nano Lett.*, **9**, 30.
58. Wang, J.J., Zhu, M.Y., Outlaw, R.A., Zhao, X., Manos, D.M., Holloway, B.C., and Mammana, V.P. (2004) *Appl. Phys. Lett.*, **85**, 1265.
59. Wang, J., Zhu, M., Outlaw, R.A., Zhao, X., Manos, D.M., and Holloway, B.C. (2004) *Carbon*, **42**, 2867.
60. Zhu, M., Wang, J., Holloway, B.C., Outlaw, R.A., Zhao, X., Hou, K., Shutthanandan, V., and Manos, D.M. (2007) *Carbon*, **45**, 2229.
61. Rao, C.N.R., Subrahmanyam, K.S., Ramakrishna Matte, H.S.S., Abdulhakeem, B., Govindaraj, A., Das, B., Kumar, P., Ghosh, A., and Late, D.J. (2010) *Sci. Technol. Adv. Mater.*, **11**, 054502.
62. Pati, S.K., Enoki, T., and Rao, C.N.R. (2010) *Graphene and its Fascinating Attributes*, World Scientific Publishing Co., Chennai.
63. Subrahmanyam, K.S., Panchakarla, L.S., Govindaraj, A., and Rao, C.N.R. (2009) *J. Phys. Chem. C*, **113**, 4257.
64. Panchakarla, L.S., Govindaraj, A., and Rao, C.N.R. (2009) *Inorg. Chim. Acta*, **363**, 4163.
65. Wu, Z.-S., Ren, W., Gao, L., Zhao, J., Chen, Z., Liu, B., Tang, D., Yu, B., Jiang, C., and Cheng, H.-M. (2009) *ACS Nano*, **3**, 411.
66. Zhiyong, W., Nan, L., Zujin, S., and Zhennan, G. (2010) *Nanotechnology*, **21**, 175602.
67. Li, N., Wang, Z., Zhao, K., Shi, Z., Gu, Z., and Xu, S. (2010) *Carbon*, **48**, 255.
68. Wu, C., Dong, G., and Guan, L. (2010) *Phys. E*, **42**, 1267.
69. Dreyer, D.R., Park, S., Bielawski, C.W., and Ruoff, R.S. (2010) *Chem. Soc. Rev.*, **39**, 228.
70. Stankovich, S., Dikin, D.A., Piner, R.D., Kohlhaas, K.A., Kleinhammes, A., Jia, Y., Wu, Y., Nguyen, S.T., and Ruoff, R.S. (2007) *Carbon*, **45**, 1558.
71. Shin, H.-J., Kim, K.K., Benayad, A., Yoon, S.-M., Park, H.K., Jung, I.-S., Jin, M.H., Jeong, H.-K., Kim, J.M., Choi, J.-Y., and Lee, Y.H. (2009) *Adv. Funct. Mater.*, **19**, 1987.
72. Pham, V.H., Cuong, T.V., Nguyen-Phan, T.-D., Pham, H.D., Kim, E.J., Hur, S.H., Shin, E.W., Kim, S., and Chung, J.S. (2010) *Chem. Commun.*, **46**, 4375.

73. Zhou, X., Zhang, J., Wu, H., Yang, H., Zhang, J., and Guo, S. (2011) *J. Phys. Chem. C*, **115**, 11957.
74. Zhu, C., Guo, S., Fang, Y., and Dong, S. (2010) *ACS Nano*, **4**, 2429.
75. Zhang, J., Yang, H., Shen, G., Cheng, P., Zhang, J., and Guo, S. (2010) *Chem. Commun.*, **46**, 1112.
76. Wang, G., Yang, J., Park, J., Gou, X., Wang, B., Liu, H., and Yao, J. (2008) *J. Phys. Chem.*, **112**, 8192.
77. Fan, X., Peng, W., Li, Y., Li, X., Wang, S., Zhang, G., and Zhang, F. (2008) *Adv. Mater.*, **20**, 4490.
78. Amarnath, C.A., Hong, C.E., Kim, N.H., Ku, B.-C., Kuila, T., and Lee, J.H. (2011) *Carbon*, **49**, 3497.
79. Guo, H.-L., Wang, X.-F., Qian, Q.-Y., Wang, F.-B., and Xia, X.-H. (2009) *ACS Nano*, **3**, 2653.
80. Sundaram, R.S., Gómez-Navarro, C., Balasubramanian, K., Burghard, M., and Kern, K. (2008) *Adv. Mater.*, **20**, 3050.
81. Compton, O.C., Jain, B., Dikin, D.A., Abouimrane, A., Amine, K., and Nguyen, S.T. (2011) *ACS Nano*, **5**, 4380.
82. Zhou, M., Wang, Y., Zhai, Y., Zhai, J., Ren, W., Wang, F., and Dong, S. (2009) *Chem. Eur. J.*, **15**, 6116.
83. Ding, Y.H. *et al.* (2011) *Nanotechnology*, **22**, 215601.
84. Cote, L.J., Cruz-Silva, R., and Huang, J. (2009) *J. Am. Chem. Soc.*, **131**, 11027.
85. Abdelsayed, V., Moussa, S., Hassan, H.M., Aluri, H.S., Collinson, M.M., and El-Shall, M.S. (2010) *J. Phys. Chem. Lett.*, **1**, 2804.
86. Kumar, P., Das, B., Chitara, B., Subrahmanyam, K.S., Gopalakrishnan, K., Krupanidhi, S.B., and Rao, C.N.R. (2012) *Macromol. Chem. Phys.*, **213**, 1146.
87. Kumar, P., Subrahmanyam, K.S., and Rao, C.N.R. (2011) *Mater. Exp.*, **1**, 252.
88. Baraket, M., Walton, S.G., Wei, Z., Lock, E.H., Robinson, J.T., and Sheehan, P. (2010) *Carbon*, **48**, 3382.
89. Ekiz, O.O., Urel, M., Guner, H., Mizrak, A.K., and Dana, A. (2011) *ACS Nano*, **5**, 2475.
90. Yao, P., Chen, P., Jiang, L., Zhao, H., Zhu, H., Zhou, D., Hu, W., Han, B.-H., and Liu, M. (2010) *Adv. Mater.*, **22**, 5008.
91. Hassan, H.M.A., Abdelsayed, V., Khder, A.E.R.S., AbouZeid, K.M., Terner, J., El-Shall, M.S., Al-Resayes, S.I., and El-Azhary, A.A. (2009) *J. Mater. Chem.*, **19**, 3832.
92. Li, Z., Yao, Y., Lin, Z., Moon, K.-S., Lin, W., and Wong, C. (2010) *J. Mater. Chem.*, **20**, 4781.
93. Li, X., Wang, X., Zhang, L., Lee, S., and Dai, H. (2008) *Science*, **319**, 1229.
94. Wang, X., Ouyang, Y., Li, X., Wang, H., Guo, J., and Dai, H. (2008) *Phys. Rev. Lett.*, **100**, 206803.
95. Cresti, A., Nemec, N., Biel, B., Niebler, G., Triozon, Fo., Cuniberti, G., and Roche, S. (2008) *Nano Res.*, **1**, 361.
96. Chen, Z., Lin, Y.-M., Rooks, M.J., and Avouris, P. (2007) *Phys. E*, **40**, 228.
97. Tapaszto, L., Dobrik, G., Lambin, P., and Biro, L.P. (2008) *Nat. Nanotechnol.*, **3**, 397.
98. Cano-Marquez, A.G., Rodrguez-Macias, F.J., Campos-Delgado, J., Espinosa-Gonzalez, C.G., Tristan-Lopez, F., Ramrez-Gonzalez, D., Cullen, D.A., Smith, D.J., Terrones, M., and Vega-Cantu, Y.I. (2009) *Nano Lett.*, **9**, 1527.
99. Kosynkin, D.V., Higginbotham, A.L., Sinitskii, A., Lomeda, J.R., Dimiev, A., Price, B.K., and Tour, J.M. (2009) *Nature*, **458**, 872.
100. Higginbotham, A.L., Kosynkin, D.V., Sinitskii, A., Sun, Z., and Tour, J.M. (2010) *ACS Nano*, **4**, 2059.
101. Jiao, L., Wang, X., Diankov, G., Wang, H., and Dai, H. (2011) *Nat. Nanotechnol.*, **6**, 132.
102. Kosynkin, D.V., Lu, W., Sinitskii, A., Pera, G., Sun, Z., and Tour, J.M. (2011) *ACS Nano*, **5**, 968.
103. Jiao, L., Zhang, L., Wang, X., Diankov, G., and Dai, H. (2009) *Nature*, **458**, 877.
104. Wu, Z.-S., Ren, W., Gao, L., Liu, B., Zhao, J., and Cheng, H.-M. (2010) *Nano Res.*, **3**, 16.
105. Kumar, P., Panchakarla, L.S., and Rao, C.N.R. (2011) *Nanoscale*, **3**, 2127.
106. Yu, W.J., Chae, S.H., Perello, D., Lee, S.Y., Han, G.H., Yun, M., and Lee, Y.H. (2010) *ACS Nano*, **4**, 5480.

107. Wang, X. and Dai, H. (2010) *Nat. Chem.*, **2**, 661.

108. Cai, J., Ruffieux, P., Jaafar, R., Bieri, M., Braun, T., Blankenburg, S., Muoth, M., Seitsonen, A.P., Saleh, M., Feng, X., Mullen, K., and Fasel, R. (2010) *Nature*, **466**, 470.

109. Yang, X., Dou, X., Rouhanipour, A., Zhi, L., Rader, H.J., and Mullen, K. (2008) *J. Am. Chem. Soc.*, **130**, 4216.

110. Yazyev, O.V. and Helm, L. (2007) *Phys. Rev. B*, **75**, 125408.

111. Bhowmick, S. and Shenoy, V.B. (2008) *J. Chem. Phys.*, **128**, 244717.

112. Joly, V.L.J., Takahara, K., Takai, K., Sugihara, K., Enoki, T., Koshino, M., and Tanaka, H. (2010) *Phys. Rev. B*, **81**, 115408.

113. Kopelevich, Y., da Silva, R.R., Torres, J.H.S., Penicaud, A., and Kyotani, T. (2003) *Phys. Rev. B*, **68**, 092408.

114. Talapatra, S., Ganesan, P.G., Kim, T., Vajtai, R., Huang, M., Shima, M., Ramanath, G., Srivastava, D., Deevi, S.C., and Ajayan, P.M. (2005) *Phys. Rev. Lett.*, **95**, 097201.

115. Shenoy, V.B. and Rao, C.N.R. (2008) *Philos. Trans. R. Soc. A*, **366**, 63.

116. Shenoy, V.B., Sarma, D.D., and Rao, C.N.R. (2006) *Chem. Phys. Chem.*, **7**, 2053.

117. Dutta, S., Lakshmi, S., and Pati, S.K. (2008) *Phys. Rev. B*, **77**, 073412.

118. Sato, H., Kawatsu, N., Enoki, T., Endo, M., Kobori, R., Maruyama, S., and Kaneko, K. (2003) *Solid State Commun.*, **125**, 641.

119. Hao, S., Takai, K., Feiyu, K., and Enoki, T. (2008) *Carbon*, **46**, 110.

120. Takai, K., Eto, S., Inaguma, M., Enoki, T., Ogata, H., Tokita, M., and Watanabe, J. (2007) *Phys. Rev. Lett.*, **98**, 017203.

121. Enoki, T. and Takai, K. (2008) *Dalton Trans.*, 3773.

122. Makarova, T. and Palacio, F. (2006) *Carbon Based Magnetism*, Elsevier, Amsterdam.

123. Subrahmanyam, K.S., Kumar, P., Maitra, U., Govindaraj, A., Hembram, K., Waghmare, U.V., and Rao, C.N.R. (2011) *Proc. Natl. Acad. Sci. U.S.A.*, **108**, 2674.

124. Late, D.J., Ghosh, A., Subrahmanyam, K.S., Panchakarla, L.S., Krupanidhi, S.B., and Rao, C.N.R. (2010) *Solid State Commun.*, **150**, 734.

125. Vasko, F.T. and Ryzhii, V. (2007) *Phys. Rev. B*, **76**, 233404.

126. Hwang, E.H. and Das Sarma, S. (2008) *Phys. Rev. B*, **77**, 115449.

127. Stauber, T., Peres, N.M.R., and Guinea, F. (2007) *Phys. Rev. B*, **76**, 205423.

128. Hwang, E.H., Adam, S., and Das Sarma, S. (2007) *Phys. Rev. Lett.*, **98**, 186806.

129. Chen, J.H., Jang, C., Adam, S., Fuhrer, M.S., Williams, E.D., and Ishigami, M. (2008) *Nat. Phys.*, **4**, 377.

130. Morozov, S.V., Novoselov, K.S., Katsnelson, M.I., Schedin, F., Elias, D.C., Jaszczak, J.A., and Geim, A.K. (2008) *Phys. Rev. Lett.*, **100**, 016602.

131. Du, X., Skachko, I., Barker, A., and Andrei, E.Y. (2008) *Nat. Nanotechnol.*, **3**, 491.

132. Schwierz, F. (2010) *Nat. Nanotechnol.*, **5**, 487.

133. Vollmer, A., Feng, X., Wang, X., Zhi, L., Mullen, K., Koch, N., and Rabe, J. (2009) *Appl. Phys. A*, **94**, 1.

134. Mak, K.F., Sfeir, M.Y., Wu, Y., Lui, C.H., Misewich, J.A., and Heinz, T.F. (2008) *Phys. Rev. Lett.*, **101**, 196405.

135. Kim, W.Y. and Kim, K.S. (2008) *Nat. Nanotechnol.*, **3**, 408.

136. Bai, J.W., Cheng, R., Xiu, F.X., Liao, L., Wang, M.S., Shailos, A., Wang, K.L., Huang, Y., and Duan, X.F. (2010) *Nat. Nanotechnol.*, **5**, 655.

137. Bai, J., Duan, X., and Huang, Y. (2009) *Nano Lett.*, **9**, 2083.

138. Heersche, H.B., Jarillo-Herrero, P., Oostinga, J.B., Vandersypen, L.M.K., and Morpurgo, A.F. (2007) *Nature*, **446**, 56.

139. Qifeng, L. and Jinming, D. (2008) *Nanotechnology*, **19**, 355706.

140. Masatsugu, S., Itsuko, S.S., and Jurgen, W. (2004) *J. Phys. Condens. Matter*, **16**, 903.

141. Subrahmanyam, K.S., Vivekchand, S.R.C., Govindaraj, A., and Rao, C.N.R. (2008) *J. Mater. Chem.*, **18**, 1517.

142. Vivekchand, S., Rout, C., Subrahmanyam, K., Govindaraj, A., and Rao, C. (2008) *J. Chem. Sci.*, **120**, 9.

143. Li, H.-Q., Wang, Y.-G., Wang, C.-X., and Xia, Y.-Y. (2008) *J. Power Source*, **185**, 1557.

144. Stoller, M.D., Park, S., Zhu, Y., An, J., and Ruoff, R.S. (2008) *Nano Lett.*, **8**, 3498.

145. Fan, Z., Yan, J., Zhi, L., Zhang, Q., Wei, T., Feng, J., Zhang, M., Qian, W., and Wei, F. (2010) *Adv. Mater.*, **22**, 3723.

146. Yan, J., Wei, T., Fan, Z., Qian, W., Zhang, M., Shen, X., and Wei, F. (2010) *J. Power Source*, **195**, 3041.

147. Yoo, E., Kim, J., Hosono, E., Zhou, H.-s., Kudo, T., and Honma, I. (2008) *Nano Lett.*, **8**, 2277.

148. Liu, Z., Liu, Q., Huang, Y., Ma, Y., Yin, S., Zhang, X., Sun, W., and Chen, Y. (2008) *Adv. Mater.*, **20**, 3924.

149. Blake, P., Brimicombe, P.D., Nair, R.R., Booth, T.J., Jiang, D., Schedin, F., Ponomarenko, L.A., Morozov, S.V., Gleeson, H.F., Hill, E.W., Geim, A.K., and Novoselov, K.S. (2008) *Nano Lett.*, **8**, 1704.

150. Wang, X., Zhi, L., Tsao, N., Tomovic, Z., Li, J., and Müllen, K. (2008) *Angew. Chem. Int. Ed.*, **120**, 3032.

151. Liu, Q., Liu, Z., Zhang, X., Yang, L., Zhang, N., Pan, G., Yin, S., Chen, Y., and Wei, J. (2009) *Adv. Funct. Mater.*, **19**, 894.

152. Wang, F., Zhang, Y., Tian, C., Girit, C., Zettl, A., Crommie, M., and Shen, Y.R. (2008) *Science*, **320**, 206.

153. Xia, F., Mueller, T., Lin, Y.-M., Valdes-Garcia, A., and Avouris, P. (2009) *Nat. Nanotechnol.*, **4**, 839.

154. Ryzhii, V., Ryzhii, M., and Otsuji, T. (2008) *Appl. Phys. Exp.*, **1**, 013001.

155. Ryzhii, V., Ryzhii, M., Ryabova, N., Mitin, V., and Otsuji, T. (2009) *Jpn. J. Appl. Phys.*, **48**, 04C144.

156. Chitara, B., Panchakarla, L.S., Krupanidhi, S.B., and Rao, C.N.R. (2011) *Adv. Mater.*, doi: 10.1002/adma.201101414

157. Rana, F., George, P.A., Strait, J.H., Dawlaty, J., Shivaraman, S., Chandrashekhar, M., and Spencer, M.G. (2009) *Phys. Rev. B*, **79**, 115447.

158. George, P.A., Strait, J., Dawlaty, J., Shivaraman, S., Chandrashekhar, M., Rana, F., and Spencer, M.G. (2008) *Nano Lett.*, **8**, 4248.

159. LeeEduardo, J.H., Balasubramanian, K., Weitz, R.T., Burghard, M., and Kern, K. (2008) *Nat. Nanotechnol.*, **3**, 486.

160. Xia, F., Mueller, T., Golizadeh-Mojarad, R., Freitag, M., Lin, Y.-m., Tsang, J., Perebeinos, V., and Avouris, P. (2009) *Nano Lett.*, **9**, 1039.

161. Mueller, T., Xia, F., Freitag, M., Tsang, J., and Avouris, P. (2009) *Phys. Rev. B*, **79**, 245430.

162. Lee, C., Kim, J.Y., Bae, S., Kim, K.S., Hong, B.H., and Choi, E.J. (2011) *Appl. Phys. Lett.*, **98**, 071905.

163. Chitara, B., Krupanidhi, S.B., and Rao, C.N.R. (2011) *Appl. Phys. Lett.*, **99**, 113114.

164. Eda, G., Unalan, H.E., Rupesinghe, N., Amaratunga, G.A.J., and Chhowalla, M. (2008) *Appl. Phys. Lett.*, **93**, 233502.

165. Malesevic, A., Kemps, R., Vanhulsel, A., Chowdhury, M.P., Volodin, A., and Van Haesendonck, C. (2008) *J. Appl. Phys.*, **104**, 084301.

166. Jung, S.M., Hahn, J., Jung, H.Y., and Suh, J.S. (2006) *Nano Lett.*, **6**, 1569.

167. Palnitkar, U.A., Kashid, R.V., More, M.A., Joag, D.S., Panchakarla, L.S., and Rao, C.N.R. (2010) *Appl. Phys. Lett.*, **97**, 063102.

168. Prashant, K., Panchakarla, L.S., Bhat, S.V., Urmimala, M., Subrahmanyam, K.S., and Rao, C.N.R. (2010) *Nanotechnology*, **21**, 385701.

169. Subrahmanyam, K.S., Kumar, P., Nag, A., and Rao, C.N.R. (2010) *Solid State Commun.*, **150**, 1774.

170. Robertson, J. and O'Reilly, E.P. (1987) *Phys. Rev. B*, **35**, 2946.

171. Bredas, J.L., Silbey, R., Boudreaux, D.S., and Chance, R.R. (1983) *J. Am. Chem. Soc.*, **105**, 6555.

172. Seshadri, R., Rao, C.N.R., Pal, H., Mukherjee, T., and Mittal, J.P. (1993) *Chem. Phys. Lett.*, **205**, 395.

173. Dong, X., Fu, D., Fang, W., Shi, Y., Chen, P., and Li, L.-J. (2009) *Small*, **5**, 1422.

174. Jung, N., Kim, N., Jockusch, S., Turro, N.J., Kim, P., and Brus, L. (2009) *Nano Lett.*, **9**, 4133.

175. Ghosh, S., Chaitanya Sharma Y, S.R.K., Pati, S.K., and Rao, C.N.R. (2012) *RSC Adv.*, **2**, 1181–1188.
176. Choudhury, D., Das, B., Sarma, D.D., and Rao, C.N.R. (2010) *Chem. Phys. Lett.*, **497**, 66.
177. Saha, S.K., Chandrakanth, R.C., Krishnamurthy, H.R., and Waghmare, U.V. (2009) *Phys. Rev. B*, **80**, 6.
178. Manna, A.K. and Pati, S.K. (2010) *Nanoscale*, **2**, 1190.
179. Wildgoose, G.G., Banks, C.E., and Compton, R.G. (2006) *Small*, **2**, 182.
180. Voggu, R., Pal, S., Pati, S.K., and Rao, C.N.R. (2008) *J. Phys. Condens. Matter*, **20**, 5.
181. Kamat, P.V. (2009) *J. Phys. Chem. Lett.*, **1**, 520.
182. Muszynski, R., Seger, B., and Kamat, P.V. (2008) *J. Phys. Chem. C*, **112**, 5263.
183. Subrahmanyam, K.S., Manna, A.K., Pati, S.K., and Rao, C.N.R. (2010) *Chem. Phys. Lett.*, **497**, 70.
184. Si, Y. and Samulski, E.T. (2008) *Chem. Mater.*, **20**, 6792.
185. Li, Y., Fan, X., Qi, J., Ji, J., Wang, S., Zhang, G., and Zhang, F. (2010) *Nano Res.*, **3**, 429.
186. Guo, S., Dong, S., and Wang, E. (2009) *ACS Nano*, **4**, 547.
187. Hong, W., Bai, H., Xu, Y., Yao, Z., Gu, Z., and Shi, G. (2010) *J. Phys. Chem. C*, **114**, 1822.
188. Wang, Y., Ni, Z., Hu, H., Hao, Y., Wong, C.P., Yu, T., Thong, J.T.L., and Shen, Z.X. (2010) *Appl. Phys. Lett.*, **97**, 163111.
189. Ma, J., Zhang, J., Xiong, Z., Yong, Y., and Zhao, X.S. (2010) *J. Mater. Chem.*, **21**, 3350.
190. Xu, C., Wang, X., and Zhu, J. (2008) *J Phys. Chem. C*, **112**, 19841.
191. Williams, G. and Kamat, P.V. (2009) *Langmuir*, **25**, 13869.
192. Das, B., Choudhury, B., Gomathi, A., Manna, A.K., Pati, S.K., and Rao, C.N.R. (2011) *Chem. Phys. Chem.*, **12**, 937.
193. Patchkovskii, S., Tse, J.S., Yurchenko, S.N., Zhechkov, L., Heine, T., and Seifert, G. (2005) *Proc. Natl. Acad. Sci. U.S.A.*, **102**, 10439.
194. Ghosh, A., Subrahmanyam, K.S., Krishna, K.S., Datta, S., Govindaraj, A., Pati, S.K., and Rao, C.N.R. (2008) *J. Phys. Chem. C*, **112**, 15704.
195. Kumar, N., Subrahmanyam, K.S., Chaturbedy, P., Raidongia, K., Govindaraj, A., Hembram, K.P.S.S., Mishra, A.K., Waghmare, U.V., and Rao, C.N.R. (2011) *Chem. Sus. Chem.*, **4**, 1662–1670.
196. Stankovich, S., Dikin, D.A., Dommett, G.H.B., Kohlhaas, K.M., Zimney, E.J., Stach, E.A., Piner, R.D., Nguyen, S.T., and Ruoff, R.S. (2006) *Nature*, **442**, 282.
197. Ramanathan, T., Abdala, A.A., Stankovich, S., Dikin, D.A., Herrera-Alonso, M., Piner, R.D., Adamson, D.H., Schniepp, H.C., Chen, X., Ruoff, R.S., Nguyen, S.T., Aksay, I.A., Prud'Homme, R.K., and Brinson, L.C. (2008) *Nat. Nanotechnol.*, **3**, 327.
198. Das, B., Prasad, K.E., Ramamurty, U., and Rao, C.N.R. (2009) *Nanotechnology*, **20**, 125705.
199. Rafiee, M.A., Rafiee, J., Wang, Z., Song, H., Yu, Z.-Z., and Koratkar, N. (2009) *ACS Nano*, **3**, 3884.
200. Prasad, K.E., Das, B., Maitra, U., Ramamurty, U., and Rao, C.N.R. (2009) *Proc. Natl. Acad. Sci. U.S.A.*, **106**, 13186.
201. Voggu, R., Das, B., Rout, C.S., and Rao, C.N.R. (2008) *J. Phys. Condens. Matter*, **20**, 472204.
202. Subrahmanyam, K.S., Voggu, R., Govindaraj, A., and Rao, C.N.R. (2009) *Chem. Phys. Lett.*, **472**, 96.
203. Varghese, N., Ghosh, A., Voggu, R., Ghosh, S., and Rao, C.N.R. (2009) *J. Phys. Chem. C*, **113**, 16855.
204. Lu, C.-H., Yang, H.-H., Zhu, C.-L., Chen, X., and Chen, G.-N. (2009) *Angew. Chem. Int. Ed.*, **48**, 4785.
205. Kim, J., Cote, L.J., Kim, F., and Huang, J. (2009) *J. Am. Chem. Soc.*, **132**, 260.
206. Xie, L., Ling, X., Fang, Y., Zhang, J., and Liu, Z. (2009) *J. Am. Chem. Soc.*, **131**, 9890.
207. Xu, Y., Liu, Z., Zhang, X., Wang, Y., Tian, J., Huang, Y., Ma, Y., Zhang, X., and Chen, Y. (2009) *Adv. Mater.*, **21**, 1275.

208. Xu, Y., Zhao, L., Bai, H., Hong, W., Li, C., and Shi, G. (2009) *J. Am. Chem. Soc.*, **131**, 13490.

209. Ramakrishna Matte, H.S.S., Subrahmanyam, K.S., Venkata Rao, K., George, S.J., and Rao, C.N.R. (2011) *Chem. Phys. Lett.*, **506**, 260.

210. Álvaro, M., Atienzar, P., Bourdelande, J.L., and García, H. (2004) *Chem. Phys. Lett.*, **384**, 119.

211. Shafirovich, V.Y., Levin, P.P., Kuzmin, V.A., Thorgeirsson, T.E., Kliger, D.S., and Geacintov, N.E. (1994) *J. Am. Chem. Soc.*, **116**, 63.

212. Elias, D.C., Nair, R.R., Mohiuddin, T.M.G., Morozov, S.V., Blake, P., Halsall, M.P., Ferrari, A.C., Boukhvalov, D.W., Katsnelson, M.I., Geim, A.K., and Novoselov, K.S. (2009) *Science*, **323**, 610.

213. Jaiswal, M., Yi Xuan Lim, C.H., Bao, Q., Toh, C.T., Loh, K.P., and Ozyilmaz, B. (2011) *ACS Nano*, **5**, 888.

214. Xie, L., Jiao, L., and Dai, H. (2010) *J. Am. Chem. Soc.*, **132**, 14751.

215. Jones, J.D., Hoffmann, W.D., Jesseph, A.V., Morris, C.J., Verbeck, G.F., and Perez, J.M. (2010) *Appl. Phys. Lett.*, **97**, 233104.

216. Zheng, L., Li, Z., Bourdo, S., Watanabe, F., Ryerson, C.C., and Biris, A.S. (2010) *Chem. Commun.*, **47**, 1213.

217. Gopalakrishnan, K., Subrahmanyam, K.S., Kumar, P., Govindaraj, A., and Rao, C.N.R., (2012) *RSC Adv.*, **2**, 1605–1608.

218. Ramakrishna Matte, H.S.S., Gomathi, A., Manna, A.K., Late, D.J., Datta, R., Pati, S.K., and Rao, C.N.R. (2010) *Angew. Chem. Int. Ed.*, **49**, 4059.

219. Yang, D., Sandoval, S.J., Divigalpitiya, W.M.R., Irwin, J.C., and Frindt, R.F. (1991) *Phys. Rev. B*, **43**, 12053.

220. Gordon, R.A., Yang, D., Crozier, E.D., Jiang, D.T., and Frindt, R.F. (2002) *Phys. Rev. B*, **65**, 125407.

221. Schumacher, A., Scandella, L., Kruse, N., and Prins, R. (1993) *Surf. Sci.*, **289**, L595.

222. Yang, D. and Frindt, R.F. (1999) *J. Phys. Chem. Solids*, **57**, 1113.

223. Matte, H.S.S.R., Plowman, B., Datta, R., and Rao, C.N.R. (2011) *Dalton Trans.*, **40**, 10322.

224. Nag, A., Raidongia, K., Hembram, K.P.S.S., Datta, R., Waghmare, U.V., and Rao, C.N.R. (2010) *ACS Nano*, **4**, 1539.

225. Kumar, N., Raidongia, K., Mishra, A.K., Waghmare, U.V., Sundaresan, A., and Rao, C.N.R. (2011) *J Solid State Chem.*, **184**, 2902.

226. Raidongia, K., Nag, A., Hembram, K.P.S.S., Waghmare, U.V., Datta, R., and Rao, C.N.R. (2010) *Chem. Eur. J.*, **16**, 149.

227. Radisavljevic, B., Radenovic, A., Brivio, J., Giacometti, V., and Kis, A. (2011) *Nat. Nanotechnol.*, **6**, 147.

228. Kiran, M.S.R.N., Raidongia, K., Ramamurty, U., and Rao, C.N.R. (2011) *Scr. Mater.*, **64**, 592.

2
Understanding Graphene via Raman Scattering

A. K. Sood and Biswanath Chakraborty

2.1
Introduction

Graphene, a two-dimensional form of carbon, is the most recently discovered allotrope in 2004 [1], following discoveries of zero-dimensional fullerenes and one-dimensional nanotubes in the past decades. This two-dimensional material has unique physical properties as compared to the three-dimensional allotropic forms of carbons, namely, diamond and graphite, as well as the above-mentioned lower dimensional forms [2–4]. Not surprising then that graphene has attracted immense interest in the past few years, both for fundamental basic science and possible applications, with the Nobel Prize in physics awarded to its discoverers, Andrei Geim and Konstantin Novoselov in 2010 [5].

Ever since the discovery of graphene, micro-Raman spectroscopy and imaging have emerged as powerful probes to characterize the quality and number of layers and their stacking order; and for quantitative measurement of electron–phonon coupling, doping levels, strain, edge chirality in nanoribbons; and for phonon dispersions near the K-point of the Brillouin zone [6, 7]. The phonons involved are the wave vector $q \approx 0$, Γ point phonons, as well as $q \neq 0$ defect-induced and higher order combination modes. As Raman scattering from phonons involves intermediate (virtual and real) electronic states, resonant Raman scattering using different excitation laser wavelengths gives unambiguous information about the electronic states of graphene. In this chapter, we review the progress made in understanding the physics of single-layer graphene (SLG), bilayer graphene (BLG), and a few-layer graphene (FLG) using Raman scattering as a probe.

2.2
Atomic Structure and Electronic Structure of Graphene

Graphene is a bipartite lattice of carbon atoms arranged in a honeycomb lattice, as shown in Figure 2.1a. The unit cell in real space (shaded portion in Figure 2.1a) contains two carbon atoms belonging to two different sublattices A and B. The

Graphene: Synthesis, Properties, and Phenomena, First Edition. Edited by C. N. R. Rao and A. K. Sood.

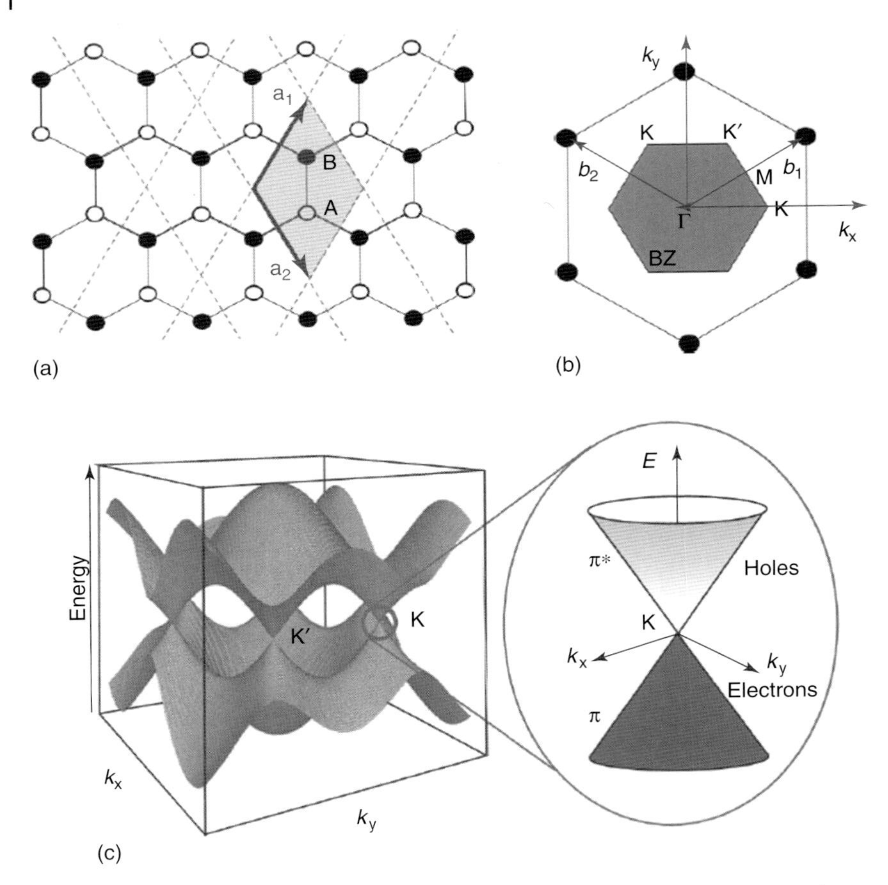

Figure 2.1 (a) Lattice structure of graphene. The unit cell is shown as the shaded rhombus. It has two atom bases (A and B). $\vec{a}_1$ and $\vec{a}_2$ are the primitive vectors. (b) The hexagonal first Brillouin zone shown in shaded area. $\vec{b}_1$ and $\vec{b}_2$ are the primitive vectors in the reciprocal space. Γ, K, K′, and M are the high-symmetry points. (c) Graphene band structure. Enlargement of the band structure close to the K or K′ point showing the Dirac cone. (Source: Adapted from [2].)

Brillouin zone is a hexagon (shaded area in Figure 2.1b) where the high-symmetry points Γ, K, K′, and M are marked. The inequivalent reciprocal lattice points K and K′ are connected by time-reversal symmetry.

One of the most important aspects of electronic dispersion of SLG realized a long time ago [8] is the linear relation between energy E and momentum $\mathbf{k}$ at the K- and K′-points

$$E(k) = \pm \hbar v_F k \tag{2.1}$$

where "+" ("−") refers to conduction (valence) band, v_F is the Fermi velocity ($\sim 10^6$ cm^{-1}), and k is measured from the K or K′ point. Figure 2.1c shows the "Dirac cone" band structure, similar to that for light. The Hamiltonian near K (K′)

can be approximated by the Dirac equation for effective massless particles

$$H_K = \hbar v_F \begin{pmatrix} 0 & k_x - ik_y \\ k_x + k_y & 0 \end{pmatrix} \tag{2.2}$$

which can be written as

$$H_K = \hbar v_F \begin{pmatrix} \sigma_x & \sigma_y \end{pmatrix} \begin{pmatrix} k_x \\ k_y \end{pmatrix} = v_F \sigma \cdot \mathbf{p} \tag{2.3}$$

where σ_x and σ_y are the Pauli spin matrices and $\mathbf{p}$ is the momentum operator. Here, σ is a pseudospin reflecting the relative contributions of A and B sublattices to the electronic wave functions near the K point. An interesting issue in recent years is the manipulation of pseudospin using elastic strain that changes the electron-hopping amplitude between carbon atoms [9] and hence induces an effective vector potential. This pseudo-magnetic field arising from the strain-induced gauge field has been experimentally shown [10] to be as large as 300 T from scanning tunneling microscopy of highly strained nanobubbles in graphene. This possibility of manipulating electronic states using pseudo-magnetic fields opens up many interesting challenges such as pseudo quantum Hall effect [11]. Recently, it has been shown that pseudospin dependence of electron–phonon matrix element results in the polarization dependence of D-band as well as its splitting for an armchair edge in graphene [12].

2.3 Phonons and Raman Modes in Graphene

The point group symmetry of SLG is D_{6h}, while that of the Bernal-stacked (AB stacking) bilayer is D_{3d}. For the trilayer, there are two different stacking sequences: ABA (Bernal) type and ABC (Rhombohedral) type with D_{3h} and D_{3d} point group symmetry, respectively [13]. The phonon dispersion relation for the SLG [14] is shown in Figure 2.2. For SLG with two atoms in the unit cell, there are three optical and three acoustic branches. At Γ-point ($q = 0$), optical phonons belong to the irreducible representation E_{2g} (Raman active) and B_{2g}. The E_{2g} mode is a degenerate transverse optical (TO) and longitudinal optical (LO) mode with in-plane vibrations of two neighboring atoms in opposite directions at a frequency of $\sim$1582 cm^{-1}. The branches ZO and ZA refer to vibrations of atoms out of the graphene plane. It may be noted that for the ZA branch, unlike the other acoustic branches for longitudinal acoustic (LA) and transverse acoustic (TA) phonons, the dispersion is quadratic ($\omega \propto q^2$). For the BLG with AB stacking, there are four atoms in the unit cell and the phonons belong to the irreducible representation $2E_{2g}$ (R) $-$ E_{1u} (IR) $+2B_{2g} + A_{2u}$ [3]. Figure 2.3 shows eigenvectors of six normal modes at Γ point of SLG (A. Das, Graphene and carbon nanotubes: Field induced doping, interaction with nucleobases, confined water and sensors. PhD thesis. Indian Institute of science, Bangalore, India, unpublished.).

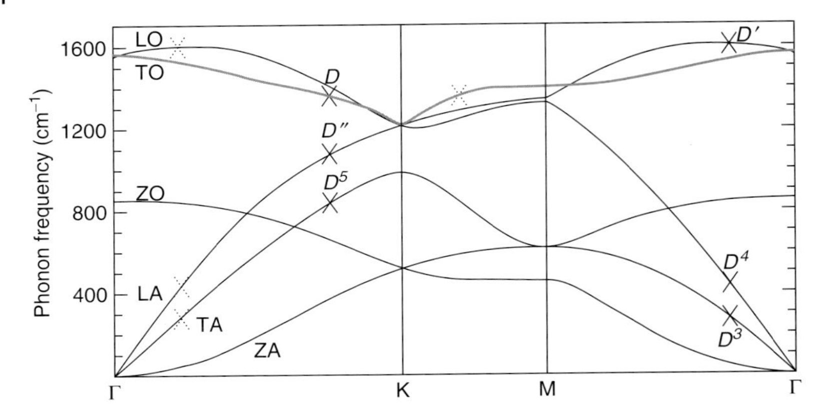

Figure 2.2 Phonon dispersion curves for monolayer graphene along the high-symmetry directions. The bold crosses indicate that the phonons along these branches are responsible for Raman modes D, D′, D″, D^3, D^4, and D^5. (Source: Taken from Ref. [14].)

The eigenvectors of six in-phase and six out-of-phase phonons of BLG are also shown.

First-order Raman scattering from crystalline medium involves phonons with wave vector $q \approx 0$. The Raman line at ~1582 cm^{-1} in SLG and multilayer graphene (MLG) as well as in graphite, usually termed as G-band, is associated with the E_{2g} phonon near the Γ-point in the Brillouin zone. Figure 2.4 shows a typical Raman spectrum of the SLG recorded in the authors' laboratory using 514.5 nm laser radiation. Besides the symmetry-allowed G-band, a number of Raman bands are observed between 1100 and 4500 cm^{-1} associated with higher order Raman processes involving phonons with $q \neq 0$ [6, 7, 14]. These can be divided into two parts: (i) defect-induced modes where the additional momentum to achieve total momentum transfer almost equal to zero is provided by elastic scattering from defects and (ii) excitation of two phonons with wave vectors **q** and -**q**, which does not require any defect-induced scattering for wave vector compensation. In the first category, defect-induced prominent Raman lines are the D line at ~ 1350 cm^{-1} (using laser excitation energy of 2.41 eV), D′ line at ~1620 cm^{-1} and the D″ line at ~1100 cm^{-1}. The other possible Raman modes (expected to be much weaker) are D^3 (~250 cm^{-1}), D^4 (~380 cm^{-1}), and D^5 (~800 cm^{-1}) marked in Figure 2.2 [14]. The D line observed in all graphitic materials with disorder is associated with phonons near the $\vec{K}$ point on the TO branch along the K-Γ direction. A weak D″ band also involves phonons near the $\vec{K}$ point, but on the LA branch (Figure 2.2). The D′ line is associated with $q \neq 0$ phonons near the Γ-point. We later point out that the D-band involves phonon-induced intervalley scattering, while D′ involves intravalley scattering.

In the second category, the prominent lines are 2D band (also called G′) at ~2700 cm^{-1}, D + D″ line at ~2400 cm^{-1}, and 2D′ at ~3200 cm^{-1}. Many interesting features of the Raman bands associated with $q \neq 0$ phonons are understood in terms of the double-resonance (DR) process. As compared to the first-order Raman

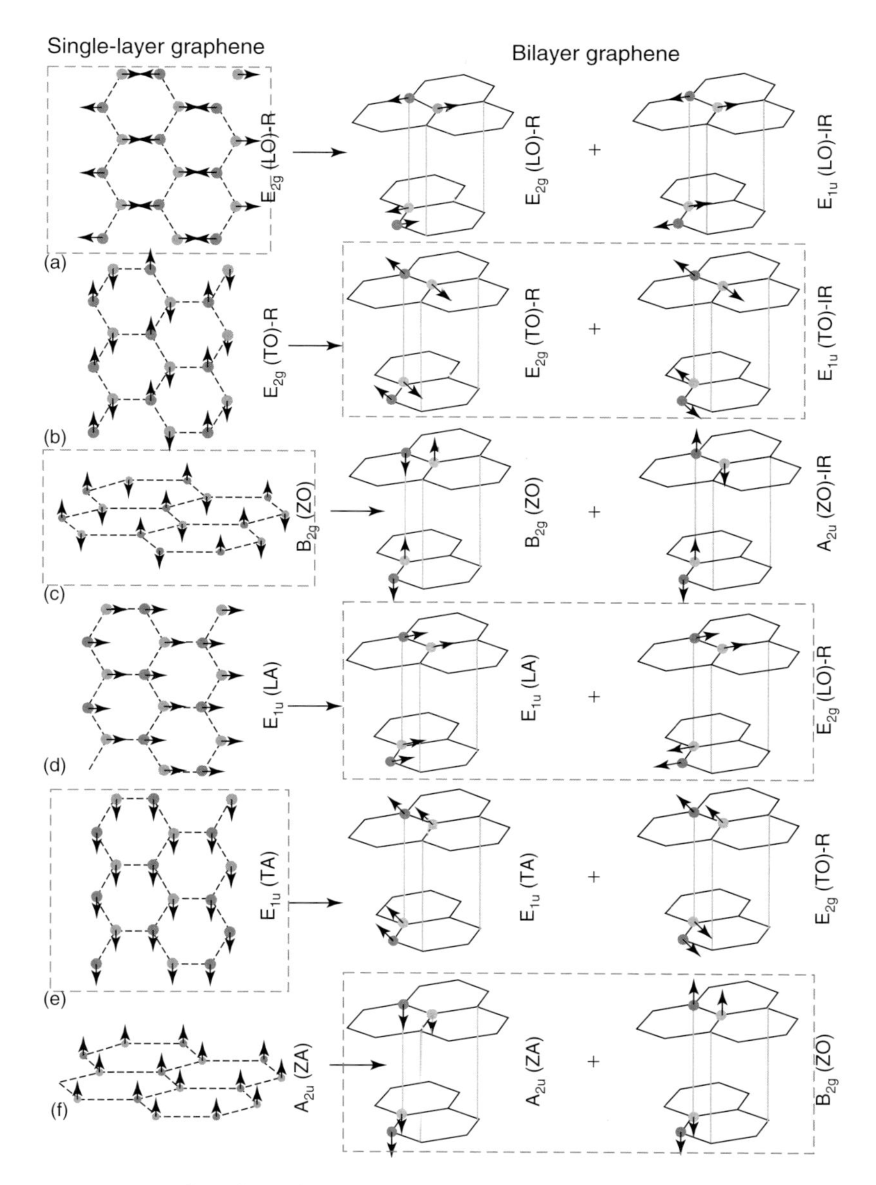

Figure 2.3 (a–f) (Left panel) Eigenvectors of six normal modes at Γ point of SLG. Eigenvectors of six in-phase (middle panel) and six out-of-phase (right panel) normal modes at Γ point of BLG. The R and IR correspond to the Raman active and infrared active modes, respectively (A. Das, Graphene and carbon nanotubes: Field induced doping, interaction with nucleobases, confined water and sensors. PhD thesis. Indian Institute of science, Bangalore, India, unpublished.).

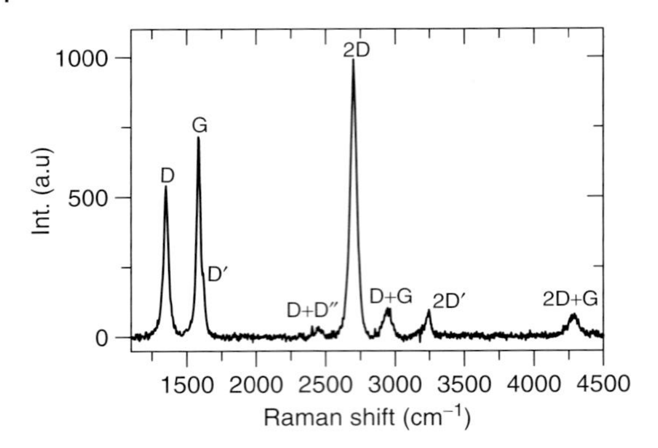

Figure 2.4 Raman spectrum of a single-layer graphene recorded with a 514.5 nm laser line. The modes are marked.

scattering (G-band), which is a third-order process, Raman scattering from the D, 2D, and other combination modes is a fourth-order process (within perturbation theory) where the cross section I for any excitation laser energy E_{L} is given by Venezuela *et al.* [14], Martin and Falicov [15]

$$I \propto \sum_{f} \left| \sum_{a,b,c} \frac{H_{fc} H_{cb} H_{ba} H_{ai}}{\left(E_L - E_c - i\frac{\gamma_c}{2}\right)\left(E_L - E_b - i\frac{\gamma_b}{2}\right)\left(E_L - E_a - i\frac{\gamma_a}{2}\right)} \right|^2 \delta(E_{\mathrm{L}} - E_f) \tag{2.4}$$

The summation is performed on the intermediate electronic states a, b, and c incorporating electronic and phononic excitations of the medium with energies E_a, E_b, and E_c. The intermediate electronic excitations serve as different quantum pathways, resulting in resonant Raman scattering and quantum interference [16]. The latter, in the context of graphene, has been seen in the nonmonotonic doping dependence of the G-band integrated intensity [17]. The final state f with energy E_f corresponds to the electronic ground state of the medium with one- or two-phonon excitations. The quantities γ_i ($i = a, b, c$) are inverse of the lifetimes of the electronic intermediate states. H_{lm} is the first-order scattering matrix element between the states l and m. The matrix element H_{ai} corresponds to electron–radiation interaction relating to absorption of light by the creation of electron–hole pair in the $\pi - \pi^*$ bands (Figure 2.5). The matrix element H_{ba} describes the scattering of carriers from state a (with momentum $\mathbf{k}$ as measuredfrom K to b (momentum $\mathbf{k} + \mathbf{q}$) by emission of a phonon (Stokes process) of momentum -$\mathbf{q}$ (electron–phonon interaction). The third step is different in the defect-induced process vis-a-vis two-phonon scattering. In the former, H_{cb} describes the elastic scattering of state b to c, returning the momentum of the carriers to $\mathbf{k}$ without a change in energy. In the two-phonon process, H_{cb} is the electron–phonon matrix element to describe the creation of a phonon of momentum $\mathbf{q}$. The final step is the electron–hole recombination in the $\mathbf{k}$ state (see vertical line marked 4 in Figure 2.5a). The intermediate matrix elements H_{ba} and H_{cb} can involve phonon- and defect-mediated scattering of electrons

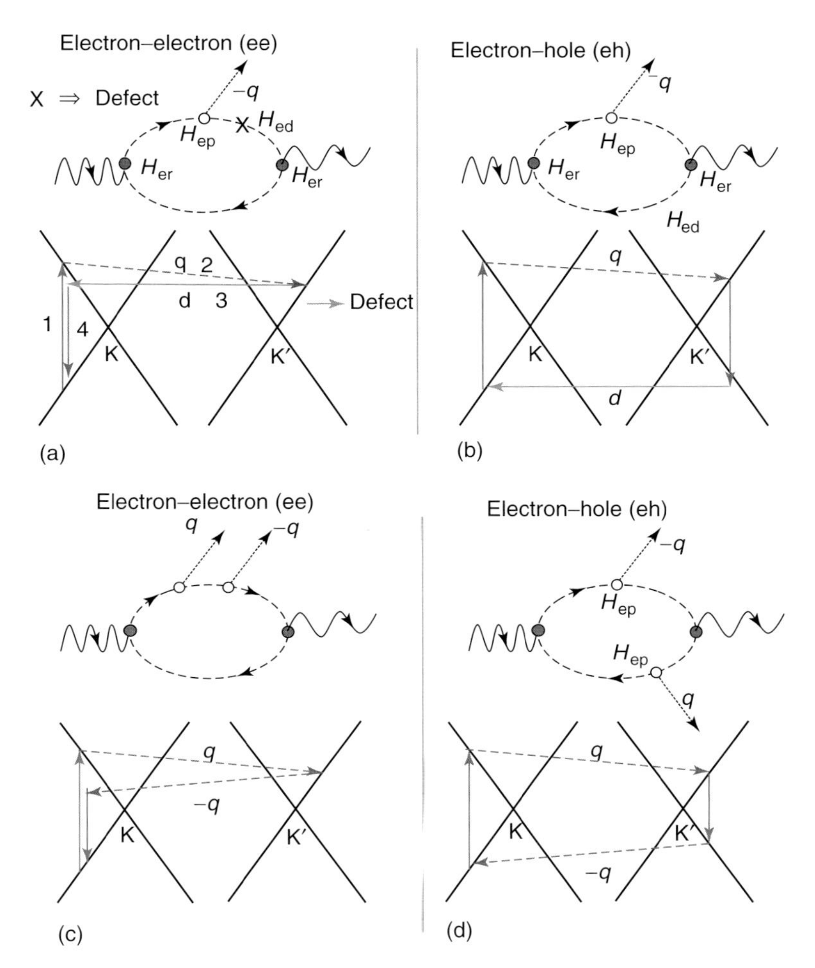

Figure 2.5 D mode double resonance in single-layer graphene considering only (a) electron–electron (e–e) and (b) electron–hole (e–h) scattering. Double resonance showing 2D mode considering (c) e–e and (d) e–h scattering. The dotted lines indicate phonon scattering and "d" indicates defect scattering. The vertices in the Feynman diagram indicate different interactions: H_{er} denotes electron–radiation interaction, H_{ep} denotes electron–phonon interaction, and H_{ed} denotes electron–defect interaction.

(e) and holes (h), and hence, Equation 2.4 will have eight different possibilities, elaborated clearly in Ref. [14] along with the comparison with other theoretical papers. Venezuela et al. [14] show that the process termed as "ab" where two intermediate scattering events are associated, one to an electron state and other to a hole state ("eh" or "he"), makes the largest contributions to the Raman intensity as compared to "aa" (intermediate scattering events involving both electrons or hole states; "ee" or "hh") processes due to quantum interference. We note that usually the higher order Raman scattering involving fourth-order process for $q \neq 0$ phonon scattering should be weaker than the third-order process because the three energy denominators are nonzero (recall that the energy need not be conserved during the intermediate virtual transition) [15, 18]. However, this is not the case for graphene, where two or more of the denominators can be equal to zero simultaneously because phonons fulfilling precisely the resonance conditions are involved, thus being termed as the DR process. For D and 2D modes, the phonons involved are near the K-point along the K - Γ direction (Figure 2.2) whose frequency depends on the q almost linearly. When different laser energies (E_L) are used (see step 1 in Figure 2.5a), the wave vector of the phonon satisfying the resonance condition in Equation 2.4 will increase with E_L, making the peak frequency of the Raman D-band dependent on E_L. Experimentally, it is seen that $\frac{d\omega_D}{dE_L} \sim 50\ \text{cm}^{-1}\,\text{eV}^{-1}$ and $\frac{d\omega_{2D}}{dE_L} \sim 100\ \text{cm}^{-1}\,\text{eV}^{-1}$ [6]. The dispersion of Raman bands involving $q \neq 0$ phonons is a helpful way to confirm the assignment of modes. Similar to the modes involving phonons that connect two valleys (intervalley), we have intravalley Raman scattering as well, where states within the same valley are connected by phonon scattering (Figure 2.6). The D′ and 2D″ are such modes where D′ is a defect-induced mode and 2D′ is a two-phonon overtone mode.

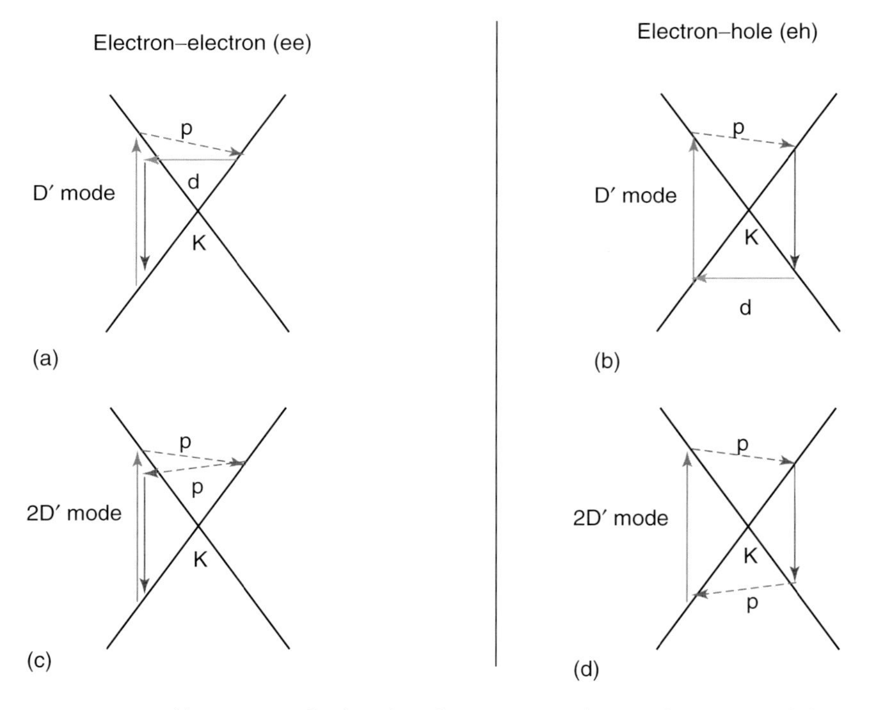

Figure 2.6 Double resonance for the D′ mode in SLG considering only (a) e–e and (b) e–h scattering. Double resonance showing 2D′ mode involving (c) e–e and (d) e–h scattering . As in D and 2D, the e–h contribution is maximum due to quantum interference effects. "p (d)" stands for phonon (defect) scattering.

2.4
Layer Dependence of Raman Spectra

2.4.1
G-Band

The G-band frequency does not depend on the number of layers. When the G-band frequency is seen to be blue shifted in graphene samples [19] (i.e., more than $\sim 1582\,cm^{-1}$), it is most likely due to unintentional doping by electrons or holes from the substrate or due to environment or functionalization of graphene. In such cases, the line width is also seen to be smaller than that for the pristine graphene. Another factor that can contribute to the G-band shift is strain introduced by the processing conditions [20].

2.4.2
2D-Band

The most distinguishing feature of the Raman spectrum depending on the number of layers is the 2D-band [22]. Figure 2.7 shows the 2D-band for layers $n = 1, 2, 3$, and 4 and the highly oriented pyrolytic graphite (HOPG) [6]. The 2D-band in SLG is a single band at $\sim 2685\,cm^{-1}$ with a full width at half maximum (FWHM) of $\sim 25\,cm^{-1}$ [22–24]. In comparison, the 2D-band in BLG can be convincingly deconvoluted into four Lorentzians marked $2D_{uu}$, $2D_{ul}$, $2D_{lu}$, and $2D_{ll}$ in Figure 2.7 [22, 25, 26]. Among these, the two central components are strong. For three layers ($n = 3$), the band is decomposed into six components. For $n > 5$, the 2D mode is indistinguishable from that of HOPG, which can be decomposed into two

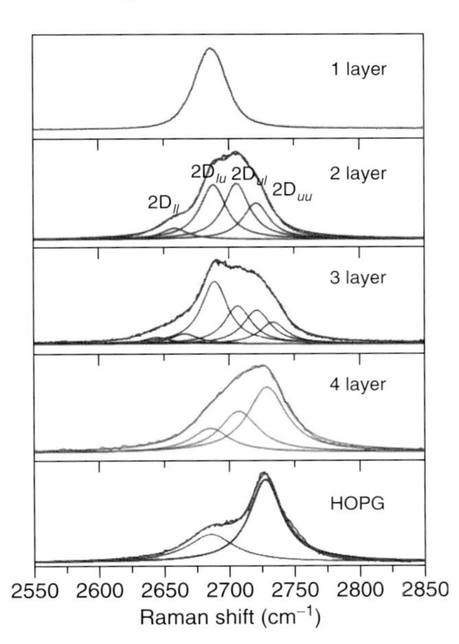

Figure 2.7 2D-band evolution with layer number. The four peaks in the bilayer spectrum are marked following the scheme of Figure 2.8. (Source: Taken from Ref. [6].).

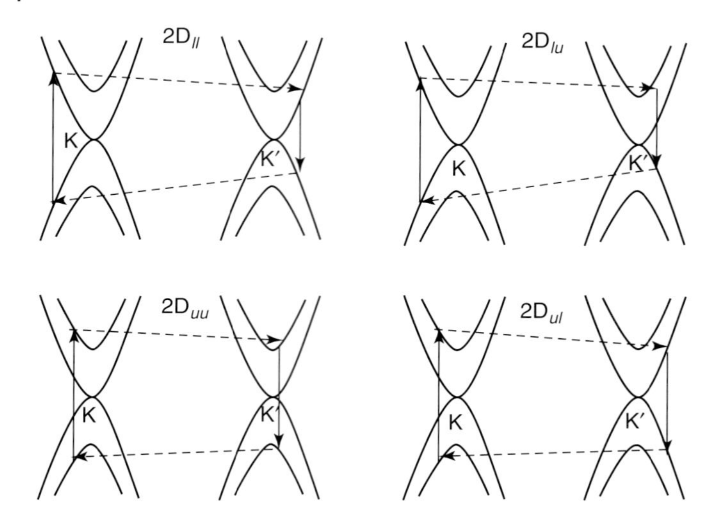

Figure 2.8 Outer processes leading to the four peaks in the bilayer 2D spectrum. Only the e–h processes are shown.

components. It is worth pointing out that the 2D mode in turbostatic graphite with AA stacking also has a single 2D band, but its peak frequency is upshifted by $\sim$50 cm^{-1} and the FWHM is more than double that of the SLG [22, 27].

The dependence of the Raman 2D-band lineshape on the number of layers can be understood from the DR mechanism that incorporates the layer-dependent electronic band structure [22]. For SLG, the DR process is shown in Figure 2.5, which explains the single 2D-band. However, for the BLG, the conduction and the valence band at the K- and K′-points are split because of interlayer coupling, as shown in Figure 2.8. The four subbands in the 2D spectrum arise because of four possible intervalley transitions, involving the upper (u) and lower (l) bands in the "eh" process as shown in Figure 2.8 [22]. Recently, it has been demonstrated [26] that the inner processes shown in Figure 2.9 are dominant as compared to the outer processes (Figure 2.8). For the inner processes, phonons along the K-Γ participate, whereas for the outer processes, the phonons involved are along K-M [14].

The 2D-band can also be used to determine the stacking sequence. For example, in the case of the ABC-stacked trilayer, the peaks are blue shifted as compared

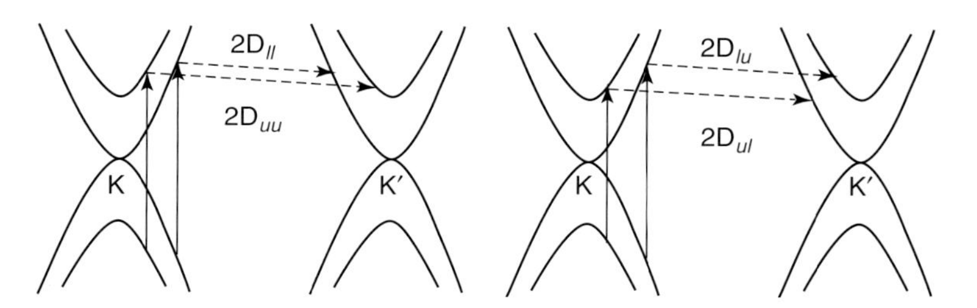

Figure 2.9 Inner processes contributing to the 2D-band.

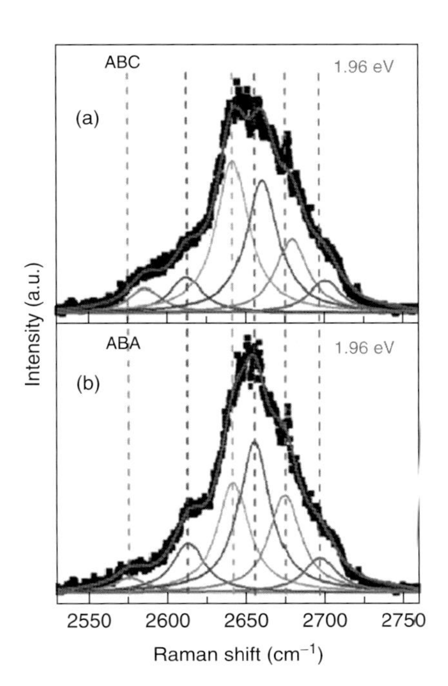

Figure 2.10 2D mode for (a) ABC- and (b) ABA-stacked trilayer graphene recorded with 1.96 eV laser energy. The Lorentzian profiles for the six subbands are shown, and the vertical dashed lines denote the positions of the subband peaks. (Source: Taken from Ref. [21].)

to ABA (Bernal stacking) stacking [21]. This is shown in Figure 2.10, where the spectrum for both the stacking types is fitted with six Lorentzians with FWHM $\sim$25 cm^{-1}.

2.4.3 D-Band

The effect of the number of layers on the defect-induced D-band is similar to that of the 2D-band, as shown in Figure 2.11, for the Raman spectra recorded from the edges of n-layer graphene ($n = 1, 2, 3, 4$, and 20) [28]. While monolayer edges produce a single-component (single Lorentzian) D-band, multicomponent (four Lorentzian) features are present for $n = 2$–4. For HOPG, only a doublet feature is seen.

In a recent study [29], it has been reported that the stacking sequence in the bilayer affects the D-band. An extra band at $\sim$1374 cm^{-1} in the incommensurate bilayer graphene (termed as *IBLG*) with AA stacking is seen, for which a quantitative explanation is lacking.

2.4.4 Combination Modes in the Range 1650–2300 cm^{-1}

Several new modes in the range 1650–2300 cm^{-1} have been reported recently, which show significant layer dependence [29, 30]. The characteristic and distinguishable peaks between 1650 and 2300 cm^{-1} are shown in Figure 2.12a [29]. These

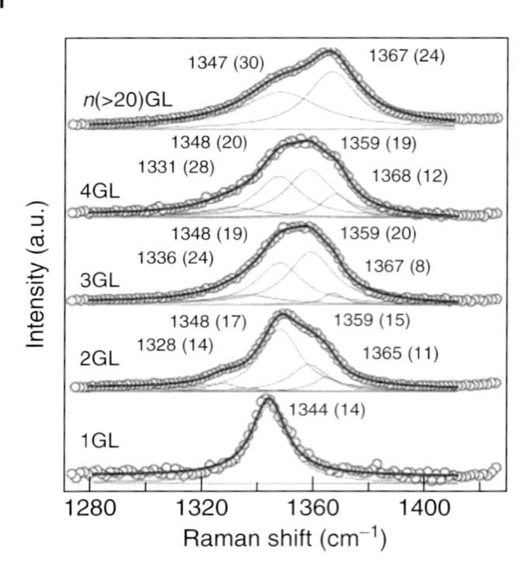

Figure 2.11 Raman spectra showing the D-band for different layers recorded with 514.5 nm laser line. (Source: Taken from Ref. [28].)

features are much lower in intensity than the G-band. The double-peak feature at $\sim 1750\,\text{cm}^{-1}$ marked "M" is an overtone of the ZO phonon and is activated by the interlayer coupling. It is thus completely suppressed in the spectra of SLG and IBLG. The lower frequency component of this doublet, "M^-," is red shifted in BLG as compared to FLG and HOPG (Figure 2.12a). The spectra in the 1840–1890 cm^{-1}

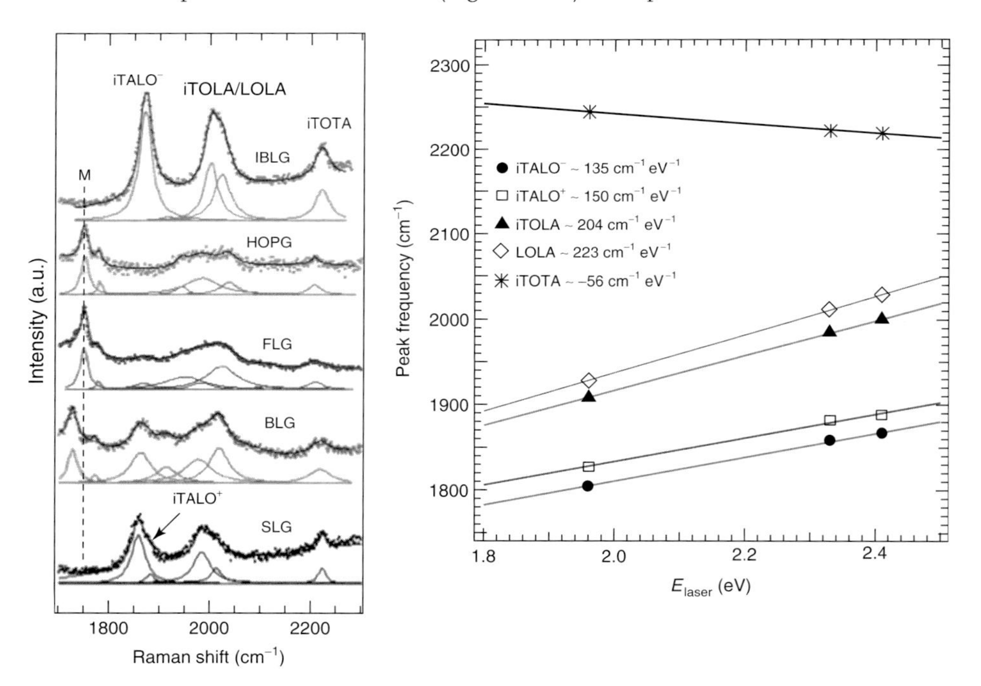

Figure 2.12 (a) Raman spectra recorded in the range 1650–2300 cm^{-1} from graphene samples using laser energy 2.33 eV. (b) Dispersion of the modes in the range 1650–2300 cm^{-1}. (Source: Taken from Ref. [29].)

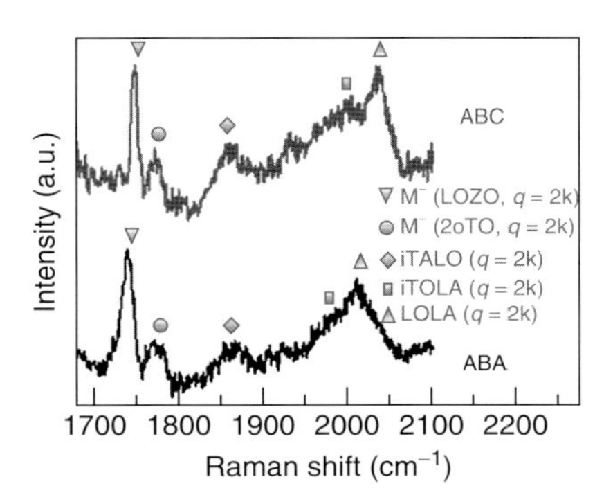

Figure 2.13 Stacking-dependent Raman spectra from trilayer graphene with ABA and ABC stackings. (Source: Taken from Ref. [21].)

range can be deconvoluted into two features, the lower one at 1860 cm^{-1} for SLG, termed *iTALO*$^-$, is a combination of iTA and LO phonons. The notation iTO and iTA in Ref. [29] refer to the TO and TA branches, respectively, in Figure 2.2. The higher frequency component iTALO$^+$ is greatly suppressed for IBLG. This mode has been assigned to a combination of ZO and LO phonons around the K-point of the Brillouin zone. In general, the intensities of both these peaks iTALO$^-$ and iTALO$^+$ decrease with the layer thickness. The third set of doublet in this range is a combination of LA and iTO (lower frequency peak at ~1983 cm^{-1} for SLG) and a combination of LO and LA (higher frequency peak at ~2012 cm^{-1} for SLG). The last distinctive feature appears at ~2200 cm^{-1} (for SLG) and is observed for all the graphene samples. This peak is a combination of iTA and iTO phonons around the K-point of the Brillouin zone. These modes also show the dispersive behavior with laser energy, which is shown in Figure 2.12b [29] where the value of dispersion is also given. It can be seen that these modes are more dispersive than the 2D mode because one component of the combination mode, LA or TA, is more dispersive in ω–q parameter space than the TO branch near the K-point (Figure 2.2). As shown for the 2D mode, the stacking sequence affects the modes in these energy regions as well and is shown for trilayer graphene with ABA and ABC stackings in Figure 2.13 [21].

2.4.5 Low-Frequency Modes

Recently, it was predicted that the low-frequency out-of-plane optical phonon near ~110 cm^{-1} is sensitive to the number of layers and could be seen in the Raman spectrum [31, 32]. Experimental verification is still to come.

2.5 Phonon Renormalization Due to Electron and Hole Doping of Graphene

The self-energy of a phonon arising from the electron-phonon coupling (EPC) renormalizes the phonon frequency and its lifetime via its real and imaginary parts, respectively. A quantitative understanding of EPC is necessary for the realization of graphene-based electronic and optoelectronic devices.

Studies have shown that the degenerate E_{2g} phonon at the Γ-point has a strong EPC leading to Kohn anomalies (KAs) [33–35]. KA refers to the anomalous screening of phonons of a particular wave vector $\mathbf{q}$ satisfying the relation $\mathbf{q} = \mathbf{k}_1 - \mathbf{k}_2$, where $\mathbf{k}_1$ and $\mathbf{k}_2$ are the wave vectors of the electron states at the Fermi level [36]. In graphene, the Fermi surface is just two points at **K** and **K**′. As $\mathbf{K}' = 2\mathbf{K}$, these are connected by vector **K**. KA in graphene thus occurs for phonon wave vector $\mathbf{q} = 0$ (for scattering between same K- or K′-point) and $\mathbf{q} = \mathbf{K}$ (electron scattering between K- and K′-points) corresponding to Γ- and **K**-point, respectively, in the phonon dispersion curve. We may recall that in a normal metal, KA occurs for phonons with wave vector $q = 2k_F$ (k_F being the Fermi wave vector) and not for $q = 0$ phonons. The Dirac cones in graphene at K and K′ result in KA for the Γ-point phonons.

Another important aspect of EPC in graphene is the breakdown of adiabatic approximation [34, 36–38]. Adiabatic approximation is mostly used when the electrons move sufficiently fast so that they can follow the slow motion of the heavy nuclei. The motion of the electrons can then be expressed as a function of atomic positions. When the atomic motion becomes faster than the time required for electronic momentum relaxation (through scattering mechanisms), the excited electrons do not have sufficient time to attain their instantaneous ground state. The electrons and phonons are thus coupled and cannot be treated separately under adiabatic approximation. This approximation is applicable when the energy gap between the ground and excited state is larger than the energy scale of the nuclear motion. Thus for metals, adiabatic approximation fails, and in gapless SLG, phenomena beyond the adiabatic Born–Oppenheimer approximation become important.

The best way to dope graphene is by gating in a field-effect transistor (FET) configuration. Usually, the gating is done in back-gate geometry [36, 37] using ~300 nm SiO_2 as a dielectric material with gate capacitance of ~12 nF cm^{-2}. In comparison, we have shown that electrolyte top gating using a solid polymer electrolyte has much higher capacitance ~1.5 μF cm^{-2} [38], and hence a very high doping of $\sim 5 \times 10^{13}$ cm^{-2} can be achieved using a modest gate voltage of ~2–3 V. Figure 2.14a shows typical Raman spectra of a monolayer graphene with positive gate voltage (electron doping) as well as negative gate voltage (hole doping) [38]. The gate voltage is quantitatively converted into carrier concentration n_c by including geometrical capacitance (C_G) as well as quantum capacitance (C_Q) [38]. Figure 2.14b,c shows that the G-band frequency increases and the FWHM decreases on doping with electrons as well as holes. These results are similar to SiO_2 back-gate graphene transistor where the maximum doping achieved was $\sim 7 \times 10^{12}$ cm^{-2} [36, 37]. The effect of chemical doping by Boron (hole) and nitrogen (electron) on the frequency shift of the G-mode is similar to electrochemical doping [39]. In comparison, the behavior of the 2D mode is quite different; the 2D mode frequency increases on hole doping and on electron doping, the mode frequency shows a very small (<1 cm^{-1}) increase till $n_c = 2 \times 10^{13}$ cm^{-2} (change in Fermi

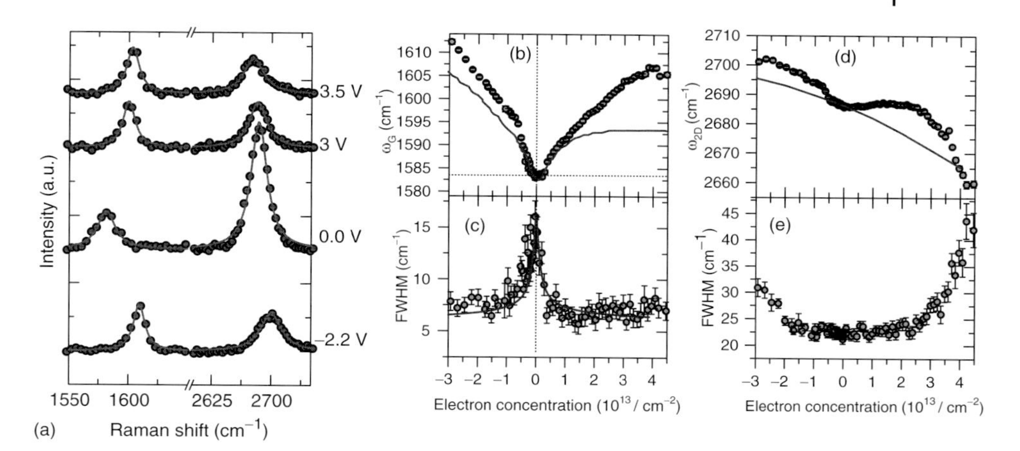

Figure 2.14 (a) The Raman spectra recorded with 514.5 nm laser wavelength. Points are experimental data and lines are Lorentzian profiles. Carrier concentration dependence of (b) G-band frequency (c) FWHM of G-band, (d) 2D-band frequency, and (e) FWHM of 2D band. (Source: Taken from [38].)

energy $\Delta E_F \sim 700$ meV) and then softens by a large amount $\sim$25 cm^{-1}. The line width does not change till $|n_c| \sim 2 \times 10^{13}$ cm^{-2}, beyond which it increases for both electron and hole doping [38].

The results for the G-band, namely, the blueshift and decrease of FWHM, can be explained as shown in Figure 2.15, where both real and virtual phonon-induced creation of electron–hole pairs are shown. The G-phonon can create electron–hole pairs as long as the Fermi energy shift (either for electron or for hole doping) is smaller than half the phonon energy (Figure 2.15a). The FWHM decreases because of the blockage of the decay channel of phonons into electron–hole pairs due to the Pauli exclusion principle when the electron–hole gap becomes higher than the

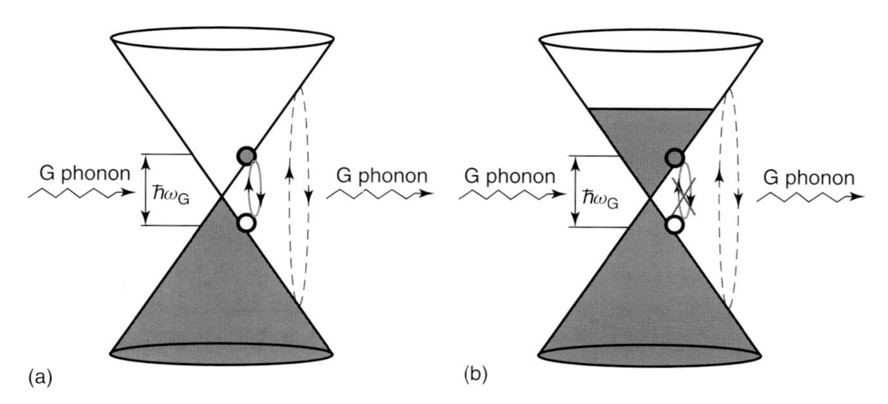

Figure 2.15 (a) Decay of phonon into electron–hole pair in an undoped SLG. (b) Blockage of phonon decay channel when $E_F > \hbar\omega_G$. Solid lines indicate real transitions and dashed lines denote virtual transitions.

phonon energy. The G-peak stiffening is due to the nonadiabatic removal of the KA from the Γ-point.

The frequency shift and the line width are given by

$$\hbar\Delta\omega_G = \text{Re}[\Pi(E_F) - \Pi(E_F = 0)]. \tag{2.5}$$

$$\text{FWHM(G)}^{\text{EPC}} = 2\text{Im}[\Pi(E_F)] \tag{2.6}$$

where Π is the phonon self-energy. The self-energy corrections are obtained from time-dependent perturbation theory (TDPT), which is expressed, for a phonon of wave vector q, as [34, 35]

$$\Pi(E_F) = \sum_{\mathbf{k},\mathbf{k}',s,s'} |W_{\mathbf{kq}}|^2 \frac{[f(\epsilon_{sk}) - f(\epsilon_{s'k'})]}{\epsilon_{s\mathbf{k}} - \epsilon_{s'\mathbf{k}'} + \hbar\omega_{\mathbf{q}} + i\delta} \tag{2.7}$$

where $W_{\mathbf{kq}}$ is the strength of the electron–phonon interaction, $f(\epsilon) = 1/[\exp(\frac{\epsilon - E_F}{k_B T}) + 1]$ is the Fermi–Dirac distribution, $\mathbf{k}' = \mathbf{k} + \mathbf{q}$, and δ is a broadening factor accounting for charge inhomogeneity. Here, $s, s' = +1$ for conduction band and -1 for valence band. The doping dependence given by Equation 2.7 is symmetric for both electron and hole doping. However, as seen in Figure 2.14b, some asymmetry is observed which is accounted for by doping-induced changes in the equilibrium lattice parameter. In later, hole doping results in frequency increase, whereas electron doping leads to softening [34].

The solid lines in Figure 2.14b are the theoretical curves taking into account the dynamic correction as well as a small contribution from the change in lattice constant [38]. A comparison of the experiment and theory for the G-mode frequency shift and line width give a quantitative estimate of the electron–phonon interaction strength or the deformation potential. It may be noted that the agreement between theory and experimental frequency shift is not good for $|n_c| \geq 1 \times 10^{13}\,\text{cm}^{-2}$ ($\Delta E_F \sim 500$ meV). One possible reason for the discrepancy between theory and experiment could be the electron –electron correlations at high doping. However, it has been shown that the deformation potential of the $\Gamma - E_{2g}$ G-mode depends negligibly on electron –electron correlations [40]. This is not the case for the phonons at the K-point, which results in the doping dependence of the D-band dispersion, as shown in Figure 2.16 [40]. This dependence has been recently verified experimentally for SLG (Chakraborty, B., Muthu, D.V.S., and A.K. Sood Experimental doping dependence of the dispersion of D and 2D modes in graphene, unpublished.).

As compared to the G-band, the 2D mode dependence on doping is different; the 2D frequency increases for hole doping and decreases for electron doping, while the line width increases for both, as shown in Figures 2.14d and e [38]. For 2D-band, the influence of the dynamic effects are negligible because the phonons giving rise to the 2D-peak are far away from the KA at **K** [41]. The 2D-peak position as a function of carrier concentration can be estimated by incorporating the lattice relaxation effect within the framework of density functional theory (DFT), (see the solid line in Figure 2.14d). The theoretical results do capture the trend but lack a quantitative agreement. Thus, we see that the G-band can be used as a readout

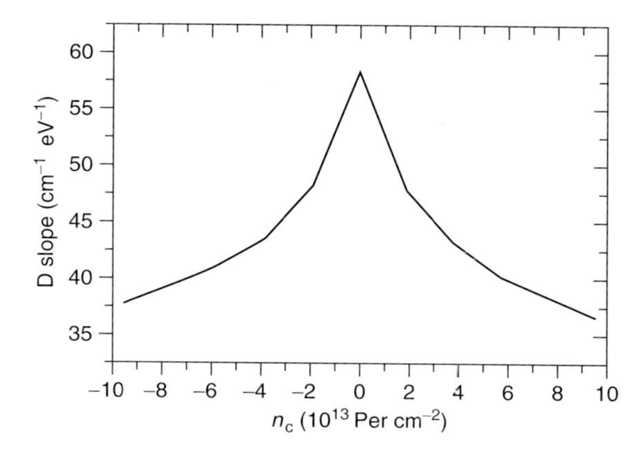

Figure 2.16 Variation of D-band dispersion on laser energy as a function of carrier concentration n_c. (Source: Data taken from [40].)

of carrier concentration, whereas the 2D-band allows to discriminate between electrons and holes.

The doping dependence of integrated intensities of the G- and 2D-modes in our experiments with gated SLG using $E_L = 2.41\,\text{eV}$ is shown in Figure 2.17. A very interesting result is seen by Chen *et al.* [17] using lower E_L ($E_L = 1.58\,\text{eV}$), where the G-band intensity is maximum at $2|E_F| = E_L - \hbar\omega_G/2$. They highlight this as a demonstration of quantum interference of different pathways involving intermediate electronic excitations in Raman scattering. In addition, hot luminescence is seen as the Fermi energy approaches $E_L/2$ because of availability of the excited-state relaxation channel. The integrated intensity of the 2D-band is also

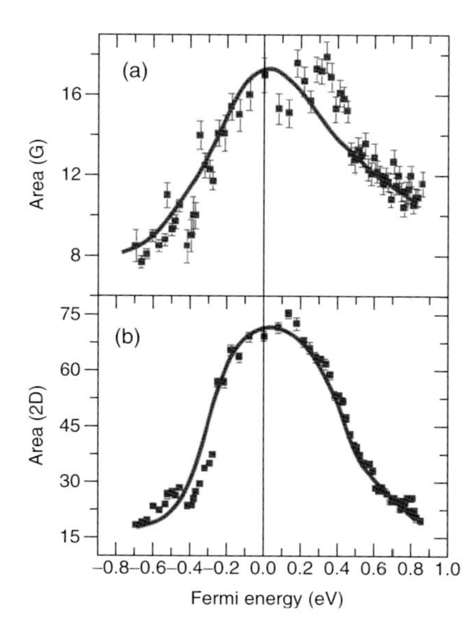

Figure 2.17 Integrated intensities of (a) G- (b) 2D-band as a function of the Fermi energy for top-gated SLG. The solid lines are guides to the eye.

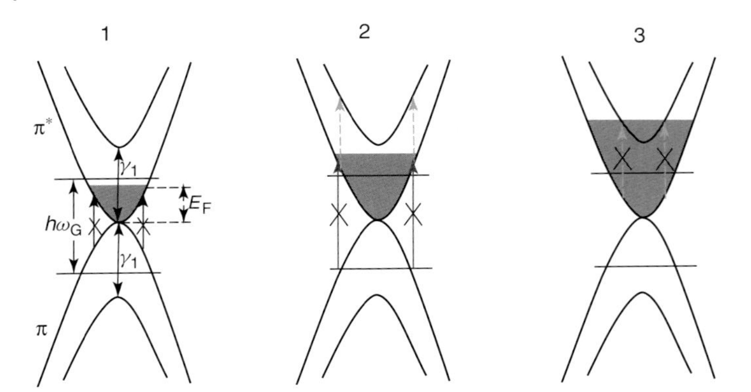

Figure 2.18 Possible transitions in doped bilayer graphene: (1) $E_F < \hbar\omega_G/2$, (2) $\hbar\omega_G/2 < E_F < \gamma_1$, and (3) $E_F > \gamma_1$. Solid lines are interband and dashed lines are intraband transitions.

influenced by quantum interference between different pathways [17]. The intensity of the 2D-band decreases as doping increases and becomes very small when all pathways are blocked.

For BLG, a similar doping dependence has been experimentally observed [42, 43]. However, the distinct electronic band structure in the bilayer results in some differences. As shown in Figure 2.18, the electronic band structure at the K-point consists of two π and two π* bands separated by an energy γ_1 (interlayer hopping integral) [43]. As a result, there will be interband $\pi \rightarrow \pi^*$ as well as the intraband transitions between two different levels in π- or π*-bands. When the Fermi energy reaches $\hbar\omega_G/2$, the phonon-induced interband transition involving the highest valence band and the lowest conduction band is forbidden. However, the intraband transitions from the filled lowest π* band to the higher energy π* band are still possible (Figure 2.18). When the gate voltage increases further and the Fermi energy reaches the second subband, intraband $\pi^* \rightarrow \pi^*$ transitions are also suppressed. A quantitative comparison of the frequency shift with the theory, including both interband and intraband transitions induced by the phonon, has been done to yield a value of interlayer coupling γ_1 in bilayer to be ~0.39 eV [43].

2.5.1
Optical Phonon Mixing in Doped Bi- and Multilayer Graphene

As discussed earlier, the zone center E_{2g} mode in monolayer graphene is degenerate. Stacking one layer above the other in AB sequence (Bernal stacking) leads to a splitting of the Raman active E_{2g} mode into two doubly degenerate modes, namely, the E_{2g} (Raman active) and E_{1u} (IR active) modes. Both these modes involve identical intralayer atomic displacements, but differ in relative interlayer motion of carbon atoms in adjacent layers: the E_{2g} mode being symmetric with respect to inversion and hence termed as *in-phase mode*, whereas the E_{1u} mode is antisymmetric and hence its motion is out-of-phase (Figure 2.3). Raman spectroscopy has been used

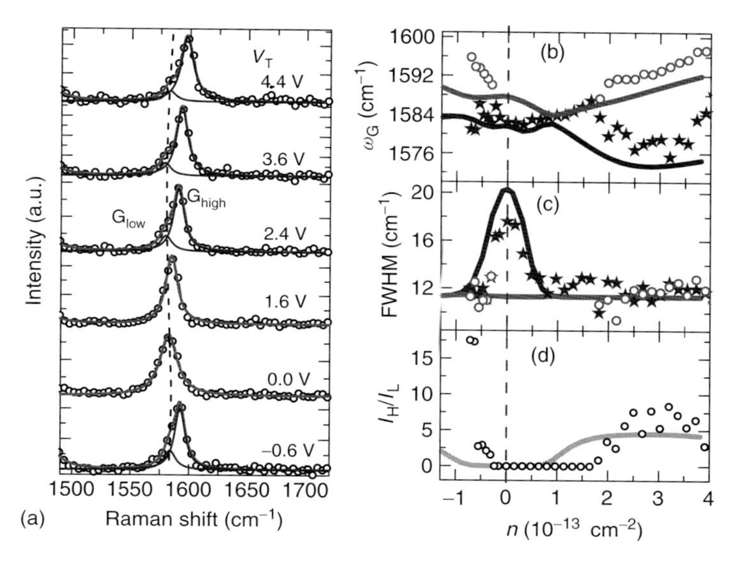

Figure 2.19 (a) G-mode Raman spectra of BLG showing G_{high} and G_{low} for different top gate voltages V_T [48]. The dashed line shows the evolution of G_{low}. Circles are experimental data points, and the lines are Lorentzian fits. Comparison with the theory for (b) frequency shifts and (c) FWHM. Open circles (G_{high}) and stars (G_{low}) represent the data. Dark gray and light gray lines represent the theoretical curves for G_{high} and G_{low} [47]. (d) The ratio between intensities I_H/I_L of the high-frequency (G_{high}) and low-frequency (G_{low}) modes. Black open circles are experimental points, and the solid line is theory [47].

to study the splitting of G-band in BLG and MLG (~10 layers) because of mixing of optical phonons induced by doping [44–49]. Figure 2.19a shows the Raman spectra of BLG at various gate voltages showing the splitting of the G-mode into the low-frequency (G_{low}) and high-frequency (G_{high}) bands [48]. It can be seen that the separation between the subbands increases with doping. The splitting is understood in terms of mode mixing due to doping-induced potential asymmetry in the two layers. At charge neutral configuration, the zone center phonons are classified as symmetric (Raman active) and antisymmetric (IR active) modes. The symmetric mode exhibiting KA has a lower frequency as compared to the antisymmetric mode. As the potential difference between the two layers builds up, the inversion symmetry breaks down and the symmetric and antisymmetric modes get mixed. They are no longer the vibrational eigenstates of the bilayer system; instead the new phonon eigenstate is a superposition of the symmetric and antisymmetric modes with the former governing the intensity of the spectrum. The frequency position and the FWHM [48] are compared with the theory [47] in Figure 2.19b,c, respectively. At the Dirac point, the line width is broadened because of strong EPC, but after splitting, the line width of G_{low} and G_{high} remains constant. The intensity ratio in Figure 2.19d shows that the crossover of intensity from G_{low} to G_{high} takes place at high doping.

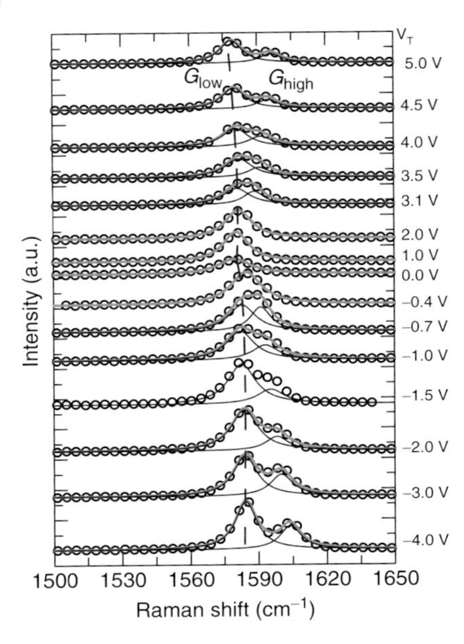

Figure 2.20 G-mode Raman spectra of multilayer graphene for different values of top gate voltages (V_T). Circles are experimental data points, and the solid lines are Lorentzian fits. The dashed line shows the evolution of the G_{low}.

The splitting of the G-band in doped MLG has also been reported [48] and explained in terms of mode mixing as well [49]. Figure 2.20 shows the results for MLG. It has been argued [49] that only the even-layered AB-stacked MLGs (essentially the repetition of the bilayer structure) with the center of inversion exhibit similar optical phonon mixing as in BLG and hence the mode splitting. The potential asymmetry in the even-layered MLG is created because of the short screening length along the *c*-axis of the stacked layers, and, therefore, the induced carriers remain mostly in layers close to the gate.

2.5.2
Charge Inhomogeneity and p–n Junction in the FET Channel Probed by Raman Spectroscopy

The characteristic signature of G- and 2D-bands in doped graphene has enabled Raman spectroscopy to probe the spatial variation of carrier concentrations in an FET [50, 51]. In an externally gated graphene transistor, a naive impression would be that the carrier profile is uniform along the graphene sample. This is correct as long as the gate voltage V_G is much larger than the drain source voltage V_{DS}. However, when $V_G \sim V_{DS}$, the channel has spatially varying carrier concentration and a p–n junction can be formed in between the source and drain [52–55]. Such a situation is shown for a BLG transistor [50] in Figure 2.21a, where the solid polymer electrolyte is used as a top gate. The corresponding electrical characteristic is shown in Figure 2.21b. To understand the nonlinear $I_{DS} - V_{DS}$ curve at a given gate voltage, we look at the schematic in Figure 2.21a. For $V_{DS} \ll V_T$ (V_T implies top gate voltage), the bilayer channel will be electron doped and doping ($n_c(x)$) will be homogeneous along the channel length. In this case, the voltage difference

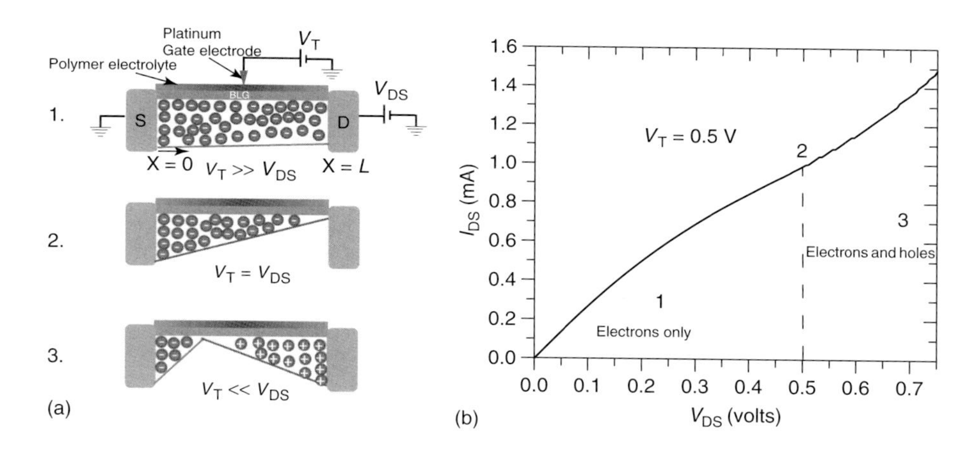

Figure 2.21 (a) Formation of a p–n junction along a bilayer graphene channel. Three situations are predicted for the variation of V_{DS} at constant V_T. (b) The corresponding electrical characteristic is shown for three conditions in (a). (Source: Adapted from [50].)

between the gate and the bilayer channel varies from V_T at the grounded source to $V_T - V_{DS}$ at the drain end. Therefore, by varying V_{DS} at fixed V_T, the doping concentration $n_c(x)$ changes from n-type to p-type along the channel length. In Figure 2.21a, we have shown the carrier distribution $n_c(x)$ along the bilayer channel for three different cases of V_{DS} at $V_T = 0.5$ V.

Case 1: $V_{DS} < V_T$, and, therefore, the carriers $n_c(x)$ will be the electrons in the bilayer channel with $n_c(x = 0) > n_c(x = L)$. In this region, I_{DS} increases sublinearly with V_{DS}.

Case 2: $V_{DS} = V_T$, $n_c(x)$ will be zero at $x = L$, and, therefore, the conduction channel gets pinched off near the drain end, making the region devoid of charge carriers. At this point, the slope of the I_{DS} versus V_{DS} curve undergoes a change (Figure 2.21).

Case 3: $V_{DS} > V_T$, the point in the channel where the pinch off occurs moves deeper into the channel drifting toward the source electrode. As $V_{DS} > V_T$, the gate is negatively biased with respect to the drain near the drain region. As a result, carriers will be holes near the drain region and the current increases because of enhanced p-channel conduction.

On the basis of this physical picture, the calculated $I_{DS} - V_{DS}$ curve in the diffusive transport model agrees well with the experiments [50]. The formation of the p–n junction in the channel can further be validated using Raman spectroscopy. Figure 2.22 shows the Raman spectra recorded at three different locations along an SLG transistor (B. Chakraborty, D.V.S. Muthu, and A.K. Sood Raman signatures of spatial dependent doping of graphene, unpublished.). The G-mode frequency along the channel shows different peak positions at different locations of the channel, and this can be used to estimate the varying carrier concentration along the channel.

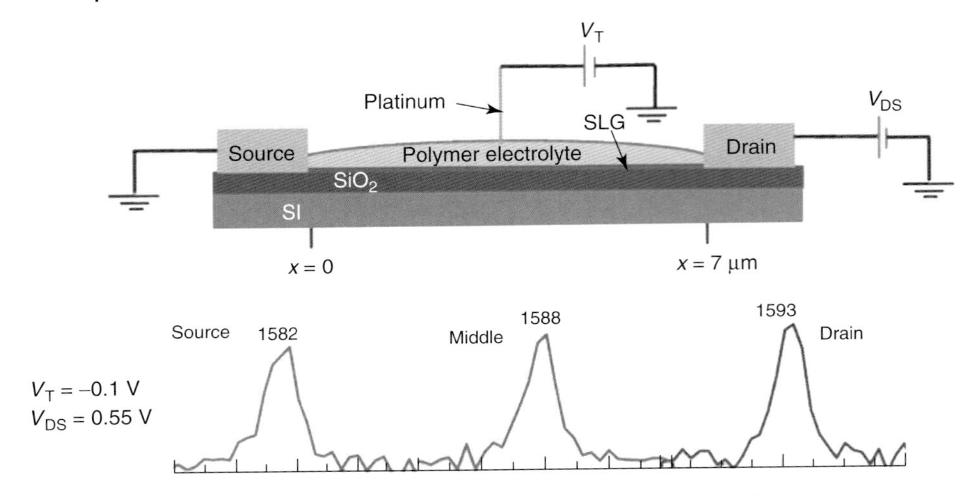

Figure 2.22 G-band Raman spectra taken at three different regions of a single-layer graphene transistor for fixed gate and drain–source voltages.

The 2D Raman spectra, which show opposite trends for electron and hole doping, can show the fingerprints of the p–n junction in between the source and the drain.

2.6 Raman Spectroscopy of Graphene Edges and Graphene Nanoribbons

The two orientations for the edge of graphene are zigzag (ZZ) and armchair (AC). In general, the edge can be a mixture of these two types. A noticeable fact is that the two similar edges form an angle of 120°, while angles 150° and 90° correspond to different edges. As a thumb rule, an AC edge and adjacent ZZ edge are separated by an angle $\theta = (2p + 1) \times 30^\circ$, while similar adjacent edges form an angle $\theta = 2p \times 30^\circ$ where p is an integer [56].

2.6.1 Effect of the Edge Orientation on the G-Band

In bulk graphene, LO and TO modes that are degenerate contribute equally to the single peak of the G-band. However, the nature of the edge influences the LO and TO modes differently. The modes LO and TO are shown in Figure 2.23 for both the AC and ZZ edges [57, 58]. The LO mode for the ZZ edge is the TO mode for the AC edge and vice versa for the TO mode. It has been shown [57, 58] that only the TO mode is Raman active at the ZZ edge, while the LO mode is Raman active at the AC edge. The phonon energy is a sum of bare phonon energy and the self-energy contribution due to the electron–phonon interaction. The energy difference between the LO and TO phonons can arise because of the self-energy correction as the TO phonons do not undergo KA at the AC and ZZ edges, whereas

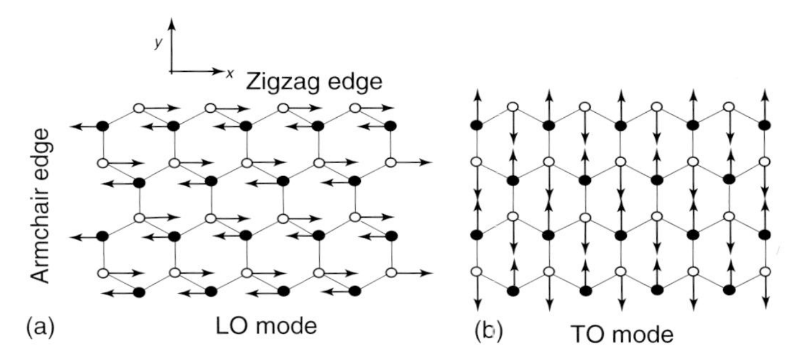

Figure 2.23 Atomic displacements for the (a) LO and (b) TO modes at zigzag edge with the LO (TO) being parallel (perpendicular) to the edge. The zigzag edge lies along the *x*-axis. The LO (TO) at the ZZ edge corresponds to the TO (LO) of the AC edge. (Source: Adapted from [57].)

LO phonons can, thereby resulting in higher TO frequency. Furthermore, only the TO (LO) mode is Raman active at ZZ (AC) edge. Therefore, the G-band for the AC edge is expected to be lower than $1582\,\mathrm{cm}^{-1}$. The self-energy correction is estimated to be $\sim 50\,\mathrm{cm}^{-1}$ [58]. The polarization dependence of the Raman G-band for two types of edges is opposite to each other. For the AC edge, the intensity is maximum when the angle θ_i between the edge and incident radiation (E_i) is 0°, whereas it is 90° for the ZZ edge. When the ZZ edges are introduced at random into part of an AC edge, the polarization dependence is lost.

Figure 2.24 shows the Raman spectra recorded at the two types of edges for different angles of incident polarization (with respect to the edge) [59]. The G-band at the AC edges is maximum for parallel and minimum for perpendicular polarization (Figure 2.24a,b), in agreement with theoretical predictions [58]. The insensitivity of the G-band to the polarization away from the edge is presented in the inset of Figure 2.24a, where there is hardly any change in the intensity with incident polarization. The solid line in the polar plot is the fit to the equation $I_G \propto \cos^2(\theta_i)$. In Figure 2.24c,d, the Raman G-band and the corresponding polar plot for the ZZ edge are shown as a function of incident polarization [59]. Unlike the AC edge, the G-band intensity is maximum for perpendicular and minimum for the parallel polarization. The line in the polar plot (Figure 2.24d) corresponds to the equation $I_G \propto \sin^2(\theta_i)$. The opposite polarization dependence of the G-mode at the AC and ZZ edges can, therefore, serve as a probe to study the edge chirality of the graphene samples.

Recent theoretical studies have also predicted [58] the softening of G-phonon energy at the AC edge, a signature of the well-studied KA that causes similar phonon softening in bulk graphene samples and can be tuned by doping. This prediction has been proved by recent experiments, where the G phonons at the ZZ edges are stiffened, while at the armchair edges softening is observed [60].

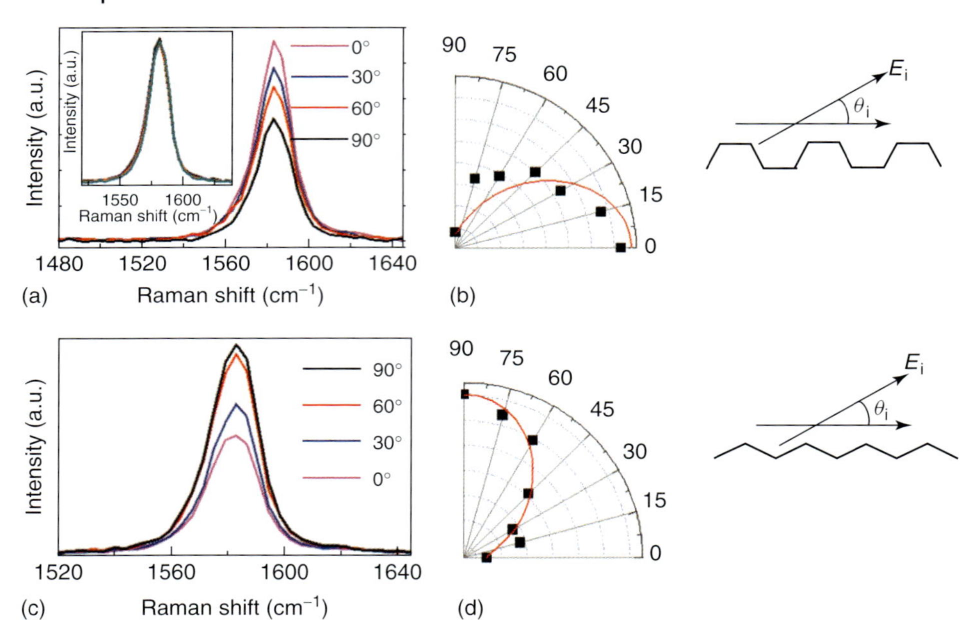

Figure 2.24 (a) G-band spectra from the armchair edge for different incident polarizations. Inset shows the Raman spectra recorded away from the edge. (b) Polar plot showing the intensity as a function of incident polarization angle θ_i. (c) G-band recorded at the zigzag edge for different θ_i. (d) Polar plot for the zigzag edge. (Source: Taken from Ref. [59].)

2.6.2
Effect of the Edge Orientation on the D-Band

The scattering of an electron at the ZZ edge is an intravalley scattering process, whereas it is intervalley at the AC edge [57, 58]. Therefore, only the AC edges participate in the DR-defect-induced process and affect the D-band. The Raman D-band intensity (I_D) is maximum with incident and scattered light being polarized along the AC edge. It has been theoretically shown that $(I_D)_{max}/(I_D)_{min} \sim 5$ to 10 [12]. Another interesting prediction is that the D-band at the AC edge is split into two due to trigonal warping effect on the electronic band structure [12]. Both these effects for the AC edge, namely, polarization dependence as well as splitting of the D-band, are shown to arise from the pseudospin dependence of the electron–phonon matrix element.

For ideal edges, the AC (ZZ) structure can be identified from the presence (absence) of the D-mode [61]. As shown in Figure 2.25, the D-band intensity is maximum for the spectrum marked "1," recorded from the predominantly AC region in a graphite edge, while regions "2" and "3" correspond to the ZZ edge and the sample interior, respectively [61].

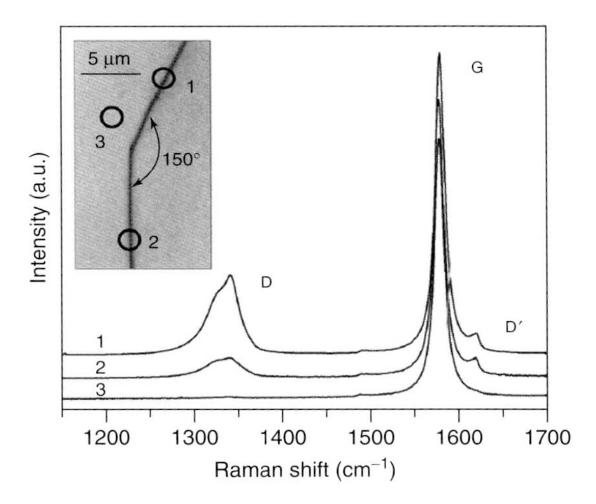

Figure 2.25 (a) Raman spectra recorded (with 1.96 eV laser energy) from three different regions in graphite marked "1," "2," and "3" in the inset. (Source: Taken from Ref. [61].)

2.6.3
Raman Spectroscopy of Graphene Nanoribbons

A small on–off ratio (~10) in graphene FETs is due to its vanishing bandgap at the Dirac point and hence limits its applications in devices. One of the ways to open up a gap is to reduce the width of the graphene (nanoribbons), bringing quantum confinement effects into play [62]. FETs using graphene nanoribbons (GNRs) of width ~ 10 nm show an on–off ratio of $\sim 10^7$ at room temperature [63], making GNR-based devices potential candidates in future digital electronics. As GNRs can be viewed as an open single-walled nanotube (SWNT) whose properties are decided by diameter and chirality, it is natural to expect that the width and nature of edges (zigzag or armchair) will influence the electronic and vibrational properties of the GNR. Indeed, the density of electronic states near the ZZ edge has a very narrow and large δ-function similar to singularity at the Dirac point, which is absent in AC edges [64].

In GNRs, the intensity ratio (I_D/I_G) of the D-peak with respect to the G-peak varies monotonically, reaching a maximum for ribbon width of ~20 nm [65]. In addition, the Raman spectra from the monolayer and few-layer nanoribbons exhibit certain characteristics depending on the ribbon width and the edge structure [66]. Figure 2.26 shows the Raman spectrum for a monolayer nanoribbon prepared by chemical route. Apart from the known bands, there are two sharp features at ~1450 and ~1530 cm^{-1}. Ren *et al.* [66] have attributed these new modes to the localized vibrations of the edge atoms. Two other weak-intensity peaks at ~1140 and ~1210 cm^{-1} are seen to correlate with the modes at ~1450 and ~1530 cm^{-1}, respectively. The frequencies of the modes at ~1450 and ~1530 cm^{-1} are insensitive to the layer numbers. Their intensity with respect to the G-band is found to decrease with the increasing width of the nanoribbon. Another intriguing feature of these

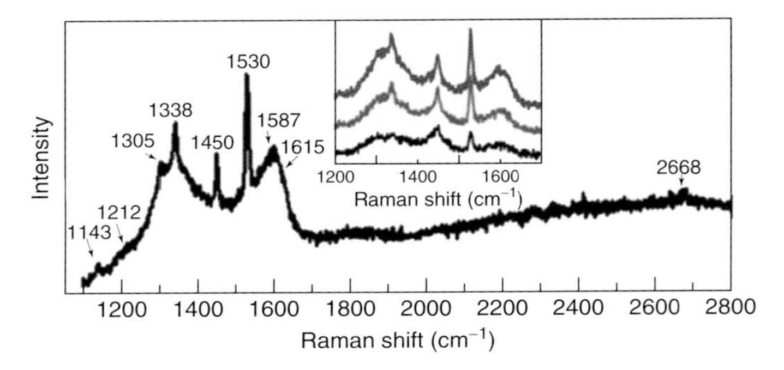

Figure 2.26 Raman spectra of monolayer GNR. Inset shows the spectra from three different GNRs. (Source: Taken from Ref. [66].)

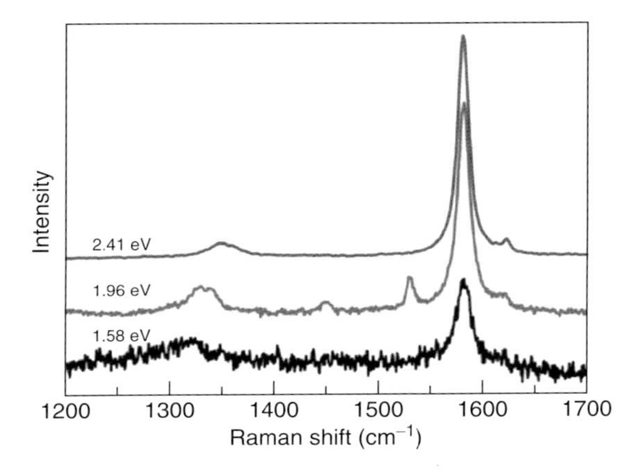

Figure 2.27 Raman spectra of few-layer GNRs recorded for three different laser energies. (Source: Taken from Ref. [66].)

modes is that they are observed only for laser energy 1.96 eV (Figure 2.27) [66]. More work is needed to understand these modes.

2.7
Effect of Disorder on the Raman Spectrum of Graphene

Here, we address how ion bombardment of graphene creating point defects influences the D-mode of SLG. Figure 2.28a shows the Raman spectra of a graphene monolayer subjected to different ion bombardment intensities [67]. The doses range from 10^{11} Ar^+ impacts per square centimeter (corresponds to one defect per 4×10^4 carbon atoms) to 10^{15} Ar^+ impacts per square centimeter (corresponds to one defect for every four carbon atoms). The D-band starts appearing as a small feature relative to the G-peak at a dose of 10^{11} $Ar^+\,cm^{-2}$. The intensity of the D-peak increases for subsequent higher doses up to 10^{13} $Ar^+\,cm^{-2}$. The

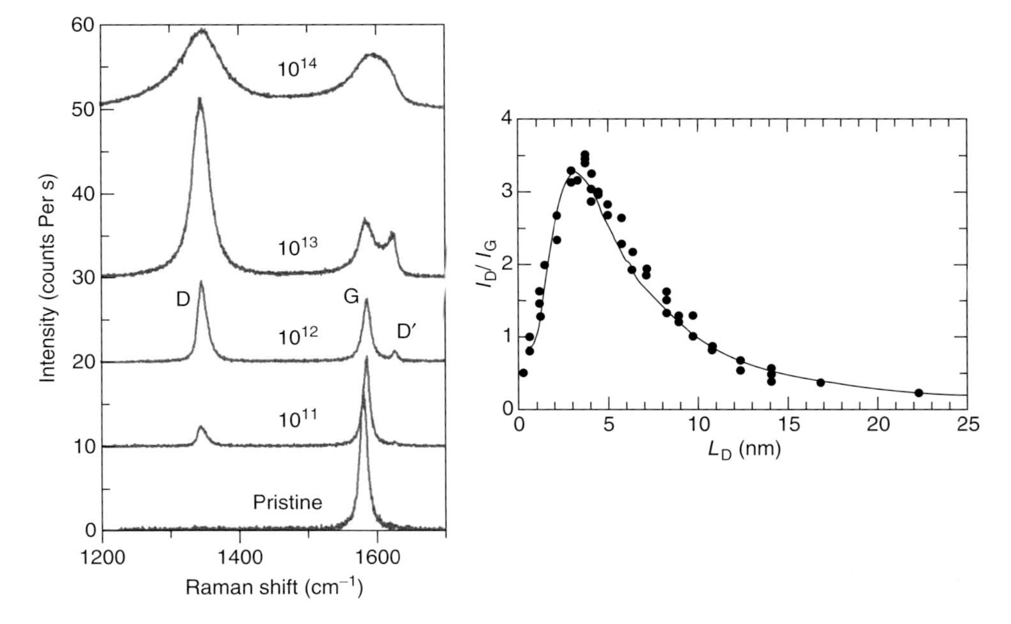

Figure 2.28 (a) Raman spectra showing the G- and D-band regions recorded using 514 nm laser wavelength in an SLG subjected to various ion doses, indicated in Ar^+ cm^{-2} next to the spectrum. (b) Intensity ratio I_D/I_G as a function of L_D. The points are experimental data, and the solid line represents the theoretical model. (Source: Taken from Ref. [67].)

other disorder-induced D′-band at ~1620 cm^{-1} also becomes evident from 10^{12} Ar^+ cm^{-2}. In Figure 2.28b, the intensity ratio between D- and G-peak (I_D/I_G) as a function of interdefect distance L_D is shown [67]. The ratio shows a nonmonotonic dependence on L_D [67, 68]: I_D/I_G increases up to $L_D \sim 3.5$ nm, attaining a maximum value at $L_D \sim 3.5$ nm and decreasing for $L_D \geq 3.5$ nm. It is to be mentioned that the determination of L_D has been done from STM measurements on HOPG, which was subjected to the same ion doses as the SLG.

The intensity trend shown in Figure 2.28b was modeled considering that ion bombardment causes modification of the SLG on two length scales, namely, r_A and r_S with $r_A > r_S$ [67] (Figure 2.29a). The area within r_S is the structurally disordered area (S region) and the region with $r_A \geq r \geq r_S$ is termed as the *activated area* (A region) where the hexagonal structure is preserved and the D-band is observed because of the mixing of electronic states near K- and K′-points. As a consequence, the selection rules break down and the D-mode intensity is enhanced. In Figure 2.29 b,c, snapshots showing the evolution of these two regions as the dose increases. Depending on the proximity to the point of ion impact, otherwise pristine regions can turn into a structurally disordered region or an activated region. With the increase in the dose, the pristine area starts getting converted to activated regions, resulting in the increase in D-peak intensity I_D. A further increase in the number of impacts could lead to the activated regions being transformed into structurally

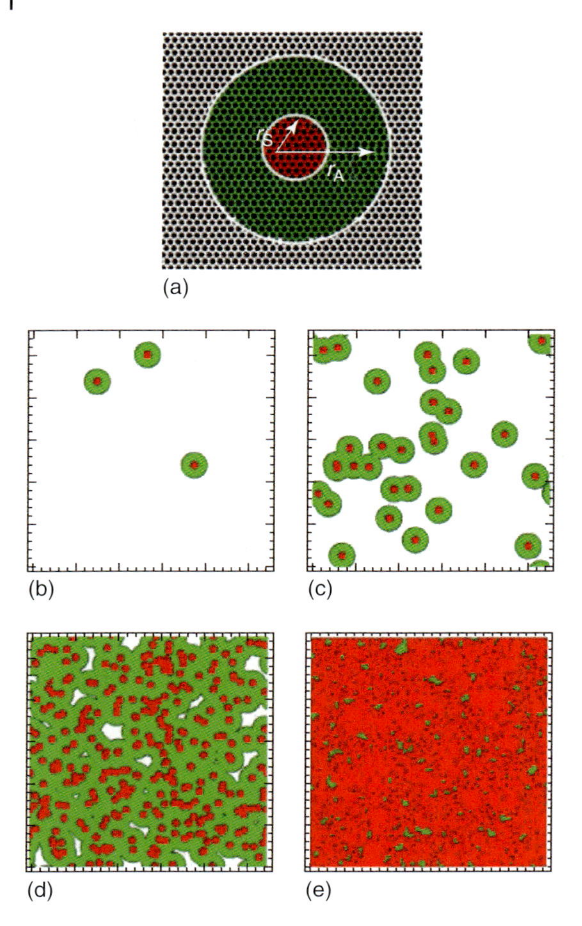

Figure 2.29 (a) A and S regions defined by r_A and r_S measured from the point of impact of the Ar^+ ion. Snapshots showing the structural evolution of the graphene sheet for different ion doses, (b) 10^{11} $Ar^+ cm^{-2}$, (c) 10^{12} $Ar^+ cm^{-2}$, (d) 10^{13} $Ar^+ cm^{-2}$, and (e) 10^{14} $Ar^+ cm^{-2}$. Situation at (d) corresponds to high D-band intensity, while (e) corresponds to a low D-band intensity. (Source: Taken from Ref. [67].)

distorted S regions that cause the reduction of I_D. The nonmonotonic nature of the intensity ratio can now be understood by considering that, in the low-defect-density regime (characterized by large L_D), the total area from which scattering occurs is proportional to the number of defects. Thus we can write $I_D/I_G \propto 1/L_D^2$. The increase in the defect density from higher doses causes the activated regions to overlap, which in turn results in maximum D-peak intensity. Any further increase in the defect density renders the graphene sheet dominated by structurally distorted regions S and hence decreasing I_D. For a given L_D, the dependence of the intensity ratio I_D/I_G on laser excitation energy has also been experimentally observed to be $\frac{I_D}{I_G} \sim E_L^{-4}$ [68].

A more microscopic picture has been given by Venezula *et al.* [14] based on the dependence of the electronic broadening parameter γ (see Equation 2.4 where γ in all the intermediate states are considered equal) on the defect concentration n_D. The parameter γ has contributions from intrinsic electron–phonon interaction (γ^{ep}) as well as extrinsic defect-induced component (γ^D), the latter increasing linearly with n_D. In addition, the Raman intensity of the D-line has a proportionality factor N_D (average number of defects in the unit cell). At low n_D, that is, large L_D, the intensity I_D is dominated by the proportionality factor N_D. But as n_D is increased further (lower L_D), the effect of increasing γ offsets the effect of the proportionality factor and hence I_D decreases.

2.8 Raman Spectroscopy of Graphene under Strain

As is expected, tensile strain can decrease phonon frequency, whereas compressive strain causes hardening of phonons. In this section, we see that the Raman spectroscopy of strained graphene not only provides information about the strain magnitude but can also determine the strain axis.

Typical experimental arrangements for applying tensile strain to graphene are shown in Figure 2.30, where graphene has been deposited on a flexible substrate [69, 70]. Figure 2.31a,b shows the G- and 2D-Raman spectra of the graphene monolayer with different values of uniaxial strain ϵ, showing the redshift [69]. Furthermore, the G-band splits into two bands, denoted by G^+ (high-frequency mode) and G^- (low-frequency mode)because of lowering of the crystal symmetry. As seen in Figure 2.31c, the mode frequencies of both the G^+ and G^- modes vary linearly with strain: for 1% strain $\Delta\omega_{G^+} \sim -10\,\text{cm}^{-1}$ and $\Delta\omega_{G^-}{\sim}-31\,\text{cm}^{-1}$. The 2D-band frequency shifts much more, $\Delta\omega_{2D}{\sim}-64\,\text{cm}^{-1}$. It was shown [69, 70] that the G^- mode eigenvector is parallel to the strain while that for G^+ is perpendicular to it, resulting in lower softening as shown in Figure 2.32. The magnitude of the phonon energy change with strain is proportional to the mode Grüneisen parameter, defined as $\gamma = -\frac{1}{\omega^0}\frac{d\omega^h}{d\epsilon_h}$, where $\epsilon_h = \epsilon_{ll} + \epsilon_{tt}$ is the hydrostatic component of the uniaxial strain, ω^0 is the unstrained G-mode frequency, and l (t) denotes the

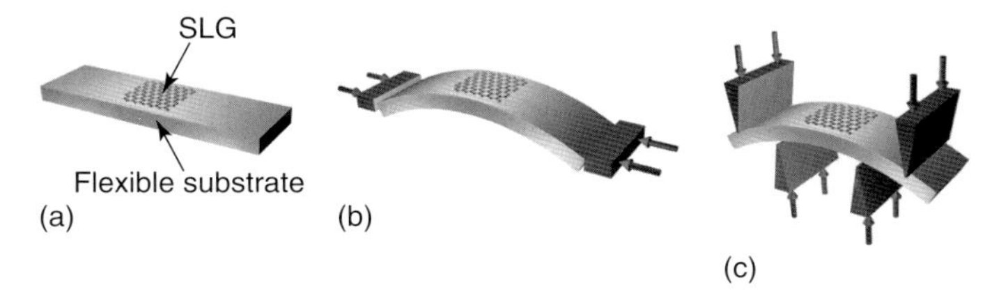

Figure 2.30 (a) A typical setup for applying strain with graphene flake deposited on a flexible substrate such as polyethylene terephthalate (PET) film or acrylic (Perspex). (b) Two-point and (c) four-point bending schemes for applying tensile strain. (Source: Taken from Ref. [69].)

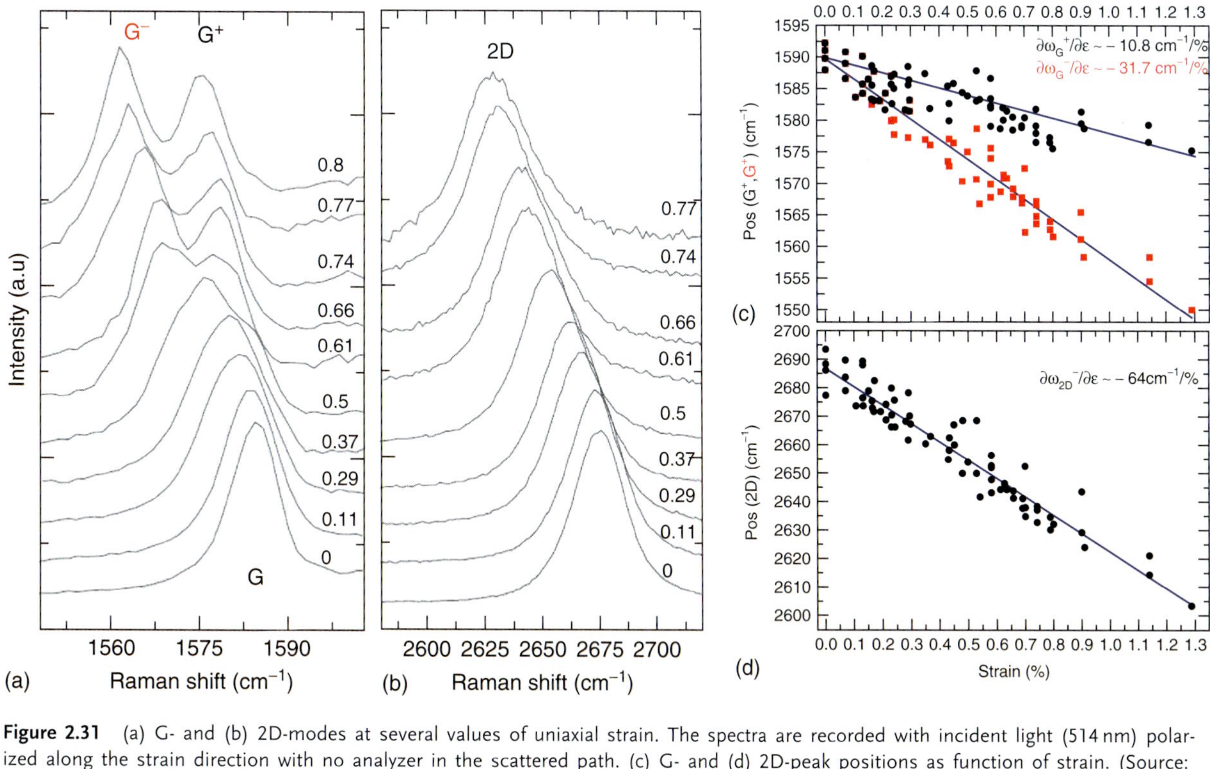

Figure 2.31 (a) G- and (b) 2D-modes at several values of uniaxial strain. The spectra are recorded with incident light (514 nm) polarized along the strain direction with no analyzer in the scattered path. (c) G- and (d) 2D-peak positions as function of strain. (Source: Taken from Ref. [69].)

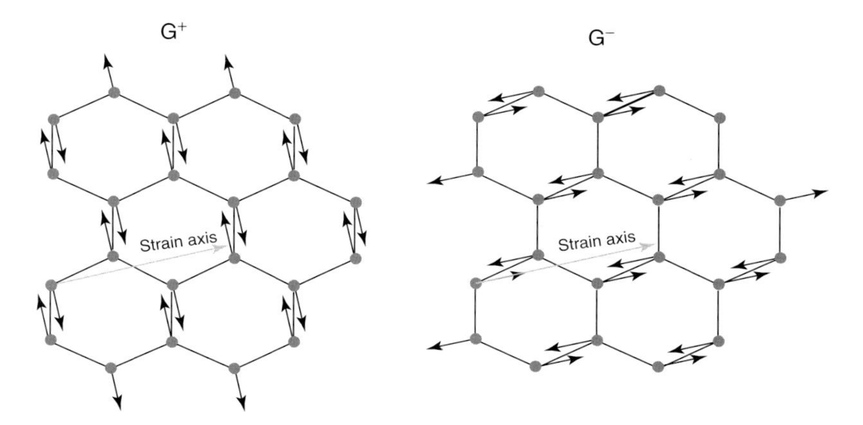

Figure 2.32 Eigenvectors representing (a) G^+- and (b) G^--mode displacements under the influence of uniaxial strain along the direction shown by "strain axis." G^--mode displacement is parallel to the strain axis, while the G^+-mode is perpendicular. (Source: Adapted from [69].)

longitudinal (transverse) direction of the strain. Similarly, the shear deformation potential β is expressed as $\beta = -\frac{1}{\omega^0}\frac{d\omega^S}{d\epsilon_S}$, where $\epsilon_s = \epsilon_{ll} - \epsilon_{tt}$ is the shear component of the uniaxial strain that causes the splitting of the G-mode. For uniaxial strain, $\epsilon_{ll} = \epsilon$ and $\epsilon_{tt} = -\nu\epsilon$ where ν is the Poisson's ratio ($\nu_{\text{substrate}} = 0.33$ of the substrate was taken assuming ideal contact between graphene and substrate). The resultant change in G-mode frequency is given by Mohiuddin *et al.* [69] $\Delta\omega^{\pm} = \Delta\omega^{\text{h}} \pm \frac{1}{2}\Delta\omega^{\text{s}}$, where $\Delta\omega^{\text{h}}$ defines the hydrostatic component and $\Delta\omega^{\text{s}}$ is the shear component of the strain. The $+(-)$ sign is for G^+ (G^-) mode. Experimental values of frequency shifts yield $\gamma \sim 2$ and $\beta \sim 1$.

The graphene crystal orientation with respect to the strain can also be determined by comparing the intensities of the G^+- and G^--modes. In the presence of an uniaxial strain, the atomic displacement (in Cartesian basis) can be expressed as a combination of longitudinal and transverse components of the strain. The intensity of the two peaks is expressed as $I_{G^-} \propto \sin^2(\theta_i + \theta_O + 3\phi)$ and $I_{G^+} \propto \cos^2(\theta_i + \theta_O + 3\phi)$, where θ_i and θ_O are the incident and scattered polarization angle,respectively, with respect to the strain axis (Figure 2.33b) [69]. ϕ is the angle between the strain axis and x-axis, which is directed along the ZZ orientation. Figure 2.33c shows the Raman spectra at a fixed value of strain with varying incident polarization keeping $\theta_O = 0$. Figure 2.33d shows the polar plot of G^+- and G^--peak intensities extracted from Figure 2.33c as a function of θ_i. The points are the experimental data, and the best fitting shown by solid lines was obtained by fitting $I_{G^-} \propto \sin^2(\theta_i + 3\phi)$ and $I_{G^+} \propto \cos^2(\theta_i + 3\phi)$, with $3\phi = 34^\circ$. Thus, strain axis relative to the sample orientation can easily be obtained by fitting the experimental ratio $I_{G^-}/I_{G^+} = \tan^2(\theta_i + \theta_O + 3\phi)$ to get the best value of ϕ.

Recent Raman experiments [71, 72] have shown the effect of both tension and compression on graphene samples, as shown in Figure 2.34, where negative (positive) strain denotes compression (tension).

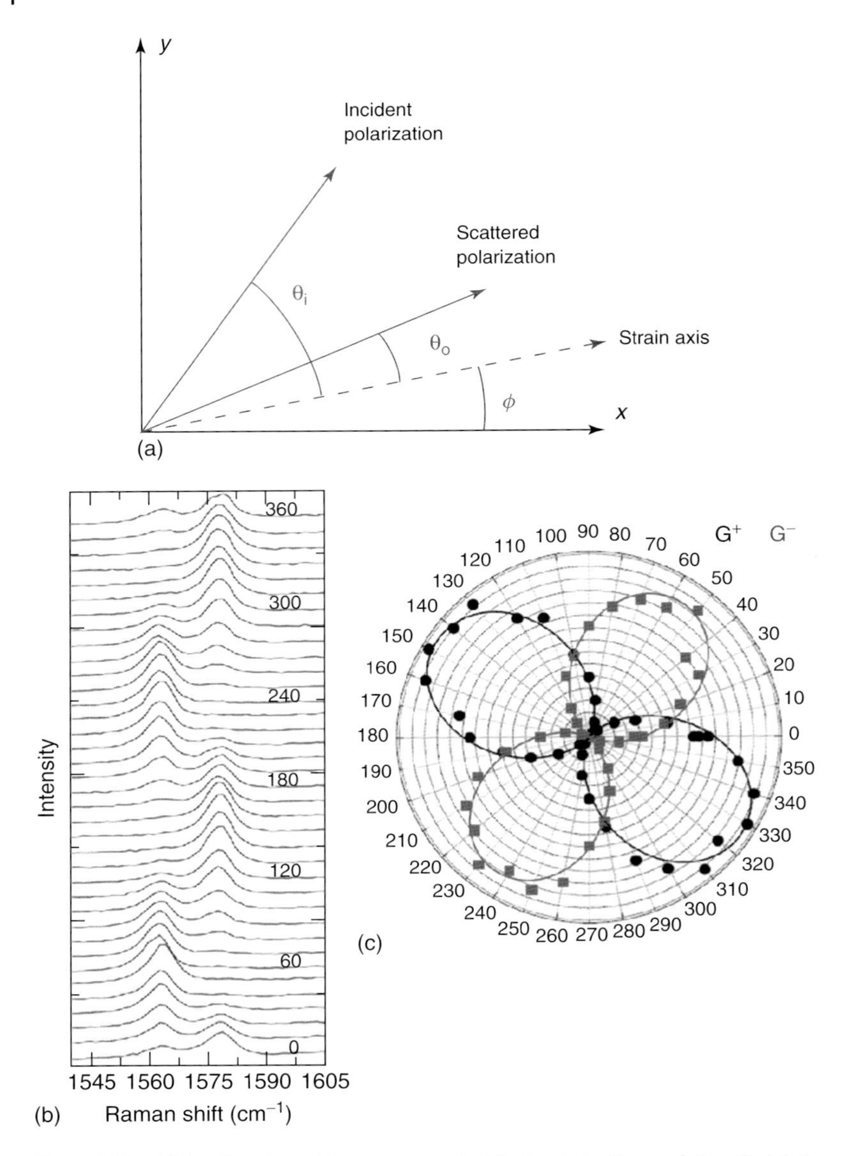

Figure 2.33 (a) Strain axis making an angle ϕ with the x (zigzag orientation of the sample)-axis. The incident and scattered polarization relative to the strain axis is denoted by θ_i and θ_0, respectively. (b) Raman spectra at a fixed value of strain with varying incident polarization and $\theta_0 = 0$. (c) Polar plot of the intensities: squares (circles) are for G^- (G^+) modes, and lines are fit to $I_{G^-} \propto \sin^2(\theta_i + 34)$ and $I_{G^+} \propto \cos^2(\theta_i + 34)$. (Source: Taken from Ref. [69].)

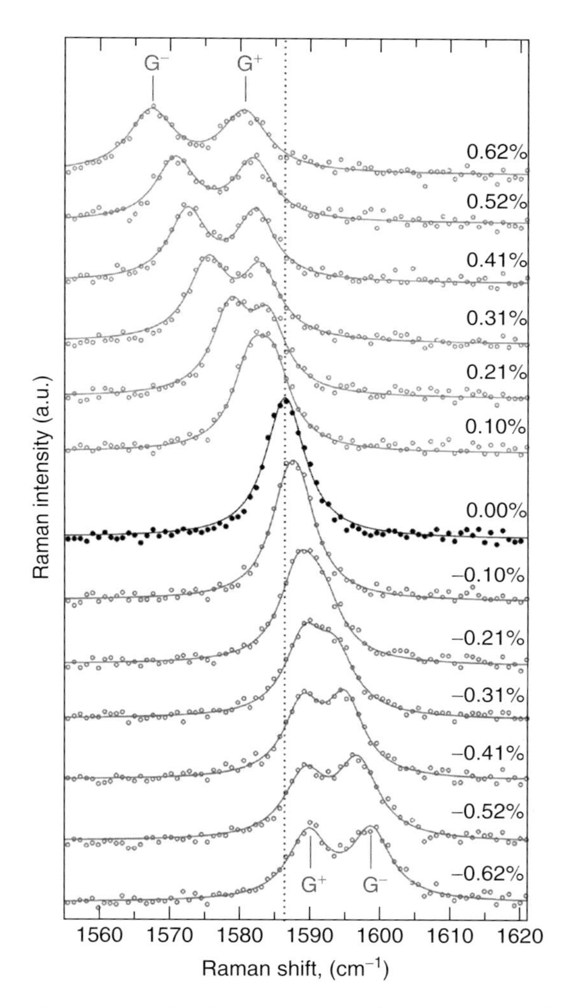

Figure 2.34 G-band Raman spectra of SLG recorded with 785 nm under uniaxial tensile strain (positive values) and uniaxial compressive strain (negative values). Points are experimental data, and solid lines are Lorentzian fit to the data. (Source: Taken from Ref. [71].)

The splitting of the 2D-mode of a monolayer graphene under strain has also been seen recently [73–75]. Figure 2.35 shows [73] the 2D Raman spectra for various levels of uniaxial strain applied along the AC (Figure 2.35a) and ZZ direction (Figure 2.35b). In the AC case, the peak-assigned $2D^-$ ($2D^+$) is obtained for incident and scattered polarization being parallel (perpendicular) to the strain axis. When the strain axis is parallel to the ZZ direction, the Raman spectra (Figure 2.35b) are recorded with both incident and scattered polarization at 50° relative to the strain axis [73]. The 2D-band red shifts with increasing strain (Figure 2.35c,d). The 2D spectra dependence on incident polarization is shown in Figure 2.36a,b for a fixed strain level [73]. For strain along the ZZ direction (Figure 2.36a), the $2D^+$-mode

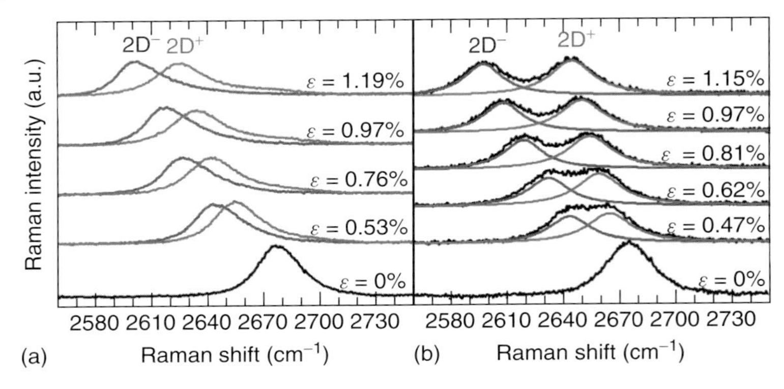

Figure 2.35 Raman spectra showing the 2D-mode recorded with 514 nm excitation at various levels of uniaxial tensile strain (ϵ) applied along (a) armchair and (b) zigzag orientation. In (a), $2D^-$ is recorded with incident and scattered polarization at 0° relative to the strain axis, while $2D^+$ is recorded with perpendicular polarization. In (b), the spectra are recorded with polarization of the incident and scattered radiation perpendicular to the strain axis. (Source: Taken from Ref. [73].)

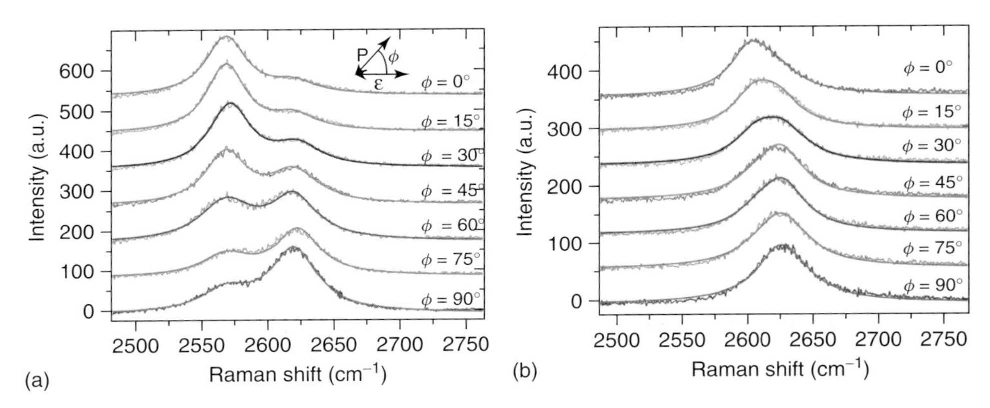

Figure 2.36 2D-mode spectra recorded with 532 nm excitation at fixed uniaxial tensile strain varying the angle ϕ between the incident polarization (P) and the strain axis (ϵ) for the (a)zigzag and (b) armchair edges. The strain for (a) is 3.8% and for (b) is 2.6%. (Source: Taken from Ref. [74].)

dominates when incident polarization P is perpendicular to the strain and $2D^-$ in parallel polarization. For the AC, the splitting is not clear: the perpendicular case spectra display a single symmetric peak, and for parallel polarization, the peak shows some asymmetry. These effects originate from changes in resonant conditions caused by the distortion of the Dirac cones and anisotropic modification of the phonon dispersion under uniaxial strain. For the unstrained graphene, the three $K \rightarrow K'$ intervalley scattering processes denoted by "1", "2", and "3" in Figure 2.37a are equivalent, leading to a single peak in the 2D Raman spectrum [73]. For strain applied along the AC direction, one of the K′-points moves further away while the other two move closer to the K-point (Figure 2.37b). The reverse

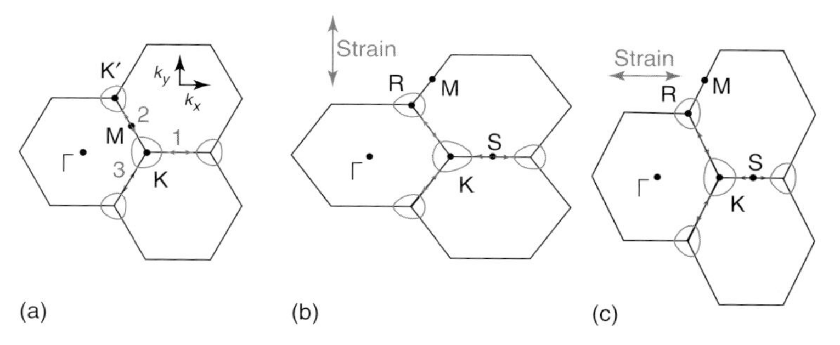

Figure 2.37 (a) Reciprocal lattice diagram for an unstrained graphene. Strain-induced distortion of the reciprocal lattice under strain along the (b) armchair edge and (c) zigzag edge. The strained K′-points are now designated as *R*-points, and the midpoints between the unstrained K- and K′-points have become inequivalent M and S points. (Source: Taken from Ref. [73].)

happens for the strain along the ZZ direction (Figure 2.37c). The phonons along ΓKS direction are involved for scattering process "1", while the phonons along ΓRM take part in processes "2" and "3." It should be pointed out that under uniaxial strain K′-points are *R* and inequivalent high-symmetry midpoints between K- and R-points are *M* and *S*, as shown in Figure 2.37b,c.

It should be noted that in the evaluation of mode Grüneisen parameter γ, knowledge of the Poisson ratio of the substrate and the degree of adhesion of graphene to it are important. Also, for the D- and 2D-modes, the uniaxial strain moves the Dirac cones, which adds to the difficulty in calculating the γ_{D} and $\gamma_{2\mathrm{D}}$. Under biaxial strain (isotropic strain), there would be no split as $\epsilon_s = 0$ since $\epsilon_{ll} = \epsilon_{tt}$, and the substrate Poisson ratio does not come into the picture. The effects of relative shifting of the Dirac cones also are not present under biaxial strain. Recent experiments have analyzed the Raman spectrum of graphene under biaxial stress imparted by piezoelectric actuator [77] and also by studying biaxial strain in graphene bubbles [76]. The values of $\gamma_{2\mathrm{D}} = 2.98$, $\gamma_{\mathrm{D}} = 2.3$ and $\gamma_{2\mathrm{D}} = 2.6$, and $\gamma_{\mathrm{D}} = 2.52$ and $\gamma_{E_{2g}} = 1.8$ are estimated. The Raman spectrum and Raman image from a graphene bubble are shown in Figure 2.38a,b, respectively [76].

2.9 Temperature and Pressure Dependence of Raman Modes in Graphene as Nanometrological Tools

In addition to determining the layer thickness and doping profile, Raman spectroscopy can well be applied as a metrological tool to estimate the temperature in graphene devices. Temperature is known to alter intrinsic carrier density and device mobility, and hence proper thermal management is an important issue for graphene devices.

Phonon frequency and line width depend on quasi-harmonic (volume change) and anharmonic phonon–phonon interactions. Temperature-dependent

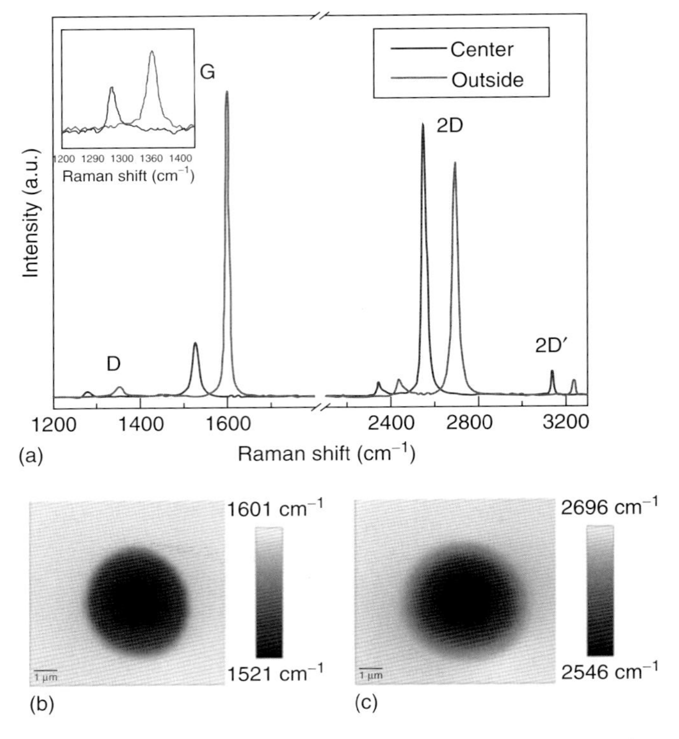

Figure 2.38 Raman spectra measured using 488 nm excitation at the center of a single-layer bubble and away from the center on the substrate. Raman map of the (b) G-mode and (c) 2D-mode. Both peak positions strongly decrease toward the center. (Source: Taken from Ref. [76].)

Raman spectroscopy on SLG and BLG flakes shows expected softening of the G- and 2D-mode frequencies (ω) with increasing temperature, expressed as $\omega(T) = \omega_0 + \chi T$, where ω_0 is the phonon frequency at $T = 0$ K. In the temperature range 100–373 K, the temperature coefficient $\chi_G = -0.016\,\text{cm}^{-1}\,\text{K}^{-1}$ and $\chi_{2D} = -0.034\,\text{cm}^{-1}\,\text{K}^{-1}$ for the SLG and $\chi_G = -0.015\,\text{cm}^{-1}\,\text{K}^{-1}$ and $\chi_{2D} = -0.066\,\text{cm}^{-1}\,\text{K}^{-1}$ for the BLG [78]. The temperature dependence of the phonon frequencies of the 2D-band has been used to estimate the laser-induced heating, leading to the measurement of thermal conductivity of graphene to be $\kappa = 1800–710\,\text{W}\,(\text{mK})^{-1}$ in the temperature range 325–500 K [79]. In another study [80], the local temperature was measured by the relative intensity ratio of Stokes (S) and anti-Stokes (AS) Raman G-line: $I_{AS}/I_S = \exp(\frac{-\hbar\omega_G}{k_B T})$, and the thermal conductivity of monolayer graphene was estimated to be $\kappa = 632\,\text{W}\,(\text{mK})^{-1}$ at 660 K. A recent review outlines the contributions of the in-plane and cross-plane phonon modes to thermal conductivity of SLG and FLG and compares the results with different experiments [81].

Recent experiments [82] subjecting graphene layers to hydrostatic pressure highlight the good adherence of monolayer graphene to the substrate (Si/SiO_2), which

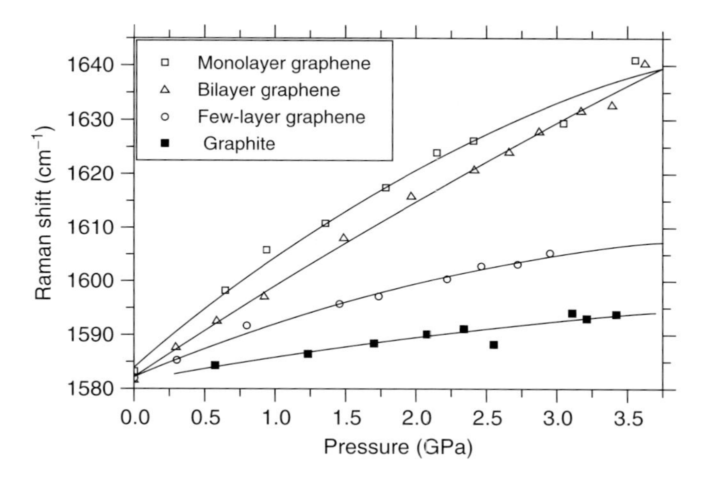

Figure 2.39 G-peak position as a function of pressure for SLG, BLG, FLG, and graphite. (Source: Taken from Ref. [82].)

may lead to several application of graphene-based pressure sensors. Figure 2.39 shows the pressure dependence of the G-band in monolayer, BLG, FLG, and graphite showing a trend that the shift of G-band with pressure is significantly higher for monolayer and BLG [82].

2.10 Tip-Enhanced Raman Spectroscopy of Graphene Layers

Tip-enhanced Raman spectroscopy (TERS) provides a better lateral resolution as compared to the diffraction-limited resolution in confocal Raman spectroscopy [85]. In TERS, a metallic tip (a few nanometers in diameter) is brought close (approximately a few nanometers) to the laser-illuminated region of the sample, resulting in a strong localized electromagnetic (EM) because of surface plasmons, similar to thr mechanism of surface-enhanced Raman scattering.

TERS has recently been applied for molecular imaging, photoluminescence mapping, and chirality determination of carbon nanotubes [86–88]. Recent TERS of graphene layers using gold cantilevers in tapping mode atomic force microscope [83, 84, 89, 90] shows a Raman enhancement of $\sim 10^6$, as shown in Figure 2.40 [83]. The intensity ratio between the D- and the G-Raman peaks (I_D/I_G) is found to be $\sim$1.4 in micro-Raman and $\sim$0.8 in TERS. A higher value of I_D/I_G in micro-Raman is due to the fact that the edges contribute to the D-band because graphene flake size ($\sim$400 nm) was smaller than the laser spot. TERS of the CVD-grown monolayer graphene has been used to image pristine, disordered, contaminated, and hydrogen-terminated regions with a spatial resolution better than 12 nm (Figure 2.41). Thus, TERS has a potential to characterize the disordered and pristine regions of graphene with nanometer resolution.

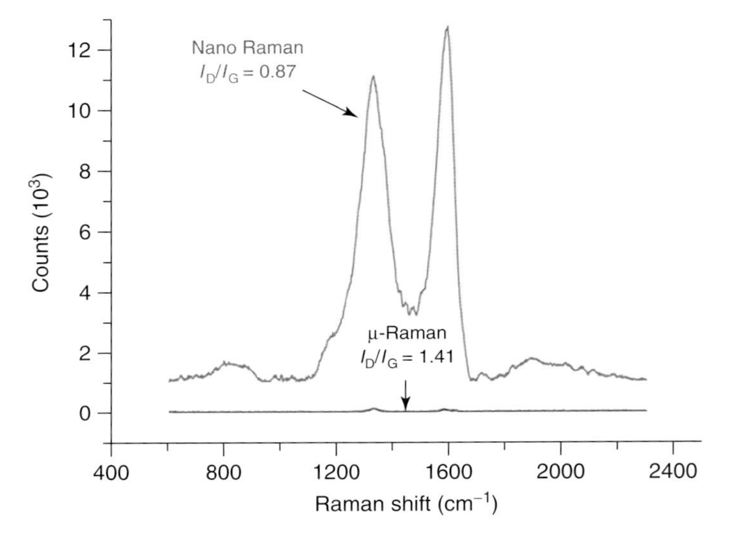

Figure 2.40 Micro-Raman and TERS spectra (termed *Nano-Raman*) of SLG. (Source: Taken from Ref. [83].)

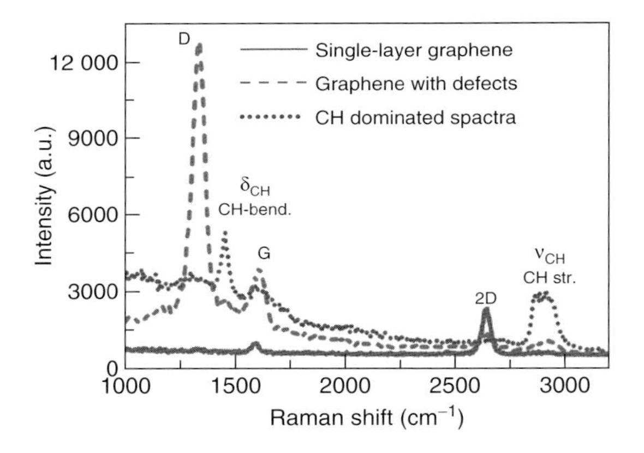

Figure 2.41 TERS recorded at different spots on the chemical vapor deposition (CVD)-grown graphene. The solid spectrum is from SLG. The dashed spectrum exhibits a pronounced D-peak signifying disorder, and the dotted spectrum is from the region with strong CH bending and stretching modes. (Source: Taken from Ref. [84].)

2.11 Conclusions

Raman spectroscopy has emerged as one of the most useful probes of characterizing graphene in terms of a number of layers, stacking order, degree of disorder, and strain. Simultaneous measurements of transport properties and Raman spectra of graphene FETs give a quantitative estimate of doping levels and the electron–phonon interaction strength, which is of critical importance in developing electronic and optoelectronic nanodevices of graphene. It will be interesting to do Raman studies of graphene devices on boron nitride layers [91, 92] and other graphene analog materials. Recently, single-layer MoS_2 transistors have been made with an on–off ratio $\sim 10^6$ [93, 94], which hold a lot of promise. Raman studies of this system as a function of electron doping show an interesting symmetry-dependent phonon softening [94], namely, the A_{1g} phonon softens on electron doping, whereas the E^1_{2g} mode is hardly affected. Future studies can include the study of functionalized graphene and graphene complexes with interesting materials such as topological insulators [95], two-dimensional electron layer in oxide heterostructures [96], and ferroelectric materials [97]. Raman studies of graphene under magnetic fields will elucidate inter-Landau level excitations [98, 99]. Electronic Raman scattering from graphene including the quantum Hall regime will be a challenge to pursue, similar to the case of 2D electron gas in semiconductor heterostructures [100]. It is clear that there are many avenues opening up in graphene research where the Raman spectroscopy in conjunction with other spectroscopies and transport measurements will play an important role.

Acknowledgments

AKS thanks the Department of Science and Technology for support under the Nanomission project. We thank our collaborators in our work on graphene.

References

1. Novoselov, K.S., Geim, A.K., Morozov, S.V., Jiang, D., Zhang, Y., Dubonos, S.V., Grigorieva, I.V., and Firsov, A.A. (2004) *Science*, **306**, 666.
2. (a) Geim, A.K. and Novoselov, K.S. (2007) *Nat. Mater.*, **6**, 183; (b) Castro Neto, A.H., Guinea, F., Peres, N.M.R., Novoselov, K.S., and Geim, A.K. (2009) *Rev. Mod. Phys.*, **81**, 109; (c) Sarma, S.D., Adam, S., Hwang, E.H., and Rossi, E. (2011) *Rev. Mod. Physics*, **83**, 407; (d) Peres, N.M.R. (2010) *Rev. Mod. Phys.*, **82**, 2673.
3. Rao, C.N.R., Sood, A.K., Subrahmanyam, K.S., and Govindaraj, A. (2009) *Angew. Chem. Int. Ed.*, **48**, 7752.
4. Rao, C.N.R., Sood, A.K., Voggu, R., and Subrahmanyam, K.S. (2010) *J. Phys. Chem. Lett.*, **1**, 572.
5. Geim, A.K. (2011) *Rev. Mod. Phys.*, **83**(3), 851.
6. Malard, L.M., Pimenta, M.A., Dresselhaus, G., and Dresselhaus, M.S. (2009) *Phys. Rep.*, **473**, 51.
7. Dresselhaus, M.S., Jorio, A., and Saito, R. (2010) *Annu. Rev. Condens. Matter Phys.*, **1**, 89.
8. Wallace, P.R. (1947) *Phys. Rev.*, **71**(9), 622.
9. Pereira, V.M., Castro Neto, A.H., and Peres, N.M.R. (2009) *Phys. Rev. B*, **80**, 045401.
10. Levy, N., Burke, S.A., Meaker, K.L., Panlasigui, M., Zettl, A., Guinea, F., Castro Neto, A.H., and Crommie, M.F. (2010) *Science*, **329**, 544.
11. Guineal, F., Katsnelson, M.I., and Geim, A.K. (2010) *Nature Physics*, **6**, 30.
12. Sasaki, K., Kato, K., Tokura, Y., Suzuki, S., and Sogawa, T. (2012) *Phys. Rev. B*, **85**, 075437.
13. Yan, J., Ruan, W.Y., and Chou, M.Y. (2008) *Phys. Rev. B*, **77**, 125401.

14. Venezuela, P., Lazzeri, M., and Mauri, F. (2011) *Phys. Rev. B*, **84**, 035433.
15. Martin, R.M. and Falicov, L.M. (1983) Resonant Raman scattering, in *Light Scattering in Solids I*, 2nd edn (ed. M. Cardona), Springer-Verlag Berlin Heidelberg, New York.
16. Ralston, J.M., Wadsack, R.L., and Chang, R.K. (1970) *Phys. Rev. Lett.*, **25**, 814.
17. Chen, C.F., Park, C.H., Boudouris, B.W., Horng, J., Geng, B., Girit, C., Zettl, A., Crommie, M.F., Segalman, R.A., Louie, S.G., and Wang, F. (2011) *Nature*, **471**, 617.
18. (a) Thomsen, C. and Reich, S. (2000) *Phys. Rev. Lett.*, **85**(24), 5214; (b) Kürti, J., Zolyomi, J., Grüneis, J., and Kuzmany, H. (2002) *Phys. Rev. B*, **65**, 165433; (c) Narula, R. and Reich, S. (2008) *Phys. Rev. B*, **78**, 165422. (d) Basko, D.M. (2008) *Phys. Rev. B*, **78**, 125418.
19. Das, A., Chakraborty, B., and Sood, A.K. (2008) *Bull. Mater. Sci.*, **31**(3), 579.
20. Yu, V., Whiteway, E., Maassen, J., and Hilke, M. (2011) *Phys. Rev. B*, **84**, 205407.
21. Cong, C., Yu, T., Sato, K., Shang, J., Saito, R., Dresselhaus, G.F., and Dresselhaus, M.S. (2011) *ACS Nano*, **5**(11), 8760.
22. Ferrari, A.C., Meyer, J.C., Scardaci, V., Casiraghi, C., Lazzeri, M., Mauri, F., Piscanec, S., Jiang, D., Novoselov, K.S., Roth, S., and Geim, A.K. (2006) *Phys. Rev. Lett.*, **97**, 187401.
23. Gupta, A., Chen, G., Joshi, P., Tadigadapa, S., and Eklund, P.C. (2006) *Nano Lett.*, **6**(12), 2667.
24. Graf, D., Molitor, F., Ensslin, K., Stampfer, C., Jungen, A., Hierold, C., and Wirtz, L. (2007) *Nano lett.*, **7**(2), 238.
25. Malard, L.M., Nilsson, J., Elias, D.C., Brant, J.C., Plentz, F., Alves, E.S., Castro Neto, A.H., and Pimenta, M.A. (2007) *Phys. Rev. B*, **76**, 201401(R).
26. Mafra, D.L., Moujaes, E.A., Doorn, S.K., Htoon, H., Nunes, R.W., and Pimenta, M.A. (2011) *Carbon*, **49**, 1511.
27. Cancado, L.G., Takai, K., Enoki, T., Endo, M., Kim, Y.A., Mizusaki, H., Speziali, N.L., Jorio, A., and Pimenta, M.A. (2008) *Carbon*, **46**, 272.
28. Gupta, A.K., Russin, T.J., Gutierrez, H.R., and Eklund, P.C. (2009) *ACS Nano*, **3**(1), 45.
29. Rao, R., Podila, R., Tsuchikawa, R., Katoch, J., Tishler, D., Rao, A.M., and Ishigami, M. (2011) *ACS Nano*, **5**(3), 1594.
30. Cong, C., Yu, T., Saito, R., Dresselhaus, G.F., and Dresselhaus, M.S. (2011) *ACS Nano*, **5**(3), 1600.
31. Saha, S.K., Waghmare, U.V., Krishnamurthy, H.R., and Sood, A.K. (2008) *Phys. Rev. B*, **78**, 165421.
32. (a) Jiang, J.W., Tang, H., Wang, B.S., and Su, Z.B. (2008) *Phys. Rev. B*, **77**, 235421; (b) Malard, L.M., Guimaraes, M.H.D., Marfa, D.L., Mazzoni, M.S.C., and Jorio, A. (2009) *Phys. Rev. B*, **79**, 125426.
33. Piscanec, P., Lazzeri, M., Mauri, F., Ferrari, A.C., and Robertson, J. (2004) *Phys. Rev. Lett.*, **93**(18), 185503.
34. Lazzeri, M. and Mauri, F. (2006) *Phys. Rev. Lett.*, **97**, 266407.
35. Ando, T. (2006) *J. Phys. Soc. Jpn.*, **75**(12), 124701.
36. (a) Pisana, S., Lazzeri, M., Casiraghi, C., Novoselov, K.S., Geim, A.K., Ferrari, A.C., and Mauri, F. (2007) *Nat. Mater.*, **6**, 198; (b) Kohn, W. (1959) *Phy. Rev. Lett.*, **2**(9), 393.
37. Yan, J., Zhang, Y., Kim, P., and Pinczuk, A. (2007) *Phy. Rev. Lett.*, **98**, 166802.
38. Das, A., Pisana, S., Chakraborty, B., Piscanec, S., Saha, S.K., Waghmare, U.V., Novoselov, K.S., Krishnamurthy, H.R., Geim, A.K., Ferrari, A.C., and Sood, A.K. (2008) *Nat. Nanotechnol.*, **3**, 210.
39. Panchakarla, L.S., Subrahmanyam, K.S., Saha, S.K., Govindaraj, A., Krishnamurthy, H.R., Waghmare, U.V., and Rao, C.N.R. (2009) *Adv. Mater.*, **21**, 4726.
40. Attaccalite, C., Wirtz, L., Lazzeri, M., Mauri, F., and Rubio, A. (2010) *Nano Lett.*, **10**, 1172.

41. Saha, S.K., Waghmare, U.V., Krishnamurthy, H.R., and Sood, A.K. (2007) *Phy. Rev. B*, **76**, 201404(R).
42. Yan, J., Henriksen, E.A., Kim, P., and Pinczuk, A. (2008) *Phys. Rev. Lett.*, **101**, 136804.
43. Das, A., Chakraborty, B., Piscanec, S., Pisana, S., Sood, A.K., and Ferrari, A.C. (2009) *Phys. Rev. B*, **79**, 155417.
44. Malard, L.M., Elias, D.C., Alves, E.S., and Pimenta, M.A. (2008) *Phys. Rev. Lett.*, **101**, 257401.
45. Ando, T. and Koshino, M. (2009) *J. Phys. Soc. Jpn.*, **78**(3), 034709.
46. Yan, J., Villarson, T., Henriksen, E.A., Kim, P., and Pinczuk, A. (2009) *Phys. Rev. B*, **80**, 241417(R).
47. Gava, P., Lazzeri, M., Saitta, A.M., and Mauri, F. (2009) *Phys. Rev. B*, **80**, 155422.
48. Chakraborty, B., Das, A., and Sood, A.K. (2011) *AIP Conf. Proceedings*, **1349**, 11.
49. Bruna, M. and Borini, S. (2010) *Phys. Rev. B*, **81**, 125421.
50. Chakraborty, B., Das, A., and Sood, A.K. (2009) *Nanotechnology*, **20**, 365203.
51. (a) Stampfer, C., Molitor, F., Graf, D., Ensslin, K., Jungen, A., Hierold, C., and Wirtz, L. (2007) *Appl. Phys. Lett.*, **91**, 241907; (b) Casiraghi, C., Pisana, S., Novoselov, K.S., Geim, A.K., and Ferrari, A.C. (2007) *Appl. Phys. Lett.*, **91**, 233108.
52. Meric, I., Han, M.Y., Young, A.F., Ozyilmaz, B., Kim, P., and Shepard, K.L. (2008) *Nat. Nanotechnol.*, **3**, 654.
53. Vasu, K.S., Chakraborty, B., Sampath, S., and Sood, A.K. (2010) *Solid State Commun.*, **150**, 1295.
54. Misewich, J.A., Martel, R., Avouris, Ph., Tsang, J.C., Heinze, S., and Tersoff, J. (2003) *Science*, **300**, 783.
55. Zaumseil, J., Donley, C.L., Kim, J., Friend, R.H., and Sirringhaus, H. (2006) *Adv. Mater.*, **18**, 2708.
56. Casiraghi, C., Hartschuh, H., Qian, H., Piscanec, S., Georgi, C., Fasoli, A., Novoselov, K.S., Basko, D.M., and Ferrari, A.C. (2009) *Nano Lett.*, **9**(4), 1433.
57. Sasaki, K., Yamamoto, M., Murakami, S., Saito, R., Dresselhaus, M.S., Takai, K., Mori, T., Enoki, T., and Wakabayashi, K. (2009) *Phys. Rev. B*, **80**, 155450.
58. Sasaki, K., Saito, R., Wakabayashi, K., and Enoki, T. (2010) *J. Phys. Soc. Jpn.*, **79**(4), 044603.
59. Cong, C., Yu, T., and Wang, H. (2010) *ACS Nano*, **4**(6), 3175.
60. Zhang, W. and Li, L. (2011) *ACS Nano*, **5**(4), 3347.
61. Cancado, L.G., Piementa, M.A., Neves, B.R.A., Dantas, M.S.S., and Jorio, A. (2004) *Phys. Rev. Lett.*, **93**, 247401.
62. Han, M.Y., Özyilmaz, B., Zhang, Y., and Kim, P. (2007) *Phys. Rev. Lett.*, **98**, 206805.
63. Li, X., Wang, X., Zhang, L., Lee, S., and Dai, H. (2008) *Science*, **319**, 1229.
64. Wakabayashi, K., Sasaki, K., Nakanishi, T., and Enoki, T. (2010) *Sci. Technol. Adv. Mater.*, **11**, 054504.
65. Ryu, S., Maultzsch, J., Han, M.Y., Kim, P., and Brus, L.E. (2011) *ACS Nano*, **5**(5), 4123.
66. Ren, W., Saito, R., Gao, L., Zheng, F., Wu, Z., Liu, B., Furukawa, M., Zhao, J., Chen, Z., and Cheng, H. (2010) *Phys. Rev. B*, **81**, 035412.
67. Lucchese, M.M., Stavale, F., Ferreira, E.H.M., Vilani, C., Moutinho, M.V.O., Capaz, R.B., Achete, C.A., and Jorio, A. (2010) *Carbon*, **48**, 1592.
68. Cancado, L.G., Jorio, A., Ferreira, E.H.M., Stavale, F., Achete, C.A., Capaz, R.B., Moutinho, M.V.O., Lombardo, A., Kulmala, T.S., and Ferrari, A.C. (2011) *Nano Lett.*, **11**, 3190.
69. Mohiuddin, T.M.G., Lombardo, A., Nair, R.R., Bonetti, A., Savini, G., Jalil, R., Bonini, N., Basko, D.M., Galiotis, C., Marzari, N., Novoselov, K.S., Geim, A.K., and Ferrari, A.C. (2009) *Phys. Rev. B*, **79**, 205433.
70. Huang, M., Yan, H., Chen, C., Song, D., Heinz, T.F., and Hone, J. (2009) *Proc. Natl. Acad. Sci. U.S.A.*, **106**(18), 7304.
71. Frank, O., Tsoukleri, G., Parthenios, J., Papagelis, K., Riaz, I., Jalil, R., Novoselov, K.S., and Galiotis, C. (2010) *ACS Nano*, **4**(6), 3131.
72. Tsoukleri, G., Parthenios, J., Papagelis, K., Jalil, R., Ferrari, A.C., Geim, A.K.,

Novoselov, K.S., and Galiotis, C. (2009) *Small*, **5**(21), 2397.

73. Yoon, D., Son, Y.W., and Cheong, H. (2011) *Phys. Rev. Lett.*, **106**, 155502.
74. Huang, M., Yan, H., Heinz, T.F., and Hone, J. (2010) *Nano Lett.*, **10**, 4074.
75. Frank, O., Mohr, M., Maultzsch, J., Thomsen, C., Riaz, I., Jalil, R., Novoselov, K.S., Tsoukleri, G., Parthenios, J., Papagelis, K., Kavan, L., and Galiotis, C. (2011) *ACS Nano*, **5**(3), 2231.
76. Zabel, J., Nair, R.R., Ott, A., Georgiou, T., Geim, A.K., Novoselov, K.S., and Casiraghi, C. (2012) *Nano Lett.*, **12**, 617.
77. Ding, F., Ji, H., Chen, Y., Herklotz, A., Dörr, K., Mei, Y., Rastelli, A., and Schmidt, O.G. (2010) *Nano Lett.*, **10**, 3453.
78. (a) Calizo, I., Balandin, A.A., Bao, W., Miao, F., and Lau, C.N. (2007) *Nano Lett.*, **7**(9), 2645; (b) Calizo, I., Miao, F., Bao, W., Lau, C.N., and Balandin, A.A. (2007) *Appl. Phys. Lett.*, **91**, 071913.
79. Lee, J.U., Yoon, D., Kim, H., Lee, S.W., and Cheong, H. (2011) *Phys. Rev. B*, **83**, 081419(R).
80. Faugeras, C., Faugeras, B., Orlita, M., Potemski, M., Nair, R.R., and Geim, A.K. (2010) *ACS Nano*, **4**(4), 1889.
81. Nika, D.L., and Balandin, A.A. (2012) *arXiv*, **1203**, 4282v1.
82. Proctor, J.E., Gregoryanz, E., Novoselov, K.S., Lotya, M., Coleman, J.N., and Halsall, M.P. (2009) *Phys. Rev. B*, **80**, 073408.
83. Snitka, V., Rodrigues, R.D., and Lendraitis, V. (2011) *Microelectron. Eng.*, **88**, 2759.
84. Stadler, J., Schmid, T., and Zenobi, R. (2011) *ACS Nano*, **5**(10), 8442.
85. Hartschuh, H. (2008) *Angew. Chem. Int. Ed.*, **47**, 8178.
86. Hartschuh, A., Sánchez, E.J., Xie, X.S., and Novotny, L. (2003) *Phys. Rev. Lett.*, **90**(9), 095503.
87. Hartschuh, A., Qian, H., Meixner, A.J., Anderson, N., and Novotny, L. (2005) *Nano Lett.*, **5**(11), 2310.
88. Cancado, L.G., Hartschuh, A., and Novotny, L. (2009) *J. Raman Spectrosc.*, **40**, 1420.
89. Saito, Y., Verma, P., Masui, K., Inouye, Y., and Kawata, S. (2009) *J. Raman Spectrosc.*, **40**, 1434.
90. Domke, K.F. and Pettinger, P. (2009) *J. Raman Spectrosc.*, **40**, 1427.
91. Ci, L., Song, L., Jin, C., Jariwala, D., Wu, D., Li, Y., Srivastava, A., Wang, Z.F., Storr, K., Balicas, L., Liu, F., and Ajayan, P.M. (2010) *Nat. Mater.*, **9**, 430.
92. Dean, C.R., Young, A.F., Meric, I., Lee, C., Wang, L., Sorgenfrei, S., Watanabe, K., Taniguch, T., Kim, P., Shepard, K.L., and Hone, J. (2010) *Nat. Nanotechnol.*, **5**, 722.
93. Radisavljevic, B., Radenovic, A., Brivio, J., Giacometti, V., and Kis, A. (2011) *Nat. Nanotechnol.*, **6**, 147.
94. Chakraborty, B., Bera, A., Muthu, D.V.S., Bhowmick, S., Waghmare, U.V., and Sood, A.K. (2012) *Phys. Rev. B*, **85**, 161403(R).
95. Hasan, M.Z. and Kane, C.L. (2010) *Rev. Mod. Phys.*, **82**, 3045.
96. Ha, S.D. and Ramanathan, S. (2011) *J. Appl. Phys.*, **110**, 071101.
97. Zheng, Y., Ni, G.X., Toh, C.T., Tan, C.Y., Yao, K., and özyilmaz, B. (2010) *Phys. Rev. Lett.*, **105**, 166602.
98. (a) Ando, T. (2007) *J. Phys. Soc. Jpn.*, **76**(2), 024712; (b) Goerbig, M.O., Fuchs, J.N., Kechedzhi, K., and Fal'ko, V.I. (2007) *Phys. Rev. Lett.*, **99**, 087402.
99. (a) Faugeras, C., Amado, M., Kossacki, P., Orlita, M., Sprinkle, M., Berger, C., de Heer, W.A., and Potemski1, M. (2009) *Phys. Rev. Lett.*, **103**, 186803; (b) Yan, J., Goler, S., Rhone, T.D., Han, M., He, R., Kim, P., Pellegrini, V., and Pinczuk, A. (2010) *Phys. Rev. Lett.*, **105**, 227401.
100. Kang, M., Pinczuk, A., Dennis, B.S., Pfeiffer, L.N., and West, K.W. (2001) *Phys. Rev. Lett.*, **86**(12), 2637.

3
Physics of Quanta and Quantum Fields in Graphene

Ganapathy Baskaran

3.1
Introduction

The concept of field is fundamental to physics. It captures the essence of a physical state. The concept that started in classical electrodynamics, elasticity, and fluid mechanics has evolved into quantum domain as quantum fields in the past one century. Bose and Fermi fields and their quanta pervade condensed matter physics – low-energy and low-temperature properties of many systems are well describable as weakly interacting bosons and fermions. In the study of phases of condensed matter, fields appear as *order parameters*.

The richness of condensed matter systems continues to pose new challenges, through new and strongly interacting quantum fields. Two-dimensional systems such as electron gas in the interface of the heterostructure in strong magnetic fields or quasi-2D cuprate superconductors exemplify this. A recent example in this category is graphene [1]. In spite of its simplicity, graphene, a honeycomb net of $p\pi$-bonded carbon atoms, supports an unexpected variety of quanta and quantum fields. Consequently, a rich low-energy physics follows. The description of quantum field theory has become a useful and, in some cases, a necessary tool to understand many of the experiments in graphene. Further, the quantum field theory approach has made important and verifiable predictions.

Several articles reviewing exciting developments in graphene have appeared in the recent past [2–6]. The aim of this chapter is to review and summarize the quantum field aspects of graphene from the point of view of physics in a qualitative manner. We start with the relativistic Dirac equation in $3+1$ dimensions and review the concepts needed to understand the $2+1$-dimensional relativistic type of behavior of electrons in graphene. Then, we summarize the results of the familiar tight-binding model of electrons in the π-orbitals of graphene, forming a honeycomb lattice. As shown by Semenoff [7] in 1984, massless Dirac-like dispersion in $2+1$-dimensions and the chirality property of the Bloch states close to the chemical potential follow from the tight-binding analysis. This theoretical appreciation of massless Dirac-like behavior of electrons in graphene happened well before an experimental observation of massless Dirac-like behavior of electrons

Graphene: Synthesis, Properties, and Phenomena, First Edition. Edited by C. N. R. Rao and A. K. Sood.

in graphene by Geim and Novoseleev [1] in 2005. Dirac cone dispersion and chirality have consequences. Some of them have been seen in experiments: (i) half-integer shift in integer Hall quantization [8, 9], (ii) antilocalization in electronic transport [10, 11], (iii) Klein tunneling [12, 13], and (iv) universal optical absorption coefficient [14]. Other predictions such as Zitterbewegung [15–20] and possibility of fractional charge excitations in the presence of Kekule dimerization [21] are yet to be seen. Massless Dirac operator in $2 + 1$ dimensions has profound field theory consequences that are yet to be seen in experiments: (i) topological mass generation, related to induced Chern-Simons term, a parity anomaly [7]; (ii) zero-field Hall effect [22]; and (iii) spin Hall effect [23]. From the mathematical point of view, massless Dirac operator in $2 + 1$ dimensions brings in certain index theorems [3, 24], which help count zero modes in the presence of nontrivial topology, topological defects, and external magnetic fields.

It turns out that one can achieve extraordinarily high mobility of electrons in graphene with less effort, for example, compared to that in semiconductor heterostructures. This has made graphene a wonderful template for experimental study of integer and fractional quantum Hall effects. The band parameters of graphene have desirable values and make the integer quantum Hall plateau visible in real experiments at temperatures as high as room temperatures [25]. We realized [26] theoretically that quantum Hall states are capable of bringing out relativistic-type effects in an experimentally accessible way. We focus on integer quantum Hall states in the presence of a crossed electric field and review our own predictions [26] of (i) vacuum collapse phenomenon, (ii) Lorentz boost, (iii) Lorentz contraction, and (iv) time dilation.

As graphene is a lower dimensional tight-binding system, one expects some important role played by Coulomb interaction. We review one consequence, namely, emergence of a spin-1 collective mode branch, a theoretical prediction made by Jafari and Baskaran [27]. This mode can be viewed as arising from presence of Pauling's resonating valence bond (RVB) correlations in the ground state of graphene.

A search for Majorana fermions in condensed matter systems have become very popular recently [28], in view of a potential role it could play in the field of topological quantum computation and quantum information. Graphene offers the possibility of Majorana fermions in a few ways. We present a new possibility of using two-channel Kondo effect to generate the Majorana zero mode in graphene. Sengupta and Baskaran [29] have theoretically predicted the gate-controlled two-channel Kondo effect in graphene. The two-channel Kondo impurity traps a fragile Majorana zero mode that could, in principle, be used for topological quantum computation. There have also been suggestions of the Majorana zero mode arising from Kekule dimerization with topological defects [30, 31] and vortices in the superconducting state in graphene.

At the end of this chapter, we summarize how lattice strain and topological defects could generate and mimic strong *pseudo-magnetic fields* that respect the overall time-reversal symmetry of graphene.

3.2
Dirac Theory in 3 + 1 Dimensions: A Review

Interest in graphene has sharply increased because of relativistic, 2 + 1 dimensional, massless, and chiral Dirac particle-like behavior of quasiparticles in graphene. To appreciate the importance and richness of Dirac fermions in general and contrast it with the 2 + 1-dimensional graphene case, we briefly review the theory of Dirac particles in 3 + 1 dimensions [32].

As is well known, Dirac discovered his equation while generalizing the nonrelativistic Schrodinger equation to the relativistic situation. Earlier attempts by Klein, Gordon, and Schrodinger himself led to the Klein–Gordon equation. It was a first *quantization* of the energy–momentum relation of the special theory of relativity, $E^2 = m^2c^4 + c^2\mathbf{p}^2$ into the Klein–Gordon equation

$$-\hbar^2 \frac{\partial^2 \psi(\mathbf{r}, t)}{\partial^2 t} = (m^2c^4 - c^2\hbar^2\nabla^2)\psi(\mathbf{r}, t) \tag{3.1}$$

Even though this equation is symmetric under Lorentz transformations, it leads to unphysical and negative values for probability density. This failure is due to the second-order character of the equation in time variable. This had Dirac worried. Guided by a strong faith in mathematical consistency, mathematical elegance, and beauty, Dirac discovered the first-order equation that bears his name:

$$-i\hbar\partial_t \Psi(\mathbf{r}, t) = (-i\hbar c\ \vec{\alpha} \cdot \nabla + mc^2\beta)\Psi(\mathbf{r}, t) \tag{3.2}$$

The price Dirac had to pay was to introduce 4×4 anticommuting gamma (Dirac) matrices, $(\vec{\alpha}, \beta)$, and generalize the single-component wave function $\psi(\mathbf{r}, t)$, a complex number appearing in the Klein–Gordon equation into a four-component spinor $\Psi(\mathbf{r}, t)$. In turn, Dirac discovered several secrets of nature in a single shot: the relativistic origin of spin degree of freedom, helicity, chirality, antiparticle (positron), vacuum as a filled sea of fermions, unexpected infinities, and so on. It is a historical fact that the Dirac equation laid the foundation for the relativistic quantum field theory of elementary particles in nature. The first successful theory was quantum electrodynamics. This has grown into an edifice, namely, the standard model of electromagnetic, weak, and strong interactions.

The four-component Dirac spinor carries rich physics [32]. To see this, let us write the Dirac equation in matrix form, in the standard representation:

$$-i\hbar\partial_t \Psi(\mathbf{r}, t) = \begin{pmatrix} mc^2\mathrm{I} & c\vec{\sigma} \cdot \mathbf{p} \\ c\vec{\sigma} \cdot \mathbf{p} & -mc^2\mathrm{I} \end{pmatrix} \Psi(\mathbf{r}, t) \tag{3.3}$$

where $\vec{\sigma}$ are the 2 × 2 Pauli matrices and I is the 2 × 2 unit matrix. The top two entries of the spinor $\Psi = \begin{pmatrix} e_\uparrow \\ e_\downarrow \\ p_\uparrow \\ p_\downarrow \end{pmatrix}$ represent the up and down spin components of the positive energy (electron) amplitude, $(e_\uparrow, e_\downarrow)$. The bottom two entries represent the negative energy (positron) amplitude $(p_\uparrow, p_\downarrow)$. What is interesting is that relativistic dynamics (momentum operator of the off-diagonal elements) mix positive and negative energies at the *one-particle level*; the amount of mixing is determined by the diagonal matrix element mc^2.

To see the mixing, let us look at the plane-wave free-particle solution. One such eigenfunction is an *electron eigenfunction*

$$\Psi(\mathbf{r}) = \sqrt{\frac{E+mc^2}{2EV}} \begin{pmatrix} \begin{pmatrix} 1 \\ 0 \end{pmatrix} \\ \frac{cp}{E+mc^2}(\vec{\sigma}\cdot\hat{\mathbf{p}}) \begin{pmatrix} 1 \\ 0 \end{pmatrix} \end{pmatrix} e^{i\frac{\mathbf{p}\cdot\mathbf{r}}{\hbar}} \tag{3.4}$$

where E is the energy of the particle and V is the volume of the system; $\hat{\mathbf{p}}$ is the unit vector along the three-momentum $\mathbf{p}$. The first two entries of the above-mentioned four-component spinor specify that the positive energy component of the electron has a spin projection along a chosen positive z-direction. The first two components alone are Pauli spinors used in nonrelativisitic description of electrons in solid-state physics, where, to a good approximation, spin and orbital motions are decoupled and independent. For a Dirac particle at nonrelativistic speeds, the first two components are called *large components* and the last two *small components*.

The last two components represent the negative energy or positron component of a physical Dirac electron. For nonrelativistic speeds, this component is small $\sim \frac{v}{c}$. In Feynman's space–time picture, an electron, during its time evolution, spends a fraction of time as a positron but moving back in time. What is interesting is that this component represents a tendency of the Dirac electron to point its spin along the direction of its motion. There are two possibilities: it could be either parallel or antiparallel to the direction of its motion. This important property is called *helicity*. This is seen by rewriting the Dirac spinor as a sum of two parts (ignoring a normalization factor):

$$\begin{pmatrix} \begin{pmatrix} 1 \\ 0 \end{pmatrix} \\ \frac{-pc}{E+mc^2} \begin{pmatrix} 1 \\ 0 \end{pmatrix} \end{pmatrix} + \frac{pc}{E+mc^2} \begin{pmatrix} \begin{pmatrix} 0 \\ 0 \end{pmatrix} \\ (1+\vec{\sigma}\cdot\hat{\mathbf{P}}) \begin{pmatrix} 1 \\ 0 \end{pmatrix} \end{pmatrix} \tag{3.5}$$

The first part is a spinor state in which a Dirac particle has both a particle and a small antiparticle content, but with spin pointing along the positive z-axis. The second part is a pure antiparticle state, where the operator $(1+\vec{\sigma}\cdot\hat{\mathbf{p}})$ rotates the spin into the direction of momentum along the unit vector $\hat{\mathbf{p}} = (\hat{p}_x, \hat{p}_y, \hat{p}_z)$. That is, the spin points exactly along the direction of motion of the particle. This property is called *helicity*. It is $+1$ if the spin points along the direction of local velocity and -1 when it is the opposite. It is easy to see that the helicity operator

$$\Sigma\cdot\mathbf{p} \equiv \begin{pmatrix} \vec{\sigma} & 0 \\ 0 & \vec{\sigma} \end{pmatrix}\cdot\mathbf{p} \tag{3.6}$$

commutes with the Dirac Hamiltonian. (Here, Σ is the *spin* operator of the Dirac particle.) However, helicity is not a Lorentz invariant quantity, as it is a product of a three vector and a pseudoscalar. The helicity property of the small component indicates that the orbital and spin motions are coupled, albeit with a small probability for nonrelativistic electrons.

In the study of electrons in solids, it is customary, because of nonrelativistic speeds, to ignore the antiparticle component of the Dirac spinor and retain only a two-component Pauli spinor. However, when a particle reaches relativistic speeds, such as a Bohr orbit passing near a heavy nucleus in heavy atoms (for example, bismuth), the orbital motion starts influencing the electron spin dynamics through the helicity character of spin of the positron component. This is the origin of spin orbit coupling. As this is a small effect, one can calculate the effect of positron component perturbatively and systematically by eliminating the small component (off-diagonal terms of the Dirac equation) by the Foldy–Wouthuysen transformation.

Massless Dirac particles are special. To see this, we consider the small mass limit, $mc^2 << \text{pc}$. Physics in this limit becomes transparent in the so-called Weyl representation:

$$-i\hbar\partial_t\Psi(\mathbf{r},t) = \begin{pmatrix} c\vec{\sigma}\cdot\mathbf{p} & mc^2\mathrm{I} \\ mc^2\mathrm{I} & -c\vec{\sigma}\cdot\mathbf{p} \end{pmatrix}\Psi(\mathbf{r},t) \tag{3.7}$$

In this representation, the mass term appears as off-diagonal elements. Thus, they cause mixing between a *particle and an antiparticle*. Recall that in the Dirac representation it is the momentum term that causes mixing between a *particle and an antiparticle*. The Dirac representation is more convenient for nearly nonrelativistic fermions, such as electrons in solids or heavy quarks. The Weyl representation is well suited for nearly relativistic particles such as neutrinos.

For $m = 0$, we get the Weyl equation where particles and antiparticles get decoupled. We get two separate two-component equations for particles and antiparticles. It is easily seen that the Weyl equation violates parity. The Dirac–Weyl Hamiltonian also has a new conserved quantity called *chirality*, defined using the γ_5 matrix. We will not go further into chirality except to point out that in the $m = 0$ case, both chirality (handedness) and helicity become the same. Further, in this limit, the helicity operator also respects Lorentz invariance. Particles and antiparticles have opposite helicity.

Nature makes use of an intrinsic *less symmetric* handedness of fermions in the world of elementary particles. It has been established that only left-handed fermions participate in weak interactions. This experimental fact, possibly due to a spontaneous symmetry breaking in nature at a deeper level (beyond the standard model), is one of the puzzles in the world of elementary particle physics.

3.3 Band Structure of Graphene: Massless Chiral Dirac Electrons in 2 + 1 Dimensions

Having briefly seen the theory of 3 + 1-dimensional Dirac particles, we will see the 2 + 1-dimensional Dirac theory through the example of graphene. Carbon atoms in graphene are connected by sp^2 bonds, resulting in a simple tight-binding behavior of π electrons. Three valence orbitals of carbon atoms, 2s, $2p_x$, and $2p_y$, undergo sp^2 hybridization. This leads to a strong σ bond formation. Three electrons from each carbon atom participate in this strong (120°) bond, resulting in a planar honeycomb lattice structure with threefold coordination for each carbon atom. We are left with one $2p_z$ orbital and one electron per carbon atom. These lone electrons undergo a nearest neighbor $p\pi$ bonding and delocalization.

In the noninteracting picture, the tight-binding model Hamiltonian is

$$H = -t \sum_{\langle i,j \rangle,\sigma} (a_{i,\sigma}^{\dagger} b_{j,\sigma} + H.c) \tag{3.8}$$

The honeycomb lattice has two sublattices and two atoms per unit cell. Here, $(a_{i,\sigma}^{\dagger}, a_{i,\sigma})$ and $(b_{j,\sigma}^{\dagger}, b_{j,\sigma})$ are creation and annihilation operators for A and B sublattices. Translation invariance and periodic boundary conditions lead to

$$H = -t \sum_{\mathbf{k},\sigma} \gamma(\mathbf{k}) a_{\mathbf{k},\sigma}^{\dagger} \; b_{\mathbf{k},\sigma} + H.c \tag{3.9}$$

Here, $\gamma(\mathbf{k}) \equiv \sum_{\Delta} e^{i\mathbf{k}\cdot\Delta}$. And $\Delta \equiv (\Delta_1, \Delta_2, \Delta_3)$ are three vectors connecting an atom in sublattice A to three of its nearest neighbors in sublattice B.

The above-mentioned single-particle Hamiltonian for a given k, $\begin{pmatrix} 0 & -t\gamma(\mathbf{k}) \\ -t\gamma(\mathbf{k})^* & 0 \end{pmatrix}$ has the following meaning. Bloch states in the two sublattices do not disperse and form two degenerate flat bands at zero energy. Nearest neighbor hopping between two sublattices mixes these bands (in a k-dependent manner) and produces dispersion. As a result, we get an electron-like and a hole-like broad bands. The resulting Hamiltonian is

$$H = \sum_{\mathbf{k},\sigma} \varepsilon_{\mathbf{k}} \left(\alpha_{\mathbf{k}\sigma}^{\dagger} \alpha_{\mathbf{k}\sigma} - \beta_{\mathbf{k}\sigma}^{\dagger} \beta_{\mathbf{k}\sigma} \right) \tag{3.10}$$

Here, $\varepsilon_{\mathbf{k}} = t\sqrt{1 + 4\cos\frac{\sqrt{3}k_x a}{2}\cos\frac{k_y a}{2} + 4\cos^2\frac{k_y a}{2}}$. The creation and annihilation operators α's and β's correspond to positive and negative energy Bloch states, respectively. As shown in Figure 3.1, two bands touch each other at the K- and K′-points in the Brillouin zone, leading to two distinct Dirac nodes. Neutral graphene has its chemical potential at zero energy, and the spectrum has particle–hole symmetry.

Semenoff [7] observed that the one-electron Bloch wave function for graphene is conveniently organized as a two-component spinor. Two components of the spinor represent amplitudes of the single-particle wave function in sublattices A and B. Notice that in graphene we have nonrelativistic electrons that move in a 2D periodic potential of carbon nuclei. The single-particle wave functions satisfy the Schrodinger equation. The Dirac equation that Semenoff obtained is only a convenient way of organizing the Schrodinger equation. As is often the case, a suitable mathematical reorganization brings in new insights and similarities, and points to new directions. If we measure wave vector $\mathbf{k}$ with respect to the K-point

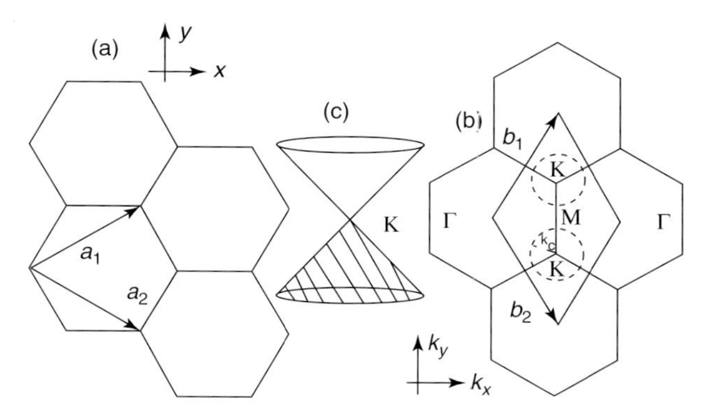

Figure 3.1 (a) Honeycomb lattice of carbon atoms in real space; $\mathbf{a}_{1,2} = \frac{a}{2}(\sqrt{3}, \pm 1)$ (b) K-space and the Brillouin zone; $\mathbf{b}_{1,2} = \frac{2\pi}{a}(\frac{1}{\sqrt{3}}, \pm 1)$ (c) Dirac cone spectrum at a K-point.

and look at the linearized dispersion, we get the k-space single-particle Hamiltonian

$$H = -i\hbar v_F \vec{\sigma} \cdot \mathbf{k} = \hbar v_F \begin{pmatrix} 0 & k_x + ik_y \\ k_x - ik_y & 0 \end{pmatrix} \tag{3.11}$$

where $v_F \approx \frac{\sqrt{3}}{2} t \approx \frac{c}{300}$ is the Fermi velocity of electrons at the Dirac nodes and c is the velocity of light. The wave vector $\mathbf{k}$ is a two-dimensional vector. Hence, only the x- and y- components of the Pauli matrices appear in the above-mentioned equation. In the approximation v_F, the Fermi velocity is the only parameter for netural graphene. We get similar massless Dirac equation for states close to the K′-point, with k_y replaced by $-k_y$.

In the absence of any external potential, four representative Bloch states are (i) *positive energy* electron state near K-point, (ii) a *negative energy* hole state near K-point, (iii) a *positive energy* electron state near K′-point, and (iv) a *negative energy* hole state near K′-point. The two-component spinor takes the following form

$$\begin{pmatrix} s \\ e^{i\tau_z \theta_{\mathbf{k}}} \end{pmatrix} e^{i\tau_z \mathbf{k}\cdot\mathbf{r}} \tag{3.12}$$

where $\tau_z \pm 1$ corresponds to valleys K and K′; s = ±1 corresponds to positive energy (electron) and negative energy (hole) states. Further, $\theta_{\mathbf{k}} \equiv \tan^{-1}(\frac{k_y}{k_x})$ is the angle subtended by the vector $\mathbf{k}$ with x-axis with K- or K′-point as the origin. Notice that states at corresponding points around K and K′ are complex conjugates of each other.

Now we are ready to discuss the meaning of the spinor. First, unlike the 3 + 1 -dimensional case, *Dirac matrices* in 2 + 1 dimensions are simply Pauli matrices. *The two components of the Spinors, however, do not represent the spin degree of freedom.* They represent particle and antiparticle degrees of freedom. The particle and antiparticle spinors are sometimes called a *pseudospin* in the context of graphene.

For graphene, the two components of the spinor (Equation 3.11) represent the amplitude of finding a particle (electron) and its antiparticle (hole). The massless character of the excitation in the case of graphene makes the physical particle to be a coherent superposition of electronlike and holelike state.

The meaning of electron and hole components is somewhat confusing! To clarify it in this context, we add a mass term by hand to the massless Dirac equation (3.11). A physical realization of such a small but finite *mass gap* is achieved by placing graphene on top of the layered hexagonal boron nitride (BN) surface in registry. As both have hexagonal structures and nearly the same lattice constant, boron (nominally negatively charged) and nitrogen (nominally positively charged) ions of BN modify the on-site potential for the carbon atoms in the two sublattices of graphene. The gap induced by this additional periodic potential is estimated to be $E_G \approx 50$ MeV. The corresponding *mass* is defined through the relation $E_G = 2mv_F^2$. A mass gap adds a term containing the third component of the Pauli matrix σ^z

$$H = -i\hbar v_F(\sigma_x k_x + \tau_z \sigma_y k_y) + m v_F^2 \sigma^z \tag{3.13}$$

Particle and hole excitations of single-layer BN itself are describable by a massive Dirac equation in 2 + 1 dimensions. However, the mass (band) gap is large, $E_G \sim 5$ eV. Consequently, it is customary to use a nonrelativistic parabolic band description for low-energy excitations.

In the presence of the mass gap, small compared to the quasiparticle energy, $\hbar v_F k \gg E_G$, the electron and hole spinors at K and K′ ($\tau_z = \pm$) take the form

$$\begin{pmatrix} 1 \\ (1 - \frac{E_G}{\hbar v_F k}) e^{i\tau_z \theta_{\mathbf{k}}} \end{pmatrix} e^{i\tau_z \mathbf{k}\cdot\mathbf{r}} \quad \& \quad \begin{pmatrix} -(1 - \frac{E_G}{\hbar v_F k}) e^{i\tau_z \theta_{\mathbf{k}}} \\ 1 \end{pmatrix} e^{i\tau_z \mathbf{k}\cdot\mathbf{r}} \tag{3.14}$$

In the other (BN) limit, when $E_G \gg \hbar v_F k$, the spinors take the form

$$\begin{pmatrix} 1 \\ \frac{\hbar v_F k}{E_G} e^{i\tau_z \theta_{\mathbf{k}}} \end{pmatrix} e^{i\tau_z \mathbf{k}\cdot\mathbf{r}} \quad \& \quad \begin{pmatrix} -\frac{\hbar v_F k}{E_G} e^{i\tau_z \theta_{\mathbf{k}}} \\ 1 \end{pmatrix} e^{i\tau_z \mathbf{k}\cdot\mathbf{r}} \tag{3.15}$$

It is interesting to see that in the presence of the mass gap the equality of the weight (modulus of the amplitude) of Bloch states in the two sublattices starts changing. Electronlike excitations lose their weight in sublattice B and holelike ones in sublattice B, respectively. In the limit, the momentum of the quasiparticle is zero (that is, bottom of the conduction band and top of the valence band) and the Bloch state has vanishing weight in one or the other sublattices

$$\begin{pmatrix} 1 \\ 0 \end{pmatrix} \quad \& \quad \begin{pmatrix} 0 \\ 1 \end{pmatrix} \tag{3.16}$$

That is, in the case of BN, the top of the valence band (at K- and K′-points) is purely nitrogen-like and the bottom of the conduction band is purely boronlike.

Massless Dirac particles in 3 + 1 dimensions possess helicity and chirality properties. What is the analog in 2 + 1 dimensions? In 2 + 1 dimensions, we have helicity in the sense that the direction of the (spinor) *pseudospin* is parallel or antiparallel to the momentum direction in the plane. This in turn leads to a chirality or vorticity property of phases of the Bloch states around K- and K′-points.

3.3.1 Phase Vortices of Bloch States in k-Space

In Equation 3.12, the relative amplitudes have a phase factor $se^{i\tau_z\theta_{\mathbf{k}}}$. This tells us that the *phase field of Bloch states* defines a vortex and an antivortex in the neighborhood of K- and K′-points in k-space. In other words, if we consider a spinor $\begin{pmatrix} 1 \\ e^{i\theta_{\mathbf{k}}} \end{pmatrix}$, as the state of the pseudospin at a point **k** in k-space, the direction of the pseduospin lies in the XY-plane and points along the direction of momentum of the Bloch wave! If the pseudospin direction is parallel to the **k** near the K-point, it is antiparallel near the K′-point. Thus, the overall time-reversal symmetry character of the graphene Hamiltonian is respected by the Bloch states.

These phase vortices mean that there is the possibility of nontrivial Berry curvature appearing as nonzero Chern number. This nontrivial topology of the phases of Bloch states in k-space enables the zero-field quantum Hall effect suggested by Haldane and spin Hall effect suggested by Kane and Mele, when perturbed suitably. The issue of Berry connection and relation to the Chern-Simons theory and quantum Hall effect is discussed in detail in Ref. [3] and references therein.

From the point of view of *pseudospin–orbit coupling* of electron degree of freedom, the above-mentioned phenomenon is remarkable. Because for physical electrons in solids, spin–orbit coupling is typically very small $\sim$100's of millielectronvolts. Only at high speeds (reaching scale c), spin–orbit coupling becomes important and spin dynamics gets coupled to orbital dynamics. What we have with the pseudospin of electrons in graphene is an extreme case of pseudospin and orbit coupling. That is, the orbital motion completely dictates the pseudospin dynamics.

It is interesting to note that spinors of massless Weyl particles in $3+1$ dimensions also have a topological property. That is, around the Weyl point in the momentum space, the phases of the two-component spinors (direction of physical spin in this case) define a magnetic monopole (hedgehog) configuration. The sign of helicity defines the sign of the monopole charge.

While there is no realization of Weyl particles in the world of elementary particles (even neutrino is believed to have a finite mass), such particles could emerge in certain quantum condensed matter.

We should also point out that by changing the mass gap in graphene systematically we have an opportunity for tunable strength for *pseudospin–orbit* coupling and study the consequences in detail.

For the physical electron in graphene, in addition to the valley pseudospin (valley spin) denoted by τ_z corresponding to K- and K′-points, there are physical spin-half degree of freedom. In what follows, unless necessary, we will not exhibit the valley spin and physical spin indices.

In the following few sections, we focus on certain consequences of the massless Dirac field theory character of low-energy fermions in graphene, including a brief introduction to anomalies in field theories.

3.4
Anomaly – A Brief Introduction

Anomaly in quantum mechanics or quantum field theory is an indication that the symmetry of a classical system fails to survive in the corresponding quantum system. This should not be a surprise as Hilbert space is much bigger and richer than the classical phase space and configuration space. Unexpected degeneracies, not necessarily related to standard spontaneous symmetry breaking, may occur in the ground state. Anomalies also manifest themselves as the calculable and somewhat anomalous response of the vacuum to external perturbations such as uniform electric and magnetic fields. This is the approach we have taken in this section when sketching anomalies.

Field theory description of physical systems often comes with mathematical ambiguities and infinities arising from uncountable continuous degrees of freedom. In order to obtain physically meaningful finite values for observables, sometimes, one has to subtract two infinities and introduce regularization procedures involving short- distance (ultraviolet) cutoffs. In discussing low-energy physics in solid-state physics, a finite lattice parameter and tight-binding approximations bring in a natural cutoff. Effective field theories of condensed matter systems have a natural lattice regularization. *Because of this, some of the anomalies that one encounters in field theories are not anomalies or surprises in condensed matter systems.* We discuss a few examples including a parity anomaly that one encounters in field theory description of graphene.

Certain anomalies in quantum field theory occur as nonvanishing expectation values of some operators that are expected to vanish in the classical context. They also occur as anomalous commutation relations. These anomalous quantities express certain reduced symmetries, reduced compared to the full symmetry of the Hamiltonian, in the physical vacuum. Emergence of zero modes is responsible for some anomalies. Evaluation of anomalies is a nontrivial problem that requires careful regularization using ultraviolet and infrared cutoffs. Anomalies play a role in phenomena such as fractionization of quantum numbers.

Spontaneous symmetry breaking in statistical mechanics and classical physics may be considered as a classical anomaly. For example, spontaneous magnetization is nonzero below T_c, in a 2D Ising model on an infinite lattice, in the limit of the external field $H \to 0^+$; that is, the magnetization $\langle s_i^z \rangle_{H=0^+}$ remains finite below T_c. However, the $H = 0$ field expectation value of spin vanishes identically, because of global spin symmetry, $\langle s_i^z \rangle_{H=0} = 0$ at all temperatures, including $T < T_c$.

Another example is a bifurcation of solutions of nonlinear differential equations from symmetric to the family of nonsymmetric solutions, as a certain parameter is varied. What is important is that quantum anomalies represent emergence of nontrivial quantum correlations, such as topological order, rather than spontaneous order in quantum systems. Zero modes of the anomalies sense topological properties of the quantum vacuum.

Anomalies in quantum field theory depend on the dimension of space. It brings out nontrivial topological structures that different space–time dimensions can support through quantum fields. As a side remark, this is the basic reason why Dirac operators that emerged in physics have profound implications for pure mathematics, in classifying manifolds and so on, in the field of topology and geometry.

In what follows, we discuss first the anomalies in $1 + 1$ and $3 + 1$ dimensions (odd space dimensions) that have common anomalies called *axial anomalies*. Then, we discuss briefly the anomaly in the $2 + 1$ dimension, which is called a *parity anomaly*. Common features [33] of all anomalies are *emergence of nonvanishing electric, magnetic fluxes, and charges, and their magnitude measures the degeneracy of zero modes in the problem.*

3.4.1
Anomalous Commutator in (1 + 1) Dimensions

One of the first anomalies to be discovered in quantum field theory is the Schwinger anomaly [34]. It occurs in $1 + 1$-dimensional massless electrodynamics. There are different manifestations of this anomaly: (i) an anomalous commutation relation between currents and (ii) dynamical mass generation. This anomaly tells us that when we consider an operator such as charge density of a $1 + 1$-dimensional theory of Dirac fermions, it has an anomalous commutation relation when projected into Hilbert space of the filled Dirac sea and excitations around it. They are anomalies in the sense that a naive evaluation of the commutator gives zero as the result. However, a derivation using a cutoff (for example, a cutoff that creates a bottom to the Dirac sea) gives a nonzero result. Further, the anomalous quantities turn out to be independent of the cutoff and survive even when the cutoff tends to infinity. This is at the heart of bosonization phenomena in models such as the Luttinger model for $1 + 1$-dimensional condensed matter systems. While Schwinger gave a very general argument for anomalies, Tomonaga [35] obtained a similar result for specific model of free fermions in $1 + 1$ dimensions. This was also used by Lieb and Mattis [36], who corrected the exact solution to the Luttinger model [37].

Let us consider a $1 + 1$-dimensional massless Dirac fermion and look at the commutation relation of the Fourier components of fermion density of left and right movers, $\rho_1(-q) \equiv \sum_i \mathrm{e}^{-\mathrm{i}qx_i}, \rho_2(q) \equiv \sum_i \mathrm{e}^{\mathrm{i}qx'_i}$ where x_i and x_i' are coordinates of left-moving and right-moving fermions.

From the above-mentioned definition, it follows that

$$[\rho_1(-q), \rho_1(q')] = 0 \tag{3.17}$$

In the second quantization form, using creation, annihilation operators $a_1^\dagger$ are a_1's

$$[\rho_1(-q), \rho_1(q')] = \sum_{k=-\infty}^{k=+\infty} a_{1k}^{\dagger} a_{1k+q-q'} - \sum_{k=-\infty}^{k=+\infty} a_{1k+p}^{\dagger} a_{1k+p} = 0 \tag{3.18}$$

Vanishing of the commutator follows from the absence of upper and lower momentum cutoffs. The last two terms of Equation 3.18, which are individually infinity, seem to cancel each other because of the limits. Suppose we introduce a finite upper and lower cutoff in the summation and evaluate the commutator in the physical vacuum with a finite k_F, we find the nontrivial commutation relation (in the sense of equivalence of operators as discussed by Tomonaga [35])

$$[\rho_1(-q), \rho_1(q')] = \delta_{q,q'} \sum_{-q<k<0} 1 = \frac{qL}{2\pi}\delta_{q,q'} \tag{3.19}$$

and $[\rho_1(-q), \rho_2(q')] = 0$. Thus, the commutation is nonzero and is independent of the short distance (large momentum) cutoff we introduced as regularization. The above equation forms the basis of bosonization of the Fermi system in the $1+1$ dimension.

3.4.2
Axial Anomaly in (1 + 1), (3 + 1) Dimensions

The anomaly that is relevant for graphene is a parity anomaly that occurs in $2+1$ dimensions. Before we move on to this, we briefly discuss two axial (also called *chiral*) anomalies: (i) Schwinger anomaly in the $1+1$-dimensional massless QED and (ii) ABJ anomaly, discovered by Adler [38], Bell, and Jackiw [39] in the $3+1$-dimensional massless electrodynamics. This anomaly is called *axial anomaly* as quantum effects produce nonconservation of an axial current. In one dimension, the time component of the axial current is the difference in density of left- and right-moving fermions.

We follow Nielson and Ninomiya [40] and calculate the response to electric and magnetic fields to understand anomalies. Let us start with the $1+1$-dimensional massless electrodynamics. Imagine applying a uniform electric field E along positive x-axis for a time t. The field acts like a chemical potential with opposite signs for the left and right movers. In a theory without cutoff, it generates current and produces equal density of right-moving particles and left-moving antiparticles. That is, the bottomless vacuum that has no cutoff produces particles and antiparticles (from its infinite resources). The increase in charge density or right and left movers is

$$\rho_{\mathrm{R}} = \frac{eE}{2\pi}t \quad \text{and} \quad \rho_{\mathrm{L}} = -\frac{eE}{2\pi}t \tag{3.20}$$

Thus, the total (vector) fermion number remains conserved

$$\partial_t \rho_V = \partial_t(\rho_R + \rho_L) = 0 \tag{3.21}$$

whereas the difference in particle and antiparticle number (axial current)

$$\partial_t \rho_A = \partial_t(\rho_R - \rho_L) = \frac{eE}{2\pi} \tag{3.22}$$

remains finite. This is the axial current anomaly.

On the other hand, if we had a cutoff at the bottom of the Dirac sea, left movers will not be affected by the Dirac sea; and right movers will produce excess particles at the expense of holes at the bottom of the Dirac sea.

It is good to pause and think about a condensed matter system such as a one-dimensional tight-binding model of metal with a finite density of electron. Let it be connected to a source and a sink at its ends. The lattice provides a natural cutoff for the present problem. It is easy to see that if we apply an electric field for a short time the Fermi momentum will shift and equal number of right-moving particles and left-moving holes (with reference to the original fermi sea) will be produced. While the total number of fermions remain conserved, the sum of the right and left movers is not conserved. So what is natural in condensed matter theory appears as an anomaly in the field theory description!

The manifestation of the axial anomaly in the $1+1$ dimension is an induced effective action for the electromagnetic field. That is, if we integrate over the massless fermion field, to leading order, we generate an induced action

$$S_{\text{ind}} = \frac{e}{2\pi}\int \mathrm{d}x\mathrm{d}t\epsilon^{\mu\nu}F_{\mu\nu} \tag{3.23}$$

The induced action also tells us that photon acquires a mass. It is a mass generation mechanism for gauge particles without spontaneous symmetry breaking. This is called the *Schwinger mechanism of dynamic mass generation*. In this case, the lowest excitation is a neutral positronium, an electron and a positron are connected by an electric flux tube. There is confinement; an electron or a positron in isolation will have infinite energy, arising from the energy of the infinitely long electric flux attached to the charge.

Next, we move onto the $3+1$ dimension and follow Nielson and Ninomiya [40]. Imagine applying a uniform magnetic field to our massless Dirac fermions along z-axis. This is a solvable problem. We get Landau quantization along the xy-plane, with Landu level quantum number n and intra-Landau level index m. Plane waves continue to be eigenfunctions for z-axis motion. The square of the Hamiltonian in the Landau basis takes the diagonal form

$$H^2 = c^2p_z^2 + (2n+1)\hbar|eB|c^2 - 2eBS_zc^2 \tag{3.24}$$

As in the 2D problem, the energy eigenvalues are independent of the intra-Landau level index m. While n takes all integer values, m varies between 1 and N_0. Here, N_0 is the total magnetic flux in units of flux quanta applied along the z-axis. For a given Landau quantum number m and n, fermions effectively behave as one-dimensional fermion with mass

$$m^2c^4 = (2n+1)|eB|c^2 - 2eBS_zc^2 \tag{3.25}$$

There is one special set of levels, $n=0$ and $S_z = \frac{1}{2}$, which behave like massless $1+1$-dimensional electrons. Now we apply an electric field E parallel to the z-axis (parallel to the applied field) for a time interval t. Particles and holes are created. As we mentioned earlier, chirality is equal to helicity in massless $3+1$-dimensional case; spin projections of the right and left movers are defined by chirality. At the end, we produce an axial current (difference in number between left and right movers along the z-direction), which is given by

$$\partial_t \rho_A = \frac{eE}{\pi}\frac{eB}{2\pi} = \frac{e^2}{2\pi^2}\vec{E}\cdot\vec{B} \tag{3.26}$$

The factor $\frac{eB}{2\pi}$ counts the number of one-dimensional channel per unit area from the degeneracy of each Landau levels.

This is the simple manifestation of the 3 + 1-dimensional axial anomaly. The induced action derived by Adler, Bell, and Jackiw using Feynman diagrams (triangular diagram) also has the same form

$$S_{\text{ind}} = \frac{e^2}{8\pi^2}\int \mathrm{d}\mathbf{r}\,\mathrm{d}t\ F_{\mu\nu}\tilde{F}_{\mu\nu} = \frac{e^2}{8\pi^2}\int \mathrm{d}\mathbf{r}\,\mathrm{d}t\vec{E}\cdot\vec{B} \tag{3.27}$$

This anomaly is exhibited by massless Dirac fermions (Weyl fermions) interacting with gauge fields in 3 + 1 dimensions. As we saw earlier, Weyl fermions in 3 + 1 dimensions have a definite chirality. The ABJ anomaly depends on the chirality of the Weyl fermions. The above-mentioned induced action explains the decay process of a pion in standard model, $\pi_0 \to 2\gamma$, which is nominally forbidden by the axial current conservation symmetry in the classical lagrangian. This action also has the form of θ term that plays an important role in issues such as confinement in non-Abelian gauge theories.

In the world of elementary particle physics, fundamental leptons are chiral. So if there is a quantum anomaly, it has consequences for various scattering and particle decay processes. It was found that it is difficult to discuss this anomaly using the lattice regularization procedure. According to Nielson and Ninomiya, a straightforward lattice regularization produces Weyl fermions in pairs. Consequently, there could be anomaly cancellations. This is called the *fermion doubling problem* and *Nielson–Ninomiya theorem* [41]. Nature seems to choose only one of the Weyl fermion species with one type of helicity.

However, Nielson and Ninomiya wanted to see [40] if there are situations in metal physics in 3 + 1 dimensions where Weyl fermion could emerge and the ABJ anomaly will leave its signature in spite of species doubling. In the process, they also provided a physical picture of the axial anomaly as a conversion of electrons of one helicity to another or flow of electrons from the Dirac node of one helicity to another Dirac node with opposite helicity in momentum space in the presence of electric and magnetic fields. They also predicted that it will lead to anomalous magnetoconductance. It is interesting that this question has surfaced again recently [42, 43] and certain three-dimensional metals with a strong spin–orbit coupling have been suggested to have Weyl fermion excitations, and fascinating predictions have been made.

3.5 Graphene and 2 + 1-Dimensional Parity Anomaly

Deser *et al.* [44], in 1982, discovered an anomaly in $2+1$-dimensional massless Dirac fermions interacting with U(1) gauge fields. Unlike the odd spatial dimensions we discussed earlier, the even spatial dimensional case, $2+1$ dimension, is special. The notion of chirality or γ_5 matrix is absent. Instead of a chiral anomaly, one gets a parity anomaly. The difference becomes clear when we discuss the effect of a magnetic field.

At this juncture, a physical picture of axial anomalies and a possible application to tight-binding condensed matter fermion system also appeared [40] on the scene. In this background, Semenoff looked for an analog of chiral anomaly in $2+1$ condensed matter systems. He identified a rather unique condensed matter system, namely, graphene, and showed the presence of massless Dirac fermion excitations. It is also remarkable that Semenoff makes the following prescient remark in his paper [7]: "... it may be possible to fabricate a graphite monolayer where the effects which we describe would be observable." Following close on his heels, Haldane [22] suggested a modification of the graphene Hamiltonian (an additional perturbation) that will bring out the parity anomaly as a *zero-field quantum Hall effect*. This anomaly paved the way for nontrivial application of the Chern-Simons theory to $2+1$-dimensional condensed matter systems.

In the presence of a vector potential and a small mass term, the Hamiltonian for electrons close to K- and K′-points is given by

$$H = v_F(\sigma_x(p_x + \frac{e}{c}A_x) + \tau_z\sigma_y(p_y + \frac{e}{c}A_y)) + mv_F^2\sigma^z \tag{3.28}$$

We focus on one of the two valleys and analyze it for two situations: (i) the case of uniform and (ii) arbitrary external magnetic field. The first one is the standard Landau level problem of massive electrons in $2+1$ dimensions. This problem can be exactly solved, as we saw before, by taking the square of the Hamiltonian and converting it to a 2D nonrelativistic Landau level problem. For $m = 0$, we have particle–hole symmetry and the Landau spectrum is symmetric and there are N_0 zero modes. Here, $N_0 = \frac{\Phi}{\frac{hc}{e}}$ is the total number of flux passing through the plane measured in units of the fundamental flux quanta. Let us focus on the zero modes. We choose radial gauge, $\mathbf{A} = \frac{1}{4}r^2 B$. The zero-mode eigenfunctions for states close to K-points (ignoring normalization factors) are $\Psi = \mathrm{e}^{-\frac{|z|^2}{4l_B^2}} z^m \begin{pmatrix} 1 \\ 0 \end{pmatrix}$, with $m = 0, 1, \ldots,$ N_0. For states close to K′-points, zero-mode eigenfunctions are $\Psi = \mathrm{e}^{-\frac{|z|^2}{4l_B^2}} z^m \begin{pmatrix} 0 \\ 1 \end{pmatrix}$.

What happens to the zero modes when we introduce a finite mass term? First, when $m \neq 0$, the Dirac Hamiltonian at the K- and K′-points separately lose parity symmetry and particle–hole symmetry. However, the overall particle–hole symmetry of the graphene Hamiltonian remains intact. Zero modes get collectively shifted in the opposite direction depending on the sign of the mass term. As the chemical potential of the Dirac sea is zero, the zero modes will be either completely filled or completely empty, depending on the sign of the mass term m $\epsilon_{\tau_z,n\pm} = \pm\{(mv_F^2)^2 + n\hbar|eB|^2 v_F^2\}^{\frac{1}{2}}$. The energy expression for the zero-mode Landau level is $\epsilon_{\tau_z,0} = \tau_z m(eB)$, where $\tau_z = \pm 1$ denotes the two valleys.

If we focus on one of the valleys, all zero modes will be either full or empty as $m \to 0^{\pm}$. Correspondingly, the vacuum will carry a macroscopic charge $\pm eN_0$. This phenomena, easily understood as filling of levels close to the chemical potential, become a *parity anomaly* in the field theory context.

The above-mentioned phenomena appear as parity-violating induced current. The current induced by the background gauge field can be calculated perturbatively to get

$$j^{\mu}(\mathbf{r}) = \frac{e}{8\pi}\epsilon^{\mu\nu\lambda}F_{\mu\nu}(\mathbf{r})\mathrm{sgn(m)} \tag{3.29}$$

Using this equation, one can calculate the total induced charge for an arbitrary background of external magnetic field as, $Q = \int \mathrm{d}\mathbf{r}j^0 = \mathrm{sgn(m)}\frac{e\Phi}{2}$, where $\Phi = \frac{1}{2\pi}\int \mathrm{d}\mathbf{r}B(\mathbf{r})$ is the magnetic flux.

For graphene, the induced current has opposite sign for the K- and K′-points because of the change in the sign of the mass term. The sum of the induced currents from the two valleys vanish

$$j_T^{\mu}(\mathbf{r}) = \frac{e}{8\pi}\epsilon^{\mu\nu\lambda}F_{\mu\nu}(\mathbf{r}) - \frac{e}{8\pi}\epsilon^{\mu\nu\lambda}F_{\mu\nu}(\mathbf{r}) = 0 \tag{3.30}$$

Thus, in graphene, *parity anomaly is hidden* or there is anomaly cancellation because of the presence of two identical species with opposite signs of the mass term.

Zero modes are at the heart of the parity anomaly. Is there a way to study occupancy of zero modes? Semenoff suggested that Zeeman splitting of the zero mode can be used as a chemical potential and the occupation of zero modes with one or other spin species can be studied using spin resonance experiments. Further, keeping Zeeman splitting the same, Landau level splitting can be controlled by rotating the magnetic field.

Earlier to Semenoff's work, parity anomaly was connected to 2D quantum Hall effect in 2D electron gas. This is reviewed in Ref. [45]. Following Semnoff's seminal work, Haldane made the concrete suggestion of a *zero-field quantum Hall effect* in graphene, using an appropriate perturbation. It involves applying a staggered flux to graphene through the phase-dependent next-nearest hopping term. The phase dependence was chosen such that net flux through each hexagonal plaquette identically vanishes. This perturbation respects the crystal symmetry of the hexagonal lattice, but violates the time-reversal symmetry. It manages to open a gap in the spectrum at K and K′ without violating parity. More importantly, it becomes an integer quantum Hall state.

Recently, Kane and Mele [23] suggested that the intrinsic spin–orbit coupling will make graphene a spin Hall insulator and bring out the parity anomaly happening within individual valleys. The effective Hamiltonian suggested by Kane and Mele is

$$H = -i\hbar v_F(\sigma_x k_x + \tau_z\sigma_y k_y) + \Delta_{so}\sigma_z\tau_z s_z \tag{3.31}$$

where $s_z = \pm 1$ stands for the spin projection of the electron. So far, we have ignored the spin indices. In the presence of a finite, spin–orbit coupling, Δ_{so}, the orbital (sublattice and valley) degree of freedom and spin degree of freedom get coupled. Kane and Mele showed by analyzing the edge states that while charge edge currents will mutually cancel each other, edge spin current will add and lead to the spin Hall effect. This makes neutral graphene a topological insulator albeit with a tiny gap. However, graphene has played a key role in opening a window to the exciting world of topological insulators [46] in condensed matter systems.

An analysis, along the lines indicated above, also leads to what is called *valley Hall effect* [47], which according to theory will show up in bilayer graphene.

If we have an arbitrary gauge field, quantum corrections due to the Dirac fermions arise as Chern-Simons terms with opposite signs for the two valleys

$$S_{cs} = \mathrm{sgn(m)}\frac{e^2}{4\pi}\int \mathbf{dr}\, dt\ \epsilon^{\mu\nu\lambda} A_\mu \partial_\nu A_\lambda \tag{3.32}$$

It is remarkable that anomalies occur in different regions in momentum space. However, in real space, they occur in the same region and add to zero. The above developments indicate that in spite of anomaly cancellation we can devise a suitable situation where Chern-Simons physics is brought out. In situations where real space and k-space get mixed up, parity anomaly could show up (G. Baskaran, unpublished.).

3.6 Zitterbewegung

According to Dirac, a physical electron has a positron (antiparticle) content. One observable consequence of this is Zitterbewegung, discovered by Schrodinger [32]. It is a kind of jittery or shaky (circular) motion, at short timescales, of the center of the mass of an electron wave packet during its free evolution. The antiparticle content, which increases with the speed of the electron, dictates the amount of Zitterbewegung. We saw in Section 3.2 that the antiparticle content has a definite helicity (spin pointing along the direction of electron motion). This means that Zitterbewegung has a small amplitude for a fast spin precision of the Dirac particle as well. Further, Huang [48] has suggested a deep connection between the intrinsic spin of an electron and Zitterbewegung. Historically, de Broglie [49] has conjectured, earlier than Schrodinger, that there should be an internal clock associated with any massive elementary particle, with a frequency and period given by $\omega_0 = \frac{2\pi}{T} = \frac{\hbar}{2mc^2}$.

For an electron at rest in vacuum, the Zitterbewegung frequency $\omega_0 = \frac{2mc^2}{\hbar} = 1.6 \times 10^{21}\mathrm{s}^{-1}$. The fluctuation in the position of the electron is given by the Compton radius $= \frac{c}{\omega_0} = \frac{\hbar}{2mc} = 1.9 \times 10^{-11}$ cm. These are prohibitively short time and distance scales. To reach experimentally manageable time and length scales, one has to reach relativistic speeds. There is a claim that in certain ion channeling experiments, Zitterbewegung or motion associated with de Broglie's internal clock is observed [50].

However, electrons confined to real solids provide a great opportunity to observe this phenomenon, by creating a filled Dirac sea like (filled valence band) situation with (i) mass (related to band gap $E_G = 2m_0\ v_F^2$) and much smaller than real electron mass and (ii) effective *light velocities* v_F, small compared to c, speed of light. Further ability to apply magnetic and electric fields in solids helps us to bring out novel consequences of the Zitterbewegung phenomenon in different situations.

Even before the advent of excitements in graphene, the Zitterbewegung phenomenon was discussed in the context of narrow gap semiconductors [18] and carbon nanotubes [19], as an interband oscillation phenomenon. At the end of this section, we introduce one of the simplest Zitterbewegung phenomena in graphene using ideas of interband oscillations.

It is customary to introduce the basic physics behind Zitterbewegung using the $1 + 1$-dimensional Dirac theory described by the Hamiltonian

$$H_0 = c\sigma_x(-i\hbar\partial_x) + \sigma_z mc^2 \tag{3.33}$$

First, we look at the time evolution of the position operator x using the Heisenberg equation of motion

$$\partial_t x(t) = \mathrm{e}^{\frac{i}{\hbar}H_0 t}\frac{i}{\hbar}[H_0, x]\mathrm{e}^{-\frac{i}{\hbar}H_0 t} = c\sigma_x(t) \tag{3.34}$$

where $\sigma_x(t) \equiv \mathrm{e}^{\frac{i}{\hbar}H_0 t}\sigma_1\mathrm{e}^{-\frac{i}{\hbar}H_0 t}$. This equation is the expression for *standard velocity*. It has a remarkable consequence for the instantaneous velocity of Dirac electron. As the eigenvalue of the matrix of RHS, $\sigma_x(t)$ is unitarily equivalent to σ_x, the instantaneous velocity of an electron is either $+$c or $-$c. This means that the *magnitude of instantaneous velocity* of even a massive Dirac electron is always equal to the velocity of light.

This is in contradiction to the classical expression for velocity of a massive particle in the theory of relativity:

$$p = \frac{mv}{\sqrt{1 - \frac{v^2}{c^2}}} \quad \text{or} \quad v = \frac{c^2 p}{\sqrt{c^2p^2 + m^2c^4}} = \frac{c^2 p}{E} \tag{3.35}$$

How do we reconcile this smooth dependence of velocity given by the above classical expression and the instantaneous velocity $\pm$ c ? To see what is going on, we take one more derivative of Equation 3.32, eliminate $\sigma_1(t)$, and integrate the equation to get the following expression. This expression has a classical component $\frac{c^2 p}{E}t$ and a deviation from it

$$x(t) = x(0) + \frac{c^2 p}{E}t + A_0\frac{i\hbar c}{2E}\mathrm{e}^{\frac{2iEt}{\hbar}} \tag{3.36}$$

where $A_0 \sim 1$ and the amplitude of oscillation $\frac{i\hbar c}{2E}$ is the Compton wave length, for speeds that are small compared to the velocity of light. The last term is extra and oscillatory. It describes the Zitterbewegung motion. It means that even though the magnitude of instantaneous velocity is c, the incessant change in direction of velocity leads to a drift velocity given by the classical expression for velocity. However, a zittery motion survives as a quantum effect in the background of a drift, when we look at the time evolution of the center of the wave packet.

Zitterbewegung becomes somewhat simpler in the case of massless Dirac particles. Let us see the physical meaning in graphene, the case of massless Dirac electrons. Let us recall that in the tight-binding model, to begin with there are two flat bands (equation) corresponding to Bloch states of A and B sublattice Wannier orbitals. They hybridize through A to B hopping (off-diagonal term) and giving rise to holelike and electronlike bands that meet at K- and K′-points.

Let us create a wave packet with net average momentum. In the language of pseudospin, it amounts to making the spinor point along a certain direction. The individual k-component of this wave packet will undergo tunneling as determined by the Hamiltonian matrix $\begin{pmatrix} 0 & -t\gamma(\mathbf{k}) \\ -t\gamma(\mathbf{k})^* & 0 \end{pmatrix}$. Whenever the momentum of the wave packet is close to K or K′, this tunneling phenomenon is simpler but special. Detailed investigation of this phenomenon exists in the literature [15–20]. But to explain the phenomenon in very simple terms, as an interband oscillation, we prepare a special initial wave packet where the pseudospin points along the z-direction. This amounts to creating a wave packet that has a finite support only in sublattice A. The 2×2 matrix determines the wave packet evolution in the form of interband oscillation.

We choose a mean value of the wave packet momentum $\hbar k$, measured with reference to and close to K or K′. In this case, it is easily seen from the structure of the massless Dirac Hamiltonian for graphene, that there will be a Rabi oscillation with frequency $\omega_k = v_F k$. Physically, the pseudospin of the wave packet tunnels back and forth between the up and down positions. Or the wave packet tunnels coherently between the two sublattices. In the language of Dirac theory, the wave packet dynamics we have been considering amounts to coherent oscillation between the particle and antiparticle states with a frequency $\omega_k = v_F k$. During each cycle of spatial oscillation, particle to antiparticle transformation takes place once.

What is interesting is that the Rabi oscillation frequency can be continuously controlled by controlling the momentum of the wave packet $\hbar k$. The experimental challenge is to inject electrons into a wave packet state with a bias on one of the sublattices and being able to measure current in the A and B sublattices.

Zitterbewegung takes an interesting form in the presence of the magnetic field in graphene [17]. We will not discuss this further.

In summary, Zitterbewegung is an interesting interband oscillation phenomenon in solids. Our ability to change parameters of Dirac-like electrons and control electron dynamics through external electric and magnetic fields makes graphene and narrowband semiconductors and quantum dots [51] a great playground to study and observe a rich variety of Zitterbewegung phenomena. From the experimental point of view, this field remains largely unexplored. As far as I know, Zitterbewegung has been seen only in one experiment, in an ion trap simulation of $1 + 1$-dimensional Dirac equation [52].

From the theory point of view, an interesting connection of the Zitterbewegung phenomenon to the ABJ anomaly that we saw in Section 3.3 has been also made [53]. There are also interesting phenomena such as superluminal propagation of parts of wave packets, and a consequence of Zitterbewegung has been discussed [54].

3.7
Klein Paradox

We are familiar with tunneling over classical barriers in nonrelativistic quantum mechanics. The wave function leaks into or penetrate through classically forbidden regions. It is a single-particle phenomenon. In Dirac theory, massive Dirac particles behave like single particles for practical purposes, without disturbing the quiscent Dirac sea. However, when there are strong perturbations, the Dirac sea gets disturbed and interesting phenomena take place. This is at the heart of Klein paradoxes and Klein tunneling. The mass gap provides a scale above which Klein paradoxes begin to occur.

A well-known example of Klein paradox is the problem of the relativistic hydrogen atom in the presence of a nucleus of charge Ze. When Z exceeds a critical value, $Z_c = 1/\alpha = 137$ (α is the fine structure constant), the Hamiltonian becomes non-self-adjoint. At Z_c, the binding energy of hydrogen 1s state just equals 2 mc^2, the mass gap (band gap). For $Z < Z_c$, the Hamiltonian acquires complex energy eigen values. It indicates an instability and a need to rearrange the Dirac vacuum locally in the vicinity of the nuclear charge. It results in local particle–hole pair production, and the problem should be analyzed by going beyond the single-particle description.

Graphene is a massless Dirac fermion system. So Klein paradoxes should be ubiquitous and occur at even for small perturbations and at low-energy scales.

From the point of view of metal or semiconductor physics, various Klein paradoxes that are being discussed are no more paradoxes; they can be explained in a natural manner using ideas of Fermi sea.

Consider a graphene sheet connected to a source and drain and bring a gate, biased at a particular negative voltage, close to an area B of the graphene sheet. From the single-electron point of view, there is a potential barrier from the gate in region B for electrons from outside this region (called region A). However, from the metallic graphene point of view, the negative gate voltage will deplete electrons from region B. This is equivalent to changing the local chemical potential and doping region B with holes.

An electron injected from region A will not see a simple barrier. It will see a region full of holes, where the group velocity of an electron changes its sign. Thus, an injected electron will resonantly tunnel to available degenerate states at the local Fermi level and behave in somewhat unexpected ways. This is because electron tunnels from one state to another state, where the initial- and final-state wave functions are not matched in sign and magnitude of wave vectors. One will get transmission and reflection coefficients that depend on the details of the shape of the boundary and angle of incidence and so on. For some geometry, there will be an unexpected perfect transmission through the barrier.

After a detailed theoretical study and prediction [12], Klein tunneling phenomena have been confirmed experimentally [13]. It is interesting to point out that to observe Klein tunneling phenomena with electron in physical vacuum, a potential of 1 MeV over a width of 0.1 Au (an electric field of 10^{16} V s^{-1}) will be needed.

A simple example of Kelin tunneling is a barrier of height V, in the form of an infinite wall extended along the y-direction, of width D. By matching the boundary condition of the Dirac waves inside and outside the barrier, we get an expression for the transmission coefficient for barrier height V $<<$ E, the particle energy

$$T = \frac{\cos^2 \phi}{1 - \cos(Dq_x) \sin^2 \phi} \tag{3.37}$$

where ϕ is the angle of incidence and q_x is the x-component of the incident wave vector. Notice the angular dependence of the transmission coefficient and also the perfect transmission for normal incidence ($\phi = 0$).

3.8
Relativistic-Type Effects and Vacuum Collapse in Graphene in Crossed Electric and Magnetic Fields

When we learn the special theory of relativity, we are fascinated by phenomena such as Lorentz contraction, time dilation, and speed dependence of the mass of a particle. These phenomena are beyond our day-to-day experience. These classical relativistic phenomena get carried to the quantum domain as well. Further, relativistic vacuum collapse or instabilities that occur under extreme conditions in quantum field theories are hard to realize in the world of elementary particles. However, graphene, in view of its relativistic-type massless Dirac spectrum, could mirror some of these phenomena in an interesting and inexpensive fashion.

Along with Lukose and Shankar [26], we predicted that graphene offers a platform to realize certain relativistic type of phenomena. We brought this out by investigating the effect of a uniform electric field, applied along the graphene sheet, on its already anomalous Landau level spectrum. We solved, within the massless Dirac fermion approximation, the problem of electrons in the presence of crossed magnetic and electric fields exactly. *We find certain novel effects of electric field on the Landau levels in graphene, which differ from the standard 2D electron gas situation.*

What we found can be termed as *analog* of (i) Lorentz boost, (ii) Lorentz contraction, (iii) time dilation, and (iv) relativistic vacuum collapse. Surprisingly, Landau level spectrum (gaps) get scaled down by a factor $(1 - \beta^2)^{\frac{3}{4}}$, where $\beta \equiv \frac{E}{v_F B}$ is an electric-field-dependent dimensionless parameter. As the value of this parameter is increased, spacings between the Landau levels decrease. This Landau level contraction is a consequence of the electric-field-induced quantum mechanical mixing of Landau levels. The entire Landau level structure collapses at a critical value of this parameter $\beta_c = 1$. The "relativistic" character of the spectrum (with Fermi velocity replacing the velocity of light) leads to a novel interpretation of our result in terms of relativistic boosts and the mixing of electric and magnetic fields in moving frames of reference.

Let us briefly recall the Landau level formation for graphene in the presence of a uniform magnetic field. The Dirac Hamiltonian at K-point becomes

$$H = v_F(\mathbf{p} - \frac{e}{c}\mathbf{A}) \cdot \vec{\sigma} \tag{3.38}$$

Here, $\mathbf{A}(\mathbf{r}) = xB\hat{y}$ is the vector potential in Landau gauge that generates a uniform magnetic field in the z-direction. It is easy to see that the square of the above Hamiltonian reduces to a problem of nonrelativistic electrons. We obtain the well-known Landau level spectrum simply by taking the square root of the nonrelativistic eigenvalues, resulting in the spectrum

$$\epsilon_{n,k_y} = \mathrm{sgn}(n)\sqrt{2|n|}\frac{\hbar v_F}{l_c} \tag{3.39}$$

Here, $l_c = \sqrt{\frac{\hbar c}{eB}}$ is the magnetic length. n is the Landau level index, $k_y = \frac{2\pi}{L_y} l$ is the quantum number corresponding to the translational symmetry along the y-axis, and both n and l are integers. Unlike the case of the nonrelativistic electron in a magnetic field, where the spectrum is equispaced and linearly dependent on the magnetic field, the graphene Landau levels are not equispaced; they have a square root dependence on both the magnetic field and the Landau level index. The degeneracy of each level is given by the number of magnetic flux quanta passing through the sample. The eigenfunctions are

$$\psi_{nk_y}(x, y) \propto e^{ik_y y} \begin{pmatrix} \mathrm{sgn}(n)\,\phi_{|n|-1}(\xi) \\ i\,\phi_{|n|}(\xi) \end{pmatrix} \tag{3.40}$$

where $\phi_n(\xi)$ are the harmonic oscillator eigen functions and $\xi \equiv \frac{1}{l_c}\left(x + l_c^2 k_y\right)$.

We now consider the above system in the presence of an additional constant electric field applied, in an open-circuit geometry, along the x-direction. The single-particle Hamiltonian is then given by

$$H = v_F(\mathbf{p} - \frac{e}{c}\mathbf{A}) \cdot \vec{\sigma} + \mathbf{I}eEx \tag{3.41}$$

The Lorentz covariant structure of the Hamiltonian, with v_F playing the role of the speed of light, can be used to solve the problem exactly [55]. It is known from special relativity that if $cB < |\mathbf{E}|$ (in our case, $v_F B < |\mathbf{E}|$), we can always boost to a frame of reference where the electric field vanishes and the magnetic field is reduced. We can then use the finite magnetic field and zero electric field solution in the Dirac equation and boost it to get the exact spectrum in the presence of crossed electric and magnetic fields. Here, the boost transformation amounts to doing a transformation on the space–time coordinate system. To implement this procedure, it is convenient to work with the manifestly covariant time-dependent Dirac equation

$$i\hbar c\gamma^\mu(\partial_\mu + i\frac{e}{\hbar c}A_\mu)\Psi(x^\mu) = 0 \tag{3.42}$$

where $x^0 = v_F t$, $x^1 = x$, $x^2 = y$, $\gamma^0 = \sigma^z$, $\gamma^1 = i\sigma^y$, $\gamma^2 = -i\sigma^x$, and $\partial_\mu = \frac{\partial}{\partial x^\mu}$. $A^0 = \phi$, the scalar potential, $A^1 = A_x$, $A^2 = A_y$, and $\Psi(x^\mu)$ is a two-component spinor. We now apply a Lorentz boost in the y-direction (perpendicular to the direction of the applied electric field)

$$\begin{pmatrix} \tilde{x}^0 \\ \tilde{x}^2 \end{pmatrix} = \begin{pmatrix} \cosh\theta & \sinh\theta \\ \sinh\theta & \cosh\theta \end{pmatrix} \begin{pmatrix} x^0 \\ x^2 \end{pmatrix} \tag{3.43}$$

and $\tilde{x}^1 = x^1$. The wave function transforms as $\tilde{\Psi}(\tilde{x}^\mu) = e^{\frac{\theta}{2}\sigma_y}\Psi(x^\mu)$. Applying the above transformations and choosing $\tanh\theta = \frac{E}{v_F B} = \beta$, we can rewrite the Dirac equation in Equation 3.42

$$\left(\gamma^0\tilde{\partial}_0 + \gamma^1\tilde{\partial}_1 + \gamma^2(\tilde{\partial}_2 + \frac{i}{l_c^2}\sqrt{1-\beta^2}\,\tilde{x}^1)\right)\tilde{\Psi}(\tilde{x}^\mu) = 0 \tag{3.44}$$

In the boosted coordinates, where $|\beta| < 1$, it is a problem of a Dirac electron in a (reduced) magnetic field, $\tilde{B} = B\sqrt{1-\beta^2}$. The time component of the 3-momentum in the boosted frame is $\tilde{\epsilon}_{n,\tilde{k}_y} = \text{sgn}(n)\sqrt{2|n|}\frac{\hbar v_F}{l_c}(1-\beta^2)^{\frac{1}{4}}$. It is important to note that this is not the physical eigenvalue spectrum of our problem. We have to apply the inverse boost transformation to obtain the physical spectrum and eigenfunctions of our problem. The result is

$$\varepsilon_n, k_y = \text{sgn}(n)\sqrt{2|n|}\frac{\hbar v_F}{l_e}(1-\beta^2)^{\frac{3}{4}} - \hbar v_F \beta k_y \tag{3.45}$$

$$\Psi_{n,k_y}(x,y) \;\propto e^{ik_y y} e^{-\frac{\theta}{2}\sigma_y} \begin{pmatrix} \text{sgn}(n)\,\phi_{|n|-1}(\xi') \\ i\,\phi_{|n|}(\xi') \end{pmatrix} \tag{3.46}$$

$$\xi' \equiv \frac{(1-\beta^2)}{l_c}\left(x + l_c^2 k_y + \text{sgn}(n)\frac{\sqrt{2|n|}l_c\beta}{(1-\beta^2)^{\frac{1}{4}}}\right) \tag{3.47}$$

The energy eigenvalues of the standard $2D$ electron gas in crossed magnetic and electric fields are given by $\epsilon_{n,k_y} = (n+\frac{1}{2})\hbar\omega_c - \hbar k_y \frac{E}{B} - \frac{m}{2}(\frac{E}{B})^2$. The main difference between the two, besides the $\sqrt{n}$ and $\sqrt{B}$ dependence, is that *the low-lying graphene Landau level spacing scales as* $(1-\beta^2)^{\frac{3}{4}}$, whereas the spacing is independent of the electric field in the nonrelativistic case. Comparing the eigenfunctions with and without the electric field, we see that the effect of the electric field is to (un)squeeze the oscillator states as well as to mix the particle and hole wave functions.

Squeezing of the wave function results from the change in l_c in an anisotropic manner. In the standard 2D electron gas, the electric field does not cause squeezing of the wave function. We can view the squeezing of the wave function as *Lorentz contraction*. Similarly, squeezing of the Landau level separation can be viewed as *time dilation*. For example, the energy level difference between n and $(n+1)$ Landau levels may be viewed as a level-dependent cyclotron frequency and a corresponding period for cyclotron motion $T_n = \frac{\hbar}{\epsilon_{n+1}-\epsilon_n}$. In the presence of an additional electric field, we have the increase in the time period to $\frac{T_n^0}{(1-\beta^2)^{\frac{3}{4}}}$. We view the increase in cyclotron motion period as *time dilation*.

As β approaches unity, we infer that to keep the Gaussian shifts within the linear extent of the system requires larger values of k_y, which takes us beyond the long wavelength approximation. Moreover, we have a collapse of the Landau level spectrum at $\beta = 1$. One may wonder if the collapse we have found is an artifact of the low-energy approximation. Interestingly, we find that in our full tight-binding calculation the collapse persists, and, in fact, it occurs at a value of β even smaller than unity.

We have performed extensive numerical computations on the tight-binding model for graphene with magnetic and electric fields, using lattice sizes ranging from 60×60 to 600×600. The magnetic field enters through the Peierls substitution, $t \to t\,e^{i\frac{2\pi e}{h}\int \mathrm{a.dl}}$. $\mathbf{A}(\mathbf{r})$ is chosen in such a way that the contribution to the phase term comes from hopping along one of the three bonds for each carbon atom. This enables us to maintain translational symmetry along the $\hat{\mathbf{e}}_2$-axis of the triangular lattice. The problem then reduces to the 1D Harper equation

$$\begin{aligned}
\epsilon\phi_{1,n_1} &= 2t\cos\left(\frac{k_2 a + n_1\varphi}{2}\right)\phi_{2,n_1} + t\phi_{2,n_1+1} \\
\epsilon\phi_{2,n_1} &= 2t\cos\left(\frac{k_2 a + n_1\varphi}{2}\right)\phi_{1,n_1} + t\phi_{1,n_1-1}
\end{aligned} \tag{3.48}$$

Here, φ is the magnetic flux passing through each plaquette, k_2 is the wave vector, and n_1 is the $\hat{\mathbf{e}}_1$ component of triangular lattice coordinate.

We choose the value of the magnetic field such that $L \gg l_c \gg a$, where L is the linear extent of the system. The condition $l_c \gg a$ ensures that we stay away from the Hofstadter butterfly kind of commensurability effects on the spectrum and $L \gg l_c$ ensures that a large number of cyclotron orbits fit in the sample. For our numerics, we expressed all energies in units of t and all lengths in units of a.

Figure 3.2 shows the results of our numerical investigation for zero and Figure 3.3 for a finite ($\beta = 0.1$) electric field. Figure 3.2 shows the spectrum at low energies, and the eigenvalues that are constant with respect to k_y are the Landau levels. They have $\sqrt{n}$ behavior and are in excellent agreement with the analytical result, and the eigenstates that vary with k_y are the chiral edge states responsible for the quantum Hall current. In our numerics, the lattice has zigzag edges at the two ends along $\hat{\mathbf{E}}_1$.

As $\beta \to 1$, the tight-binding results show a faster collapse. Figure 3.4 shows that the collapse has already occurred at $\beta = 0.9$, near one of the Dirac points.

Is it possible to observe these relativistic-type effects in the laboratory?. We have shown in our paper that the relativistic-type effects show up in "dielectric breakdown" of the quantum Hall state of graphene. In the light of new (cyclotron resonance) spectroscopic experiments [56], we claim that the contraction in Landau level spacing and the collapse can be observed at fields attainable in laboratories. The gap between $n = 0$ and $n = 1$ for $B \sim 1\,\mathrm{T}$ is ~ 35 MeV, for $E \sim 3 \times 10^5\,\mathrm{Vm}^{-1}$, a 10% reduction in the gap is expected. And the collapse of the Landau levels should also be observable by applying $E \sim 10^6\,\mathrm{Vm}^{-1}$.

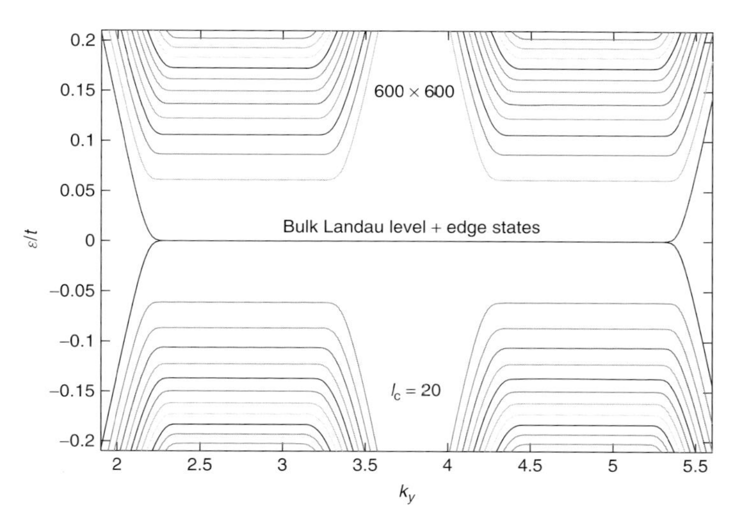

Figure 3.2 Energy eigenvalues ϵ_{n,k_y} for electrons in graphene computed from the tight-binding model for a hexagonal lattice subjected to a magnetic field $B = 27.3$ T (or $l_c = 20\,a$ where a is the triangular lattice spacing) for a system size of $600 \times 600\ a$. The plot shows ϵ_{n,k_y} in the units of t as a function of k_y, where k_y is the wave vector in the y-direction. Two sets of horizontal lines are Landau levels corresponding to the two valleys and $n = 0$ Landau level and the edge states are degenerate.

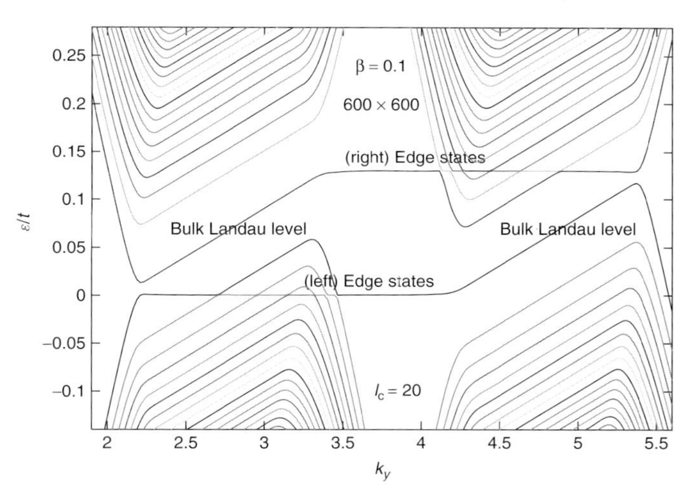

Figure 3.3 Energy eigenvalues ϵ_{n,k_y} for electrons computed for the tight-binding model for parameters given in Figure 3.2 and an external electric field E applied along the x-axis, given by the parameter $\beta = \frac{E}{v_F B} = 0.1$. The electric field gives a linear k_y dependence to the bulk Landau levels, whereas it gives a constant shift to the edge states. The part of the solid line labeled "bulk Landau level" are $n = 0$ Landau levels and the parallel lines above and below them are Landau levels corresponding to positive and negative n, respectively. The set of points parallel to k_y labeled "edge states" are surface states localized at the zigzag boundary.

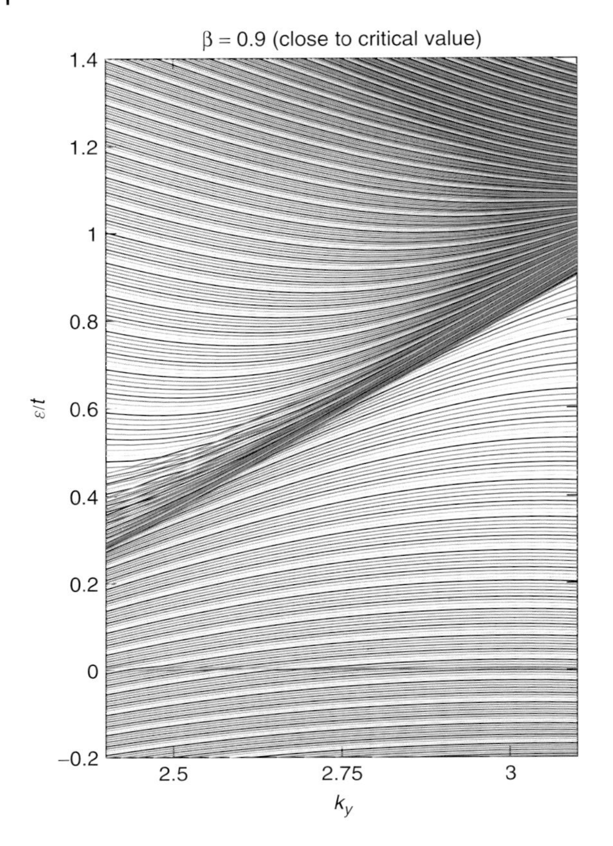

Figure 3.4 The energy eigenvalues around the Dirac point plotted as a function of k_y for $\beta = 0.9$. The collapse can be clearly seen.

3.9
Prediction of Spin-1 Quanta from Resonating Valence Bond Correlations

So far, we have seen spin-half fermionic excitations in graphene. Bosonic excitations in graphene are the standard phonons, flexural modes, and acoustic plasmons. These are nonmagnetic spin-0 excitations. Are there low-energy electronic spin-1 excitations in the diamagnetic graphene? Such low-energy excitations could arise from certain hidden *magnetic* correlations in graphene through electron–electron interaction effects. It was predicted by Jafari and Baskaran [27] that the resonating valence bond or spin singlet correlations that Pauling suggested for graphene does have a nontrivial consequence – it generates a new spin-1 collective mode branch. There has been an interesting exchange about the validity of our theoretical approach in Ref. [57].

Graphene is a broadband semimetal. The band width is in the order of 16 eV. Electron correlation in graphene is believed to be weak or moderate. In the repulsive Hubbard model description of graphene, on-site repulsion or Hubbard U $\sim$ 5 to 8 eV and band width $\sim$15 eV. Recent estimates of U put it around 8 eV or a little higher. Electron correlation effects are known to get amplified in lower dimensional systems. Are there nontrivial electron correlation effects in two-dimensional graphene? In view of the linearly vanishing density of states at the Fermi level, the available interaction strength U is not sufficient to cause Fermi surface instabilities or create a Mott insulating state.

Various theoretical approaches and experimental results suggest that a Fermi-liquid-like description works well for graphene. So far, there are no dramatic experimental anomalies that demand spin–charge decoupling or quantum number fractionization. It should be pointed out, however, that in ARPES there are certain anomalous self-energy features that seem to imply certain (less dramatic?) non-Fermi liquid behavior.

If we view graphene as a limit of a sequence of $p\pi$-bonded systems such as benzene, naphthalene, anthracene, carotene, and C_{60}, there is an interesting phenomenon that arises from electron interaction effects. For benzene, the first excited state is a triplet lying at an energy $\sim$2.2 eV above the singlet ground state. The first excited singlet state lies at $\sim$5 eV above the ground state. Thus, there is a $\sim$3 eV energy difference between the first excited triplet and excited singlet states. This large splitting gets naturally explained by electron correlation and exchange effects and the RVB picture of the $p\pi$-bonded system. It is also known that as we go to larger $p\pi$- bonded molecules such as anthracene, pentacene, and fullerene, the singlet triplet splitting survives. A natural question is, what happens to the spin-1 exciton as we add more and more carbon atoms and reach graphene limit? Does it survive as a separate spin-1 branch excitation in the entire or part of the Brillouin zone (BZ); or does it partly or completely disappear into the particle–hole continuum of the semimetal?

Along with Jafari [27], we examined this question. We showed that graphene possesses a new, unsuspected gapless branch of a spin-1 and a charge neutral collective mode. This branch lies below the electron–hole continuum (Figure 3.6); its energy vanishes linearly with momenta as $\hbar\omega_s \approx \hbar v_F q(1 - \alpha q^2)$ about three-symmetry points (Γ,K K$'$) in the BZ (Figure 3.6).

As graphene interpolates a metal and insulator, our collective mode can be viewed both from the metallic and insulating stand point. In paramagnetic metals, "zero sound" is a Fermi surface collective mode [58]. The "charge" as well as the "spin" degree of a Fermi sea can undergo independent oscillations because of the long-range Coulomb interaction. The electron–electron interactions in normal metals do not usually manage to develop a low-energy spin collective mode branch because of the nature of the particle–hole spectrum. However, *the particle–hole spectrum of 2 d graphene with a "window"* (Figure 3.6) *provides a unique opportunity for a spin-1 collective mode branch to emerge in the entire BZ*. From this point of view, our spin-1 collective mode is a *spin-1 zero sound* (SZS) of a 2+1 dimensional *massless Dirac sea*, rather than a *Fermi sea*.

From an insulator point of view, our collective mode defines a spin-triplet exciton branch. Triplet excitons are well known in insulators and semiconductors. Usually, the difference in energy between a triplet and singlet exciton in a broadband insulator such as diamond is of a few tens of million electronvolts.

Our spin-1 collective mode may also be thought of as a manifestation of Pauling's [59] RVB state of graphene: the spin-1 quantum is a delocalized triplet bond in a sea of resonating singlets. The gaplessness makes it a long-range RVB rather than Pauling's short-range RVB. Later we present an argument to suggest that at low energies the neutral spin-1 excitation might undergo quantum number fractionization into two spin-$\frac{1}{2}$ spinons.

In our study, we start with a two-dimensional Hubbard model for graphene and calculate particle–hole susceptibility in the spin triplet channel using random phase approximation (RPA). RPA is known to work well for metals and semiconductors. In this approximation, a given quasiparticle quasihole pair carrying a fixed total (center of mass) momentum undergoes scattering into various relative momentum states through Hubbard U. *The Coulomb repulsion between an up spin and down spin electron on a given site appears as an attraction between an up spin electron and a down spin hole.* It is this attraction that causes a bound state formation in the spin-triplet channel. The end result is that a collective mode splits off the bottom of the particle–hole continuum, as shown in Figure 3.6.

The particle–hole continuum of excitations for standard 2D free Fermi gas is given in Figure 3.5. The continuum has gapless excitations for a range of momentum up to $\pm 2k_F$. In contrast, the particle–hole continuum for graphene is considerably modified, and as shown in Figure 3.6, it exhibits *a window*. The *"window" is a characteristic of a 1D free Fermi gas system.*

The density of states of particle–hole continuum is obtained by looking at the imaginary part of the noninteracting particle–hole susceptibility Im $\chi^0(\omega, \mathbf{q})$. For a finite range of q and ω, Im $\chi^0(\omega, \mathbf{q})$ can be evaluated analytically

$$\mathrm{Im}\chi^0(\mathbf{q}, \omega) = \frac{1}{16 v_F^2} \frac{2\omega^2 - (q v_F)^2}{\sqrt{\omega^2 - (q v_F)^2}} \sim \frac{1}{16\sqrt{2 v_F}} \frac{q^{3/2}}{\sqrt{\omega - q v_F}}$$

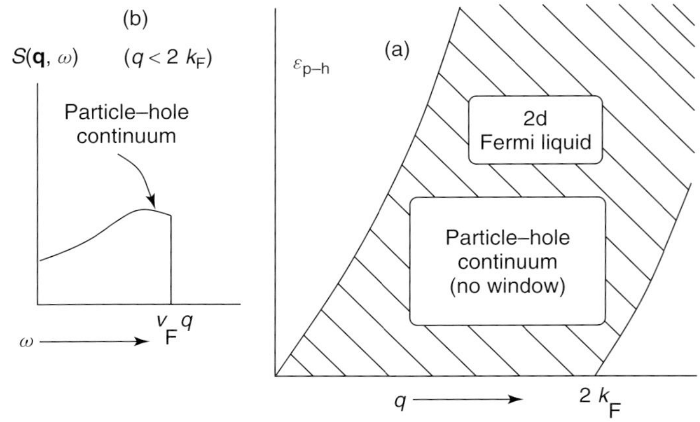

Figure 3.5 (a) Particle–hole continuum without a "window" for a 2D Fermi gas. (b) $S(\mathbf{q}, \omega)$ for $q < 2k_F$.

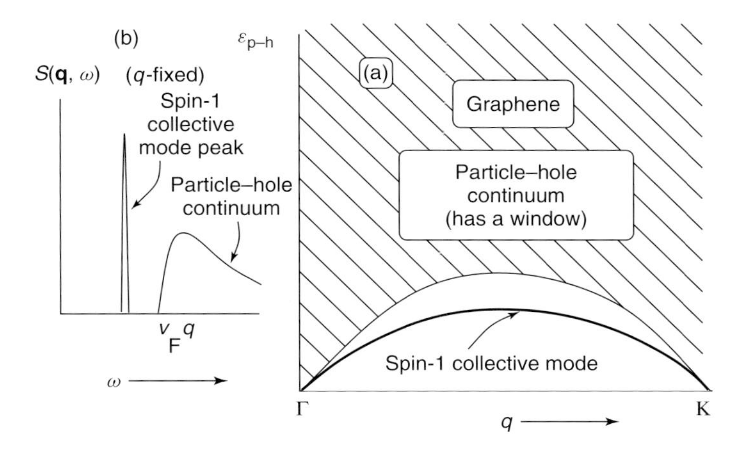

Figure 3.6 (a) Particle–hole continuum with a "window" for graphite. (b) $S(\mathbf{q}, \omega)$

with a square root divergence at the edge of the particle–hole continuum in $(\omega, \mathbf{q})$ space. This expression has the same form as density of states of a particle in 1D (with energy measured from $v_F q$). Note that in fact Im $chi^0(q, \omega) = \pi \rho_q(\omega)$, where $\rho_q(\omega)$ is a free particle–hole pair *DOS* for a fixed center of mass momentum q. *That is, the particle–hole pair has a phase space for scattering, which is effectively one dimensional.* Thus, we have a particle–hole bound state in the spin-triplet channel for arbitrarily small U. However, we also have a prefactor $q^{3/2}$ that scales the density of states. This, together with the square root divergence of the density of states at the bottom of the particle–hole continuum, gives us a bound state for every q as $q \to 0$, with the binding energy vanishing as αq^3, as shown in the following.

The collective mode in *magnetic* channel is the solution of

$$1 - U\chi^0(\mathbf{q}, \omega) = 0$$

or equivalently, $\mathrm{Im}\chi^0(\mathbf{q}, \omega) = 0$ and $\mathrm{Re}\chi^0(\mathbf{q}, \omega) = \frac{1}{U}$. The asymptotic expression for $Re\chi^0(\mathbf{q}, \omega)$ is found to be

$$\mathrm{Re}\chi^0(\mathbf{q}, \omega) \approx \frac{1}{4\pi^2 v_F}\left(k_c + \frac{\sqrt{2}}{\sqrt{1-z}} \arctan\left(\frac{\sqrt{2}}{\sqrt{1-z}}\right)\right)$$

where $z \equiv \frac{\omega}{q v_F}$. Using the above expression, we obtain the following dispersion relation for the collective mode

$$\omega = q v_F - \frac{q^3}{32\pi^2 v_F (\frac{1}{U} - \frac{k_c}{4\pi v_F})^2} \equiv q v_F - E_B(q)$$

as $\omega \to q \to 0$. Here, $E_B(q)$ is the *binding* energy of the particle–hole pair of momentum q around the Γ point. The binding energy around the K-points is roughly half of this.

We mentioned earlier that our collective mode is a "magnetic zero sound." While magnetic zero sound is difficult to get in normal metals, graphite manages to get it in the entire BZ because of the window in the particle–hole spectrum (Figure 3.6).

Within our RPA analysis, the collective mode frequency becomes negative at the Γ point for $U > U_c \sim 2t$. Because there are two atoms per unit cell, this could be either an antiferromagnetic or ferromagnetic instability. Other studies have indicated an AFM instability for $U > U_c \sim 2t$.

The square root divergence of density of states at the bottom edge of the particle–hole continuum tells us that *the low-energy spin physics is effectively one-dimensional*. To this extent, *in a final theory*, we may expect our spin-1 excitation to be a triplet bound state of "two neutral spin-half spinons" rather than "e^+e^- electron–hole pairs." Further, as the energy of the spin-1 quantum approaches zero, the binding energy also approaches zero and the electron–hole bound state wave function becomes elliptical, with diverging size. We may then view the low-energy spin-1 quanta as a "critically (loosely) bound" two-spinon state, very much like the quantum number fractionization of the des Cloizeaux–Pearson spin-1 excitation in the 1D spin-$\frac{1}{2}$ antiferromagnetic Heisenberg model. Our result also suggests a nonlinear sigma model and novel 2 + 1-dimensional bosonization scheme for graphite.

3.10 Majorana Zero Mode from Two-Channel Kondo Effect in Graphene

It is well recognized theoretically, in some cases even experimentally, that certain quantum many-body systems in two spatial dimensions support anyon excitations. Is there room for anyons in graphene? In this section, we briefly review our recent theoretical proposal (G. Baskaran, unpublished.) of using the two-channel Kondo effect in graphene to create a zero-energy Majorana fermion. A Majorana zero mode in 2 + 1 dimension is a non-Abelian Ising anyon. Our suggestion is that an array of suitably placed two-channel Kondo impurities could be manipulated using STM to perform quantum computation. Further, a tunable two-channel Kondo effect in graphene was theoretically predicted by Sengupta and Baskaran [29].

Anyons are classified as Abelian and non-Abelian. Crudely, Abelian anyons are fermions or bosons that carry Abelian U(1) flux, characterized by the complex number $e^{i\theta}$. Non-Abelian anyons carry non-Abelian fluxes characterized by a matrix – this leads to braiding properties that are noncommutative.

In recent times, the Majorana fermion has become an actively discussed topic in condensed matter physics. The Majorana fermion has a nominal *fractional Hilbert space dimension* of $\sqrt{2}$. Further, it has been suggested by Kitaev [60] that Majorana fermions could be used as *topological qubits* for building quantum computers. The Majorana fermion, in view of its topological character, enjoys a *topological protection*. It makes it less susceptible to standard decoherence processes.

Soon after Dirac's theory of electron and positron, Majorana created a relativistic field theory where fermions are their own antiparticles. It is a theory of *real*

Majorana fermion as opposed to the *complex* Dirac fermion. So far, there is no realization of Majorana fermions in the world of elementary particles.

Some exactly solvable models have Majorana zero modes: (i) boundaries of Ising chain in a traverse field, (ii) Kitaev's p-wave superconducting chains, and (iii) Emery and Kivelson solution[61] of the anisotropic two-channel Kondo problem. There is also a good hope of finding Majorana fermions in real systems, based on theoretical predictions: (i) half vortices [62, 63] of Sr_2RuO_4, a p + ip superconductor [64–66], (ii) charge $\frac{e}{4}$ quasiparticles of paired quantum Hall states [67, 68], (iii) some hybrid heterostructures or wires of s-wave superconductors and spin-orbit-coupled systems in Zeeman field [69], and (iv) domain wall and edge states in p + ip superconductors[70].

There has also been a suggestion that Majorana zero modes might occur in graphene in the presence of certain topological defects [30, 31]. Some special fractional quantum Hall states, such as paired Hall state [66] or Read Rezayi state [71], if they occur in graphene, could support Majorana fermions and Fibonacci anyons.

To define a Majorana fermion, consider $c_i^\dagger$, creation operator of an electron at site i. The real and imaginary parts of a (complex) Dirac fermion constitute two (real) Majorana fermions ζ_i and η_i. That is, $c_i^\dagger \equiv \frac{1}{2}(\zeta_i + i\eta_i)$ and $c_i \equiv \frac{1}{2}(\zeta_i - i\eta_i)$. Thus $\zeta_i = c_i^\dagger + c_i$ and $\eta_i = i(c_i^\dagger - c_i)$. This definition implies the following anticommutation relation for Majorana fermions

$$\{\zeta_i, \zeta_j\} = \{\eta_i, \eta_j\} = \{\eta_i, \zeta_j\} = 0 \quad \text{and} \quad \zeta_i^2 = \eta_i^2 = 1 \tag{3.49}$$

From this definition, nominally, an electron has a real and an imaginary part. Under normal conditions, real and imaginary parts *occupy the same single-particle state* and are inseparable. Under what conditions is a complex fermion able to isolate its real and imaginary parts and keep them spatially apart?

In the case of the p-wave 1D wire, the boundary traps a localized Bogoliubov quasiparticle with zero energy – it is an equal amplitude superposition of a particle and an antiparticle: $\alpha_b \equiv c_b^\dagger + c_b$. Here, "$b$" stands for a localized bound state wave function ψ_i at an edge. In the case of the 2D p-wave superconductor, Ivanov [62] showed that a half vortex in a p+ip superconductor in 2D, the trapped Majorana zero mode, is a non-Abelian (Ising) anyon.

One of the first appearances of localized Majorana fermion in condensed matter physics was in two-channel Kondo effect [72]. Here, a spin-half moment is Kondo coupled symmetrically to two identical channels (for example, two identical bands with identical Fermi surfaces). Emery and Kivelson [61] showed that a spin-half moment, in its attempt to form a singlet with both channels, creates a *pair resonance*, resulting in trapping of a Majorana mode in the vicinity of the Kondo impurity.

As shown by Sengupta and Baskaran [29], doped graphene offers a natural platform for two-channel Kondo effect. To enable symmetric coupling of the Kondo spin to the two channels (two identical bands around K- and K′-points), one has to place the magnetic impurity at special positions in the graphene lattice. In what follows we show exactly, for the case of SU(2) spin symmetric Kondo coupling, without actually solving the two-channel Kondo problem that a zero-energy Majorana mode is created exactly at the impurity spin site. This *ultralocalization of Majorana mode* was already present in the work of Emery and Kivelson at an exactly solvable limit of anisotropic Kondo spin coupling.

We start with a spin-half moment placed on the graphene sheet at the origin. By symmetry s-wave, $l = 0$ radial channel couples to the impurity spin. We follow the standard procedure and combine the incoming and outgoing $l = 0$ channel into a one-dimensional right-moving (or left-moving)electron system. In the case of graphene, there is a subtlety arising from the chiral character of the Dirac fermions. In this chapter, we do not focus on this aspect. Details are found in our paper. Our analysis is equally applicable to Kondo spins on a metal surface that contains two identical Fermi surfaces, for example, from the surface metallic bands.

There are four one-dimensional channels – two spin and two valley (pseudospin) species. In the absence of magnetic field and spin–orbit coupling, the four channels are identical, characterized by a single Fermi velocity v_F. At low energies, the four channels represent the four right-moving 1D Dirac particles. The effective two-channel Kondo Hamiltonian is given by

$$H = H_0 + H_I \tag{3.50}$$

$$H_0 = \hbar \frac{v_F}{2\pi} \sum_{j=1}^{2} \sum_{\sigma=\uparrow,\downarrow} \int_{-\infty}^{\infty} dx \psi_{j,\sigma}^{\dagger}(x)(i\partial_x)\psi_{j,\sigma}(x)$$

$$H_I = J \sum_{\alpha\beta} \mathbf{S} \cdot \psi_{j,\alpha}^{\dagger}(0)\vec{\sigma}_{\alpha,\beta}\psi_{j,\beta}(0)$$

Following Affleck and Ludwig [72], one uses The Sugawara construction and gets an exact Majorana representation involving eight independent $1 + 1$ right-moving Majorana fermions for the relevant free electron degree of freedom. The single derivative kinetic energy term in H_0 is written as bilinears of Sugawara currents. Because all channels are identical, currents get naturally separated into (i) $J_c(x) \equiv \sum_{j,\sigma} \psi_{j,\sigma}^{\dagger}(x)\psi_{j,\sigma}(x)$, charge current; (ii) $t\vec{J}_s(x) \equiv \sum_{j,\alpha,\beta} \psi_{j,\sigma}^{\dagger}(x)\vec{\sigma}_{\alpha,\beta}\psi_{j,\beta}(x)$, spin current; and (iii) flavor current, $\vec{J}_f(x) \equiv \sum_{j,j',\sigma} \psi_{j,\sigma}^{\dagger}(x)\vec{\sigma}_{j,j'}\psi_{j,\sigma}(x)$. The free part of the Hamiltonian takes the form

$$H_0 = \frac{v_F}{2\pi} \int_{-\infty}^{\infty} dx \left[\frac{1}{8} : J_c(x)J_c(x) : + \right.$$
$$\left. + \frac{1}{4} : \vec{J}_f(x) \cdot \vec{J}_f(x) : + \frac{1}{4} : \vec{J}_s(x) \cdot \vec{J}_s(x) : \right] \tag{3.51}$$

Further, the charge current satisfies the U(1) current commutation relation and the spin and flavor currents satisfy $SU(2)_2$ (SU(2) level-2) Kac-Moody algebra

$$[J_c(x), J_c(x')] = i\delta'(x - x')$$
$$\left[J_s^a(x), J_s^b(x')\right] = i\epsilon^{abc} J_s^c \delta'(x - x') + \frac{i}{2\pi}\delta^{ab}\delta'(x - x')$$
$$\left[J_f^a(x), J_f^b(x')\right] = i\epsilon^{abc} J_f^c \delta'(x - x') + \frac{i}{2\pi}\delta^{ab}\delta'(x - x') \tag{3.52}$$

Incidentally, the above current algebra is a manifestation of the $1 + 1$-dimensional Schwinger anomaly we saw in Section 3.4.

Following Ludwig and Maldacena [74], we introduce exact Majorana fermion representation for the Derac fermion currents. Corresponding to the four Dirac fields, there are eight Majorana fields: three, $(\chi_x(x), \chi_y(x), \chi(x))$, for spin degree of freedom; three, $(\chi_1(x), \chi_2(x), \chi_3(x))$, for flavor (isospin) degree of freedom; and two for charge degree of freedom, $(\chi_4(x), \chi_5(x))$. Currents have the bilinear form in terms of Majorana fermions $\vec{J}_s(x) = i\{\chi_y(x)\chi_z(x), \chi_z(x)\chi_x(x), \chi_x(x)\chi_y(x)\}$, $\vec{J}_f(x) = i\{\chi_2(x)\chi_3(x), \chi_3(x)\chi_1(x), \chi_2(x)\chi_3(x)\}$, and $J_c(x) = 2i\chi_4(x)\chi_5(x)$.

Substituting for currents in terms of Majorana fermions, the full Hamiltonian becomes

$$
\begin{aligned}
H_c + H_f &= \frac{v_F}{4\pi} \sum_{\alpha=1}^{5} \int_{-\infty}^{\infty} dx : \chi_\alpha(x)(i\partial_x)\chi_\alpha(x) : \\
H_s &= \frac{v_F}{4\pi} \sum_{\alpha=x,y,z} \int_{-\infty}^{\infty} dx : \chi_\alpha(x)(i\partial_x)\chi_\alpha(x) : \\
&+ \frac{iJ}{2} \int_{-\infty}^{\infty} dx\delta(x)\mathbf{S} \cdot (\vec{\chi}(x) \times \vec{\chi}(x))
\end{aligned}
\tag{3.53}
$$

H_s is the interaction term. *Impurity spin gets coupled only to the spin current (charge neutral and flavor singlet) built from electrons of the two valleys.* The flavor and charge part of the Hamiltonian H_c and H_f remains unaffected by the Kondo spin.

Now we provide a simple and new identity, which helps bring out the existence of a zero-energy Majorana mode without actually solving the problem

$$
i\mathbf{S} \cdot (\vec{\chi}(x) \times \vec{\chi}(x)) \equiv (\mathbf{S} \cdot \vec{\chi}(x))^2 + \text{constant} \tag{3.54}
$$

This identity follows by expanding the square on the right-hand side and using, $\chi_\alpha^2 = 1, (S^\alpha)^2 = 1$ and $S^\alpha S^\beta \sim i\epsilon^{\alpha\beta\gamma} S^\gamma$.

The above identity already tells us that only a Majorana fermion, rather than a physical electron, forms a Kondo singlet. To this extent, some *fractional degeneracy* should remain with the Kondo spin! In what Follows, we bring this out explicitly and exactly. To achieve this, we use the Majorana fermion $(\eta_0, \eta_x, \eta_y, \eta_z)$ representation for the spin operator $\mathbf{S} \equiv i\eta_0\vec{\eta}$. Here, η_0 is called the *scalar component*. The other three are vector components. Four Majorana fermions enlarge the Hilbert space dimension to $(\sqrt{2})^4 = 4$. To project to two-dimensional physical Hilbert space of Kondo spin, we use the constraint $\hat{D} \equiv \eta_0\eta_x\eta_y\eta_z = 1$. Further, the operator $\hat{D}$ has eigenvalues ± 1 because $\hat{D}^2 = 1$. The projection operator that projects out the unphysical part of the Hilbert space of the Kondo spin is $\hat{P} = \frac{1}{2}(\hat{D} + 1)$.

Using the above Majorana fermion representation for the Kondo spin operator, it follows that

$$
(\mathbf{S} \cdot \vec{\chi})^2 = (\eta_0\vec{\eta} \cdot \vec{\chi})^2 = (\vec{\eta} \cdot \vec{\chi})^2 \tag{3.55}
$$

Remarkably, the scalar component Majorana fermion η_0 disappears from the Kondo interaction term. Thus, the interaction term simplifies even further

$$H_s = \frac{v_F}{4\pi} \sum_{\alpha=x,y,z} \int_{-\infty}^{\infty} \mathrm{d}x : \chi_\alpha(x)(i\partial_x)\chi_\alpha(x) :$$
$$+ \frac{J}{2} \int_{-\infty}^{\infty} \mathrm{d}x \delta(x)(\vec{\eta} \cdot \vec{\chi}(x))^2 \tag{3.56}$$

Theorem: We have a Majorana zero mode for the above Hamiltonian. Proof: Absence of the scalar Majorana fermion η_0 in the interaction Hamiltonian makes it an exact zero mode.

For the above Hamiltonian, restriction to physical Hilbert space also becomes very simple. This is because the constraint operator $\hat{D} = (\eta_0\eta_x\eta_y\eta_z + 1)$ commutes with our Hamiltonian (Equation 3.56). As the eigenvalue of the constraint operator is ± 1, the unphysical Hilbert space becomes Z_2 gauge redundancy. That is, *the spectrum of the problem is identical in both sectors corresponding to eigenvalues of* $\hat{D} = \pm 1$. This is reminiscent of Kitaev's honeycomb lattice spin model where such a gauge redundancy makes the problem a Z_2 local gauge theory.

Thus, a gauge invariant quantity like the energy spectrum remains unaltered by the enlargement of the Hilbert space. *That is, the zero mode that we have found survives.* It is not an artifact of the enlargement of the Hilbert space. However, one has to perform a suitable gauge average (projection) when we calculate the expectation value of physical operators using wave functions $|\Psi\rangle$ from the enlarged Hilbert space

$$\langle \hat{A} \rangle = \langle \Psi | \hat{P}\hat{A}\hat{P} | \Psi \rangle \tag{3.57}$$

From the above result, it follows that the physical electrons of the two channels, with a competing desire to form a spin singlet with the Kondo spin, agree to isolate a vector Majorana fermion and generate a Kondo coupling. In the process, Kondo spin leaves behind a scalar Majorana fermion η_0 untouched. This is our zero Majorana mode. In other words, the two-dimensional Hilbert space of the Kondo spin gets factorized into $\sqrt{2} \times \sqrt{2}$, a vector part and a scalar part. The vector part creates a disturbance to the left-moving vector Majorana fermion through the Kondo coupling. The scalar part remains untouched. As the left-moving Majorana fermion system generates a single Majorana fermion mode and forms a Kondo singlet, it has to create a Majorana fermion at the boundary at infinity. This makes the overal fermion number an integer, and the total Hilbert space dimension of the many-body system remains an integer.

What is special about our Majorana mode is that it is ultralocalized – it is localized on the Kondo spin. Other known Majorana fermions have a finite spatial extent. For example, the Majorana mode in the vortices of the p-wave superconductor is localized within the coherence length, the size of the normal core of the vortex. The coherence length is determined by the superconducting gap. In low-temperature superconductors or proximity-induced superconductivity, the gap is small. Hence, the size of the Majorana mode is typically large ~100 Å unit. While the large size may have certain advantage, our ultralocalized Majorana mode makes it special from decoherence point of view. The Kondo spin, rather than the electron spin cloud around the Kondo spin, feels the decohering effects.

For implementing topological quantum computation, our ideas are as follows. We place an even number N of spin-half Kondo impurity spins, well separated from each other on the graphene sheet or metallic carbon nanotube or a suitable metal surface. We have a near degeneracy of ground-state degeneracy of $(\sqrt{2})^N$ arising from N Majorana fermions. Quantum computation amounts to an excursion in the $(\sqrt{2})^N$ dimensional ground-state Hilbert space manifold, through sequence of unitary operators, which in this case is a sequence of braiding operations. Our suggestion is to use STM tips to move the impurity spins around as part of the braiding of the Majorana fermions induced by the two-channel Kondo impurities put on the graphene.

However, there is a serious issue with respect to physics. Even though the zero-energy Majorana mode is exactly localized on the Kondo spin, it is surrounded by a medium that supports gapless excitations. Thus, any physical motion (adiabatic transport) of the Kondo spin creating low-energy excitations will become a source of decoherence. One way to avoid this is to induce superconductivity on the graphene or metal surface through proximity effect. As long as the superconducting gap is small compared to the Kondo energy scale, the two-channel symmetry survives and the zero-energy Majorana mode also survives.

One can also avoid adiabatic transport of the Kondo spins and use a method suggested by Bonderson *et al.* [75]. In their method, braiding is mimicked without physical displacement of positions of Majorana mode by a sequence of pairwise accessing (measurement) of Majorana fermions through proper unitary operators.

In two-channel Kondo effect, one of the key issues is the stability of zero-energy mode with respect to channel asymmetry, due to various perturbations. These issues need further investigation.

3.11 Lattice Deformation as Gauge Fields

A special property of the electronic band structure of the single-band honeycomb lattice in graphene makes lattice displacements and topological defects very special for single-particle electronic states in the vicinity of K- and K′-points. Perturbations, such as lattice strain or topological defects that respect time-reversal symmetry, start behaving like time-reversal breaking pseudo-magnetic fields! However, the pseudo-magnetic fields have opposite signs for states near K- and K′-points. Thereby, the overall time-reversal symmetry of the graphene Hamiltonian is respected by a full set of single-electron states. Further, nontrivial consequences follow.

This remarkable magnetic-field-like behavior of lattice strain was theoretically suggested by Suzumura and Ando [76] in 2002. Recently, with the advent of new developments in graphene, this issue has been analyzed theoretically in great detail [77]. In an important development, the pseudo-magnetic field has been observed experimentally [78] in strained graphene using STM. The density of states as measured by STM resembles the Landau level spectrum with spacing varying as $\sqrt{n}$, where n is the Landau level index. The experimentally estimated pseudo-magnetic field is a few hundred Tesla and is in reasonable agreement with the theoretically estimated pseudo-magnetic field.

It has been shown [76], by expanding the hopping matrix elements as a function of lattice displacements, that the long wavelength lattice distortion essentially adds a *vector potential* to the momentum term. This also follows from symmetry arguments. We consider one of the valleys. For convenience, we rewrite the pseudo-vector potential as a vector potential term containing electric charge e and the velocity of light c as

$$H = v_F \left(\mathbf{p} - \frac{e\mathbf{A}}{c} \right) \cdot \vec{\sigma} \tag{3.58}$$

Here, $\mathbf{A}$ is an effective U(1) gauge field. In the presence of lattice strain, the hopping matrix element t gets modulated spatially. The long wavelength modulation can be expressed in terms of the strain fields u_{xy}, u_{xx}, and u_{yy}. They are defined by the in-plane displacements u_x and u_y and out-of-plane displacements h through the relation

$$u_{\alpha\beta} = \frac{\partial_\alpha u_\beta + \partial_\beta u_\alpha}{2} + \frac{\partial_\alpha h \partial_\beta h}{2} \tag{3.59}$$

The relation between the gauge field and the lattice strain is given by

$$eA_x = g_2 \hbar (u_{xx} - u_{yy}) \text{ and } eA_y = 2\hbar g_2 u_{xy} \tag{3.60}$$

with $g_2 = \frac{3\kappa\beta t}{4}$, $\kappa = \frac{\sqrt{2}\mu}{2B}$ and $\beta = \frac{\partial ln(t)}{\partial ln(a)}$. For graphene, the shear modulus and bulk modulus are $\mu \approx 10$ eV Au^{-2} and $B \approx 12.5$ eV Au^{-2}, respectively. The Gruneisen parameter is $\beta \approx 0.56$. The pseudo-magnetic field is calculated using the relation $\mathbf{B} = \nabla \times \mathbf{A}$. For experimentally realizable strain, one finds that the pseudo-magnetic field is $\sim$10 T.

What we have seen so far is that states close to K feel the lattice strain as a magnetic field of a specific sign. What is important is that states close to the K′-point feels a gauge field with opposite sign.

Let us discuss briefly how topological defects mimic an effective gauge field. Important topological point defects are five-membered pentagonal and seven-membered heptagonal rings. When they occur as a pair, they are called *Stone–Wales defects*. The presence of an isolated pentagon or heptagon in an otherwise perfect hexagonal lattice destroys the bipartite character of the lattice – the point defect fuses the two sublattices along a line. That is, if we take a closed circuit by moving along the same sublattice, we will come back to the same site only after encircling the defect twice. It means that the spinor eigenfunctions (say at the K- or K′-points) have to be given a suitable phase twist along the line. In a continuum theory, this twist is enabled through a fictitious π flux that penetrates through the pentagon or hexagon. It was shown earlier [79] that similar phenomena occur in the presence of Rivier lines in random tetrahedral network (tight- binding model for amorphous semiconductor). That is, the Rivier line acts as a solenoid that carries a Bohm–Aharonov π-flux for certain band edge states.

3.12 Summary

In this chapter, we have tried to view the richness of quantum phenomena exhibited by graphene through quanta and quanta fields. The simple looking graphene, a 2D hexagonal net of carbon atoms, supports 2 + 1-dimensional massless Dirac fermions, flexural modes, gauge-field-like phonons, spin-1 collective quanta, relativistic-type effects such as Lorentz boost, Lorentz contraction, and time dilation. One of the most exciting possibilities is our prediction of finding zero-mode Majorana fermion through two-channel Kondo effect. From the point of view of topological quantum computation, this is very important and worth pursuing experimentally and theoretically.

This chapter is in the form of a qualitative summary. We have not covered certain developments. There are very good developments such as connection of the graphene problem to some problem of quantum mechanics in curved space–time [80], conformal symmetry of the Weyl equation [81] in graphene and some consequences, and emergent non-Abelian gauge fields in bilayer graphene [82]; we have not discussed them. There are also remarkable phenomena such as the universal value of optical absorption coefficient [14] of graphene, antilocalization [10] arising from absence of coherent back scattering due to the chiral character of low-energy electrons. A whole variety of unusual and perhaps unexpected fractional quantum Hall states that could support novel excitations such as Fibonacci anyons are possible. Our own suggestion of high T^c superconductivity in graphite has evolved into a distinct possibility of very high temperature superconductivity in graphene [83] with d + id symmetry [84] in doped graphene. It has been suggested that they could support Majorana edge modes [85] under suitable conditions.

Acknowledgment

I thank the DAE, India, for the Raja Ramanna Fellowship. This research was supported by the Perimeter Institute for Theoretical Physics. It is a pleasure to thank my colloborators Akbar Jafari, R. Shankar, Vinu Lukose, Krishnendu Sengupta, Sandeep Pathak and Vijay Shenoy.

References

1. (a) Novoselov, K.S., Geim, A.K., Morozov, S.V., Jiang, D., Zhang, Y., Dubonos, S.V., Grigorieva, I.V., and Firsov, A.A. (2004) *Science*, **306**, 666; (b) Geim, A.K. and Novoselov, K.S. (2007) *Nat. Mater.*, **6**, 183.
2. Castro Neto, A.H., Guinea, F., Peres, N.M.R., Novoselov, K.S., and Geim, A.K. (2009) *Rev. Mod. Phys.*, **81**, 109.
3. (a) Vozmediano, M.A.H., Katsnelson, M.I., and Guinea, F. (2010) *Phys. Reports*, **493**; (b) Cortijo, A., Guinea, F., and Vozmediano, M.A.H. arXiv:1112.2054 (Review to appear in J. Phys. A).
4. Katsnelson, M.I. and Novoselov, K.S. (2007) *Solid State Commun.*, **143**, 3.
5. Abergel, D.S.L., Apalkov, V., Berashevich, J., Ziegler, K., and

Chakraborty, T. (2010) *Adv. Phys.*, **59**, 261.
6. (a) Baskaran, G. (2011) *Graphene and Its Fascinating Attributes*, (eds Pati, S.K., Enoki, T., and Rao, C.N.R.) World Scientific, Singapore; (b) (2011) *Mod. Phys. Lett.*, **9**, 605.
7. Semenoff, G.W. (1984) *Phys. Rev. Lett.*, **53**, 2449.
8. Novoselov, K.S., Geim, A.K., Morozov, S.V., Jiang, D., Katsnelson, M.I., Grigorieva, I.V., Dubonos, S.V., and Firsov, A.A. (2005) *Nature*, **438**, 197.
9. Zhang, Y., Tan, Y.-W., Stormer, H.L., and Kim, P. (2005) *Nature*, **438**, 201.
10. Morozov, S.V., Novoselov, K.S., Katsnelson, M.I., Schedin, F., Ponomarenko, L.A., Jiang, D., and Geim, A.K. (2006) *Phys. Rev. Lett.*, **97**, 016801.
11. Wu, X. *et al.* (2007) *Phys. Rev. Lett.*, **98**, 136801.
12. Katsnelson1, M.I., Novoselov, K.S., and Geim, A.K. (2006) *Nat. Phys.*, **2**, 620.
13. Young, A.F. and Kim, P. (2009) *Nat. Phys.*, **5**, 222–226.
14. Nair, R.R., Blake, P., Grigorenko, A.N., Novoselov, K.S., Booth, T.J., Stauber, T., Peres, N.M.R., and Geim, A.K. (2008) *Science*, **320**, 1308.
15. Katsnelson, M.I. (2006) *Eur. Phys. J.*, **B51**, 157–160.
16. Zawadzki, W. and Rusin, T.M. (2011) *J. Phys.: Condens. Matter*, **23**, 143201.
17. (a) Rusin, T.M. and Zawadzki, W. (2008) *Phys. Rev.*, **B 78**, 125419; (b) Schliemann, J. (2008) *New J. Phys.*, **10**, 043024.
18. (a) Ferrari, L. and Russo, G. (1990) *Phys. Rev.*, **B42**, 7454; (b) Zawadzki, W. (2005) *Phys. Rev.*, **B72**, 085217; (c) Schliemann, J., Loss, D., and Westervelt, R.M. (2005) *Phys. Rev. Lett.*, **94**, 206801.
19. Zawadzki, W. (2006) *Phys. Rev.*, **B74**, 205439.
20. (a) Katsnelson, M.I. (2006) *Eur. Phys. J.*, **B51**, 157; (b) Cserti, J. and Dávid, G. (2006) *Phys. Rev.*, **B74**, 172305.
21. Hou, C.-Yu., Chamon, C., and Mudry, C. (2010) *Phys. Rev.*, **B81**, 075427.
22. Haldane, F.D.M. (1988) *Phys. Rev. Lett.*, **61**, 2015.
23. Kane, C. and Mele, E. (2005) *Phys. Rev. Lett.*, **95**, 226801.
24. Pachos, J.K. and Stone, M. (2007) *Int. J. Mod. Phys.*, **B21**, 5113.
25. Novoselov, K.S., Jiang, Z., Zhang, Y., Morozov, S.V., Stormer, H.L., Zeitler, U., Maan, J.C., Boebinger, G.S., Kim, P., and Geim, A.K. (2007) *Science*, **315**, 1379.
26. Lukose, V., Shankar, R., and Baskaran, G. (2007) *Phys. Rev. Lett.*, **98**, 116802.
27. Baskaran, G. and Jafari, S.A. (2002) *Phys. Rev. Lett.*, **89**, 016402.
28. Wilczek, F. (2009) *Nat. Phys.*, **5**, 614.
29. Sengupta, K. and Baskaran, G. (2008) *Phys. Rev.*, **B77**, 045417.
30. Chamon, C., Hou, C.-Y., Mudry, C., Ryu, S., and Santos, L. (2012) *Phys. Scr.*, **T146**, 014013.
31. Ghaemi, P. and Wilczek, F. (2012) *Phys. Scr.*, **T146**, 014019.
32. Thaller, B. (1993) *The Dirac Equation*, Springer.
33. Jackiw, R. (1984) *Phys. Rev.*, **D29**, 2375.
34. Schwinger, J. (1962) *Phys. Rev.*, **125**, 397.
35. Tomonaga, S. (1950) *Prog. Theor. Phys.*, **5**, 544.
36. Mattis, D.C. and Lieb, E.H. (1965) *J. Math. Phys.*, **6**, 304.
37. Luttinger, J.M. (1963) *J. Math. Phys.*, **4**, 1154.
38. Adler, S. (1969) *Phys. Rev.*, **177**, 2426.
39. Bell, J.S. and Jackiw, R. (1969) *Nuovo Cimento*, **60A**, 4.
40. Nielsen, H.B. and Ninomiya, M. (1983) *Phys. Lett.*, **B130**, 389.
41. Nielsen, H.B. and Ninomiya, M. (1981) *Phys. Lett.*, **B105**, 219.
42. Wan, X., Turner, A.M., Vishwanath, A., and Savrasov1, S.Y. (2011) *Phys. Rev.*, **B83**, 205101.
43. Burkov, A.A. and Balents, L. (2011) *Phys. Rev. Lett.*, **107**, 127205.
44. Deser, S., Jackiw, R., and Templeton, S. (1982) *Phys. Rev. Lett.*, **48**, 975.
45. Fradkin, E., Dagotto, E., and Bohanovsky, D. (1986) *Phys. Rev. Lett.*, **57**, 2967.
46. (a) Hasan, M.Z. and Kane, C.L. (2010) *Rev. Mod. Phys.*, **82**, 3045; (b) Qi, X.-L. and Zhang, S.-C. (2011) *Rev. Mod. Phys.*, **83**, 1057.
47. Ghaemi, P., Ryu, S., and Lee, D.H. (2010) *Phys. Rev.*, **B81**, 081403(R).
48. Huang, K. (1952) *Am. J. Phys.*, **20**, 479.

49. de Broglie, L. (1924) Thése de Doctorat. Une Tentative d'Interprétation Causale et non Linéaire de la Mécanique Ondulatoire. Gauthier-Villars, Paris (1956).
50. Catillon, P., Cue, N., Gaillard, M.J., Genre, R., Gouanure, M., Kirsch, R.G., Poizat, J.-C., Remillieux, J., Roussel, L., and Spighel, M. (2008) *Found Phys.*, **38**, 659.
51. Biswas, T. and Ghosh, T.K. (2012) *J. Phys.: Condens. Matter*, **24**, 185304. arXiv:1201.5252.
52. Gerritsma, R., Kirchmair, G., Zahringer, F., Solano, E., Blatt, R., and Roos, C.F. (2010) *Nature*, **463**,
53. Arunagiri, S. arXiv:0911.0975.
54. Thaller, B. arXiv:quant-ph/0409079.
55. MacDonald, A.H. (1983) *Phys. Rev.*, **B28**, 2235.
56. Sadowski, M.L. *et al.* (2007) *Sol. St. Commn.* **145**, 123.
57. (a) Peres, N.M.R., Arajo, M.A.N., and Castro Neto, A.H. (2004) *Phys. Rev. Lett.*, **92**, 199701; (b) Baskaran, G. and Jafari, S.A. (2004) *Phys. Rev. Lett.*, **92**, 99702.
58. Pines, D. (1962) *The Many Body Theory*, W.A. Benjamin.
59. Pauling, L. (1960) *The Nature of Chemical Bond*, Cornell Univ. Press.
60. (a) Kitaev, A.Yu. arXiv:quant-ph/9707021; (b) Kitaev, A.Yu. (2001) *Phys. Usp. (suppl.)*, **44**, 131; (c) Nayak, C., Simon, S.H., Stern, A., Freedman, M., and Sarma, S.D. (2008) *Rev. Mod. Phys.*, **80**, 1083.
61. Emery, V.J. and Kivelson, S.A. (1996) *Phys. Rev.*, **46**, 10812.
62. (a) Volovik, G.E. (1999) *JETP Lett.*, **70**, 609; (b) Ivanov, D.A. (2001) *Phys. Rev. Lett.*, **86**, 268.
63. Jang, J., Ferguson, D.G., Vakaryuk, V., Budakian, R., Chung, S.B., Goldbart1, P.M., and Maeno, Y. (2011) *Science*, **331**, 186.
64. Mackenzie, A.P. and Maeno, Y. (2003) *Rev. Mod. Phys.*, **75**, 657.
65. Rice, T.M. and Sigrist, M. (1995) *J. Phys.: Condens. Matter*, **7**, L643.
66. Baskaran, G. (1996) *Physica*, **B223–224**, 490.
67. Moore, G. and Read, N. (1991) *Nucl. Phys.*, **B360**, 362.
68. (a) Radu, I.P., Miller, J.B., Marcus, C.M., Kastner, M.A., Pfeiffer, L.N., and West, K.W. (2008) *Science*, **320**, 899; (b) Willett, R.L., Pfeiffer, L.N., and West, K.W. (2009) *Proc. Natl. Acad. Sci. U.S.A.*, **106**, 8853.
69. (a) Fu, L. and Kane, C.L. (2008) *Phys. Rev. Lett.*, **100**, 096407; (b) Lutchyn, R., Sau, J., and Das Sarma, S. (2010) *Phys. Rev. Lett.*, **105**, 077001; (d) Potter, A.C. and Lee, P.A. (2010) *Phys. Rev. Lett.*, **105**, 227003; (e) Alicea, J. arXiv:1202.1293; (f) Beenakker, C.W.J. arXiv:1112.1950.
70. Kwon, H.-J., Sengupta, K., and Yakovenko, V.M. (2004) *Eur. Phys. J.*, **B37**, 349–361.
71. (a) Read, N. and Rezayi, E. (1999) *Phys. Rev.*, **B59**, 8084; (b) Rezayi, E.H. and Read, N. (2009) *Phys. Rev.*, **B79**, 075306.
72. Hewson, A.C. (1993) *The Kondo Problem to Heavy Fermions*, Cambridge University Press.
73. Affleck, I. and Ludwig, A.W.W. (1991) *Nucl. Phys.*, **B360**, 641; (1993) *Phys. Rev.*, **B 48**, 7297.
74. Maldacena, J.M. and Ludwig, A.W.W. (1997) *Nucl. Phys.*, **B506**, 565.
75. Bonderson, P., Freedman, M., and Nayak, Chetan. (2008) *Phys. Rev. Lett.*, **101**, 010501.
76. Suzumura, H. and Ando, T. (2002) *Phys. Rev.*, **B65**, 235412.
77. Guinea, F., Katsnelson, M.I., and Geim, A.K. (2010) *Nat. Phys.*, **6**, 30.
78. Levy, N., Burke, S.A., Meaker, K.L., Panlasigui, M., Zettl, A., Guinea, F., Castro Neto, A.H., and Crommie, M.F. (2010) *Science*, **329**, 544.
79. Baskaran, G. (1986) *Phys. Rev.*, **B33**, 7594.
80. Cortijo, A. and Vozmediano, M.A.H. (2007) *Euro. Phys. Lett.*, **77**, 47002.
81. Iorio, A. (2011) *Ann. Phys.*, **3261**, 334.
82. San-Josea, P., Gonzalez, J., and Guinea, F. arXiv:1110.2883.
83. (a) Baskaran, G. (2002) *Phys. Rev.*, **B65**, 212505; (b) Pathak, S., Shenoy, V., and Baskaran, G. (2010) *Phys. Rev.*, **B 81**, 085431.
84. Black-Schaffer, A.M. and Doniach, S. (2007) *Phys. Rev.*, **B75**, 134512.
85. Black-Schaffer, A.M. arXiv:1204.2425.

4
Magnetism of Nanographene

Toshiaki Enoki

4.1
Introduction

Graphene (single sheet of graphite) has been a central issue in basic materials science and device applications in the decade after the successful isolation of graphene sheet by Novoselov and Geim in 2004 [1]. Owing to this historically important contribution to graphene, the 2010 Nobel Prize in physics was awarded to these two scientists. What is important in graphene is its unconventional electronic structure, which does not exist in traditional materials. It is described in terms of massless Dirac fermion moving on the two-dimensional (2D) honeycomb bipartite lattice (two-sublattice system; Figure 4.1a) as given in the following relativistic Weyl equation [2]:

$$\hat{H} = \sigma \nu_F \mathbf{p} \tag{4.1}$$

where $\mathbf{p}$ and ν_F are the momentum and Fermi velocity, respectively, and σ is the spin operator of the pseudospin, which are effectively converted from the structural degree of freedom 2 in the bipartite lattice consisting of two independent sublattices. Accordingly, the linear $\mathbf{p}$-dependent kinetic energy gives cone-shaped electronic bands for the π-valence and π^*-conduction states (Dirac cone) that touch each other at a point called the *Dirac point*, resulting in a zero-gap semiconducting feature, as shown in Figure 4.2. This unconventional electronic structure brings about unprecedented electronic phenomena such as anomalous quantum Hall effect [3, 4], spin Hall effect [5], and Klein tunneling [6].

The electronic structure of graphene, which physicists employ, is understood differently in chemistry, although the conclusion is the same. Here, in chemistry, we have an important phenomenological rule called *Clar's aromatic sextet rule* with the benzene ring (aromatic sextet) singly bonded to the environment as the fundamental unit [7]. This rule is utilized to discuss the electronic stability of aromatic molecules, which are extrapolated to graphene when the size becomes infinite. The problem we should handle with the Clar's rule is to tile aromatic sextets onto polycyclic aromatic hydrocarbon molecules or graphene. According to this rule, the most stable molecule is the one having a maximal number of sextets. On the basis of this

Graphene: Synthesis, Properties, and Phenomena, First Edition. Edited by C. N. R. Rao and A. K. Sood.

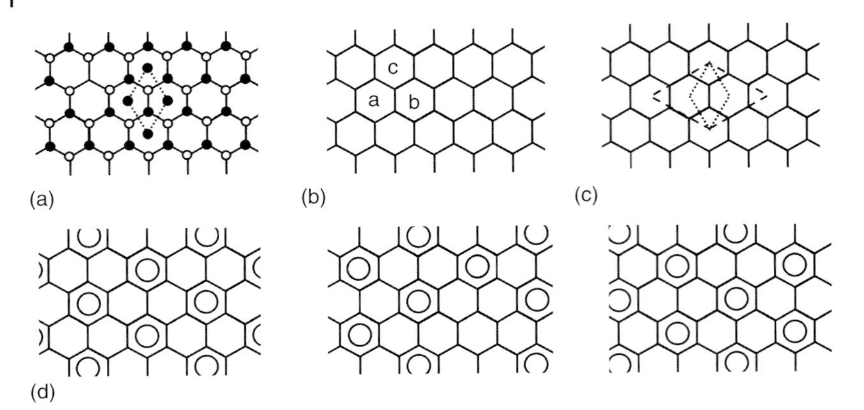

Figure 4.1 (a) A 2D honeycomb bipartite lattice of graphene sheet. The unit cell given with the dotted parallelogram has two independent sublattice sites (A(•), B(○)). (b) Three independent hexagons; a, b, c. (c) $\sqrt{3} \times \sqrt{3}$ superlattice given with the large dashed parallelogram and the unit cell (small dotted parallelogram). (d) Triply degenerate Clar's representations of graphene having $\sqrt{3} \times \sqrt{3}$ superlattices, with a sextet at hexagon a, b, or c.

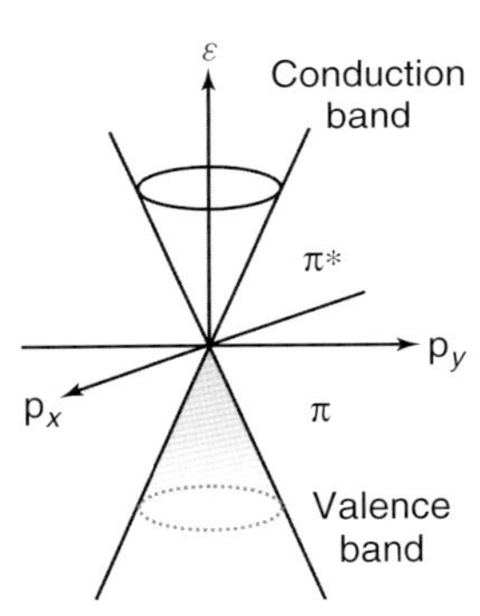

Figure 4.2 The electronic structure of graphene described in terms of two Dirac cones of the valence π- and conduction π*-bands, which touch each other at the Dirac point, at which the Fermi level is positioned.

rule, infinite-size graphene has a sextet in every three hexagons in the honeycomb lattice, forming a $\sqrt{3} \times \sqrt{3}$ superlattice triply degenerate as shown in Figure 4.1d. The fact that sextets are placed only on one-third of the hexagon rings tells us that graphene is less stable, resulting in the electronic activity of the graphene sheet.

When a graphene sheet is cut into fragments, the edges created in these fragments seriously modify the electronic structure of graphene depending on the geometry of the created edges. There are two independent cutting directions: armchair and zigzag directions. By cutting a graphene sheet and terminating the edge carbon atoms by foreign atoms such as hydrogen, zigzag and armchair edges are created without σ-dangling bonds. According to theoretical and experimental works [8–14], the nonbonding π-electron state (edge state) is created at the zigzag edge, in spite of the absence of such a state in an armchair edge. Moreover, the edge state localized around the zigzag edge is strongly spin polarized, and consequently the presence of the magnetic edge state makes finite size graphene magnetic.

Let us discuss the origin of edge state from the point of view of both physics and chemistry. In physics, where the electron in graphene has the feature of massless Dirac fermion, the presence of the edge state is a consequence of the broken symmetry of its pseudospin, as shown in Figure 4.3a, that is, in the zigzag edge, sites belonging to only one of the two sublattices exist, although sites of A and B sublattices are always paired in the armchair edge. Clar's representations in chemistry can give the same thing in magnetism. In armchair-edged molecules, we can see that the molecules are tiled with a large number of sextets, as exhibited in Figure 4.3b. This suggests that the armchair-edged molecules are energetically stable. In contrast, the zigzag-edged molecules shown in Figure 4.3c and d with cases of triangular and linear molecules are less stable because of the

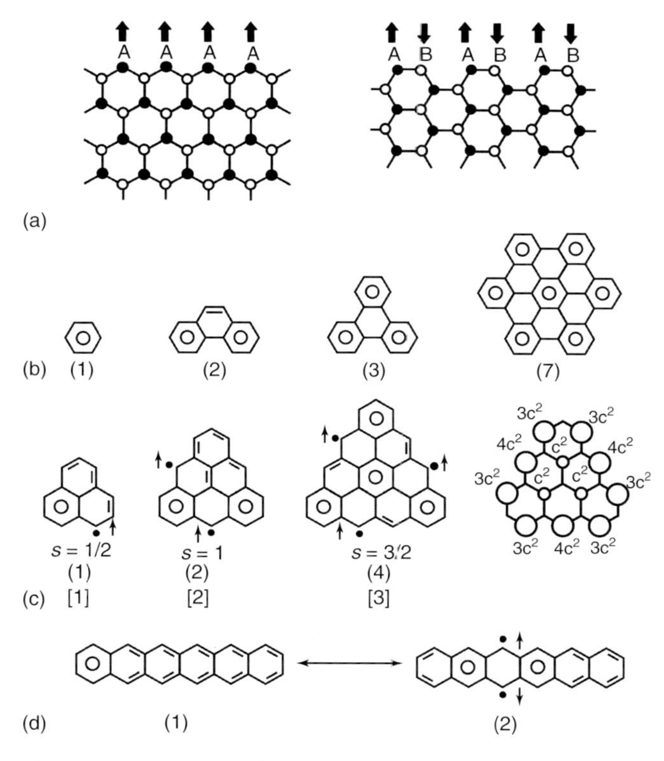

Figure 4.3 (a) Zigzag and armchair edges together with sublattices A and B. The arrow denotes pseudospin. (b–d) Clar's representations of armchair-edged molecules (b), zigzag-edged triangular molecules (c), and zigzag-edged linear molecules (d). The right end of (c) shows the spatial distribution of the local density of states of the edge state in a triangular molecule consisting of six hexagon rings. The numbers in ( ) and [ ] are the number of sextets and the number of unpaired electrons, respectively. The spin state is given by *s*.

small number of constituting sextets. More importantly, unpaired electrons having $s = 1/2$ in nonbonding π-orbitals can exist, making these molecules magnetic. Owing to the Hund rule, triangular molecules consisting of 3, 6, and 10 hexagon rings are ferromagnetic with $s = 1/2$, 1, and 3/2 for 1, 2, and 3 unpaired electrons, respectively, while spins are coupled in an antiparallel manner in linear molecules, forming an antiferromagnetic state or an open-shell singlet state in other words. Here, it should be noted that the spatial distribution of the magnetic nonbonding orbital is mostly localized in the zigzag edge region as displayed in Figure 4.3c. This is why the nonbonding state is called *edge state*.

From the above discussion, it is inferred that the important origin of graphene-based spin magnetism comes from the strongly spin-polarized edge state inherent to the π-electron. However, in addition to this, we should comment that defects of σ-electron origin can also cause localized spin. When the conjugated π-electron network of a graphene sheet is destroyed by a strong external perturbation such as ion bombardment or reaction with active chemical species, carbon atoms that are not bonded to the neighboring atoms are created. Here, the naked carbon atom with a σ-dangling bond has a localized spin of $s = 1/2$, which makes graphene magnetic. Consequently, we expect a variety of magnetic behavior with the coexistence of edge-state spins and σ-dangling bond spins in the magnetism of a defective graphene sheet. When we go back to the early stage of history in carbon-based magnetism [15, 16], which started in the 1950s [17], the origin of localized spins responsible for magnetism has not been clearly understood for more than four decades. The discovery of edge state and the characteristics that distinguish it from the σ-dangling bond have solved the confusion that arises in understanding carbon-based magnetism and, importantly, contributed to understanding its essential origin [18–23].

In this chapter, the magnetism of nanographene is reviewed on the basis of the presence of two types of localized spins with the impetus mentioned above. In addition, unconventional magnetic phenomena observed in nanographene-based nanoporous carbon are discussed in relation to the interaction of nanographene with guest molecular species physisorbed in nanopores. The chapter is organized as follows: in Section 4.2, the theoretical background of magnetism in nanographene and graphene edges is discussed; in Section 4.3, the experimental approach to magnetism of nanographene is discussed; in Section 4.4, magnetic phenomena that arise in the interaction with guest molecules in nanographene-based nanoporous carbon is addressed; and Section 4.5 provides the summary of this chapter.

4.2
Theoretical Background of Magnetism in Nanographene and Graphene Edges

As we have seen in Section 4.1, a nonbonding edge state is created at a zigzag edge as a consequence of the broken symmetry of pseudospin. This means that all the edge carbon sites at the same zigzag edge belong to only one of the two sublattices, resulting in the ferromagnetic arrangement of edge-state spins within the zigzag edge. The strength of the intra-zigzag-edge ferromagnetic exchange interaction is considerably strong and is in the range of several 10^3 K [23, 24]. The inter-zigzag interaction has a strength of $10^{-1}-10^{-2}$ times the intra-zigzag-edge interaction and a sign of + or − (ferromagnetic or antiferromagnetic) depending on the mutual geometrical relationship between the relevant zigzag edges. These interactions determine the magnetic structure of nanographene. Here is a general rule called *Lieb's theorem* [25], with which the spin state is calculated. The theorem indicates that the spin state is given by the following simple relation:

$$s = \frac{1}{2}(N_A - N_B) \tag{4.2}$$

where N_A and N_B are the numbers of sites belonging to sublattices A and B, respectively. Using this theorem, the armchair-edged molecules shown in Figure 4.3b are confirmed to be nonmagnetic, which is consistent with the Clar rule. In contrast, the ferromagnetic structure of the zigzag-edged triangular molecules shown in Figure 4.3c is obtained, while the zigzag-edged linear molecules shown in Figure 4.3d form an antiferromagnetic structure with the net magnetic moment being zero, apparently similar to the spin state of armchair-edged molecules.

Many theoretical works have examined edge-state spin magnetism [18–24], which is concluded to be consistent with the Lieb theorem. Here, we discuss the magnetic structures of zigzag-edged triangular and hexagonal nanographene because comparing the two is instructive regarding the way in which the geometry affects the magnetism [19]. The magnetism of the edge-state spins can be described in terms of the mean-field Hubbard Hamiltonian:

$$\hat{H} = \hat{H}_0 + U\sum_i \left(n_{i\uparrow}\langle n_{i\downarrow}\rangle + n_{i\downarrow}\langle n_{i\uparrow}\rangle\right), \quad \hat{H}_0 = -t\sum_{\langle i,j\rangle,\sigma}\left(c_{i\sigma}^{+}c_{j\sigma} + H.C.\right) \tag{4.3}$$

$\hat{H}_0$ is the single-orbital Hamiltonian with a transfer integral t connecting neighboring sites i and j belonging to sublattices A and B, respectively, in the bipartite lattice. The second term is the on-site Coulomb interaction with U, where the summation is taken with i running over the number of sites N_η, η = A,B (A, B; A, B sublattices). This term plays an essential role in creating magnetic moments. Figure 4.4 shows the energy spectra and the distribution of the local magnetic moments obtained by the Hubbard model and density functional theory (DFT) calculations. A spin gap appears around the Fermi level, producing spin polarization in the edge-state spins, as shown in Figure 4.4b for the triangular nanographene. The local magnetic moments, which are arranged in an antiparallel manner between sublattices A and B, are well localized near the zigzag-shaped edges, and their strength decays sharply from the edge to the interior. All the three zigzag edges constituting the triangular nanographene consist of atoms belonging to sublattice A (B), which have up (down) spins arranged in parallel, producing a ferromagnetic structure with a net nonzero spontaneous magnetic moment in triangular nanographene. This is consistent with Lieb's theorem. For hexagonal nanographene that consists of six zigzag edges, three edges belonging to sublattice A (B) have up (down) spins, whereas three edges

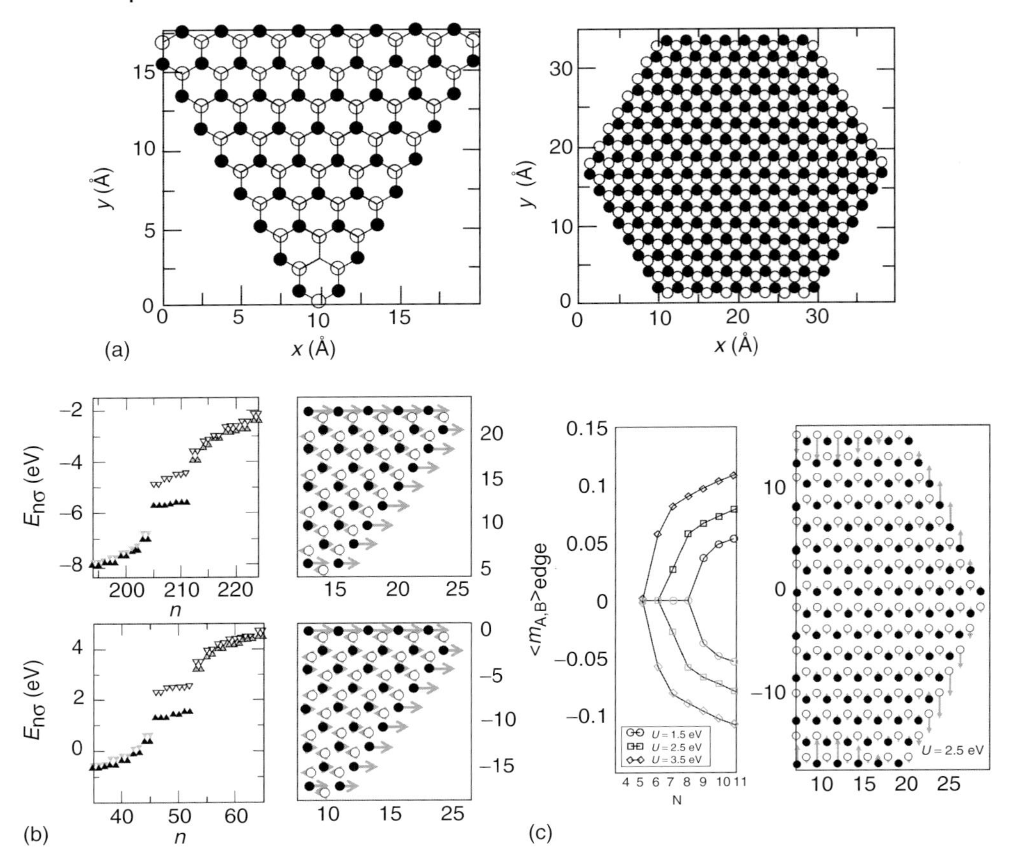

Figure 4.4 (a) (Left) Triangular and (right) hexagonal nanographene ($N = 8$) with carbon atoms in A (filled circles) and B (open circles) sublattices. The size of a nanographene is given by the number N of edge carbon atoms. (b) (Left) Energy spectra and (right) distributions of local magnetic moments for the triangular nanographene ($N = 8$), calculated by a mean-field Hubbard model with $t = 2.5$ eV and $U = 3.85$ eV (lower row) and DFT calculation (upper row). Filled triangles (open triangles) in the energy spectra correspond to the filled (empty) single particle state. Triangles (inverted triangles) represent up (down) spin states. The arrow represents the local moment whose strength is given by the length. (c) (Left) Sublattice-resolved average magnetic moment as a function of N for hexagonal nanographene with $t = 2.5$ eV and $U = 1.5$, 2.5, and 3.5 eV calculated by a mean-field Hubbard model and (right) distributions of local magnetic moments for $N = 8$ and $U = 2.5$ eV [19].

belonging to sublattice B (A) have down (up) spins. Accordingly, the compensation between the up and down spins stabilizes an antiferromagnetic structure having no spontaneous magnetization. In discussing the magnetism of edge-state spins, it should be noted that a weak spin–orbit interaction of carbon, which is estimated as $\sim 5\ \text{cm}^{-1}$ [26], makes the edge-state spins to be weakly anisotropic. They are

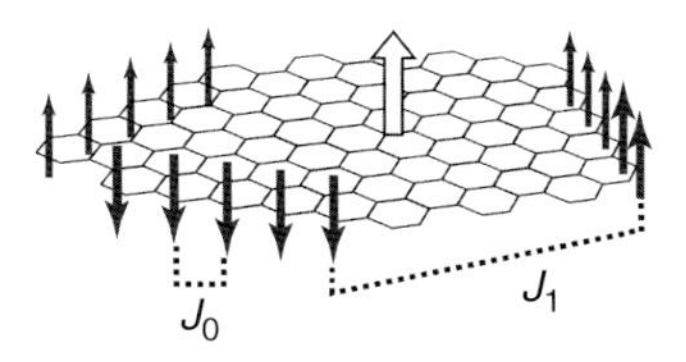

Figure 4.5 Schematic model of an arbitrarily shaped nanographene sheet and spatial distribution of edge-state spins (filled arrow). J_0 and J_1 are the intra- and inter-zigzag-edge interactions, respectively. The large open arrow is the net magnetic moment after compensation between ferromagnetic clusters in the zigzag edges.

described in terms of a low-dimensional (low D) anisotropic Heisenberg spin system in a nanographene sheet [24].

An arbitrarily shaped nanographene sheet whose periphery is described by a combination of zigzag and armchair edges has a ferrimagnetic structure, as shown in Figure 4.5 [27, 28]. The localized edge-state spins within a zigzag edge are ferromagnetically arranged through a strong ferromagnetic intra-zigzag-edge exchange interaction ($J_0 \sim 10^3$ K) [19, 23, 24]. The ferromagnetically arranged edge-state spins in two zigzag edge regions separated by an armchair region interact with each other with the aid of conduction-π-electron-mediated inter-zigzag-edge interaction J_1, whose sign varies from positive (ferromagnetic) to negative (antiferromagnetic) and strength varies depending on the mutual relationship between the two zigzag edges. J_1 has a moderate strength ($J_1 \sim 10$–100 K). Consequently, depending on the shape of the nanographene sheet, the interplay between strong intra-zigzag-edge ferromagnetic interaction J_0 and moderate inter-zigzag-edge ferromagnetic/antiferromagnetic exchange interaction J_1 is expected to produce a ferrimagnetic structure with a nonzero net magnetic moment that remains after compensation between the antiferromagnetically coupled ferromagnetic spin clusters.

Chemical modifications of edges make the magnetism of edge state change seriously [29, 30]. In graphene nanoribbons having monohydrogenated zigzag edges on both sides, the compensation between up and down spins on opposite sides produces an antiferromagnetic structure with no spontaneous magnetization. However, when one side is dihydrogenated and the opposite side is monohydrogenated, uncompensation between the up and down spins yields a ferromagnetic structure in which even the carbon atoms in the interior are strongly spin polarized, as shown in Figure 4.6a. The creation of ferromagnetic spontaneous magnetization is a consequence of the broken symmetry between the two sides. The fluorinated version of a graphene nanoribbon having the same structure is also in a ferromagnetic state, although the edge-state spins disappear on the difluorinated side, as illustrated in Figure 4.6b.

Making defects in the interior of a graphene sheet also creates a spin-polarized edge state. In this case, the magnetism has a geometrical dependence similar to that discussed above [31, 32]. In particular, when the presence of a defect makes sublattice A nonequivalent to sublattice B, a net magnetic moment is

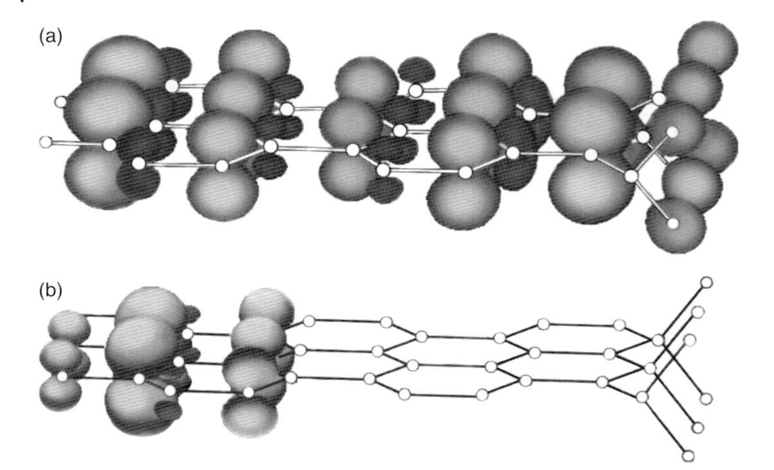

Figure 4.6 (a) Ferromagnetic structure of graphene nanoribbon having monohydrogenated zigzag edge on one side and dihydrogenated zigzag edge on the opposite side. Wave function represented by light (dark) color is polarized with up (down) spin. (b) Ferromagnetic structure of graphene nanoribbon having the same structure as (a) but with fluorine-terminated edges [30].

created; that is, the defect has the spin state expected from Lieb's theorem (Equation 4.2). However, we should remember the following two issues in discussing the magnetism of defects. The first is the deformation of the graphene lattice. When a foreign atom is bonded to a carbon atom in the interior of a graphene sheet, a defect created at the site to which the foreign atom is bonded is subjected to a local lattice deformation in the vicinity of the site, resulting in a slight contribution of the σ-state admixed to the π-state. This results in the modification of the edge state from the conventional one. The second is the contribution of σ-dangling bonds. When defects possess unterminated carbon sites with σ-dangling bonds, nonbonding states of σ electron origin are also responsible for magnetism. Figure 4.7 shows the density of states plots for a graphene sheet with hydrogen-bonded defects (Figure 4.7a) and vacancy defects (Figure 4.7b), where the interdefect distance is $12a_{CC}$ (C–C bond length; $a_{CC} = 0.142$ nm) [31]. In the former, edge states of only π-electron origin are created, whereas in the latter σ-dangling bond states are formed in addition to the edge state. In the hydrogen-terminated defect, edge state spin having $\sim 1\ \mu_B$ shows an exchange splitting of 0.23 eV as shown in Figure 4.7c. In the vacancy defect, the edge-state spins of π-electron origin and the σ-dangling bond spins coexist. The former is itinerant, with fractional magnetic moments, whereas the latter is localized, with a magnetic moment of $\sim 1\ \mu_B$. The exchange coupling is estimated as 0.14 eV, as shown in Figure 4.7d. In addition, it should be noted that local charge transfer makes the minority spin state self-doped, causing the partial spin polarization.

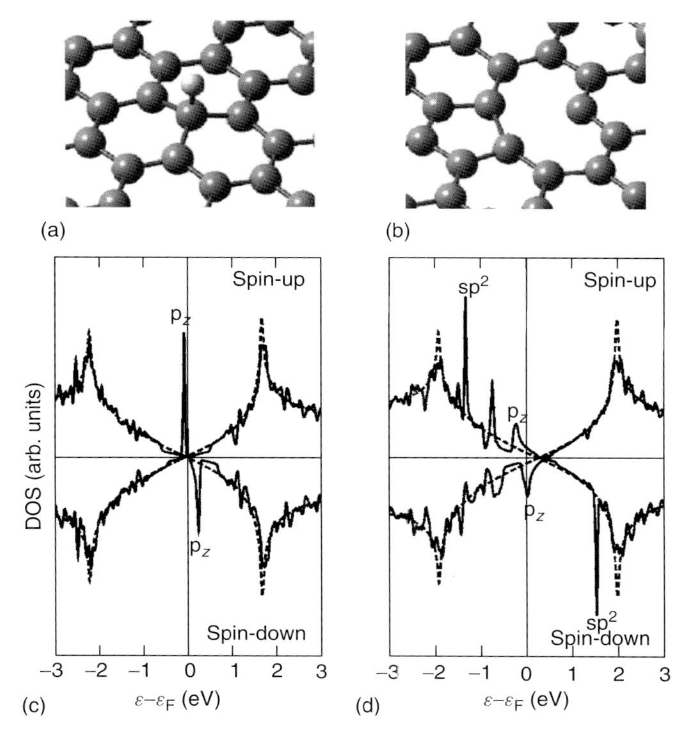

Figure 4.7 Structures of a hydrogen-bonded defect (a) and a vacancy defect (b). The small light gray sphere in (a) represents a hydrogen atom. (c and d) Density of states plots for the systems with (c) the hydrogen-bonded defect and (d) the vacancy defect. The interdefect distance is $12a_{CC}$ (C–C bond length; $a_{CC} = 0.142$ nm). The dashed line shows the density of states of the ideal graphene. Labels indicate the character of the defect states: p_z, edge state; sp^2, σ-dangling bond [31].

4.3 Experimental Approach to Magnetism of Nanographene

4.3.1 Magnetic Structure of Edge-State Spins in Nanographene

Here, we employ a nanographene-based nanoporous carbon (activated carbon fiber (ACF)) as a model system in discussing the magnetism of edge-state spins. ACFs [27, 28, 33–36] consist of a 3D disordered network of nanographite domains, each of which is a loose stack of three to four nanographene sheets with a mean in-plane size of 2–3 nm, in which 200 – 300 carbon atoms are involved, as illustrated in Figure 4.8. The concentration of edge-state spins corresponds to several spins per nanographene sheet and has the magnitude predicted by Clar's aromatic sextet rule. The spins constituting an individual arbitrarily shaped nanographene sheet form a ferrimagnetic structure, as illustrated in Figure 4.5.

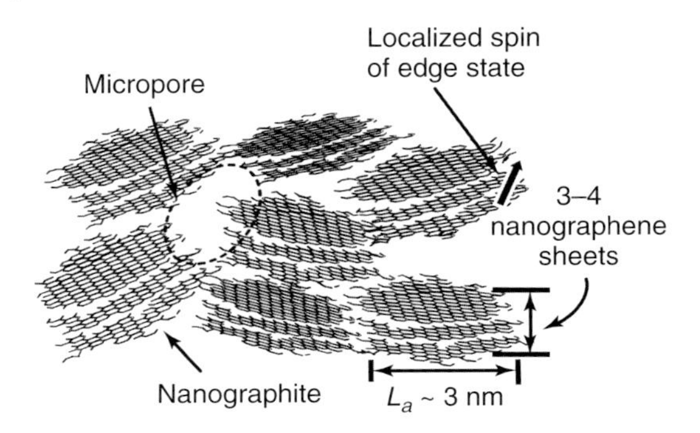

Figure 4.8 Structural model of nanoporous activated carbon fiber (ACF) with localized edge-state spins at the zigzag edges. A nanopore is created in a space surrounded by nanographite domains.

In the network of nanographite domains, two more exchange interactions should be taken into account owing to the structural hierarchy; a weak antiferromagnetic inter-nanographene-sheet interaction (J_2) and a weak antiferromagnetic inter-nanographite-domain interaction (J_3). The order of the strengths is $|J_0| > |J_1| \gg |J_2| > |J_3|$. The magnetism of ACFs is ultimately a consequence of the cooperation of J_0, J_1, J_2, and J_3.

Before discussing the magnetism, let us examine the electron transport properties of ACFs. Pristine ACFs exhibit insulating behavior (Anderson insulator [37]), as shown in Figure 4.9d [36]. The conductivity obeys the formula of Coulomb-gap-type variable range hopping in an Anderson insulator,

$$\sigma\,(T) = \sigma_0 \exp\left[-\left(\frac{T_0}{T}\right)^{1/2}\right] \tag{4.4}$$

suggesting that the electron transport is governed by electron hopping between conductive nanographene sheets under the influence of a charging process in the 3D disordered network of nanographene sheets [33–36]. The magnetism of ACFs reflects the electron transport features. Figure 4.9a–c exhibits the temperature dependence of the electron spin resonance (ESR) intensity, ESR line width ΔH_{PP}, and static magnetic susceptibility χ of the edge-state spins in pristine ACFs [36]. The ESR intensity, which represents the dynamic magnetic susceptibility, obeys the Curie law down to 30 K, and suddenly drops by 50% below 20 K, unlike the static susceptibility, which obeys the Curie law in the entire temperature range. ΔH_{PP} follows a trend corresponding to the behavior of the intensity. It decreases linearly from 6.2 to 2.2 mT as the temperature decreases to 30 K and suddenly increases by 30% (0.6 mT) below 20 K. These experimental findings are therefore understood as the representation of a transition from the high-temperature homogeneous spin state to the low-temperature inhomogeneous state at ~20 K. This is clearly evident in changes in the ESR line profile and saturation trend that appear near the transition temperature ($T_c \sim 20$ K). The line shape deviates from a Lorentzian and seriously saturates as the microwave power is elevated in the low-temperature range

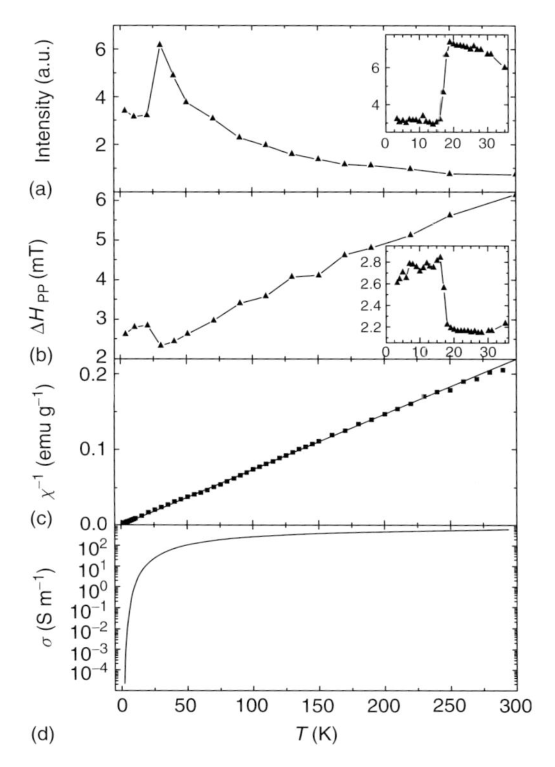

Figure 4.9 Temperature dependence of the (a) electronic spin resonance (ESR) intensity, (b) ESR line width (ΔH_{PP}), (c) reciprocal of the static susceptibility (χ), and (d) electrical conductivity (σ) of the pristine ACFs. ESR signals were observed with a microwave power of 1 μW. Insets in (a) and (b) are close-ups in the low-temperature range [36].

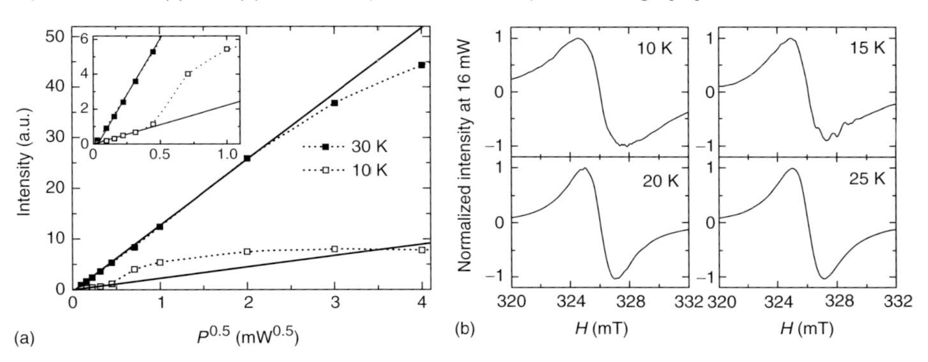

Figure 4.10 (a) Microwave power dependence of ESR intensity in vacuum, at 30 K (■) and 10 K (□). The blown-up view of low-power region is given in the inset. The solid lines are straight-line fit to the low-power region and dashed lines are guides for the eyes. (b) The ESR signals in vacuum at 10, 15, 20, and 25 K measured with 16 mW. The irregular feature typically observed in the signal at 15 K is a consequence of the hole-burning effect in the strong microwave excitation [36].

below 30 K (Figure 4.10a). More importantly, the ESR signal at a strong microwave power of 16 mW shows a prominent hole-burning effect (irregular feature in the ESR signal) just below T_c, indicating a serious inhomogeneity in this temperature range (Figure 4.10b). Above 25 K, the line shape becomes sharpened and purely Lorentzian, indicating exchange/motional narrowing.

The experimental findings presented above allow us to understand the magnetic structure and dynamics of the edge-state spin system. In electron hopping in the 3D random network of metallic nanographene sheets, π-electron carriers interact strongly with the edge-state-localized spins of a given sheet. At high temperatures, a rapid electron hopping process revealed by the high conductivity (Figure 4.9d) subjects the edge-state spins to motional narrowing, yielding a homogeneous spin system in the entire network. Therefore, the spin system is modeled merely as a metallic system with an interaction between localized edge-state spins and conduction π-electron carriers [36, 38, 39]. The strong carrier-localized spin interaction is demonstrated by the linear temperature dependence of the line width (Figure 4.9b); that is, according to the Korringa relation [40], the ESR line width, which is inversely proportional to the relaxation time $T_{\text{edge-}\pi}$ in the energy relaxation from the edge-state spins to the conduction carriers, is given as follows:

$$\Delta H \propto \frac{1}{T_{\text{edge-}\pi}} = \left(\frac{4\pi}{\hbar}\right) J^2_{\text{edge-}\pi} D(\varepsilon_F)^2 k_B T \qquad (4.5)$$

where $J_{\text{edge-}\pi}$ and $D(\varepsilon_F)$ are the exchange interaction between the edge-state spin and the conduction π carrier, and the density of states at the Fermi level ε_F, respectively.

The situation becomes different in the low-temperature regime below T_c. The conductivity suggests strong electron localization due to Coulomb interaction at lower temperatures. The appearance of hole-burning near the transition temperature demonstrates the considerable effect of structural inhomogeneity on the edge-state spins. Indeed, an inhomogeneous distribution of on-resonance fields arises from the structural inhomogeneity originating in the hierarchical features of a 3D disordered network with irregularly shaped nanographene sheets as the fundamental unit (Figure 4.5). In the individual arbitrarily shaped nanographene sheets, a ferrimagnetic structure is formed, having a net nonzero magnetic moment with a strength that varies depending on its shape, causing an inhomogeneous static distribution of the ESR on-resonance fields. The inhomogeneous distribution of the on-resonance fields does not appear at high temperatures owing to motional narrowing caused by rapid inter-nanographene-sheet electron hopping. However, as the temperature decreases, electron localization develops, and nanographene sheets become independent of each other because electron hopping becomes slower. Finally, the inhomogeneity survives at lower temperatures below T_c, at which the electron hopping frequency becomes small enough to reveal the inhomogeneous line width. This magnetic behavior unveils the ferrimagnetic structure of an arbitrarily shaped nanographene sheet.

4.3.2
Magnetism of σ-Dangling Bond Defects in Graphene

Defects can be an ingredient of magnetism in graphene, as pointed out in Section 4.2. Ion bombardment [32, 39, 41–43] and reaction with foreign species that have strong chemical activities [44–49] allow us to create defects that make graphene magnetic. Ion bombardment with hydrogen and helium ions demonstrates the creation of magnetic defects, which form a ferromagnetic state in some cases. In this section, the magnetism of defective nanographene created by the reaction with strongly chemically active fluorine is discussed. Here, the reaction of fluorine atoms with carbon atoms of a nanographene sheet destroys the π-conjugated electron system by creating defects having sp^3 bonds, resulting in the formation of σ-dangling bonds [44–47].

In nanographene, the reaction process of fluorine takes place in a two-step manner owing to the presence of edges as shown in Figure 4.11, different from that in infinite-size graphene and bulk graphite, in which all the constituent carbon atoms are subjected to fluorination with no preferential sites. In the initial stage of fluorination, a fluorine atom attacks an edge carbon atom of nanographene, as the edge carbon atom is more chemically active than a carbon atom in the interior of nanographene sheet. Accordingly, the reaction of fluorine atoms makes the edge carbon atoms difluorinated, converting the sp^2/π configuration to the sp^3 one. After all the edge carbon atoms are difluorinated, fluorine atoms attack the carbon atoms in the interior of the nanographene sheet. In this second process, a σ-dangling bond having $s = 1/2$ is created at the carbon site adjacent to the carbon site attacked by the fluorine atom. In the case of the nanographene sheet with an in-plane size of 2–3 nm, which is discussed in this section, the total number of carbon atoms and the number of edge carbon atoms are estimated to be about

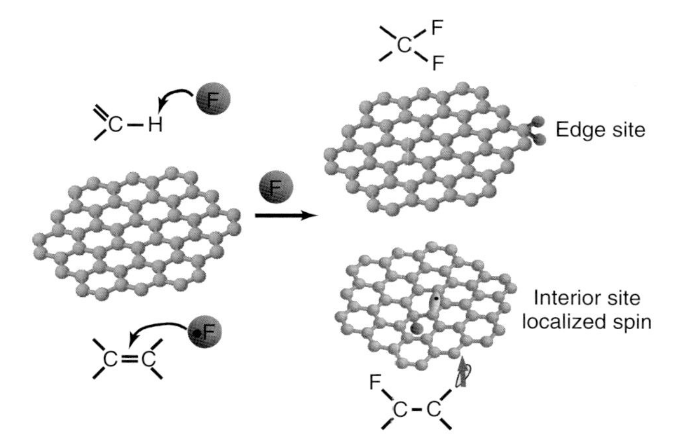

Figure 4.11 A schematic model of two-step fluorination of a nanographene sheet. In the first step (upper), edge carbon atoms are difluorinated, while carbon atoms in the interior of a nanographene sheet are monofluorinated in the second step (lower).

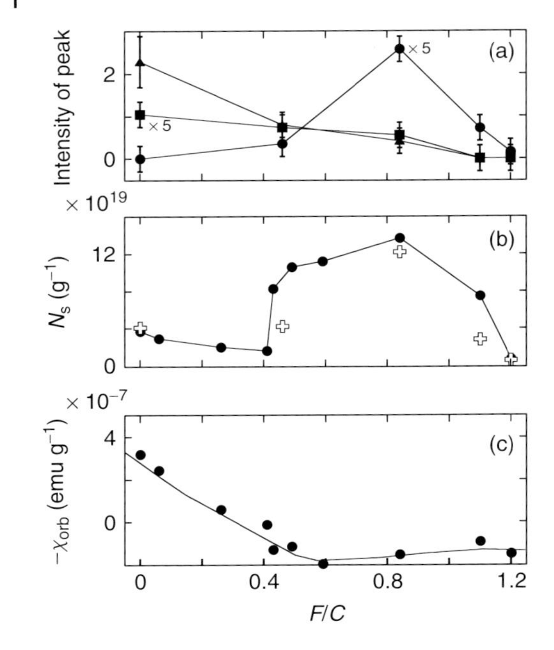

Figure 4.12 (a) The intensities of the π-edge state (squares), σ-dangling bond state (circles), and π* state (triangles) peaks as a function of fluorine concentration, obtained from NEXAFS experiments. (b) The total localized spin concentration *N*s (circles) as a function of fluorine concentration. The expected total density of magnetic moments in the F-ACF (crosses) obtained by multivariable analysis of the NEXAFS spectra with the contributions of the edge state and σ-dangling bond state. (c) The absolute value of the orbital susceptibility χ_{orb} [45].

200–300 and 60–70, respectively. From this estimate, the fluorine concentration at which all the edge carbon atoms are difluorinated is F/C ~ 0.4, and fluorination is completed at F/C ~ 1.2. Note the 0.2 excess from the concentration (F/C = 1), at which bulk graphene is completely fluorinated. This means that the excess fluorine atoms are consumed in difluorinating edge carbon atoms, while carbon atoms in the interior are monoflourinated.

Let us discuss the magnetism of the fluorinated nanographene having two types of magnetic species: edge-state spins and σ-dangling bond spins. Magnetic susceptibility and near edge X-ray absorption spectra (NEXAFS) experiments have been carried out as a function of fluorine concentration. The experimental findings of the magnetism together with the electronic structure are shown in Figure 4.12 for the fluorinated nanographene of ACFs. The orbital susceptibility, whose absolute value represents the extension of the conjugated π-electron state, decreases monotonically as the fluorine concentration is elevated. In the NEXAFS of the C K edge, the peak assigned to the π*-conduction band is observed at 285.5 eV. As shown in Figure 4.12a, the intensity of this π*-conduction band peak decreases as the fluorine concentration is elevated. These findings of the orbital susceptibility and the NEXAFS π^* peak intensity prove a successive degradation of the conjugated π-electron structure on fluorination. The localized spin concentration shown in Figure 4.12b decreases with the increase in the fluorine concentration up to F/C $\sim$ 0.4, at which all the edge carbon atoms are difluorinated. Taking into account the degradation of the conjugated π-electron structure on fluorination, the decrease in the spin concentration in this fluorination range is caused by the decrease in the edge-state spin contribution. The NEXAFS exhibits a peak at 284.5 eV assigned to the edge state near the Fermi level. The large negative chemical shift from the π^* peak is a consequence of the large screening effect of the edge state [50] because the edge state has a large local density of states. The fluorination-induced decrease in the intensity of the edge-state peak shown in Figure 4.12a confirms the decrease in the edge-state spin concentration on fluorination.

There is a change in magnetism when the fluorine concentration ranges between F/C $\sim$ 0.4 and 1.2. The spin concentration increases above F/C $\sim$ 0.4, it shows a broad hump around F/C $\sim$ 0.8, and it finally becomes negligible at the saturated fluorine concentration of F/C $\sim$ 1.2. We can discuss the change in magnetism on the basis of the aforementioned scenario of fluorination in the second step; that is, the increase in the spin concentration with a broad hump around F/C $\sim$ 0.8 suggests that the σ-dangling bond spins are responsible for this change in the fluorine concentrations of 0.4 $<$ F/C $<$ 1.2 in the second step of fluorination. Indeed, the concentration of the σ-dangling bond spins is maximized around F/C $\sim$ 0.8, at which half of the interior carbon sites are fluorinated. In the region where F/C is 0.4–0.8, fluorination takes place resulting in the formation of σ-dangling bond spins, while σ-dangling bond spins are destroyed through fluorination of the σ-dangling bonds when F/C is 0.8–1.2. In the NEXAFS, the peak assigned to the σ-dangling bond is observed at 284.9 eV. The less negative chemical shift from the π^*-peak is suggestive of a small screening effect of the σ-dangling bond. The intensity of the σ-dangling bond peak exhibited in Figure 4.12a well tracks the behavior of the σ-dangling bond spin concentration with a broad hump at F/C $\sim$ 0.8.

Finally, let us discuss the difference in the magnetic properties between the edge-state spins and the σ-dangling bond spins. Figure 4.13 shows the dependence of the mean molecular field $<\alpha>_{\mathrm{av}}$ on fluorine concentration, representing the exchange field acting between the spins in the whole fluorine concentrations. At concentrations below F/C $\sim$ 0.4, in which the spins are assigned to those from the edge state, the exchange field decreases monotonically and it becomes negligible at F/C $\sim$ 0.4, at which all the edge-state spins disappear. This importantly proves that the edge-state spins are interacting with each other through the exchange interaction mediated by the π-conduction carriers [23, 24, 27, 28, 38, 39]. This is consistent with what is discussed in Section 4.1. Above F/C $\sim$ 0.4, where the magnetism is governed by the σ-dangling bond spins, the localized spins behave independent of each other, with the exchange interaction being negligibly small. This suggests that the σ-dangling bond spins are isolated from the surroundings.

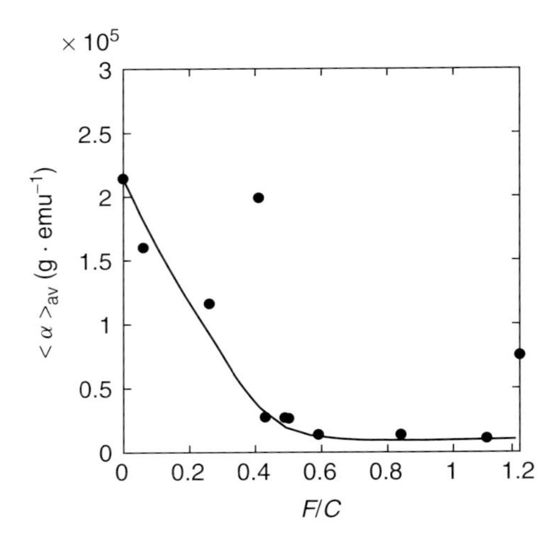

Figure 4.13 The mean molecular field coefficient $<\alpha>_{av}$ as a function of F/C in the fluorinated ACFs [44].

4.4 Magnetic Phenomena Arising in the Interaction with Guest Molecules in Nanographene-Based Nanoporous Carbon

The edge-state spins are an interesting building block in designing molecular magnetism [15, 16]. They can also be utilized to design magnetic functions. Nanoporous ACFs, which consist of a 3D-disordered network of nanographite domains, offer opportunities for finding interesting magnetic phenomena and functionality with the employment of guest species adsorbed physisorptively onto the nanopores. In this section, magnetic switching phenomenon and helium sensor are discussed with the employment of edge-state spins in nanoporous ACFs.

4.4.1 Magnetic Switching Phenomenon

The network structure of ACFs with a large fraction of nanopores in volume is soft and flexible due to the loose connections of nanographite domains. As a consequence, guest adsorption/desorption can expand/contract the nanopore volume, resulting in mechanically compressing nanographite domains that surround the nanopore space [51]. This motivates us to explore the possibility of controlling the magnetism of edge-state spins on the adsorption/desorption of guest molecules in the nanopores. The magnetic investigations together with guest molecule adsorption are carried out using various kinds of nonmagnetic chemically inert guest molecules, which do not hinder the change that takes place in the magnetism of edge-state spins during adsorption/desorption.

Figure 4.14 shows the adsorption isotherm, the spin susceptibility χ_s, and the ESR line width ΔH of ACFs during adsorption/desorption of H_2O molecules

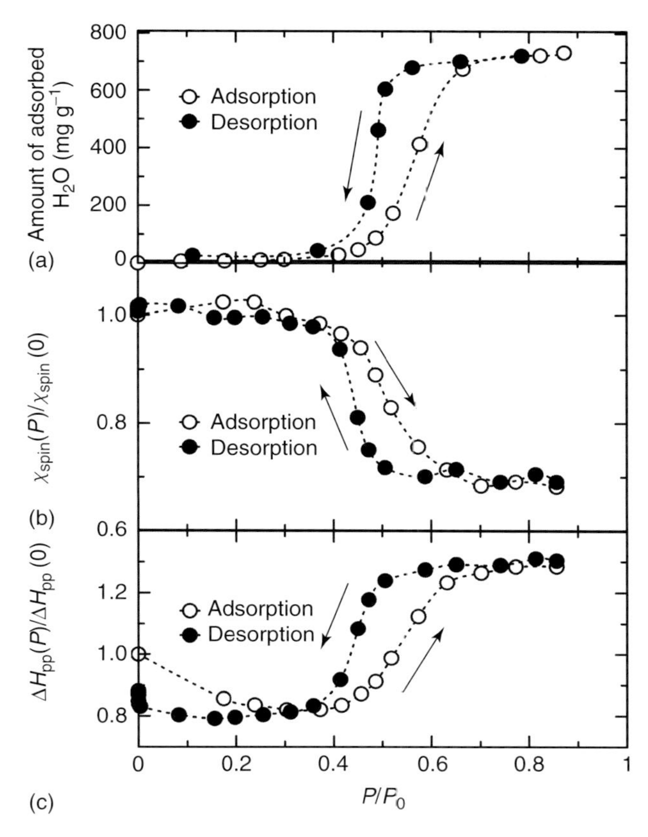

Figure 4.14 (a) The H_2O adsorption isotherm, (b) the magnetic susceptibility χ_s, and (c) the ESR line width ΔH as a function of vapor pressure P of H_2O at room temperature. P_0 is the saturation vapor pressure [52].

[52]. The adsorption isotherm shows an on/off-type feature with a threshold vapor pressure of $P_t/P_0 \sim 0.5$ (P_0, the saturation vapor pressure); no adsorption takes place when the vapor pressure range is below P_t owing to the hydrophobic nature of the nanographene sheet, whereas adsorption starts when the vapor pressure is above P_t and it rapidly saturates just after P_t. The magnetic susceptibility, which is proportional to the strength of the magnetic moment of the edge-state spin, tracks the behavior of the isotherm with fidelity. ACFs have a large magnetic moment in the absence of the adsorbed H_2O molecules in the low vapor pressure regime, whereas it decreases by 30–40% when the nanopores are completely filled with H_2O molecules above the P_t. This can be explained in terms of mechanical compression effect on the magnetism of the edge-state spins, as illustrated in Figure 4.15. In the absence of H_2O molecules in the nanopores, the inter-nanographene-sheet distance is 0.38 nm, which is considerably longer than the inter-graphene-sheet distance of 0.335 nm in the bulk graphite [51]. Thus the inter-nanographene-sheet antiferromagnetic exchange interaction J_2 is so small that the net magnetic

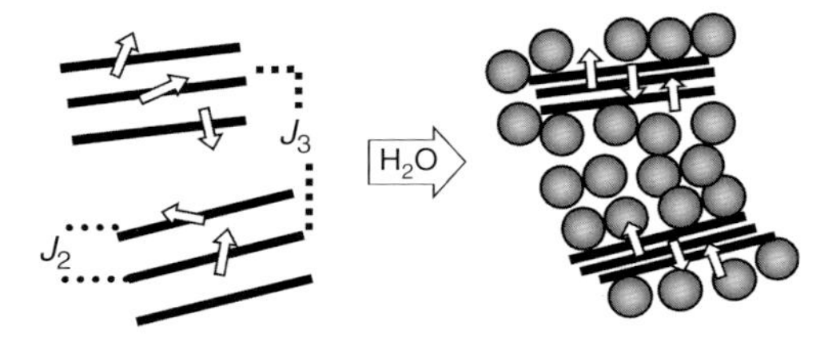

Figure 4.15 Magnetic switching phenomenon. J_2 and J_3 are the inter-nanographene-sheet antiferromagnetic interaction in a nanographite domain and the inter-nanographite-domain antiferromagnetic interaction, respectively. The compression of nanographite domains by H_2O molecules (shaded circle) condensed in the nanopores strengthens J_2 at the expense of J_3. The arrow indicates the net magnetic moment of edge-state spins in an individual nanographene sheet.

moment of the individual nanographene sheet behaves independently of that of the adjacent nanographene sheet. When the nanopores are filled with H_2O molecules above P_t, the condensed H_2O molecules surrounding the nanographite domain counterwork to compress mechanically the nanographite domain, resulting in the decrease in the inter-nanographene-sheet distance to 0.34 nm. This enhances the strength of the inter-nanographene-sheet antiferromagnetic interaction J_2 with the inter-nanographite-domain interaction J_3 being reduced, and then the magnetic moments of the nanographene sheets are arranged in an antiparallel manner in the individual nanographite domain, the net magnetic moment of ACFs being reduced. This is the high-spin/low-spin magnetic switching phenomenon, in which the mechanical compression by the physisorbed guest molecules induces a change in the magnetism of the edge-state spins. The mechanism of the magnetic switching phenomenon can be explained theoretically on the basis of the Hubbard model for a loose stack of nanographene sheets [53]. The behavior of the ESR line width confirms this mechanical-compression-induced change in the magnetic structure. The ESR line width abruptly increases by 60% above the threshold pressure P_t on adsorption of H_2O. The compression of the nanographite domain by the H_2O molecules condensed in the nanopores separates the nanographite domains from each other, working to elongate the inter-nanographite-domain distance. Accordingly, the H_2O adsorption weakens the strength of the inter-nanographite-domain antiferromagnetic interaction J_3. On the basis of the change in the exchange interactions J_2 and J_3, the H_2O-adsorption-induced increase in the ESR line width is understood by the exchange narrowing mechanism given as

$$\Delta H = \frac{\langle \Delta M^2 \rangle}{H_{ex}} \tag{4.6}$$

where $\langle \Delta M^2 \rangle$ and H_{ex} are the second moment of the dipole–dipole interaction and the exchange field, respectively. Within the individual nanographite domain, the exchange narrowing mechanism works to make the intradomain contribution to the width negligibly small under the operation of the inter-nanographene-sheet interaction J_2, which is considerably larger than the inter-nanographite-domain interaction J_3 at a vapor pressure range above P_t. Eventually, $\langle \Delta M^2 \rangle$ is governed by the inter-nanographite-domain dipole–dipole interaction. The decrease in the strength of J_3, which represents H_{ex} in Equation (4.6), broadens the line width on adsorption of H_2O.

The magnetic switching phenomenon is universal, and can be observed for other guest molecules although the behavior is modified depending on their chemical features [54–57]. Here, the key point is the presence of the hydroxyl group as we can see by comparing various guest molecules with and without the hydroxyl group, as summarized in Figure 4.16 with H_2O, CH_3OH, C_2H_5OH, $(CH_3)_2CO$, C_6H_6, $CHCl_3$, CCl_4, C_6H_{14}, C_6H_{12}, as physisorbed guests [54]. The repulsive force

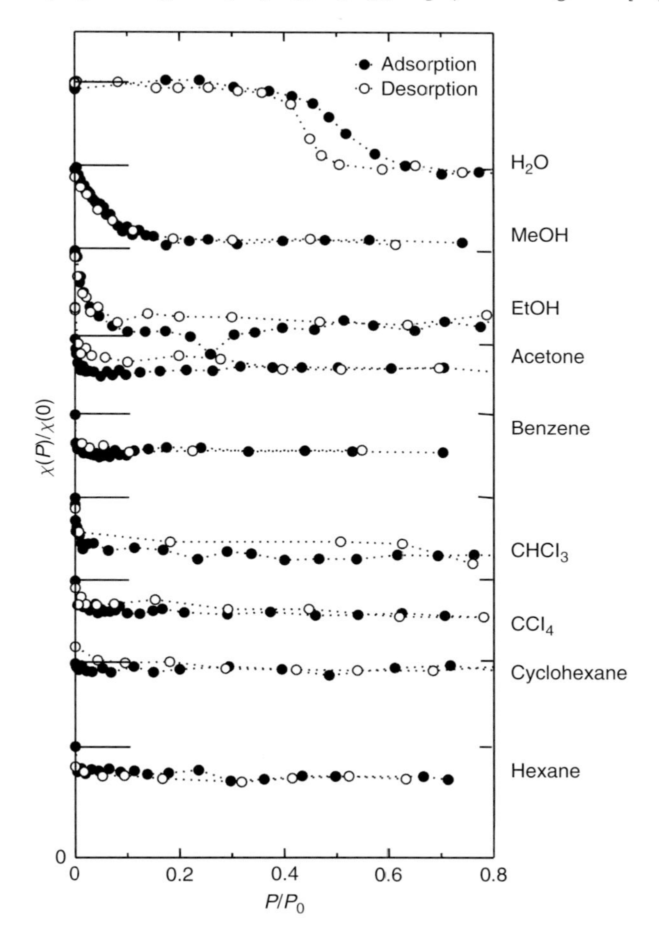

Figure 4.16 The spin susceptibilities of ACFs under the vapor pressure P of various guest molecules; H_2O, CH_3OH, C_2H_5OH, $(CH_3)_2CO$, C_6H_6, $CHCl_3$, CCl_4, C_6H_{12}, C_6H_{14} at room temperature. P_0 is the saturation vapor pressure at room temperature. The vertical axis is shifted for each plot for clarity, where the distance between the bars is 0.3 except for the top and bottom plots [54].

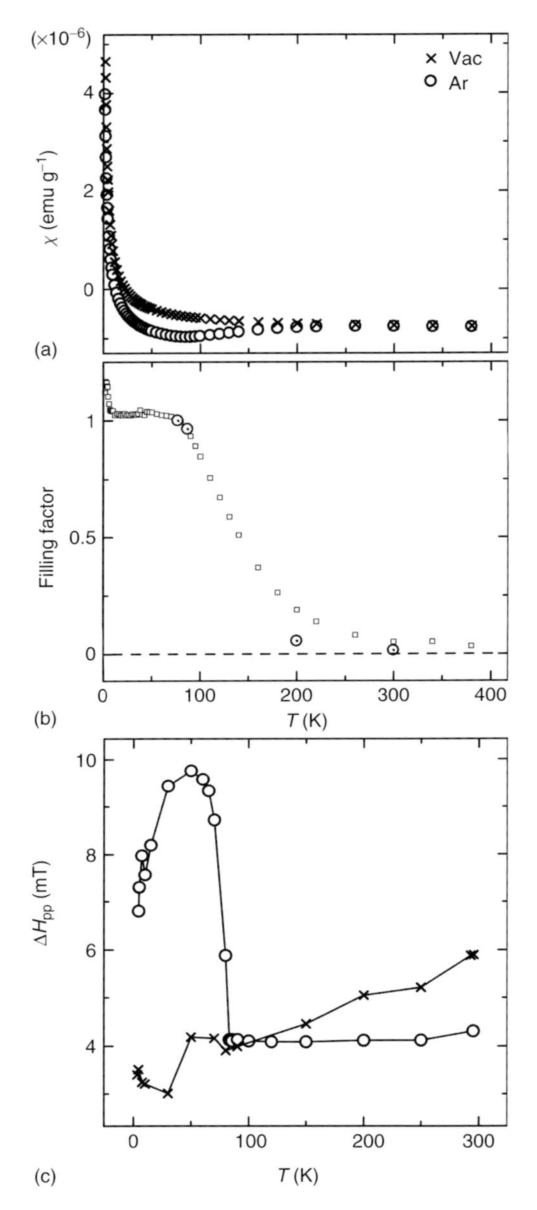

Figure 4.17 (a) Temperature dependence of the susceptibilities χ for the nonadsorbed (vac) and Ar-adsorbed (Ar) ACF samples. (b) Filling factor (circles) of Ar in the nanopores as a function of temperature from the adsorption isotherm. The squares are the absolute value of the difference in the susceptibilities between the nonadsorbed and Ar-adsorbed samples. The value is normalized with respect to the value at 77 K. (c) Temperature dependence of the ESR linewidth ΔH_{pp} for the nonadsorbed (×) and Ar-adsorbed (◯) samples. Solid lines are only guides for the eye [57].

works between a hydrophobic nanographene sheet and a molecule having hydroxyl group, whereas organic molecules with no hydroxyl group attractively interact with the nanographene sheet. Accordingly, the threshold vapor pressure for the magnetic switching phenomenon tends to decrease as the weight of the hydroxyl group in a molecule is lowered; P_t decreases on the order of $H_2O > CH_3OH > C_2H_5OH$. In the molecules having no OH groups –$(CH_3)_2CO$, C_6H_6, $CHCl_3$, CCl_4, C_6H_{14}, C_6H_{12}, the threshold pressure is almost negligible, the magnetic switching phenomenon being not obvious.

Ar adsorption shows an interesting magnetic switching phenomenon, in which the magnetic short-range ordering can be mechanically enhanced [57]. Figure 4.17 shows the change in the magnetism of the edge-state spins on adsorption of Ar molecules. Ar molecules in the nanopores have the gas–liquid transition around 85 K, similar to those in the bulk state. The change in the volume induced by the transition in condensed Ar molecules can be responsible for the magnetic switching. Figure 4.17a indicates the difference in the susceptibility between the ACFs in vacuum and that in Ar atmosphere. The difference, which suggests that the susceptibility is smaller in the Ar-ACFs than in the pristine ACFs, is visible below about 250 K. A part of the difference comes from the diamagnetic core contribution (negative sign) of the Ar molecules adsorbed. However, as shown in Figure 4.17b, the absolute value of the susceptibility difference becomes larger than the amount of the adsorbed Ar obtained from the isotherm at a temperature range below 200–250 K. This verifies the decrease in the spin susceptibility in the Ar-adsorbed ACFs below 200–250 K, suggesting the development of antiferromagnetic short-range ordering of the edge-state spins. This effect is clearly seen in the temperature-dependent behavior of the ESR line width exhibited in Figure 4.17c. The line width discontinuously increases just below the boiling point of Ar molecules, in spite of the absence of the anomalous change in the ESR line width of the pristine ACFs. The line width broadening is a consequence of the development of antiferromagnetic short range ordering induced by the mechanical compression of nanographite domains. Here, the ESR line width is governed by the following term:

$$\sum_k |F_k|^2 \left\langle S_k^z(\tau)\, S_{-k}^z(0)\right\rangle^2 \qquad (4.7)$$

where the static contribution $|F_k|$ is the k-component of the Fourier transform of the second moment for which the dipole–dipole interaction is responsible, and the dynamic contribution $\left\langle S_k^z(\tau)\, S_{-k}^z(0)\right\rangle$ representing the spin fluctuations is the correlation function of S_k^z governed by the isotropic exchange interaction. The decrease in the inter-nanographene-sheet distance on Ar adsorption enhances the dimensionality of the edge-state spin system to three dimensional owing to the strengthening of the inter-nanographene-sheet antiferromagnetic interaction J_2, resulting in the development of the spin correlation $\left\langle S_k^z(\tau)\, S_{-k}^z(0)\right\rangle$. This brings about antiferromangetic short-range ordering as observed in the low-temperature region below the Ar boiling point.

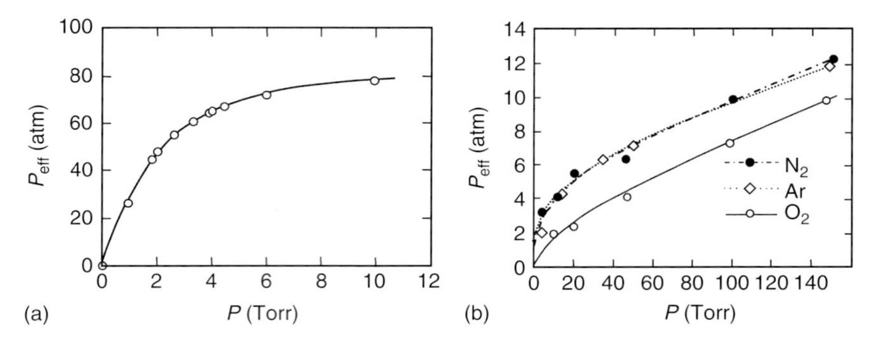

Figure 4.18 The effective pressure in the nanopores as a function of apparent pressure for (a) He, and (b) N_2, O_2, and Ar. The solid, dashed, and dotted dash lines are guides for the eye [58].

4.4.2 Helium Sensor

Next, we show the anomalous behavior in the interaction of the edge-state spins with He atoms on physisorption [58, 59]. Figure 4.18 shows the pressure dependence of the effective pressure in the nanopores of ACFs for N_2, O_2, He, and Ar adsorption at room temperature. Here, the effective pressure is obtained from the adsorption isotherm. The effective pressure of guest gases inside the nanopores is orders of magnitude larger than the pressure of the gaseous state in the environment. Note that around a gaseous pressure of 100 Torr, the effective pressure inside the nanopores reaches 7–10 atm for N_2, Ar, and O_2. This means that the guest gaseous species condense significantly (capillary condensation) in the nanopores owing to the enhanced condensation energy associated with the gas–nanographene interaction because the accommodated gaseous molecules surrounded by the surfaces of nanographene sheets are affected by that interaction in every direction. Interestingly, in the case of He, an exceptionally significant condensation takes place; that is, the pressure inside the nanopores reaches almost 80 atm, which corresponds to about 1/10th of the density of liquid helium, even when 10 Torr of He is introduced. Here, the condensation rate is over 6000.

Gas adsorption phenomena in the nanopores, investigated using gas adsorption isotherms, can be tracked by using edge-state spins as a probe. Here, the ESR saturation technique gives information on the spin-lattice relaxation of the nanographene edge-state spins. The energy that the spin system gains from the microwave is relaxed to the environment through the spin-lattice relaxation process, which is mediated by phonons in general. Figure 4.19 presents the dependence of the ESR intensity (saturation curve) on microwave power taken at room temperature in vacuum and in the atmosphere of 10 Torr He, O_2, Ar, and N_2. In vacuum, the ESR signal tends to be easily saturated in the low microwave power range, suggesting that the spin-lattice relaxation rate $1/T_1$ is considerably small. The relaxation is driven by the acoustic in-plane phonons, whose Debye temperature

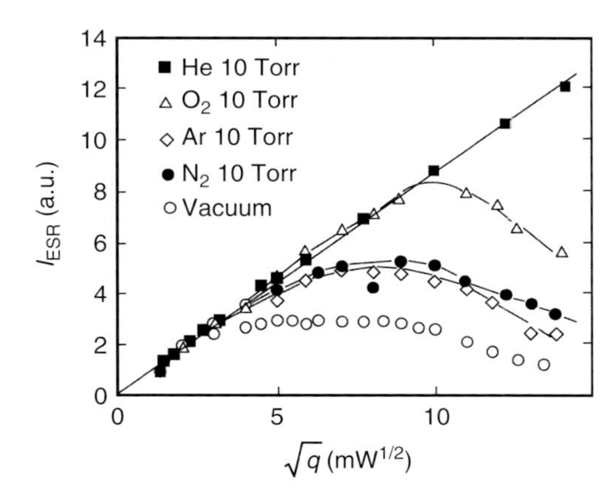

Figure 4.19 The microwave power (q) dependence of the ESR intensity (I_{ESR}) under the presence of gaseous materials at 10 Torr at room temperature. The behavior of the sample with H_2 or Ne is almost identical to that with N_2 or Ar [58].

is considerably high in the range of 2000 K for bulk graphite [60]. When the size of a graphene sheet is reduced, the phonon energy becomes discrete because of the quantum size effect. Indeed, in the case of a nanographene sheet having an in-plane size of 2–3 nm, the discreteness of phonons reaches about 200 K, making the contribution of the phonons less effective owing to the presence of an energy gap of ~200 K in the spin-lattice relaxation process, particularly at low temperatures. This is the reason the ESR signal is easily saturated in the moderate microwave power range due to the elongated T_1. In contrast, the introduction of gaseous species revives the spin-lattice relaxation paths, giving a less saturated feature of the saturation curves in the atmosphere of gaseous materials, as shown in Figure 4.19. The extreme case occurs in He atmosphere, where no saturation appears in the whole microwave power range investigated. In other words, the energy absorbed in the edge-state spins from the microwave can be easily relaxed to the environment in the atmosphere of He. These experimental findings strongly suggest an important contribution of guest gaseous species to the spin-lattice relaxation mechanism.

In the nanopores, guest molecules collide with edge-state spin sites around the peripheries of nanographene sheets. The collisional process, which is governed by van der Waals interaction between the guest molecule and nanographene, works effectively to accelerate the spin-lattice relaxation rate particularly for the case in which the phonon-assisted process is not at work, if the relaxation is incorporated with the spin-reversal event [59]. In this case, relaxation rate $1/T_1$ in the collision-induced relaxation process is described in the following equation:

$$1/T_1 = n\nu\sigma_X \tag{4.8}$$

where n and σ_X are the density of guest molecules and the collisional cross section related to the spin-flip process, respectively, and ν is the mean velocity of the molecules

$$\nu = \sqrt{3RT/M} \tag{4.9}$$

where M is the mass of the guest molecule. A more detailed analytical expression can be obtained on the basis of the van der Waals interaction between the gas molecule and nanographene, and it is given for He gas as follows:

$$1/T_1 = \left[\frac{6.05}{9\,(2\pi)^{1/2}\,\hbar^4}\right] nM^{3/2}\left(\frac{p}{\Delta E}\right)^2\left[\frac{\left(e^2 IJ\right)^4}{R_0^6}\right] w\left(k_{\mathrm{B}}T\right)^{1/2} \tag{4.10}$$

where R_0 is the minimum distance at which a He atom can access an edge-state spin site; I, J, and ΔE are parameters related to the electronic structure of the He atom [59]—I is the matrix elements of z between the 1s and $2p_z$ states, J is that of x between 2s and $2p_x$ states, and $\Delta E = E(^1\mathrm{P}) - E(^1\ \mathrm{S})$; p is given to be λ/Δ, where λ is the spin–orbit interaction of carbon p-state and Δ is the energy difference between the excited and ground states of edge-state spins; w is the probability of the carbon $2p_z$ state being vacant. A comparison with the experimental results indicates that Equation (4.10) well represents the experimental results, suggesting that the edge-state spins work as a probe of guest molecules. In other words, the edge-state spins are useful as a gas sensor probe, in which the van der Waals-interaction-assisted collisional process plays an essential role. An important issue, which should be pointed out, is the exceptionally large He condensation in the ACF nanopores. Although the origin of this remains unsolved, it suggests the presence of ultramicropores, which only the smallest diameter He atom (0.257 nm) can access.

4.5
Summary

Graphene has an unconventional electronic structure described in terms of massless Dirac fermion in a 2D hexagonal bipartite lattice. When a graphene sheet is cut into nanofragments (nanographene), the electronic structure is seriously modified depending on their shapes. In this context, there are two specific shapes: armchair and zigzag; in the zigzag-shaped edges, nonbonding π-electron state (edge state) is created, in spite of the absence of such a state in the armchair-shaped edges. This is understood on the basis of the broken symmetry of pseudospin in massless Dirac fermion. The creation of the edge state in the zigzag edges is due to the degradation of aromaticity from the aspect of chemistry. The edge state localized in the zigzag edge is strongly spin polarized, making nanographene magnetic with localized spins in contrast to the nonmagnetic structure of infinite-size graphene and graphite.

Localized edge-state spins in a zigzag edge are arranged ferromagnetically through a strong ferromagnetic exchange interaction, with its strength being several 10^3 K. In addition to this, the edge-state spins are subjected to weak magnetic anisotropy owing to the weak spin–orbit interaction of carbon. Accordingly, the magnetic structure of the edge-state spin system can be described in terms of the weakly anisotropic Heisenberg spin system in a low-dimensional lattice. In an arbitrarily shaped nanographene sheet, whose periphery consists of a combination of magnetic zigzag edges and nonmagnetic armchair edges, inter-zigzag-edge interaction operates with its strengths being $10^{-1} - 10^{-2}$ of the intra-zigzag-edge interaction. The sign of the inter-zigzag-edge interaction changes between ferromagnetic and antiferromagnetic and its strength varies depending on the mutual geometrical relation between two zigzag edges. Accordingly, under the cooperation of strong intra-zigzag-edge and intermediate-strength inter-zigzag-edge interactions, the magnetic structure of an arbitrarily shaped nanographene sheet becomes ferrimagnetic with a net nonzero magnetic moment. Chemical modifications of edges modify the magnetic structure as well.

An sp^3-defect with a σ-dangling bond is another magnetic ingredient. In contrast to the edge-state spins, which are subjected to exchange interaction, the σ-dangling bond spins are free from the exchange interaction. The coexistence of the two types of spins brings about a variety of magnetic structures in nanographene. This is verified by investigating fluorinated nanographene sheets in which fluorination creates σ-dangling bond spins.

The unconventional magnetic phenomena are observed in nanoporous ACFs consisting of a 3D-disordered network of nanographite domains in relation to the interaction with physisorbed guest molecules. Physisorbed guest molecules condensed in the nanopore space surrounded by nanographite domains, each of which is a loose stack of three to four nanographene sheets, compress nanographite domains, resulting in the decrease in the inter-nanographene-sheet distance. This brings about a magnetic switching phenomenon, in which mechanical compression by the physisorbed molecules changes the magnetic state from high spin to low spin. Guest gas molecule adsorption accelerates the spin-lattice relaxation rate owing to the collisional process. In particular, helium molecules, which are condensed to a huge extent into the nanopores, change the rate to an exceptionally strong extent. This phenomenon can be utilized as a probe for helium sensor.

Acknowledgment

The author acknowledges support from Grant-in-Aid for Scientific Research No. **20001006** from the Ministry of Education, Culture, Sports, Science, and Technology, Japan. He expresses his sincere gratitude to Kazuyuki Takai and Manabu Kiguchi for the fruitful discussion.

References

1. Novoselov, K.S., Geim, A.K., Morozov, S.V., Jiang, D., Zhang, Y., Dubonos, S.V., Grigorieva, I.V., and Firsov, A.A. (2004) *Science*, **306**, 666–669.
2. Castro Neto, A.H., Guinea, F., Peres, N.M.R., Novoselov, K.S., and Geim, A.K. (2009) *Rev. Mod. Phys.*, **81**, 109–162.
3. Novoselov, K.S., Geim, A.K., Morozov, S.V., Jiang, D., Katsnelson, M.I., Grigorieva, I.V., Dubonos, S.V., and Firsov, A.A. (2005) *Nature*, **438**, 197–200.
4. Zhang, Y., Tan, Y.-W., Stormer, H.L., and Kim, P. (2005) *Nature*, **438**, 201–204.

5. Kane, C.L. and Mele, E.J. (2005) *Phys. Rev. Lett.*, **95**, 226801/1–226801/4.
6. Katsnelson, M.I., Novoselov, K.S., and Geim, A.K. (2006) *Nat. Phys.*, **5**, 620–625.
7. Clar, E. (1972) *The Aromatic Sextet*, John Wiley & Sons, Ltd, London.
8. Stein, S.E. and Brown, R.L. (1987) *J. Am. Chem. Soc.*, **109**, 3721–3729.
9. Tanaka, K., Yamashita, S., Yamabe, H., and Yamabe, T. (1987) *Synth. Met.*, **17** (1–3), 143–148.
10. Fujita, M., Wakabayashi, K., Nakada, K., and Kusakabe, K. (1996) *J. Phys. Soc. Jpn.*, **65**, 1920–1923.
11. Kobayashi, Y., Fukui, K., Enoki, T., Kusakabe, K., and Kaburagi, Y. (2005) *Phys. Rev. B*, **71**, 193406/1–193406/4.
12. Niimi, Y., Matsui, T., Kambara, H., Tagami, K., Tsukada, M., and Fukuyama, H. (2005) *Appl. Surf. Sci.*, **241**, 43–48.
13. Kobayashi, Y., Fukui, K., Enoki, T., and Kusakabe, K. (2006) *Phys. Rev. B*, **73**, 125415/1–125415/8.
14. Enoki, T., Kobayashi, Y., and Fukui, K. (2007) *Int. Rev. Phys. Chem.*, **26**, 609–645.
15. Makarova, T.L. and Palacio, F. (2006) *Carbon Based Magnetism: An Overview of the Magnetism of Metal Free Carbon-based Compounds and Materials*, Elsevier, Ansterdam.
16. Ito, K. and Kinoshita, M. (2000) *Molecular Magnetism*, Kodansha, Tokyo.
17. Ingram, D.J.E. and Bennett, J.E. (1954) *J. Am. Chem. Soc.*, **126**, 7416–7417.
18. Bendikov, M., Duong, H.M., Starkey, K., Houk, K.N., Carter, E.A., and Wudl, F. (2004) *J. Am. Chem. Soc.*, **126**, 7416–7417.
19. Fernández-Rossier, J. and Palacios, J.J. (2007) *Phys. Rev. Lett.*, **99**, 177204/1–177204/4.
20. Jiang, D.-E., Sumpter, B.G., and Dai, S. (2007) *J. Chem. Phys.*, **127**, 124703/1–124703/5.
21. Jiang, D.-E. and Dai, S. (2008) *J. Phys. Chem. A*, **112**, 332–335.
22. Hod, O., Barone, V., and Scuseria, G.E. (2008) *Phys. Rev. B*, **77**, 035411/1–035411/6.
23. Wakabayashi, K., Sigrist, M., and Fujita, M. (1998) *J. Phys. Soc. Jpn.*, **67**, 2089–2093.
24. Yazyev, O.V. and Katsnelson, M.I. (2008) *Phys. Rev. Lett.*, **100**, 047209/1–047209/4.
25. Lieb, E. (1989) *Phys. Rev. Lett.*, **62**, 1201–1204.
26. Matsubara, K., Tsuzuku, T., and Sugihara, K. (1991) *Phys. Rev. B*, **44**, 11845–11851.
27. Enoki, T. and Takai, K. (2009) *Solid State Commun.*, **149**, 1144–1150.
28. Enoki, T. and Takai, K. (2006) *Carbon Based Magnetism: An Overview of the Magnetism of Metal Free Carbon-based Compounds and Materials*, (eds T.L. Makarova and F. Palacio), Chapter 17.
29. Kusakabe, K. and Maruyama, M. (2003) *Phys. Rev. B*, **67**, 092406/1–092406/4.
30. Maruyama, M. and Kusakabe, K. (2004) *J. Phys. Soc. Jpn.*, **73**, 656–663.
31. Yazyev, O.V. and Helm, L. (2007) *Phys. Rev. B*, **75**, 125408/1–125408/5.
32. Yazyev, O.V. (2008) *Phys. Rev. Lett.*, **101**, 037203/1–037203/4.
33. Fung, A.W.P., Wang, Z.H., Dresselhaus, M.S., Dresselhaus, G., Pekala, R.W., and Endo, M. (1994) *Phys. Rev. B*, **49**, 17325–17335.
34. Shibayama, Y., Sato, H., Enoki, T., Bi, X.X., Dresselhaus, M.S., and Endo, M. (2000) *J. Phys. Soc. Jpn.*, **69**, 754–767.
35. Shibayama, Y., Sato, H., Enoki, T., and Endo, M. (2000) *Phys. Rev. Lett.*, **84**, 1744–1747.
36. Joly, V.L.J., Takahara, K., Takai, K., Sugihara, K., Enoki, T., Koshino, M., and Tanaka, H. (2010) *Phys. Rev. B*, **81**, 115408/1–115408/6.
37. Shklovskii, B. and Efros, A. (1984) *Electronic Properties of Doped Semiconductors*, Springer-Verlag, Berlin.
38. Mrozowski, S. (1979) *J. Low Temp. Phys.*, **35**, 231–298.
39. Chen, J.-H., Li, L., Cullen, W.G., Williams, E.D., and Fuher, M.S. (2011) *Nat. Phys.*, **7**, 535–538.
40. Korringa, J. (1950) *Physica*, **16**, 601–610.
41. Lehtinen, P.O., Foster, A.S., Ma, Y., Krasheninnikov, A.V., and Nieminen, R.M. (2004) *Phys. Rev. Lett.*, **93**, 187202/1–187202/4.

42. Ohldag, H., Tyliszczak, T., Höhne, R., Spemann, D., Esquinazi, P., Ungureanu, M., and Butz, T. (2007) *Phys. Rev. Lett.*, **98**, 187204/1–187204/4.
43. Barzola-Quiquia, J., Esquinazi, P., Rothermel, M., Spemann, D., Butz, T., and García, N. (2007) *Phys. Rev. B*, **76**, 161403/1–161403/4.
44. Takai, K., Sato, H., Enoki, T., Yoshida, N., Okino, F., Touhara, H., and Endo, M. (2001) *J. Phys. Soc. Jpn.*, **70**, 175–185.
45. Kiguchi, M., Takai, K., Joly, V.L.J., Enoki, T., Sumii, R., and Amemiya, K. (2011) *Phys. Rev. B*, **84**, 045421/1–045421/6.
46. Nakadaira, M., Saito, R., Kimura, T., Dresselhaus, G., and Dresselhaus, M.S. (1977) *J. Mater. Res.*, **12**, 1367–1375.
47. Yagi, M., Saito, R., Kimura, T., Dresselhaus, G., and Dresselhaus, M.S. (1999) *J. Mater. Res.*, **14**, 3799–3804.
48. Elias, D.C., Nair, R.R., Mohiuddin, T.M.G., Morozov, S.V., Blake, P., Halsall, M.P., Ferrari, A.C., Boukhvalov, D.W., Katsnelson, M.I., Geim, A.K., and Novoselov, K.S. (2009) *Science*, **323**, 610–613.
49. Zhou, J., Wang, Q., Sun, Q., Chen, X.S., Kawazoe, Y., and Jena, P. (2009) *Nano Lett.*, **9**, 3867–3870.
50. Hou, Z., Wang, X., Ikeda, T., Huang, S.-F., Terakura, K., Boero, M., Oshima, M., Kakimoto, M., and Miyata, S. (2011) *J. Phys. Chem. C*, **115**, 5392–5403.
51. Suzuki, T. and Kaneko, K. (1988) *Carbon*, **26**, 743–745.
52. Sato, H., Kawatsu, N., Enoki, T., Endo, M., Kobori, R., Maruyama, S., and Kaneko, K. (2003) *Solid State Commun.*, **125** (11/12), 641–645.
53. Harigaya, K. and Enoki, T. (2002) *Chem. Phys. Lett.*, **351**, 128–134.
54. Sato, H., Kawatsu, N., Enoki, T., Endo, M., Kobori, R., Maruyama, S., and Kaneko, K. (2006) *Carbon*, **46**, 110–116.
55. Hao, S., Takai, K., Kang, F., and Enoki, T. (2008) *Carbon*, **46**, 110–116.
56. Takai, K., Kumagai, H., Sato, H., and Enoki, T. (2006) *Phys. Rev. B*, **73**, 035435/1–03543513.
57. Takahara, K., Hao, S., Tanaka, H., Kadono, T., Hara, M., Takai, K., and Enoki, T. (2010) *Phys. Rev. B*, **82** (12), 121417/1–121417/4.
58. Nakayama, A., Suzuki, K., Enoki, T., Ishii, C., Kaneko, K., Endo, M., and Shindo, N. (1995) *Solid State Commun.*, **93** (4), 323–326.
59. Sugihara, K., Nakayama, A., and Enoki, T. (1995) *J. Phys. Soc. Jpn.*, **64** (7), 2614–2620.
60. Komatsu, K. (1955) *J. Phys. Soc. Jpn.*, **10**, 346–356.

5
Physics of Electrical Noise in Graphene

Vidya Kochat, Srijit Goswami, Atindra Nath Pal, and Arindam Ghosh

5.1
Introduction

5.1.1
Single-Layer Graphene

Graphene consists of hexagonal arrays of carbon atoms as shown in Figure 5.1a. Each carbon atom has six electrons: two electrons fill the inner core shell 1s, whereas the other four electrons occupy the energy levels 2s, $2p_x$, $2p_y$, and $2p_z$. In graphene, these orbitals are combined to form sp^2-hybridized orbitals and give rise to the planar structure. The fourth electron is in the $2p_z$ orbital and can hop to the other neighboring $2p_z$ orbitals, leading to the formation of π-band (Figure 5.1a). The tight-binding model of graphene was originally calculated by P. R. Wallace in 1947 [1]. From the crystallographic point of view, the honeycomb lattice is not a Bravais lattice as it consists of two inequivalent neighboring lattice sites. However, one can form a triangular lattice with two atoms (A and B) per unit cell. Hence, the eigenvector can be represented by

$$\psi_k = \begin{pmatrix} \psi_k^{\mathrm{A}} \\ \psi_k^{\mathrm{B}} \end{pmatrix} \tag{5.1}$$

Using the nearest neighbor tight-binding model, the eigenvalue equation for π-electrons can be written as [1–4]

$$\gamma_0 \begin{pmatrix} 0 & f(k) \\ f^*(k) & 0 \end{pmatrix} \begin{pmatrix} \psi_k^{\mathrm{A}} \\ \psi_k^{\mathrm{B}} \end{pmatrix} = \varepsilon_k \begin{pmatrix} \psi_k^{\mathrm{A}} \\ \psi_k^{\mathrm{B}} \end{pmatrix} \tag{5.2}$$

where γ_0 is the nearest neighbor hopping energy and $f(k)$ is given by

$$f(k) = 1 + 4\cos\left(\frac{3k_x a}{2}\right)\cos\left(\frac{\sqrt{3}k_y a}{2}\right) + 4\cos^2\left(\frac{\sqrt{3}k_y a}{2}\right) \tag{5.3}$$

There are two solutions to this eigenvalue equation ($\lambda = \pm$)

$$\varepsilon_k^{\gamma} = \lambda\gamma_0 f(k) \tag{5.4}$$

Graphene: Synthesis, Properties, and Phenomena, First Edition. Edited by C. N. R. Rao and A. K. Sood.

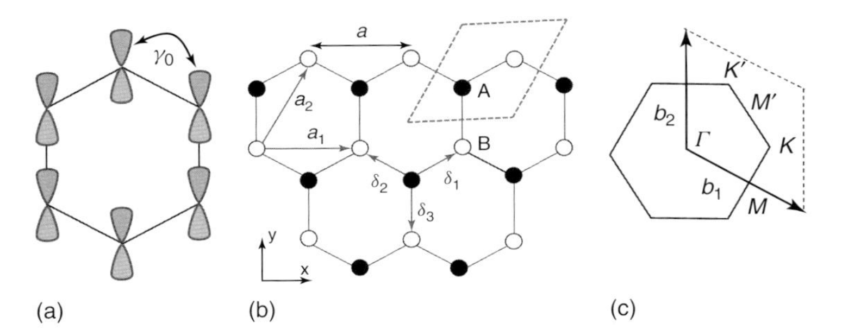

Figure 5.1 (a) $2p_z$ orbital of graphene with γ_0 being the hopping element between the two nearest carbon atoms. (b) Lattice structure of graphene. The unit cell is shown as the dashed rhombus. It has two atom bases (A and B), $\boldsymbol{a_1}$ and $\boldsymbol{a_2}$ are the primitive vectors. Three nearest neighbors are represented by the vectors $\boldsymbol{\delta}_1$, $\boldsymbol{\delta}_2$, and $\boldsymbol{\delta}_3$. (c) The hexagonal first Brillouin zone. The dashed rhombus is the unit cell with primitive vectors $\boldsymbol{b_1}$ and $\boldsymbol{b_2}$ in the reciprocal space. Γ, M, M′, K, and K′ are the high-symmetry points.

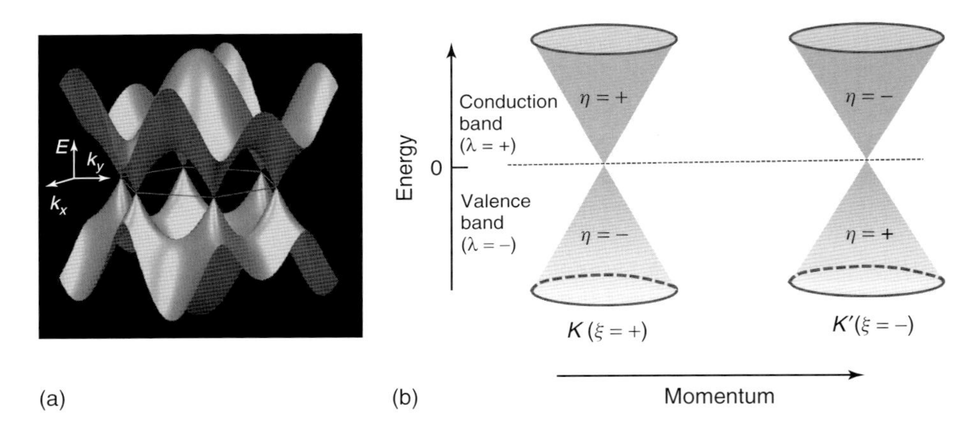

Figure 5.2 (a) Band structure of single-layer graphene using the nearest neighbor tight-binding approximation. The conduction and valence bands touch at the six corners of the BZ. (Source: Image was obtained from the personal homepage of Dr. Thomas Szkopek, McGill University.) (b) Relation between band index λ, valley pseudospin ξ, and chirality η in graphene.

where $\lambda = +$ corresponds to the upper π^*-band (conduction band), whereas $\lambda = -$ signifies the lower π-band (valence band). Figure 5.2a shows the band structure of graphene for $\gamma_0 = 3$ eV. The points where the π-band touches the π^*-band are called the *Dirac points*. They are situated at the points where the energy dispersion is zero, that is, $\varepsilon_k^\lambda = 0$ (Figure 5.2b). It can be shown that there are two inequivalent Dirac points D and D', which are situated at the points K and K′, respectively,

$$\boldsymbol{k}^D = \pm\boldsymbol{K} = \pm\frac{4\pi}{3\sqrt{3}a}(1,0) \tag{5.5}$$

where the points K and K′ are defined as the corners of the first Brillouin zone (BZ) as shown in Figure 5.1c.

5.1.1.1 Effective Tight-Binding Hamiltonian: Sublattice and Valley Symmetry

In order to discuss the low-energy excitations in graphene, we need to concentrate on the dispersion relation near the Dirac points. The wave vector can be decomposed to $\boldsymbol{k} = \pm\boldsymbol{K} + \boldsymbol{q}$, where, $|\boldsymbol{q}| \ll |\boldsymbol{K}| \sim 1/a$. Calculating $f(k)$ near the Dirac points and considering only the first-order terms, the effective low-energy Hamiltonian can be expressed as [2, 3]

$$H_q^{\xi} = \xi\hbar v_{\rm F}\left(q_x\sigma_x + q_y\sigma_y\right) \tag{5.6}$$

where we have defined the Fermi velocity as

$$v_{\rm F} = \frac{3|\gamma_0|a}{2\hbar} \tag{5.7}$$

and the Pauli matrices as

$$\sigma_x = \begin{pmatrix} 0 & 1 \\ 1 & 0 \end{pmatrix} \sigma_y = \begin{pmatrix} 0 & -i \\ i & 0 \end{pmatrix} \tag{5.8}$$

Furthermore, we have introduced the *valley pseudospin* $\xi = \pm$, where $\xi = +$ denotes the K-point; and $\xi = -$, the K′-point. The energy dispersion relation becomes

$$\varepsilon_{\boldsymbol{q},\xi=\pm}^{\lambda} = \lambda\hbar v_{\rm F}|\boldsymbol{q}| \tag{5.9}$$

which is independent of the *valley pseudospin* ξ, leading to the twofold "valley degeneracy."

The Hamiltonian (Equation 5.6) can also be represented by

$$H_q^{\xi} = \hbar v_{\rm F}\tau_z \otimes \boldsymbol{q}\cdot\boldsymbol{\sigma} \tag{5.10}$$

with the wave function as a four-component vector [3]

$$\psi_q = \begin{pmatrix} \psi_{\boldsymbol{q},+}^{\rm A} \\ \psi_{\boldsymbol{q},+}^{\rm B} \\ \psi_{\boldsymbol{q},-}^{\rm B} \\ \psi_{\boldsymbol{q},-}^{\rm A} \end{pmatrix} \tag{5.11}$$

via the 4×4 matrices

$$\tau_z \otimes \boldsymbol{\sigma} = \begin{pmatrix} \boldsymbol{\sigma} & 0 \\ 0 & -\boldsymbol{\sigma} \end{pmatrix} \text{ and } \boldsymbol{\sigma} = \left(\sigma_x, \sigma_y\right) \tag{5.12}$$

5.1.1.2 Valley and Sublattice Pseudospin

In the four-component spinor representation, the first two components indicate the sublattice wavefunction amplitudes at the K-point whereas the last two represent the amplitudes at the K′-point. For graphene, we can define two different types of pseudospin: First, the *sublattice pseudospin* represented by the Pauli matrices

σ_j, corresponding to the A and B sublattices and secondly, the *valley pseudospin* described by another set of Pauli matrices τ_j representing the K and K′ valleys in graphene. This leads to a four-fold degeneracy in the Dirac Hamiltonian arising from the presence of the two sublattices and valleys.

The eigenstates of the Hamiltonian (Equation 5.10) can be written as

$$\psi_{q,\lambda}^{\xi=+} = \begin{pmatrix} e^{\frac{-i\theta_q}{2}} \\ \lambda e^{\frac{i\theta_q}{2}} \\ 0 \\ 0 \end{pmatrix} \quad \text{and} \quad \psi_{q,\lambda}^{\xi=-} = \begin{pmatrix} 0 \\ 0 \\ e^{\frac{i\theta_q}{2}} \\ \lambda e^{\frac{-i\theta_q}{2}} \end{pmatrix} \tag{5.13}$$

where θ_q is defined as

$$\theta_q = \tan^{-1}\left(\frac{q_y}{q_x}\right) \tag{5.14}$$

5.1.1.3 Chirality

One can define the *chirality or helicity operator* (η_q) as the projection of pseudospin along the momentum direction

$$\eta_q = \frac{\boldsymbol{q}\cdot\boldsymbol{\sigma}}{|\boldsymbol{q}|} \tag{5.15}$$

which is the Hermitian and unitary operator with eigenvalue $\eta = \pm 1$. It commutes with the 2D Dirac Hamiltonian and can even be expressed as [3]

$$H_q^{\xi} = \xi\hbar v_{\mathrm{F}}|\boldsymbol{q}|\eta_q \tag{5.16}$$

Combining Equations 5.10 and 5.16, one can write $\lambda = \xi\eta$. Hence, the band index λ is completely determined by the chirality and the valley pseudospin. Near the K-point ($\xi = +$), for electron, $\eta = +$, which means the pseudospin has the same direction as the momentum. Similarly, for hole, $\eta = -$, and hence the pseudospin has the direction opposite to the momentum. The band structure near the K and K′ points are shown in Figure 5.2b.

5.1.1.4 Berry Phase and Absence of Backscattering

If we look at the eigenstates in Equation 5.13, we find that if the wave vector $\boldsymbol{q}$ is rotated adiabatically with time around the origin, effectively θ_q changes by 2π, and hence the wave function gains a phase of π. In the literature, this is commonly known as *Berry phase* of π. It has been shown by Ando [5] that owing to this Berry phase π, any kind of backscattering $\boldsymbol{q} \to -\boldsymbol{q}$ will change the sign of the wave function. Hence, to conserve the chirality of the carriers, the intravalley and intervalley backscattering are suppressed, provided the scatterers are remote from the graphene sheet by more than the lattice constant. As a consequence, graphene can have very high carrier mobility [6–8].

5.1.2
Bilayer Graphene

Similar to single-layer graphene (SLG), bilayer graphene (BLG) is also an interesting 2D material. The lattice structure of a BLG is shown in Figure 5.3a, where the bottom and top layers are represented by the dashed line and solid line, respectively. The indices 1 and 2 in Figure 5.3a,b label the sublattices of the bottom and top layers, respectively. The unit cell of BLG has a basis of four atoms (A_1, B_1, A_2, and B_2). In the so-called Bernal stacking of layers in BLG (Figure 5.3b), the A_2 sublattice of the top layer is exactly on top of the sublattice B_1 of the bottom layer. In the tight-binding approximation, the band structure of BLG has been calculated considering only the in-plane nearest neighbor hopping energy ($A_1 - B_1$ or $A_2 - B_2$), γ_0, and the interlayer hopping energy ($A_2 - B_1$), γ_1. Under these assumptions, the K-point Hamiltonian can be written as [1, 2, 9]

$$H = \begin{pmatrix} 0 & \gamma\pi & 0 & 0 \\ \gamma\pi^* & 0 & \gamma_1 & 0 \\ 0 & \gamma_1 & 0 & \gamma\pi \\ 0 & 0 & \gamma\pi^* & 0 \end{pmatrix} \tag{5.17}$$

where $\pi = \left(k_x - ik_y\right)$ and $\gamma = 3\gamma_0 a/2$. At zero magnetic field, BLG consists of four bands whose energy eigenvalues are given by

$$\varepsilon_{sj}^{k} = s\left[\left(\sqrt{\left(\frac{\gamma_1}{2}\right)^2 + (\gamma k)^2}\right) + (-1)^j \frac{\gamma}{2}\right] \tag{5.18}$$

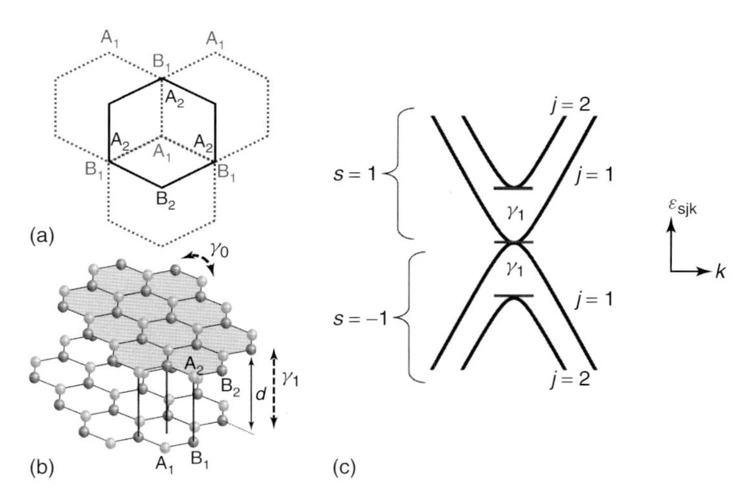

Figure 5.3 (a) Top view and (b) side view of the lattice structure of a BLG. A_1 and B_1 are the sublattices of the bottom layer, and A_2 and B_2 are the sublattices of the top layer. γ_1 is the nearest neighbor interlayer hopping term and d is the separation between the two layers. (c) Energy dispersion of a BLG. γ_1 denotes the energy separation between the two subbands ($j = 1$ and $j = 2$). $s = +1$ and $s = -1$ for the conduction and valence bands, respectively.

where s is the band index, $s = +1$ for conduction band and $s = -1$ for valence band. j is the subband index with values 1 and 2, as shown in Figure 5.3c. Even though the energy dispersion of bilayer graphene is parabolic like conventional semiconductors, the presence of A/B sublattice symmetry leads to chirality which gives rise to the conserved pseudospin index. The chiral four-component wave function for bilayer graphene which consists of both the layer and the sublattice degrees of freedom, can be found in a number of earlier publications [2, 9–11].

5.1.2.1 **Biased Bilayer Graphene**

Recently, interest in BLG has grown because of the ability to open and control the energy gap between the conduction and valence bands by external means [2, 11–13]. If the equivalence of the A_1 and B_2 sites is destroyed, a gap Δ_g opens up between the conduction and valence bands. BLG has been modeled theoretically by assuming two independent parallel plates. If the potential energy between the two layers is $\Delta V = eE_{ext}d$, E_{ext} and d being the external electric field and distance between the two layers, respectively, the Hamiltonian can be written as [10–13]

$$H = \begin{pmatrix} -\frac{\Delta V}{2} & \gamma\pi & 0 & 0 \\ \gamma\pi^* & -\frac{\Delta V}{2} & \gamma_1 & 0 \\ 0 & \gamma_1 & \frac{\Delta V}{2} & \gamma\pi \\ 0 & 0 & \gamma\pi^* & \frac{\Delta V}{2} \end{pmatrix} \quad (5.19)$$

The eigenvalues of this Hamiltonian are given by

$$\varepsilon_{sj}^{k} = s\sqrt{\left(\frac{(\Delta V)^2}{4} + \frac{\gamma_1^2}{2} + \gamma^2k^2\right) + (-1)^j\sqrt{\frac{\gamma_1^2}{4} + \gamma^2k^2\left(\gamma_1^2 + (\Delta V)^2\right)}} \quad (5.20)$$

The band structure for biased BLG is shown in Figure 5.4b for $\Delta V = 0.4$ V, having a "Mexican hat" structure. However, there is an opening of a bandgap (Δ_g) between the conductance and valence bands. Interestingly, Δ_g is not minimum at the charge neutrality point (CNP) as shown in Figure 5.4b. The value of the bandgap is given by

$$\Delta_g = \sqrt{\frac{\gamma_1^2(\Delta V)^2}{\gamma_1^2 + (\Delta V)^2}} \qquad \Delta V < \gamma_1 \quad (5.21)$$

Equation 5.21 shows that the maximum value of the bandgap can be $\approx\gamma_1$. Experimentally, this has been achieved through chemical doping [15, 16], as well as application of an external electric field [17] that sets a potential difference between the two layers of BLG. In Figure 5.4a, we describe a realistic situation where we consider the graphene bilayer, with interlayer separation d, to be located at a distance d_{ox} from a parallel metallic gate. The application of an external gate voltage (V_g) induces a total excess electronic density $n = n_1 + n_2$ on the bilayer system where n_1 (n_2) is the excess density on the layer closest to (farthest from) the gate. Practically, $n_1 \neq n_2$ because of asymmetric screening in the two layers. The excess charge n_2 will give rise to the change in potential energy $\Delta U = e^2n_2^2/\varepsilon_0\varepsilon_r$ between

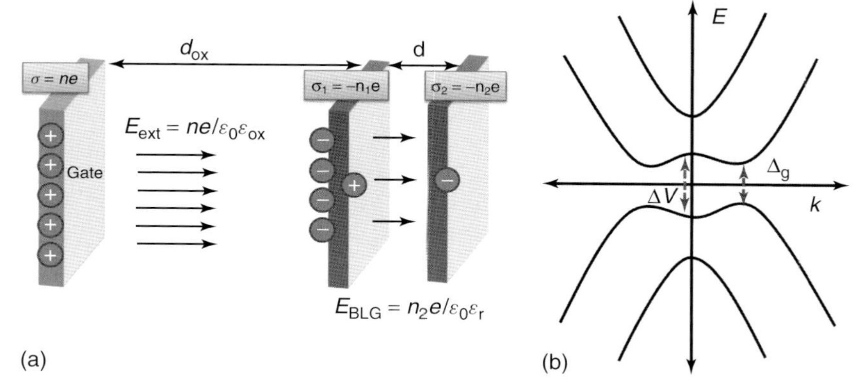

Figure 5.4 (a) Schematic of the graphene bilayer, with interlayer separation d, located at distance d_{ox} from a parallel metallic gate. The gate voltage V_g induces a total excess electronic density $n = n_1 + n_2$ on the bilayer system, where n_1 (n_2) is the excess density on the layer closest to (farthest from) the gate. E_{ext} is the external electric field due to the gate voltage, whereas, $E_{BLG} = n_2e/\varepsilon_0\varepsilon_r$ is the internal electric field generated because of the excess density n_2 in the upper layer [14]. (b) Band structure of a BLG in presence of perpendicular electric field. ΔV is the potential drop between the two layers of graphene and Δ_g is the bandgap [10].

the layers and the expression for bandgap (Δ_g) is given as [10, 14],

$$\Delta_g = \Delta_0 + \frac{e^2 n_2^2}{\varepsilon_0 \varepsilon_r} \tag{5.22}$$

where ε_r is the bilayer dielectric constant and Δ_0 is the bare asymmetry parameter due to an additional transverse electric field producing finite asymmetry $\Delta(0)$ at zero density. Recently, there was the observation of a bandgap of 250 meV in a dual-gated bilayer device [18], making BLG a promising candidate for nanoelectronics.

5.1.3 Multilayer Graphene

Similar to SLG and BLG, multilayer graphene (MLG) is also expected to exhibit novel phenomena at low charge densities owing to enhanced electronic interactions and competing symmetries [9]. The energy dispersion relation is parabolic for MLG, $E \sim |k|^2$. With increasing layer numbers, the screening properties increase, and hence it is difficult to induce carriers by applying a gate voltage in thicker layers. Beyond two layers, the electronic properties are highly dependent on the stacking properties of the layers. In the most common version of bulk graphite, the stacking order is ABA ... (Bernal stacking) or ABC ... (rhombohedral stacking). Theoretically, it has already been predicted that there can be a spontaneous breaking of layer equivalence even in the absence of an electric field, and hence there can be an opening of the bandgap [19, 20] (Figure 5.5b,c). Recently, it has been observed that TLG exhibits different transport properties with different stacking orders: there was an unexpected spontaneous gap opening in charge-neutral rhombohedral TLG

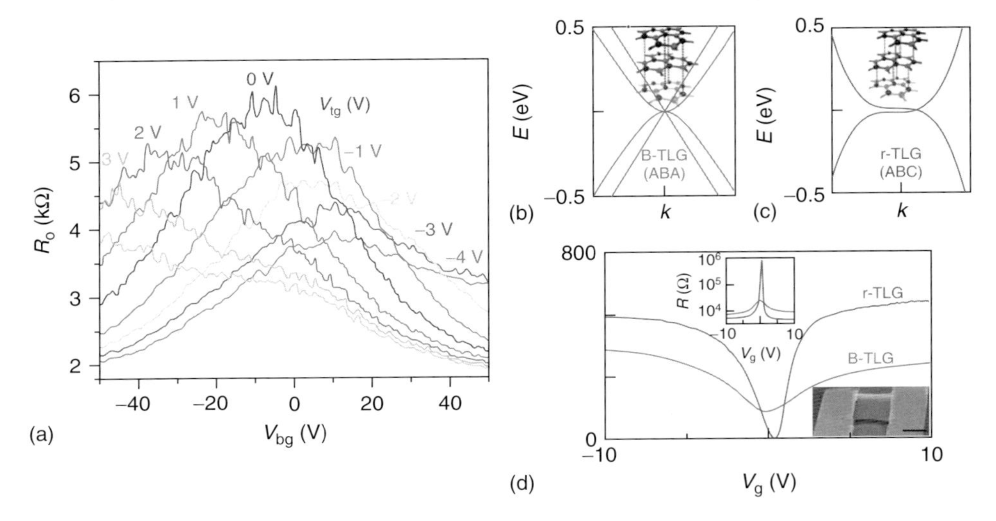

Figure 5.5 (a) Square resistance of a tri-layer device versus back gate voltage V_{bg} for different values of the top gate voltage V_{tg} at 50 mK. The value of maximum resistance systematically decreases when the perpendicular electric field applied is increased, revealing the semimetallic behavior. (Source: Adapted from Ref. [21].) (b) Band structures (main panel) and schematics (inset) of B-TLG (ABA) and r-TLG (ABC), respectively. (c) $G(V_g)$ for two different suspended TLG devices at $T =$ 1.5 K. (d) Upper inset: $R(V_g)$ in log-linear scales for the same devices. Lower inset: SEM image of a suspended graphene device. Scale bar, 2 μm. (Source: Adapted from Ref. [22].)

(Figure 5.5d), whereas the Bernal-stacked TLG exhibits semimetallic behavior [21–24] (Figure 5.5a). The quantum Hall effect has been studied in TLG [25, 26], and there was the absence of $\nu = \pm 2$ plateaus, as predicted by the theory. All these theoretical predictions and experimental investigations have instigated the research on MLG from both the application and fundamental point of view.

5.1.4 Disorder and Scattering Mechanism in Graphene

One of the most important aspects of todays' graphene research is to understand the effect of disorder in electronic transport of graphene based field effect transistors. Disorder can come from various sources:

1) There are always intrinsic lattice imperfections, point defects, dislocations or rough edges, which are the major source of short-range scattering.
2) Formation of ripples [27] on the surface of a supported graphene can create local curvature [28] and hence fluctuations in local chemical potential and effective gauge fields [2].
3) One of the most important sources is the trapped charge in the graphene-substrate interface [29–32] that can give rise to the long-range Coulomb scattering and limit the mobility significantly. Other than these,

the adsorbed atoms on the graphene surface, such as the water molecules, or resist residues (during device processing) can also limit the performance of graphene transistors. We will now discuss about the various scattering mechanism in graphene devices.

5.1.4.1 Coulomb Impurity Scattering

It has already been discussed that because of the suppressed intervalley and intravalley backscattering, carriers in graphene can show very high mobility [6–8]. However, experimentally it has been shown that the conductivity of graphene increases linearly with gate voltage, except when very close to the CNP [33]. This linear dependence of experimental conductivity on density has been explained by the charge impurity scattering due to the long-range Coulomb disorder from the substrate [32]. The effect of these charge impurities was investigated by intentionally adding potassium ions to graphene in ultrahigh vacuum [29]. It was shown that the CNP shifts more toward the higher gate voltages by increasing the doping level and that the conductivity becomes more linear with density (Figure 5.6A). In graphene, the typical ballistic length observed in experiments is of the order of 100 nm [34]. As the typical sample size used in most of the experiments is more than a few micrometers (5–10 μm), one needs to consider the diffusive transport theory to explain the observed results. The linear density dependence of conductivity was explained by the scattering from the long-range Coulomb interaction from the charge impurities in the SiO_2 substrate [30, 31]. The transport properties of SLG and BLG have been calculated using the semiclassical Boltzmann transport theory [30, 32]. The Boltzmann conductivity for two-dimensional electron gas (2DEG) can be written as

$$\sigma = \frac{2e^2}{h} k_F v_F \tau \tag{5.23}$$

where v_F is the Fermi velocity and τ is the scattering time. The scattering time has been calculated for SLG and BLG considering both the Coulomb scattering and short-range scattering. Table 5.1 compares the dependence of conductivity (σ) with density (n) for SLG and BLG with conventional 2DEG in the experimentally relevant limit $q_{TF}/2k_F > 1$ [31], where q_{TF} is the Thomas–Fermi inverse screening length. At lower density, the scattering is dominated by the Coulomb scattering, giving rise to the linear density dependence on conductivity, whereas the effect of short-range scattering becomes important at higher density, where the conductivity becomes sublinear with density [36]. In BLG, the density dependence of the

Table 5.1 Dependence of σ for various scattering mechanisms in an experimentally relevant situation for $q_{TF}/2\,k_F > 1$ [31].

	2DEG	SLG	BLG/MLG
Bare Coulomb scattering	$\sigma \sim n^2$	$\sigma \sim n$	$\sigma \sim n^2$
Screened Coulomb	$\sigma \sim n$	$\sigma \sim n$	$\sigma \sim n$
Short-range scattering	$\sigma \sim n$	$\sigma \sim n^0$	$\sigma \sim n$

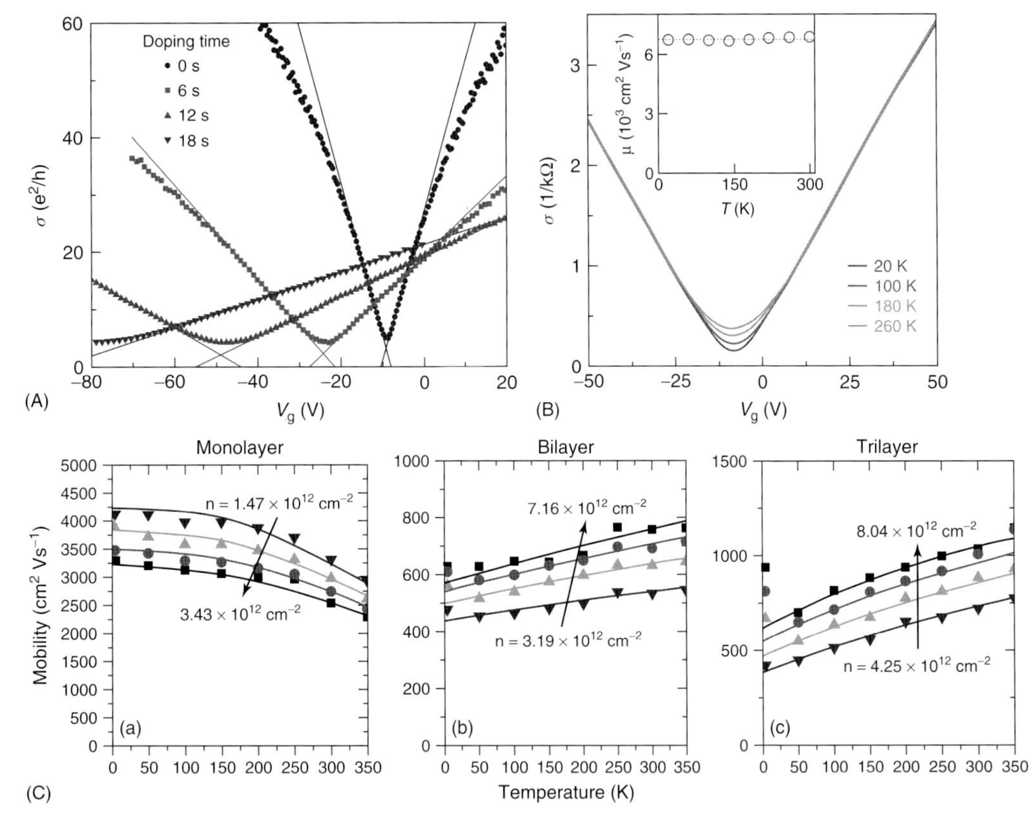

Figure 5.6 (A) Conductivity (σ) versus gate voltage (V_g) for the pristine sample and three different doping concentrations of potassium. (Source: Adapted from Ref. [29].) (B) The measured conductivity of BLG as a function of gate voltage V_g (or carrier density). The measured conductivity increases linearly with the density. Inset shows the temperature dependence of mobility. (Source: Adapted from Ref. [7].) (C) Hall mobility as a function of temperature for different hole densities in monolayer graphene, BLG, and trilayer graphene respectively (a–c). The symbols are the measured data and the lines are fits. (Source: Adapted from Ref. [35].)

conductivity shows linear behavior with density [7] (Figure 5.6B), which indicates that the BLG carrier transport is controlled by two distinct and independent physical scattering mechanisms, that is, the screened Coulomb disorder due to random charged impurities in the environment and the short-range disorder [32].

5.1.4.2 Phonon Scattering

Phonons remain as an inevitable source of scattering even when all the extrinsic sources of disorder are excluded. They represent an intrinsic source of scattering mechanism, influencing the transport in graphene near room temperature [32]. For graphene, the phonon scattering can either be intravalley acoustic/optical phonon scattering where the carriers are scattered within a valley or intervalley scattering which scatters the carriers between the K and K′ valleys. However, recent experiments reveal that the remote polar optical phonons in the substrate (i.e., SiO_2) have an adverse effect on carrier mobility [29, 35, 37]. The detailed temperature dependence of Hall mobility at different densities for SLG, BLG, and TLG were studied in Ref. [35], as shown in Figure 5.6c, where the results were explained by taking both charge impurity and phonon scattering into account.

5.1.4.3 Electron–Hole Puddles at Low Density

In spite of having zero density of states (DOS) at the Dirac point, experimentally, it is almost impossible to approach $n \rightarrow 0$ because of the formation of these electron–hole puddles arising mainly due to the lack of screening [38]. This is primarily due to the static charged impurity centers which are randomly distributed in the environment, in particular at the substrate-graphene interface. The inhomogeneity can also be due to the formation of ripples on graphene as a result of intrinsic structural wrinkles or due to extrinsic factors like the roughness of the substrate. Formation of these puddles has been directly observed by various imaging experiments in both SLG [39, 40] and BLG [41]. At low density, the inhomogeneous puddles control transport phenomena in graphene as well as in 2D semiconductors [40].

5.2 Flicker Noise or "1/*f*" Noise in Electrical Conductivity of Graphene

The conventional time-averaged transport measurements provide us with information regarding the scattering by the static disorder and the various interactions present in the system. But the practical applications of these systems require a thorough understanding of the dynamics of the disorder present in the system in addition to the various scattering mechanisms. The disorder dynamics is manifested in the form of conductance fluctuations in a current-carrying conductor when measured in the time domain. The Fourier transform of the autocorrelation function of these fluctuations is called the power spectral density (PSD), which has the form $S(f) \propto 1/f^{v}$, where f is the frequency and $v \approx 1$. A typical time series showing these conductance fluctuations and the corresponding power spectrum is shown in Figure 5.7b. These conductance fluctuations are attributed to the atomic scale motion of scatterers, which gives rise to flicker

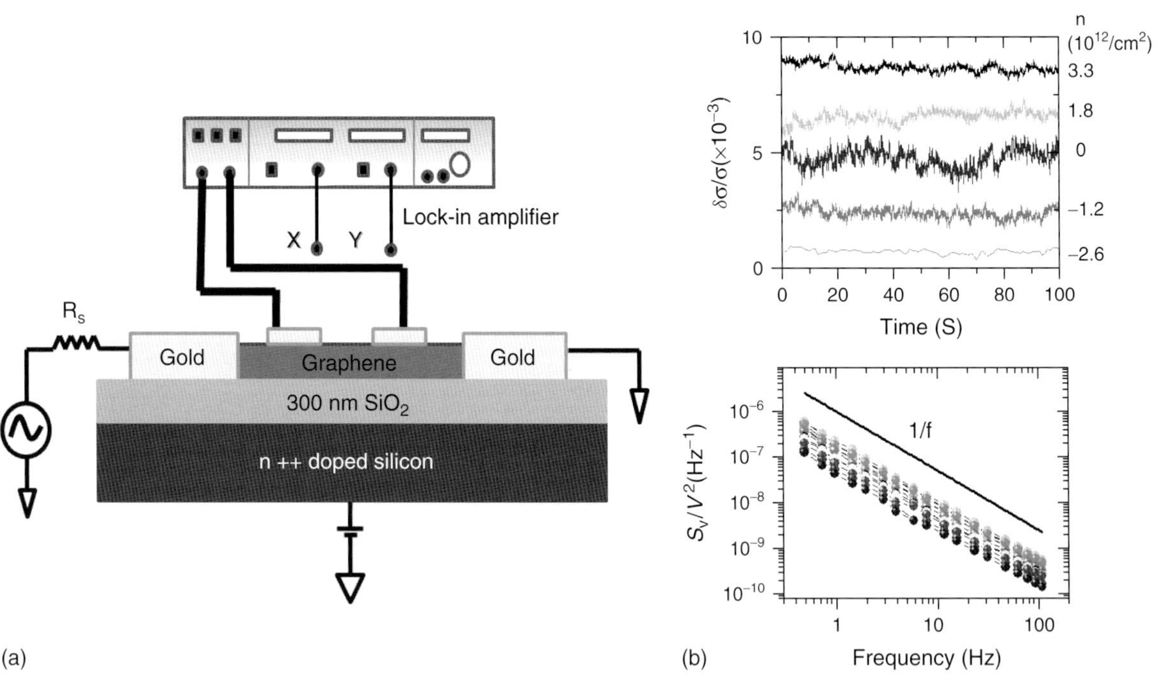

Figure 5.7 (a) Schematic of four-probe noise measurement technique using lock-in amplifier. (b) Typical time series (top) of conductance fluctuations in graphene at different carrier densities and the corresponding power spectrum (bottom).

noise or the $1/f$ noise. This forms a more sensitive transport-based probe than the standard resistance/conductance measurements to study the coupling of mobile disorder to electronic transport. Hence, for any new emerging material, it is an absolute necessity to investigate the $1/f$ noise in order to understand the intrinsic performance limits, which in turn determine the practical utility of the material.

The $1/f$ noise is measured by the low-frequency AC four-probe method using a lock-in amplifier in a high-vacuum environment, as shown in Figure 5.7a. The sample is biased by an AC carrier signal having a carrier frequency much larger than the bandwidth of the power spectrum measured. The output signal from the sample is recorded by the lock-in amplifier at two phases, where the in-phase (X) component is the sum of the signal and the background, whereas the out-of phase (Y) component involves only the background. The time-dependent output of the lock-in amplifier is then digitized, followed by a multistage decimation of the signal that eliminates effects of higher harmonics of the power line and other unwanted frequencies and aliasing effects. The data after digitization and decimation is then processed to estimate the PSD. The background noise measured simultaneously from the Y component is subtracted from the total noise to obtain the power spectrum of noise from the sample [42].

The origin of the $1/f$ noise in graphene was attributed to the trapping and detrapping of charge carriers at the graphene–oxide interface, which is similar to the mechanism giving rise to $1/f$ noise in metal-oxide-semiconductor field effect transistor (MOSFETs) [43] and carbon nanotube (CNT) field effect transistors (FETs) [44]. The experiments of Liu *et al.* [45] on double-gated SLG devices showed that the dominant contribution to noise did not come from the channel region under the top gate but was a result of the charge transfer from the oxide surface underneath. The $1/f$ noise also reflects the screening ability of the charge carriers against the external potential fluctuations, which in turn is related to the low-energy band structure of graphene. Thus, $1/f$ noise forms a robust probe to distinguish between SLG having linear energy dispersion and MLG having parabolic bands. Lin *et al.* [46] showed that there is a strong suppression of $1/f$ noise in BLG nanoribbons and that the noise in SLG and BLG devices showed the opposite behavior. The noise amplitude for bilayer nanoribbon was minimal at the CNP and increased with the increase in the carrier density (n), whereas the noise amplitude was maximum at the Dirac point and decreased with increasing n for SLG. For SLG, at high n, the Coulomb potential of the charged traps in the oxide was effectively screened and hence noise decreased at large carrier densities. For BLG, the increase in noise at large n was attributed to a bandgap opening at high electric fields, which weakens the ability to screen external potential fluctuations. Noise measurements on few-layer graphene (FLG) and MLG also exhibited a similar gate voltage dependence as for BLG, as shown in Figure 5.8 [47]. This was again explained to be a consequence of the reduction in the DOS with increasing n, due to field-induced gap opening at the Fermi energy. In addition, FLG and MLG exhibited extremely low values of the Hooge parameter defined as $\gamma_H = nfS_\sigma a_G/\sigma^2$ (which is a parameter that characterizes the noise levels in different systems) of the order 10^{-6} to 10^{-5} compared to SLG where $\gamma_H \sim 10^{-3}$ to 10^{-2}. The FLG and MLG,

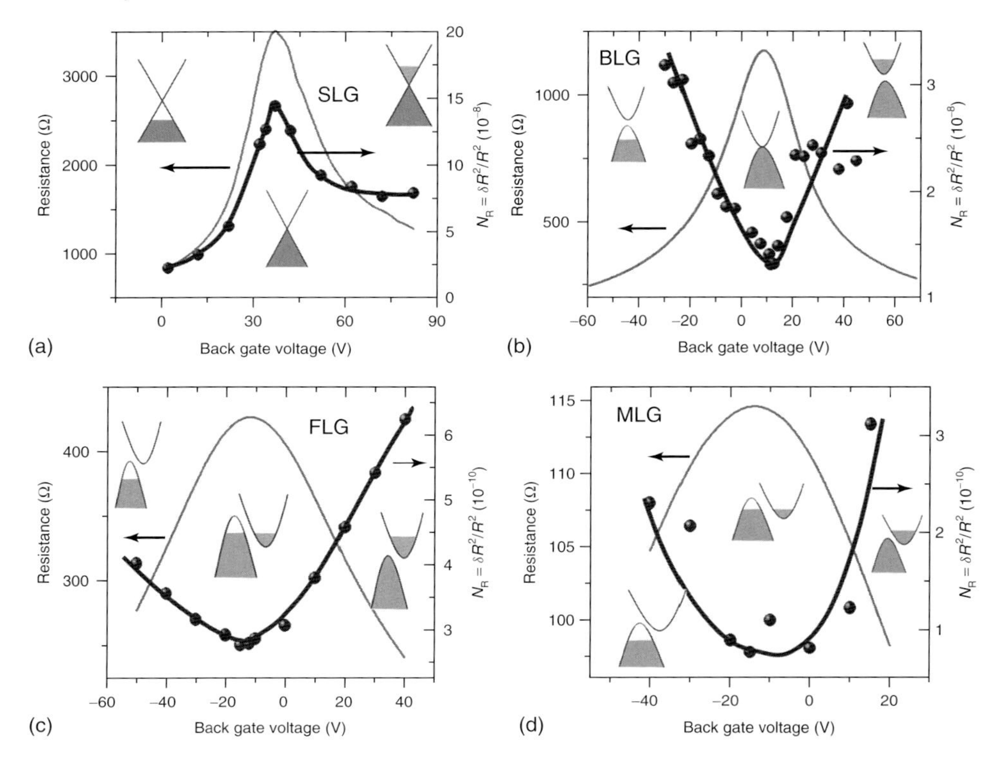

Figure 5.8 The resistance and the normalized noise PSD (N_R) as functions of back gate voltages are shown for: (a) SLG, (b) BLG, (c) FLG, and (d) MLG devices at $T = 100$ K. The insets show the bandstructure at $\varepsilon_F = 0$ and in the metallic regimes. (Source: Adapted from Ref. [47].)

owing to their near three-dimensional character, have parabolic bands that provide finite DOS at low energies, which screens the trap potentials very efficiently. Thus, FLG and MLG can be very promising for low-noise nanoelectronic applications.

Anomalies in the dependence of noise on carrier density in SLG were reported by Xu *et al.* [48] and Heller *et al.* [49]. They observed a dip in noise near the Dirac point contrary to the expected increase, which is shown in Figure 5.9.

Xu *et al.* [48] attributed this M-shaped dependence of noise on n to the spatial charge inhomogenity, which breaks the system into conducting puddles of electrons and holes at low carrier densities. As one moves away from the Dirac point, one of the puddles (electrons or holes) reduces and becomes homogeneous. The other type of puddle still contributes to noise, and this leads to an increase in noise before reaching a noise maximum, after which the noise reduces because the trap potentials are effectively screened by the majority carriers. Heller *et al.* [49] performed noise studies on liquid-gated SLG and BLG devices and proposed an augmented charge-noise model that identifies two separate mechanisms that give rise to noise in graphene FETs. Near the Dirac point, the noise is dominated by

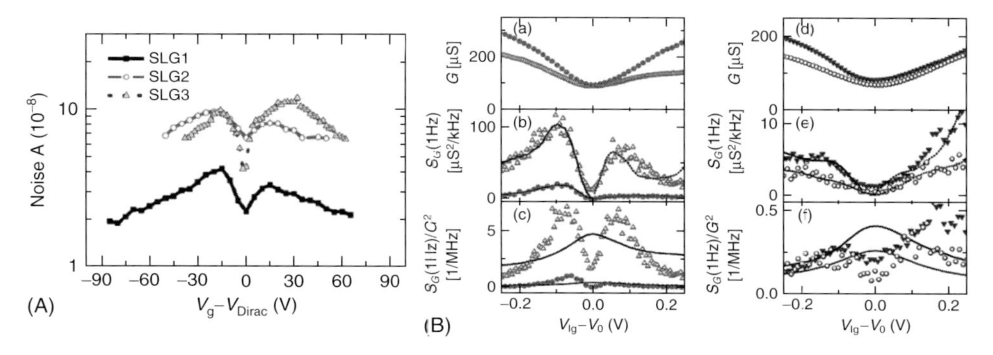

Figure 5.9 (A) M-shaped dependence of noise on n obtained by Xu *et al.* [48] for three different SLG. (B) Conductance and noise as a function of n for two SLG (a–c) and two BLG (d–f) devices reported by Heller *et al.* [49] showing a dip in noise at CNP.

the random charge fluctuations from the traps in the substrate in close proximity to graphene. At high carrier densities, a gate-independent noise source comes into play, which is due to the mobility fluctuations caused by the scatterers evenly distributed in the graphene channel modeled as a series resistor term.

5.2.1
Microscopic Origin of 1/f Noise in Graphene

A comprehensive theoretical noise model based on the fluctuating charge distribution (FCD) model was proposed by Pal *et al.* [50], which related the band structure to the microscopic origin of $1/f$ noise in graphene. At low temperatures, carrier density dependence of noise magnitude was found to have opposite behaviors for linear and parabolic band structures as shown in Figure 5.10. The FCD model identifies two processes giving rise to the $1/f$ noise that arises from the fluctuating charges in the local environment of graphene consisting of the substrate, top dielectric, adsorbates, acrylic residues, and so on.

1) ***Charge exchange noise*** due to exchange of charge carriers between graphene and the environment through trapping–detrapping processes. This *charge exchange noise* can be defined as the change $\delta\sigma$ in conductivity when charge δn is exchanged between graphene and its surroundings. As $\sigma = \sigma(n, \mu)$, where n is the carrier density and μ is the mobility

$$\delta\sigma(t) = \frac{\partial\sigma}{\partial n}\delta n(t) + \frac{\partial\sigma}{\partial\mu}\delta\mu(t) \tag{5.24}$$

According to the correlated carrier density–mobility fluctuation model [43], $\delta\mu \sim \mu_{\text{avg}}^2 \sum \delta n$, where Σ is related to the scattering rate $\left(\tau_{\text{C}}^{-1}\right)$ entirely due to the Coulomb potential of the trapped charge δn located inside the substrate. The n dependence of Σ can be calculated using the semiclassical Boltzmann transport equation for graphene, $\sigma = \frac{2e^2}{h} k_{\text{F}} v_{\text{F}} \tau$, where k_{F} is the Fermi wave vector and v_{F} is the Fermi velocity of the charge carriers in graphene, and is

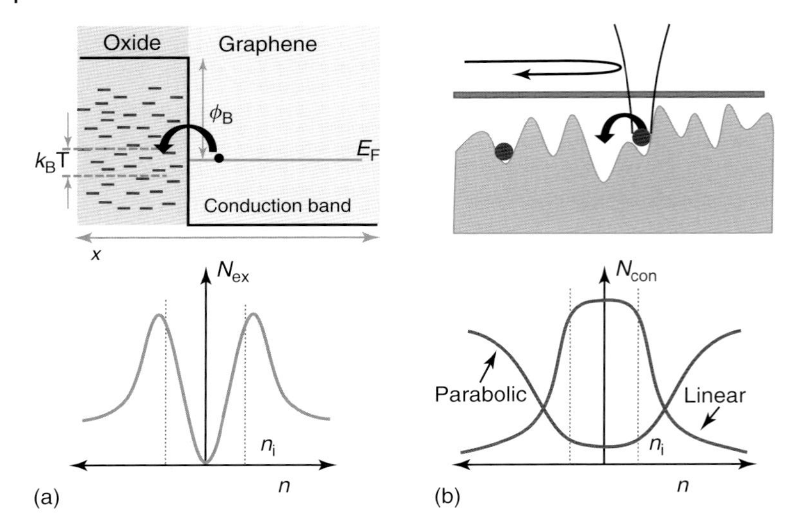

Figure 5.10 (a) Charge exchange noise due to trapping–detrapping of carriers between graphene and substrate and the corresponding noise behavior with n. (b) Charge configuration noise that distinguishes between linear and parabolic band structures.

given by

$$\Sigma = \frac{\pi}{e}\frac{n}{k_F v_F}\frac{\hbar/\tau_C}{\delta n} \tag{5.25}$$

For SLG, $1/\tau_{C|SLG} \sim 1/\sqrt{n}$ [30], while for BLG, $1/\tau_{C|SLG} \sim n^0$ [31]. Incorporating this in Equation 5.25 makes $\sum$ independent of density to the leading order for both linear and parabolic bands. The PSD of conductivity noise can be calculated from Equation 5.24 where the dominant contribution was found to be due to mobility fluctuations alone and is given by

$$N_{ex} = A(T)\left(\frac{d\sigma}{dn}\right)^2 \tag{5.26}$$

2) ***Charge configuration noise*** arises from a slow rearrangement of charges within the local environment of graphene. This leads to the alteration of the disorder landscape due to the Coulomb potential from the trapped charges, resulting in random fluctuations in the scattering cross-section (Λ_C). Within a local interference model [51], $\delta\Lambda_C \sim \Lambda_C$, which makes $\delta\sigma_C \sim \sigma l \Lambda_C$. Here l is the mean scattering length and $\Lambda_C \propto |v_q|^2$, where v_q is the screened Coulomb potential due to the trapped charges, making the configuration noise sensitive to the graphene band structure. Thus,

$$\frac{\delta\sigma_C^2}{\sigma^2} \sim l^2\Lambda_C^2 \sim l^2|v_q|^4 \tag{5.27}$$

For linear bands (SLG), Thomas–Fermi wave vector, $q_{TF} \sim k_F \sim \sqrt{n}$, which makes $v_q \sim 1/q_{TF} \sim 1/\sqrt{n}$. For parabolic bands (BLG/MLG),

$q_{TF} \sim n^0$, making v_q independent of carrier density. Hence, at large $|n|$, the configuration noise can be approximated to be $N_{con} \sim |n|^{\gamma}$, where $\gamma = -2(1-\varepsilon)$ for SLG and 2ε for BLG. Here, ε represents the n dependence of l, which is $\varepsilon \sim 0 - 0.5$ for the screened Coulomb [30, 31] and the interface phonon scattering [52] which also depends on T.

The total normalized noise PSD can be written as

$$\frac{S_{\sigma}(f)}{\sigma^2} = A(T)\left(\frac{\mathrm{d}\sigma}{\mathrm{d}n}\right)^2 + B(T)|n|^{\gamma} \qquad (5.28)$$

Figure 5.10 shows the two processes and their dependence on n.

Fitting Equation 5.28 yielded excellent agreement with the experimental noise data in all the devices, as can be seen from Figure 5.11, even though the relative contributions from N_{con} and N_{ex} varied significantly in these devices.

In the case of substrated SLG devices, N_{con} exceeds N_{ex} at all $|n|$ and, in particular, near low $|n|$ where the charge exchange noise is minimal and graphene has enhanced sensitivity to changes in the disorder landscape. The fit in the case of exfoliated and chemical vapor deposition (CVD)-grown SLG on SiO_2 yields $\gamma \sim -1.0$, which reflects screening by the linear bands of SLG. The nonmonotonic behavior in the case of suspended SLG can easily be understood from a much smaller contribution of N_{con} because of the absence of the substrate and hence is dominated by the charge exchange noise between graphene and residues/adsorbates on the surface. The fact that $\gamma \sim -2.0$ for suspended SLG near room temperature indicates that l becomes nearly independent of n. In the case of BLG, S_{σ}/σ^2 increases monotonically with $|n|$ but becomes nonmonotonic at higher T, similar to the results reported by Heller *et al.* [49]. Equation 5.28 describes the BLG noise data rather well over the entire n range with a positive γ for all T, which reflects the parabolic energy bands in BLG. For BLG, $\gamma \approx 1.0$ at low T, but decreases to $\sim$0.1 at higher T indicating a nearly constant l, possibly due to competing scattering from the longitudinal acoustic phonons, causing the measured noise to follow the $(\mathrm{d}\sigma/\mathrm{d}n)^2$ alone. Thus, the FCD model can explain the microscopic mechanism of $1/f$ noise in graphene and the results from the various other groups [46, 48, 49] can also be understood well in this framework. Figure 5.12 summarizes the noise levels in these graphene FETS in terms of the Hooge parameter, γ_H, which shows that at all n, the substrated SLG devices, exfoliated and CVD grown, are most noisy, whereas suspended SLG and thicker graphene films are a 100 times quieter.

5.2.2
Effect of Bandgap on Low-Frequency Noise in Bilayer Graphene

Low-frequency resistance noise measurements on double-gated BLG devices shown schematically in Figure 5.13a allow independent tunability of bandgap Δ_g with both n and the transverse field (ε), which serves to separate the influence of band structure and carrier density on screening [53]. A finite ε is established when the voltages between the top (V_{tg}) and bottom (V_{bg}) gates are different. This opens a bandgap

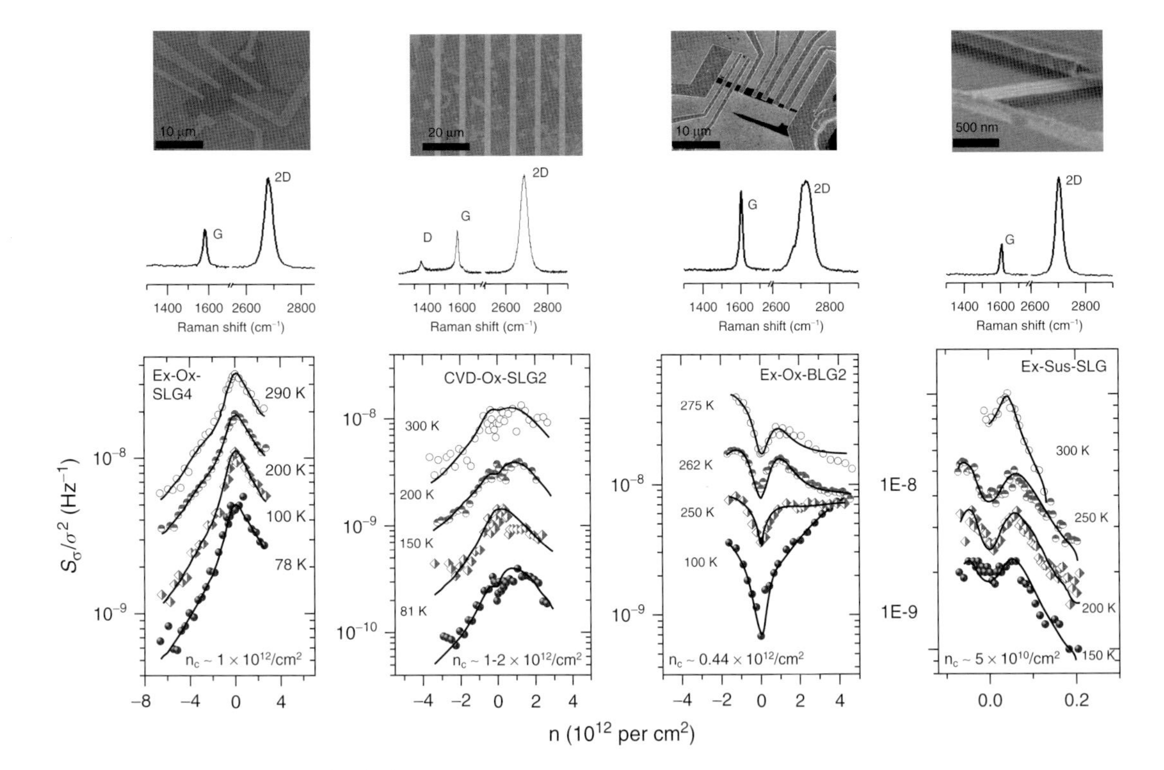

Figure 5.11 Noise magnitude $\left(S_\sigma/\sigma^2\right)$ versus n at different temperatures fitted with the FCD model for (a) substrated SLG, (b) substrated CVD-grown SLG, (c) substrated BLG, and (d) suspended SLG. The top and the middle panels show the SEM images and the Raman spectrum, respectively, for the corresponding cases. (Source: Adapted from Ref. [50].)

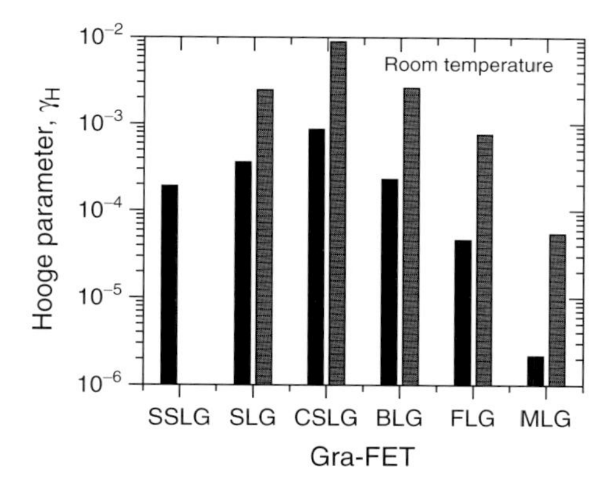

Figure 5.12 Comparison of the Hooge parameter (γ_H) for various GraFET devices at room temperature for two different *n*: suspended single-layer graphene (SSLG); substrated single-layer (SLG), bilayer (BLG), few-layer (FLG), multilayer (MLG), and CVD-grown single-layer graphene (CSLG). (Source: Adapted from Ref. [50].)

that becomes increasingly prominent at higher V_{bg}, resulting in larger resistance at the CNP as illustrated in Figure 5.13b. The normalized variance, N_R, and the average resistances as a function of V_{tg}, plotted for different V_{bg} in Figure 5.13c, show that N_R has a minimum at a particular V_{tg} and increases monotonically at high n, similar to what was observed in the case of BLG nanoribbons [46]. The V_{tg} for the noise minimum does not coincide with CNP, indicating that the noise minimum has been shifted away from CNP in the presence of finite ε and that the minimum in noise corresponds to a zero bandgap situation. Figure 5.13d shows the top gate voltages at CNP $\left(V_{tg}^{R\max}\right)$ and noise minimum point $\left(V_{tg}^{N\min}\right)$, as the $V_{bg}$is varied. From these, the dependence of charge density (n_Δ) on electric field (ε_0) can be calculated as

$$n_\Delta = \left(\frac{\in_0 \in_{cp}}{ed_{cp}}\right)\left(V_{tg}^{N\min} - V_{tg}^{R\max}\right) \tag{5.29}$$

$$\varepsilon_0 = \frac{\left(V_{bg} - V_{tg}^{N\min}\right)}{(d_{ox} + d_{cp})} \tag{5.30}$$

where d_{ox} and d_{cp} are the thicknesses of the oxide layer (BG) and the cross-linked polymer layer (TG), respectively, and $\in_{cp}$ represents the dielectric constant of TG. The vertical dashed line at $\varepsilon_0 = 0$ implies that the noise minimum occurs when the system is charge-neutral, that is, the system is electron-doped by the same amount as the intrinsic hole doping, which corresponds to zero bandgap. Hence, $1/f$ noise can be a very powerful technique to identify the zero bandgap situations in double-gated BLG devices.

In conclusion, the $1/f$ noise can be a direct electrical-transport-based probe to the band structure of graphene and hence can be used to readily identify

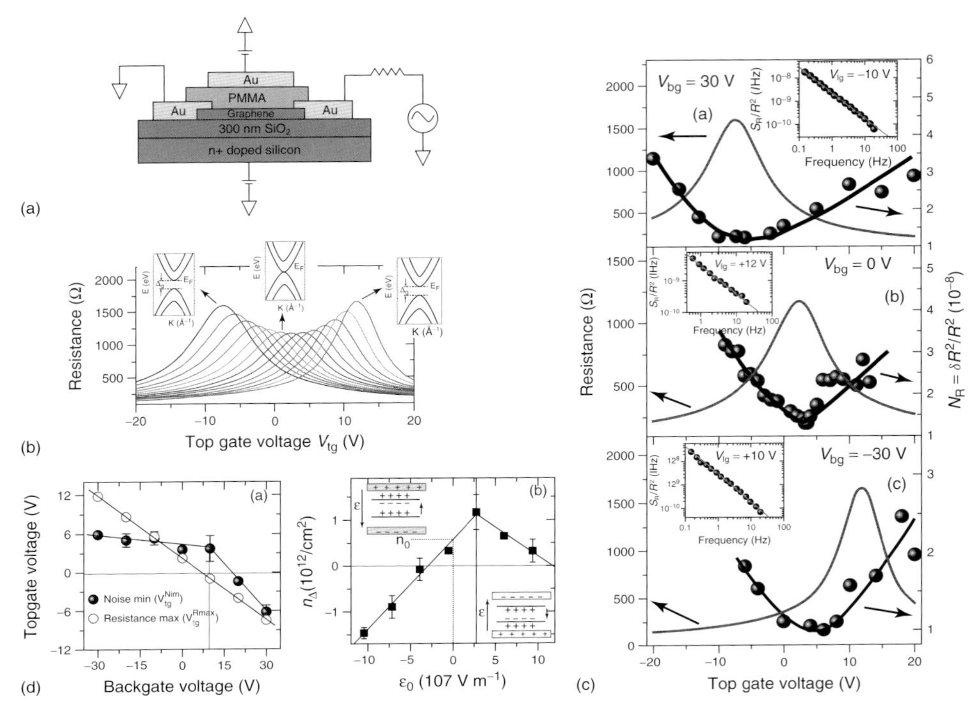

Figure 5.13 (a) Schematic of double-gated BLG device structure. (b) Resistance versus V_{tg} for various V_{bg}. Insets show schematics of corresponding band structures. (c) The resistance and the normalized noise power spectral density (N_R) as functions of V_{tg} for various V_{bg}. (d) Top gate voltages at charge neutrality point ($V_{tg}^{R_{max}}$) and noise minimum point ($V_{tg}^{N_{min}}$), plotted as functions of V_{bg}. Dependence of charge density (n_Δ) on the external electric field (ε_0) at the noise minimum point. (Source: Adapted from Ref. [53].)

SLG from multilayered ones that can be a very powerful technique particularly suited for nanostructured graphene. This study shows the need for substrates with lower trap DOSs that would impede significant charge rearrangement. Also, the extremely low noise levels of FLG and MLG films make them suitable as interconnects in spite of their limitedness in active electronics because of reduced gating ability.

5.2.3
Shot Noise in Graphene

Shot noise is the temporal fluctuation in electric current caused by discreteness of charge carriers and from the randomness in their arrival times at the drain electrode. Shot noise has been used to study the quantum of charge in superconductors [54] and in fractional quantum Hall regime [55, 56], in order to understand the statistics and interactions in mesoscopic devices. It is quantified by the Fano factor, which is the current noise, S_I normalized to the Poissonian noise and is given by

$$\mathcal{F} \equiv \frac{\Sigma_n [T_n (1 - T_n)]}{\Sigma_n T_n} = \frac{S_I}{2q < I >} \tag{5.31}$$

where T_n is the transmission probability of the charge carriers. When $T_n \ll 1$, as in the case of a tunnel junction, the transmission of electrons is totally random and is governed by Poisson statistics and in this limit $\mathcal{F} \rightarrow 1$. In the case of a disordered phase-coherent conductor, sub-Poissonian shot noise has been predicted with $\mathcal{F} = \frac{1}{3}$. This kind of sub-Poissonian shot noise with $\mathcal{F} = \frac{1}{3}$ was predicted in the case of short and wide graphene strips at the Dirac point [57]. This was attributed to the jittering motion of confined Dirac fermions called *Zitterbewegung*, which is the result of interference between positive and negative energy states and produces the same shot noise as classical diffusion. In the case of shot noise across a ballistic p–n junction in graphene, $\mathcal{F} = 1 - 1/\sqrt{2}$ was predicted [58]. The experimental measurements of shot noise in graphene by Danneau *et al.* showed gate-dependent $\mathcal{F}$ in ballistic graphene with large and small W/L. For large W/L, $\mathcal{F}$ reaches a universal value of 1/3 at the Dirac point, whereas for $W/L < 3$, $\mathcal{F}$ is lowered at the Dirac point [59]. This dependence of $\mathcal{F}$ on chemical potential has been explained using the theory of evanescent mode transport in graphene. At large n, the number of propagating evanescent waves increases, thereby reducing the backscattering and the Fano factor. Contradictory results were obtained by DiCarlo *et al.*, which showed that $\mathcal{F} \sim 0.35$–0.37 in both electron- and hole-doped regions with no dependence on n. The values obtained for SLG p–n junctions were also similar [60]. This is in accordance with some recent numerical simulations that studied the increasing disorder strength as the cause of vanishing $\mathcal{F}$ dependence. But in the case of MLG, $\mathcal{F}$ changes from 0.33 at CNP to 0.25 at $n \sim 6 \times 10^{12}$ cm^{-2}. These results are shown in Figure 5.14. In spite of these theoretical and experimental efforts, a proper understanding of shot noise in SLG and MLG is still lacking.

5.3 Noise in Quantum Transport in Graphene at Low Temperature

5.3.1 Quantum Transport in Mesoscopic Graphene

At low temperatures, in samples having phase coherence lengths much larger than the mean free path, the quantum interference effects introduce an additional correction to conductivity leading to weak localization (WL). The interference between the electron waves scattered by disorder and their time-reversed paths forming closed trajectories lead to their constructive interference and result in enhanced backscattering. In the presence of a magnetic field (B), the interfering paths gain an additional phase factor that destroys the WL, leading to negative magnetoresistance (MR). In graphene, the processes responsible for electron dephasing are very

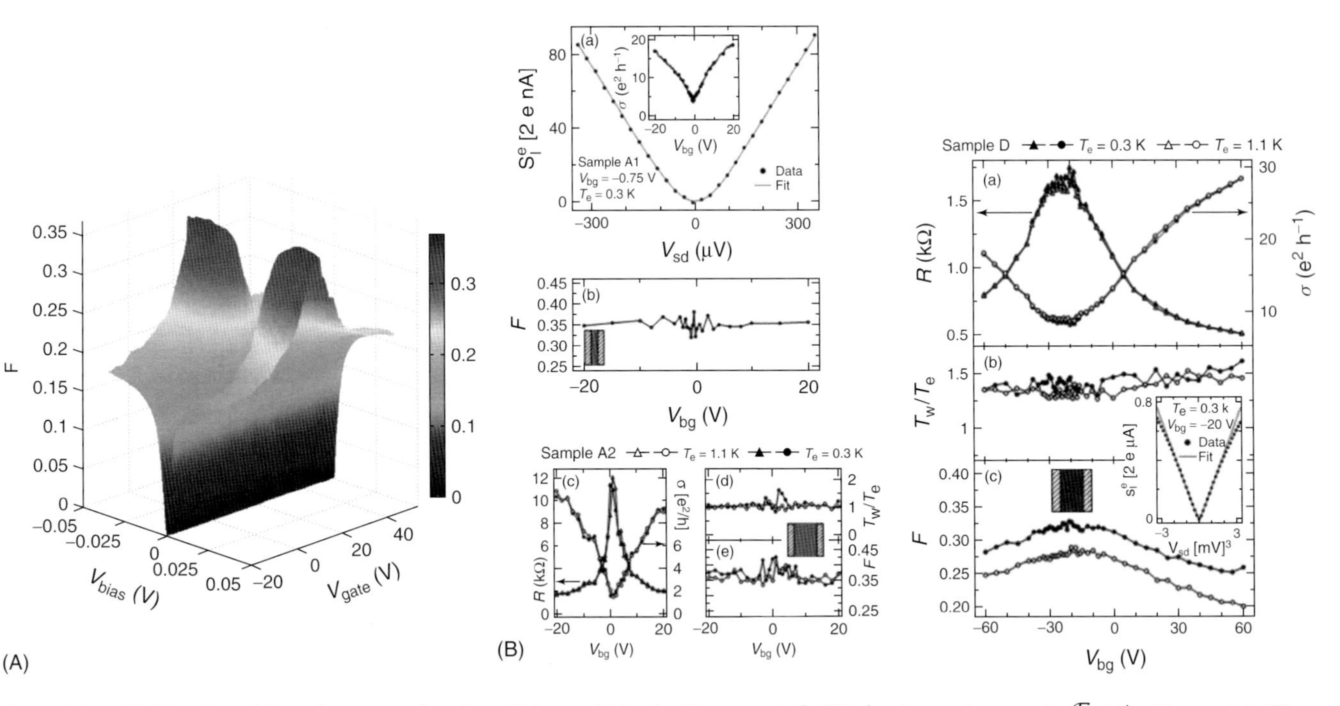

Figure 5.14 (A) Mapping of Fano factor as a function of V_{bias} and V_{gate} by Danneau *et al.* [61] showing an increase in $\mathcal{F}$ at the Dirac point. (B) Shot noise results obtained by DiCarlo *et al.* [62] showing no change in $\mathcal{F}$ as a function of n in the case of the single-layer graphene (left panel), whereas in the case of the multilayer graphene (right panel), $\mathcal{F}$ is maximum at CNP and decreases with increase in n.

different from that in metals and other 2DEGs because of the chiral nature of the charge carriers. This gives them an additional quantum number of pseudospin, owing to contributions to their wave functions from both the sublattices. The carriers in graphene completing closed trajectories acquire a Berry phase of π that leads to destructive interference and weak antilocalization (WAL) instead of WL. The quantum interference effects in graphene are sensitive not only to the inelastic phase-breaking processes but also to the various elastic-scattering mechanisms. Intravalley scattering, which breaks the chirality, destroys the interference within the K and K' valleys and suppresses the WL. This is caused by ripples, dislocations, atomically sharp defects, and also the trigonal warping of the Fermi surface. But the intervalley scattering, which occurs because of the defects of the size of the order of the lattice constant a, allows interference of carriers from different valleys by negating the chirality breaking and warping effects and restores the WL in graphene.

The WL correction to the Boltzmann conductivity of disordered carriers in a 2D honeycomb lattice, first calculated by H. Suzuura and T. Ando, can be either positive or negative depending on the interaction range of the scattering potentials [61]. The universality class of graphene, which determines the symmetries present in the system, was also studied under the influence of long-range and short-range disorder. The honeycomb lattice has two sublattices corresponding to atoms at A and B sites, and this plays the role of the pseudospin. In addition, there is another set of pseudospin specified by the valley index of the two inequivalent valleys K and K' in the first BZ of graphene.

The low energy dispersion at each of these points, K and K', is described by the Dirac Hamiltonian for massless chiral Dirac fermions, $\mathcal{H} = \boldsymbol{\sigma} \cdot \boldsymbol{p}$. In an ideal graphene, with no short-range disorder, the pseudospin corresponding to the valley symmetry is conserved. In this case, where the two Dirac cones are decoupled because of the absence of intervalley scattering, the system represents that of a half-odd integer spin where the pseudospin rotational symmetry for the sublattice can be broken by intravalley scattering. Therefore, in the absence of a short-range disorder, the system belongs to the symplectic class having pseudospin-dependent interaction with time-reversal invariance. The correction to the Boltzmann conductivity from Cooperons (which originate from the maximally crossed diagrams in the diagrammatic perturbation theory) in the absence of intervalley scattering is proportional to $e^{i(\varphi_{k_\alpha} - \varphi_{k_\beta})}$, which is almost equal to -1 (due to the Berry phase π). This gives rise to a positive correction to conductivity, $\Delta\sigma_{\mathrm{LR}} = (2e^2/\pi^2\hbar)\ln(l_\varphi/l)$, and results in antilocalization (WAL). In contrast, short-range disorder originating from lattice defects can scatter the charge carriers with large enough change in momentum, so as to impart intervalley scattering. This breaks the valley symmetry and thus the system can be viewed as an integer-spin system containing two separate sets of pseudospins with $S = 1/2$, which correspond to valley and sublattice spaces. As the only symmetry that the system now has is the time-reversal invariance, it can be classified in the orthogonal symmetry class. In the presence of intervalley scattering, the Cooperon contribution to conductivity $\sim j_\alpha j_\beta e^{i(\varphi_{k_\alpha} - \varphi_{k_\beta})}$, where j represents the valley index, ± 1 for K and K' respectively. This negates the effect of Berry phase and restores the WL correction to graphene conductivity $\Delta\sigma_{SR} = -\left(e^2/2\pi^2\hbar\right)\ln\left(l_\varphi/l\right)$, which is consistent with the orthogonal symmetry class. At low carrier densities (n), where the scattering is predominantly due to the Coulomb scatterers, one expects graphene to be in the symplectic class characterized by WAL, whereas at high n, the Coulomb disorder is heavily screened and the short-range scattering dominates. In this regime, intervalley scattering causes WL resulting in a crossover to orthogonal symmetry class. This crossover from symplectic to orthogonal class was experimentally observed by Tikhonenko *et al.* [62] in their SLG samples. The observation of WAL was realized by increasing the temperature, which decreases the phase decoherence time, and by lowering the carrier density, which increases the intervalley scattering time. The evolution of magnetoconductance (MC) with decreasing n at three different T is shown in Figure 5.15. It can be seen that at large n, WL behavior is observed even at high T.

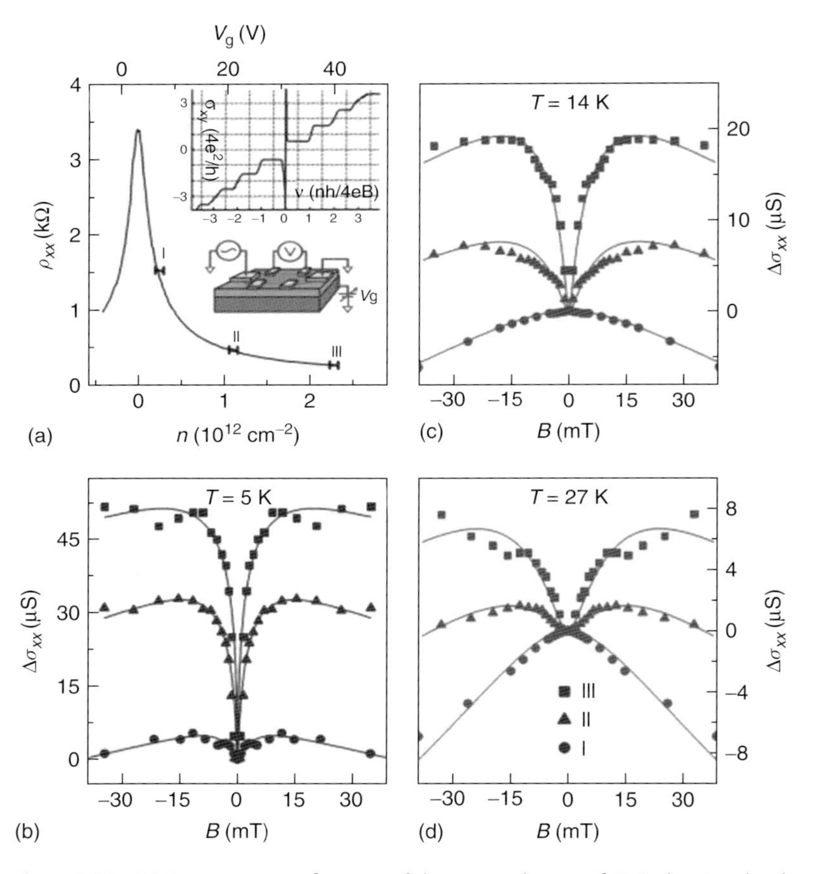

Figure 5.15 (a) Resistivity as a function of the carrier density of SLG showing the three regions where the MC is studied. (b–d) Evolution of the MC with decreasing n, T = 5, 14, and 27 K. (Source: Adapted from Ref. [62].)

Morpurgo and Guinea [63] studied another crossover in graphene from symplectic class as a function of magnetic field. When $B > B_{\varphi}$, the phase-breaking field, the time-reversal (TR) symmetry of the system is broken and the Cooperon contribution to conductivity vanishes. Thus there is a transition from symplectic class to unitary class characterized by broken TR symmetry as a function of magnetic field. They also studied whether the presence of other defects could cause spontaneous breaking of TR symmetry even in the absence of an external magnetic field, thus leading to a suppression of the WL characteristic of unitary class. These defects include the topological lattice defects and ripples that can generate vector potentials giving rise to pseudomagnetic fields and hence destroy TR invariance. In fact, in the first experiment, which studied the WL in graphene, by Morozov *et al.* [64], a strong suppression of WL in graphene near the Dirac point as well as in the metallic regime was observed, as shown in Figure 5.16. This behavior was exhibited in >80% of the samples they studied and observed in a wide temperature range from 0.3 to 77 K. AFM measurements confirmed that graphene prepared by micromechanical cleavage is not absolutely flat and contains a lot of wrinkles/ripples that cause local elastic distortions, resulting in random gauge fields. This gauge field breaks the TR symmetry and suppresses the WL in graphene. In MLG, WL behavior was restored because the ripples are expected to reduce in the thicker and more rigid multilayers.

Taking into consideration all the intervalley/intravalley scattering rates $(\tau_i^{-1}/\tau_*^{-1})$, the expression for magnetoconductivity was obtained by McCann *et al.*

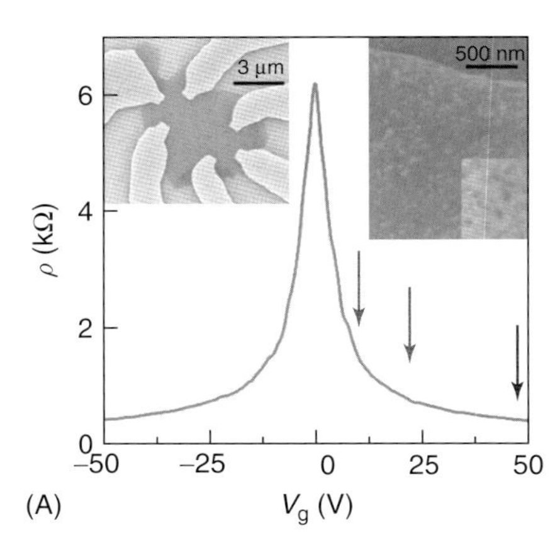

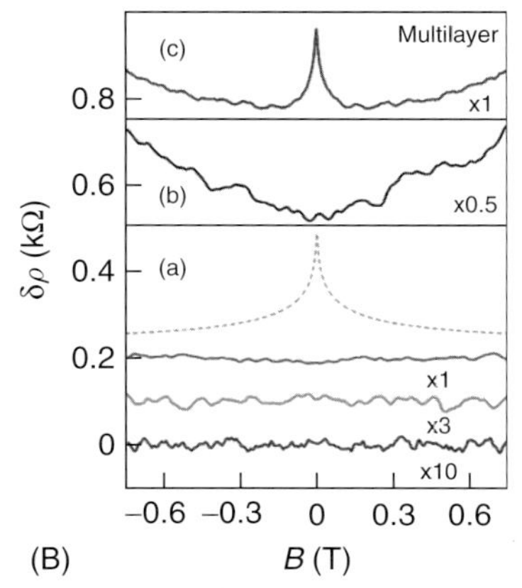

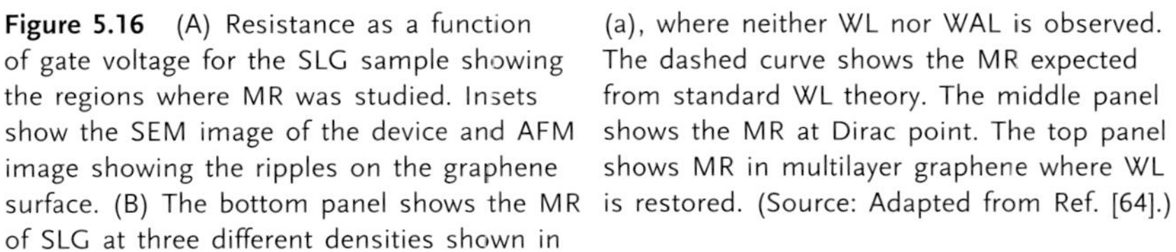
Figure 5.16 (A) Resistance as a function of gate voltage for the SLG sample showing the regions where MR was studied. Insets show the SEM image of the device and AFM image showing the ripples on the graphene surface. (B) The bottom panel shows the MR of SLG at three different densities shown in (a), where neither WL nor WAL is observed. The dashed curve shows the MR expected from standard WL theory. The middle panel shows the MR at Dirac point. The top panel shows MR in multilayer graphene where WL is restored. (Source: Adapted from Ref. [64].)

[65] as

$$\frac{\pi h}{e^2} \times \Delta\sigma(B) = F\left(\frac{\tau_B^{-1}}{\tau_\varphi^{-1}}\right) - F\left(\frac{\tau_B^{-1}}{\tau_\varphi^{-1} + 2\tau_i^{-1}}\right) - 2F\left(\frac{\tau_B^{-1}}{\tau_\varphi^{-1} + \tau_i^{-1} + \tau_*^{-1}}\right) \tag{5.32}$$

where $(z) = \ln z + \psi(0.5 + z^{-1})$, ψ being the Digamma function and $\tau_{B,\varphi,i*}^{-1} = 4eDB_{B,\varphi,i,*}/\hbar$. For $B_* \gg B_i$, the MR displays a distinct WL behavior. This is expected in graphene tightly coupled to the substrate, which generates atomically sharp defects. In the case of suspended graphene or graphene loosely attached to the substrate, the intervalley scattering time may be large. Now if $B_i < B_\varphi < B_*$ then $\delta g = 0$ and MR displays neither WL nor WAL.

The sensitivity of WL to inelastic as well as elastic-scattering processes in different carrier density regimes and in samples with different shapes and mobilities was studied experimentally by Tikhonenko *et al.* [66] The narrowest sample has strong intervalley scattering from the edges and hence a large intervalley scattering rate τ_i^{-1} compared with the wider samples (square and rectangle shaped). The MC for all the three samples is shown in Figure 5.17.

The presence of WL in all n regimes indicates a strong intervalley scattering rate, τ_i^{-1}. Figure 5.17B (a,b) shows the MC at three different temperatures in the Dirac region and metallic region, respectively, for the square sample. The stronger downturn in the Dirac region indicates reduced intervalley scattering and hence the contribution from WAL term (third term in Equation 5.32) increases. For the rectangular sample, which also had the largest mobility, the decrease in MC is sharper because of smaller τ_φ^{-1} and has a rapid downturn at large B indicating smaller τ_i^{-1}. The narrowest sample does not show any WAL feature at all, indicating large intervalley scattering from the edges. The role of atomically sharp defects as grain boundaries in producing intervalley scattering and WL was further established by the experiments on polycrystalline graphene grown by chemical vapor deposition by Yu *et al.* [67]. The MR measurements across a single grain boundary in graphene showed significant WL behavior, which was absent in measurements involving only a single grain.

5.3.2
Universal Conductance Fluctuations in Graphene

Universal conductance fluctuations (UCF) refer to the conductance fluctuations observed in disordered metals and 2DEGs as the magnetic field or the chemical potential is varied in the mesoscopic transport regime. The magnitude of these fluctuations was found to be $\delta G \approx e^2/h$, which was independent of sample size and degree of disorder [68]. Kechedzhi *et al.* studied the correlation function thermometry from the quantum transport and demonstrated its applicability in calculating the temperature of the carriers in graphene. The width of the correlation function of the conductance fluctuations is determined by the thermal broadening of the Fermi–Dirac distribution, which allows a direct extraction of electron

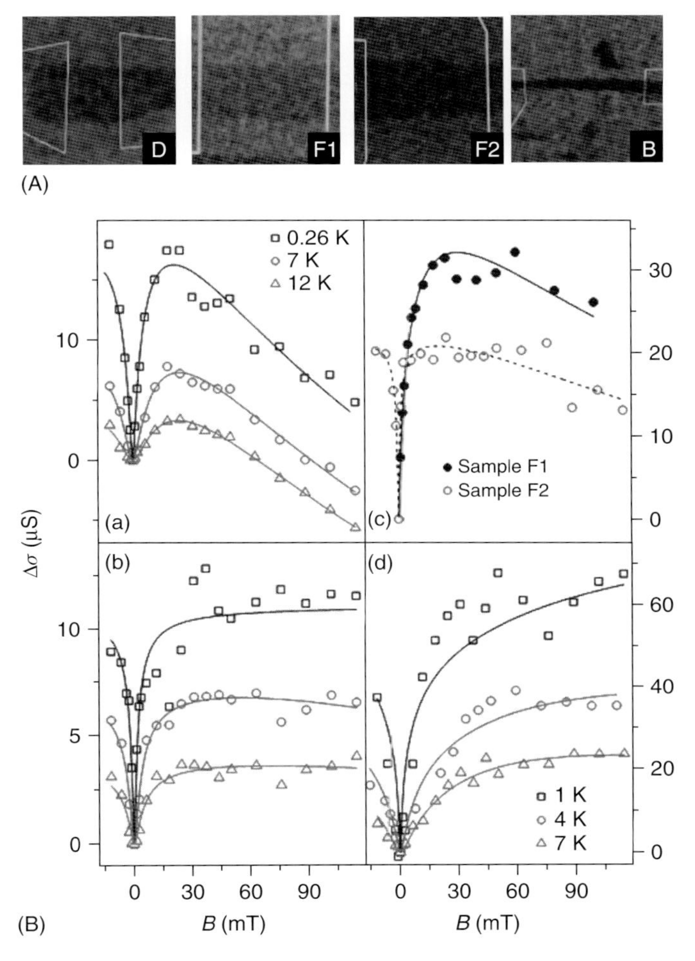

Figure 5.17 (A) The panel shows the SEM micrographs of ($4 \times 4\ \mu m^2$) of the measured devices. (B) Magnetoconductivity of (a) sample D in the Dirac region, (b) sample D at $n \sim 10^{12}\ cm^{-2}$, (c) samples F1 and F2 at $n \sim 7 \times 10^{11} cm^{-2}$, and (d) sample B at $n \sim 8 \times 10^{11} cm^{-2}$. The solid lines are the best fits of Equation 5.32 (Source: Adapted from Ref. [66])

temperature from the width at half-maximum of the correlation function as shown in Figure 5.18 [69].

The UCF in graphene has been found to be dependent on the elastic scattering in addition to other inelastic scattering mechanisms unlike conventional disordered metals. The UCF theory for graphene was proposed by Kharitonov and Efetov [70], and is extended from the conductivity calculations that were done previously for obtaining the WL correction in graphene. The conductance fluctuation magnitude, considering the various valley-symmetry-breaking scattering processes, is given

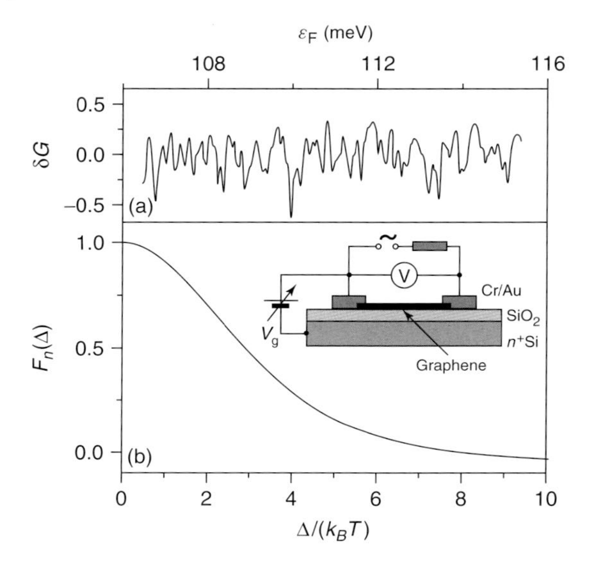

Figure 5.18 (a) The mesoscopic conductance fluctuations in graphene as a function of the Fermi energy. (b) The correlation function of the above fluctuations. (Source: Adapted from Ref. [69].)

by

$$\langle[\delta G]^2\rangle = a_\gamma \langle[\delta G]^2\rangle_{\mathrm{m}} \tag{5.33}$$

where $\langle[\delta G]^2\rangle_{\mathrm{m}}$ gives the conductance variance for conventional metals and a_γ counts the total number of gapless Cooperon and diffusion modes i that contribute to the conductance fluctuations. We consider here only the isospin/valley contributions from the low-energy pseudospin (sublattice) singlet modes because the pseudospin triplet modes are gapped out. The isospin singlet mode ($i = 0$) is unaffected by any other scattering mechanisms, whereas the isospin triplet mode, $i = 1$ (isospin $= 0$), is suppressed by intervalley scattering and the isospin triplet modes, $i = 2, 3$ (isospin $= \pm 1$), are suppressed by intervalley/intravalley scattering and trigonal warping.

1) When all the scattering effects are negligible, all the four modes contribute to conductance fluctuations and $\langle[\delta G]^2\rangle$ becomes four times that in conventional metals.
2) In the case of weak intervalley scattering, but strong intravalley scattering or warping, the modes $i = 0, 1$ alone contribute to conductance fluctuations, whereas the modes $i = 2, 3$ are gapped. Hence, $\langle[\delta G]^2\rangle$ becomes twice greater than that in conventional metals.
3) When intervalley scattering is also strong, along with intravalley scattering and warping, all the triplet modes are gapped and only the singlet mode contributes to $\langle[\delta G]^2\rangle$.

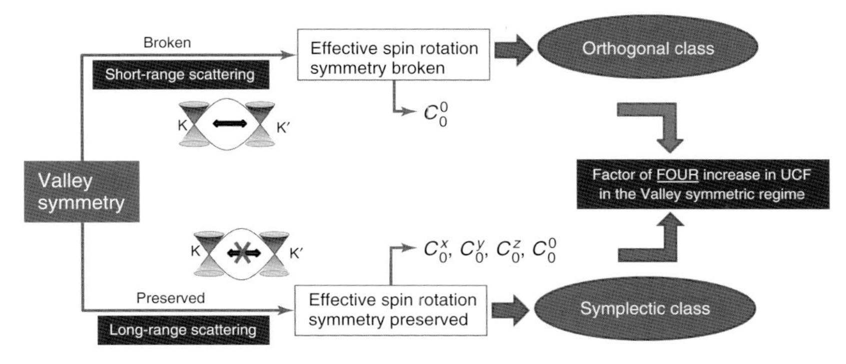

Figure 5.19 Schematic of the valley symmetry breaking leading to crossover from symplectic to orthogonal symmetry class leading to a factor of 4 increase in UCF in the valley symmetric regime.

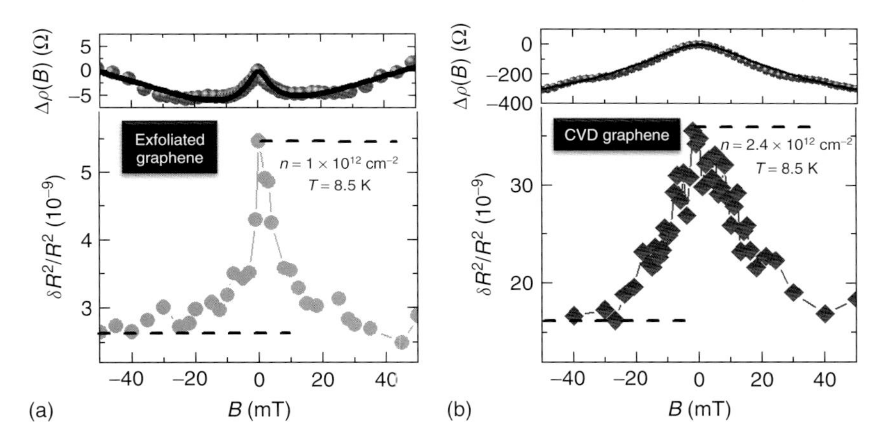

Figure 5.20 Factor of 2 reduction in UCF noise as a function of magnetic field for (a) exfoliated and (b) CVD-grown single-layer graphene. (Source: Adapted from Ref. [71].)

Hence, a factor of 4 reduction in conductance variance can be inferred as a direct evidence of valley symmetry breaking in graphene, which leads to a universality class crossover from symplectic to orthogonal symmetry class as shown schematically in Figure 5.19.

UCF can also be used as a probe to TR symmetry in graphene, which in turn can establish the effects of pseudomagnetic fields generated by ripples and local moment formation at defects and edges that lead to spontaneous TR symmetry breaking [63]. Pal *et al.* [71] studied the effect of magnetic field on the magnitude of conductance fluctuations and found that for $B > B_\varphi$, the UCF magnitude reduced by an exact factor of 2 because of vanishing contribution from Cooperons, implying a crossover to unitary class, which has broken the TR symmetry as shown in Figure 5.20. This study proves that TR symmetry is not broken spontaneously by the short-range disorder in graphene.

Pal *et al.* [71] studied the dependence of mesoscopic conductance fluctuations in graphene on the symmetry properties of the Hamiltonian, which can be a direct probe to the nontrivial universality class of graphene. The first unambiguous observation of the effect of valley symmetry on UCF in monolayer disordered graphene was also reported. Figure 5.21a shows the conductivity fluctuations as a function of gate voltage at various temperatures. With increasing temperature, the interference effect diminishes as the inelastic phase coherence length decreases, and hence the magnitude of fluctuations also goes down. This temperature effect on quantum interference can also be seen in Figure 5.21b, where the factor of 2 reduction in UCF variance as a function of B goes away at higher temperatures. Figure 5.21c shows the magnitude of UCF within each phase-coherent box for two SLG samples. While the absolute magnitude of UCF variance is expectedly on the order of e^2/h in all cases, the most striking aspect is its suppression by a factor of $\sim$4 as gate voltage is swept from the Dirac point toward the high electron or hole density regime. The density scale of such a suppression seems to follow the crossover of linear (the Coulomb scattering) to sublinear (short-range scattering) density dependence of conductivity. This demonstrates the observation of valley coherent states near the Dirac region, which can be of significance toward realizing valleytronic devices.

To conclude, studies of WL and UCFs in graphene give valuable information about the valley coherent states and the various scattering processes that break the symmetries in graphene. These also determine the universality class of graphene, and the crossover of symmetry classes have been discussed from both theoretical and experimental considerations.

5.4 Quantum-Confined Graphene

The semimetallic behavior (zero bandgap) nature of graphene imposes a serious technological bottleneck. One of the results of this is low on/off ratio in a graphene-based FET, which may limit its use in digital logic circuits. Even in very high-mobility substrate-supported graphene, the ratio of on-state current to the same in off-state current remains around 100, which is at least 2 orders of magnitude below that required for the contemporary digital circuit elements [72]. A possible route toward introducing a gap in the DOSs of two-dimensional graphene is via quantum confinement to lower dimensions. This may lead to a new class of nanostructured graphene devices that display unique transport and noise properties.

5.4.1 1D Graphene–Nanoribbons (GNRs)

Nanoribbons, as the name suggests, are narrow strips (approximately a few tens of nanometers or less) of graphene. The width of the nanoribbon (extent of confinement) should determine the gap in the DOS. Therefore, in principle, nanoribbons offer the possibility of fabricating devices with tunable bandgaps.

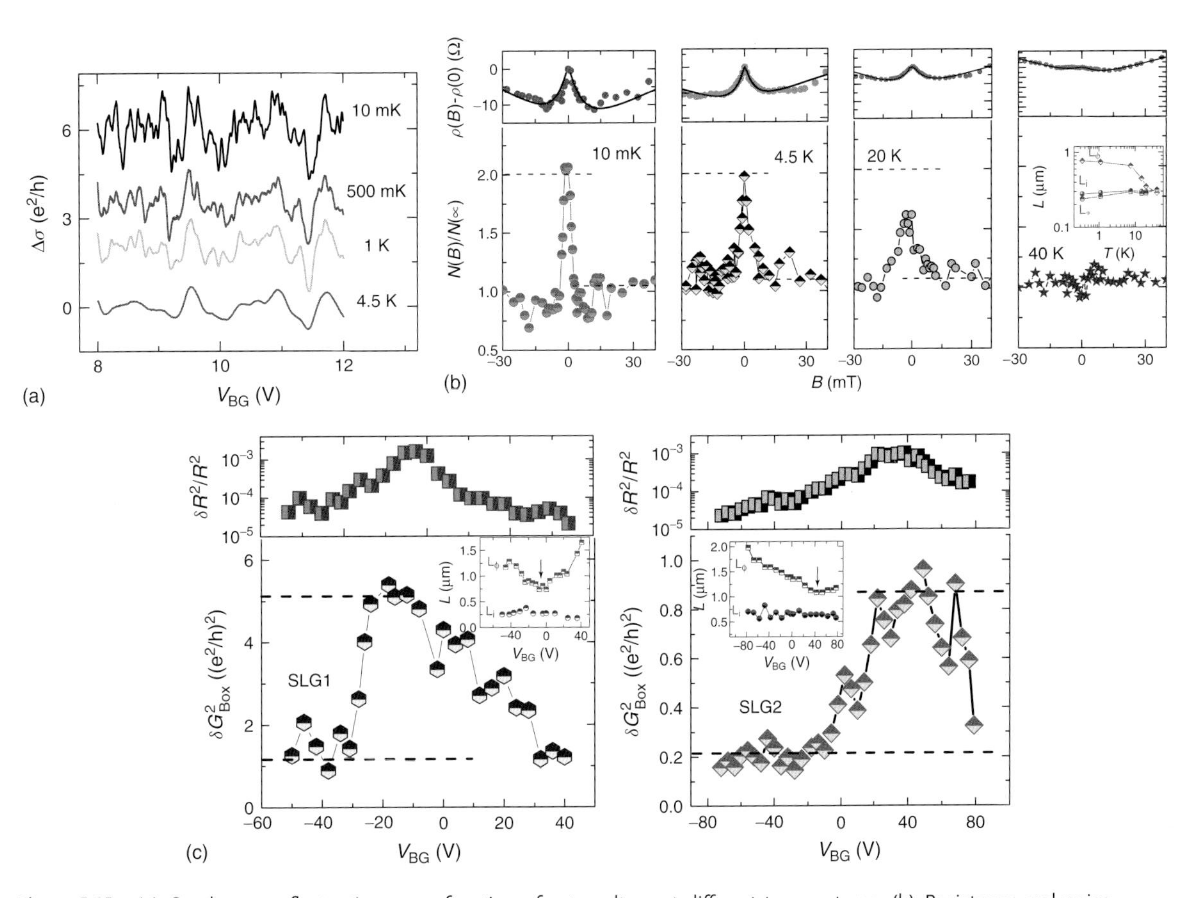

Figure 5.21 (a) Conductance fluctuations as a function of gate voltage at different temperatures. (b) Resistance and noise magnitude as a function of magnetic field are plotted for different temperatures. (c) The conductance variance of a single phase-coherent region in single-layer graphene (δG^2_{box}) shows a factor of 4 increase near the Dirac point, signifying universality class crossover. (Source: Adapted from Ref. [71].)

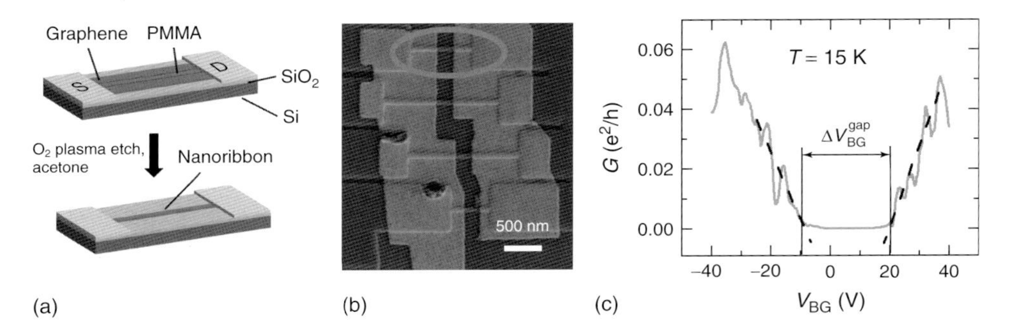

Figure 5.22 (a) Fabrication of graphene nanoribbon (GNR) using a poly(methyl methacrylate) (PMMA) etch mask. (b) SEM image of GNR devices consisting of top gate. The device highlighted by a grey circle with $W = 70$ nm and $L = 800$ nm, was measured in this present work. (c) Typical conductance spectroscopy for such a GNR showing a transport gap.

Figure 5.22a outlines the most basic method for the preparation of nanoribbons. Here, the 2D graphene is coated with an electron-beam resist, and the appropriate exposure leaves only a small portion of graphene covered with the resist. This serves as a mask for subsequent oxygen plasma etching, where graphene below the resist is protected, thereby forming a nanoribbon. An SEM image of typical graphene nanoribbon (GNR) devices are shown in Figure 5.22b. Figure 5.22c shows studies of nonequilibrium transport through a GNR device of $W = 70$ nm and $L = 800$ nm, indicating the presence of a gap at low temperature. However, the reported values of this gap are significantly higher than those expected from pure confinement effects [73–77]. A large body of experimental work has shown that this unexpectedly large *transport gap* arises because of the disorder at the edges, as well as the bulk [74–76]. The strong disorder potential causes the nanoribbon to fragment into electron and hole puddles, when it is close to the CNP. Thus, the gap is influenced more strongly by disorder and interactions, as compared to quantum confinement effects.

In an effort to reduce disorder, other fabrication methods have been explored. One of the popular alternatives is to use nanowires as etch masks [78, 79], shown in Figure 5.23a. This has two significant advantages: (i) the inherently small size of the nanowires allows for the fabrication of narrower nanoribbons and (ii) a single electron-beam lithography step reduces organic contaminants on the nanoribbons. Such nanoribbons not only show quantized conductance but also exhibit lower noise magnitude [80] (Figure 5.23b).

The low-frequency noise in these systems has been used to probe the DOS in SLG and BLG. Xu *et al.* [81] have shown that $1/f$ noise is much more sensitive to the band structure than conductance. Figure 5.23c shows that the noise magnitude (A) shows peaks, even when the quantization in conductance is smeared out. In addition, measurement of shot noise in nanoribbons has shown that it is particularly sensitive to the hopping through localized states [82]. Although reports on electrical noise are limited at present, the above-mentioned studies indicate that it could be an important tool to probe the conduction mechanism in GNRs.

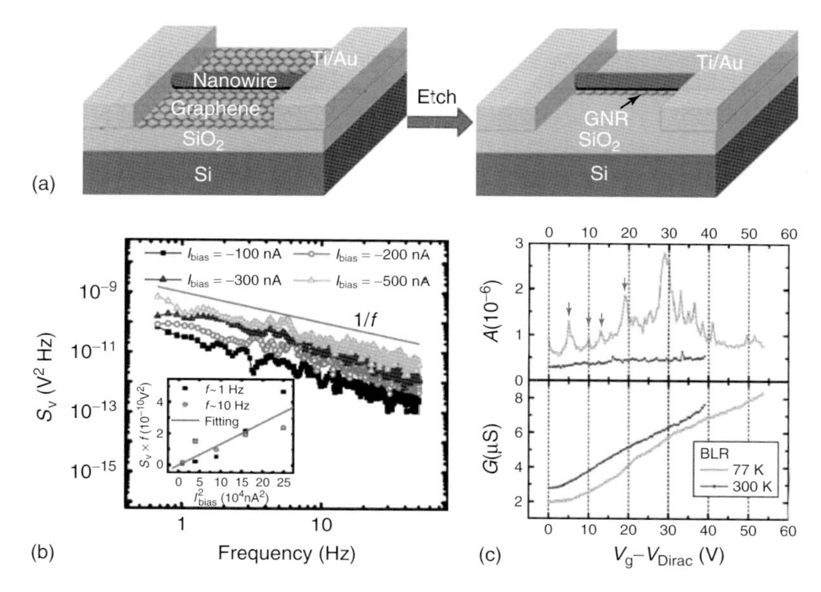

Figure 5.23 (a) Schematic showing the preparation of a GNR device using nanowire mask. (Source: Adapted from Ref. [78].) (b) The power spectral density (S_V) versus frequency (f) for a GNR showing a characteristic $1/f$ spectrum. (Source: Adapted from Ref. [80].) (c) Peaks in the noise magnitude (A) persist in a bilayer GNR (top panel) even when conductance plateaus are hardly visible (bottom panel). (Source: Adapted from Ref. [81].)

The low-frequency noise in a lithographically fabricated GNR device (Figure 5.24b) at various temperatures has been studied by (A.N. Pal and A. Ghosh, unpublished). Figure 5.24a shows the PSD at various gate voltages at a fixed temperature ($T = 97$ K). It shows the $1/f$ nature of the power spectrum, which is generally observed at higher temperatures ($>\sim 50$ K).

In Figure 5.24b, the noise magnitude, $\delta G^2/G^2 = \int S_G(f)/G^2 df$, is shown as a function of gate voltage for various temperatures. It is clear that the noise magnitude decreases with increasing gate voltage away from the neutrality region, similar to the unconfined graphene transistors. However, with decreasing temperature, the noise magnitude increases, in contrast with the mesoscopic graphene transistors described earlier. The temperature dependence of noise magnitude at three gate voltages are shown in Figure 5.24c. The increase in noise with decreasing temperature was observed in localized systems with transport governed by variable range hopping [83]. This suggests that with decreasing temperature, the carriers in nanoribbons become localized and the noise is originated mainly due to the hopping between two localized sites. However, if the temperature is decreased below ~ 50 K, the nature of the spectrum changes; nevertheless, it remains $1/f$. The time series shows a series of sharp peaks in conductance as a function of time even when the gate voltage is held constant (see $T = 0.3$ K data in Figure 5.24d). At this temperature, thermal energy is sufficiently small and the transport occurs mainly due to the

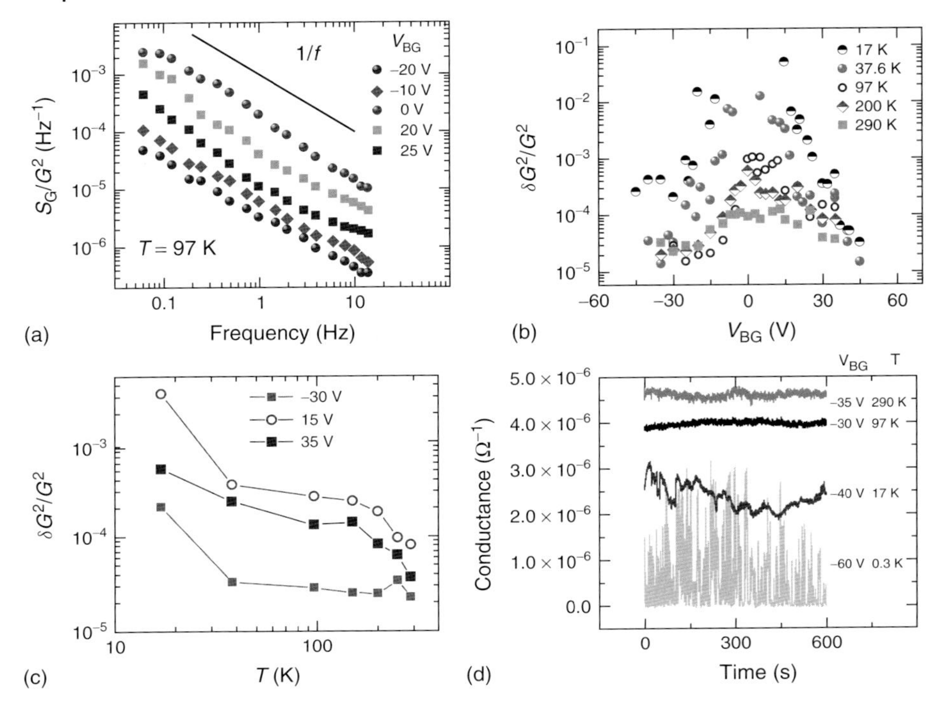

Figure 5.24 (a) Noise power spectra at various gate voltages at $T = 97$ K. (b) Noise magnitude ($\delta G^2/G^2$) as a function of gate voltage plotted at different temperatures. (c) Temperature dependence of noise at three different gate voltages extracted from (b). (d) Time series of conductance at different temperatures away from the gap region. (Source: Adapted from A.N. Pal and A. Ghosh, unpublished.)

resonant tunneling between the localized states. Any time-dependent change in the disorder landscape can alter the resonant transmission, which naturally manifests in the series of irreproducible peaks in conductance as a function of time.

Another route toward the fabrication of GNRs involves the *unzipping* of CNTs [84–86]. This allows the formation of a large number of GNRs with significantly smaller widths. Figure 5.25a outlines the basic procedure used to obtain such GNRs. Multiwalled CNTs are first treated thermally (calcinated), resulting in a preferential oxidation at defect sites. This is followed by a chemical treatment and sonication, which results in the unzipping of the tubes. Figure 5.25b compares the ratio of intensities of the D- and G-peaks in the Raman spectrum for different GNR fabrication methods. It is clear that such unzipped GNRs have significantly lower disorder than those fabricated using lithography. Furthermore, transport measurements have confirmed that such GNRs are extremely clean and have much smoother edges [87]. This makes them particularly attractive for studying a variety of exciting spintronic and electronic properties such as magnetism at the edges and the effects of quantum confinement.

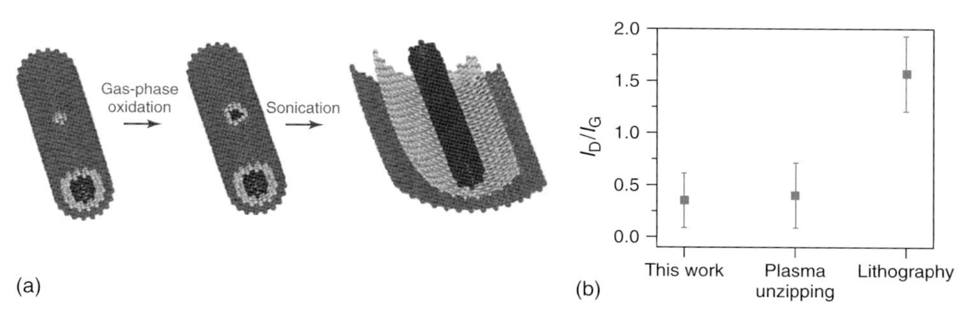

Figure 5.25 (a) Schematic showing the unzipping of carbon nanotubes resulting in the formation of GNRs. (b) A comparison of disorder levels (using Raman spectroscopy) in different GNRs fabricated via different routes. (Source: Adapted from Ref. [86].)

5.5 Conclusions and Outlook

The understanding of noise in graphene FETs thus not only provides a unique characterization tool but also can be exploited as a physical resource. The crucial effect of substrate on the noise performance appears to be generic to graphene-like atomically thin membranes, and hence requires a systematic effort to engineer the appropriate substrate. Recent investigations on graphene on h-boron nitride substrates would not only lead to high carrier mobility but also to very low noise, although this has not been verified yet. The suspended graphene flakes are also expected to exhibit very low noise magnitude. On the other hand, the sensitivity of the UCFs to the underlying symmetry of the Hamiltonian provides an elegant way to understand the ground state of various phases of graphene and its derivatives. This is particularly applicable to the many-body phenomena observed in graphene, such as the broken symmetry states in BLG and SLG, or the $\nu = 0$ Landau level and the possibility of an order–disorder transition. The grain boundaries in chemical-vapor-deposited graphene have been shown to cause backscattering in graphene, but their roles in mesoscopic fluctuations remain unanswered. It has been predicted that one can use these defects as valley filter based on scattering off a line defect, and noise can be an ideal candidate to explore such possibilities.

References

1. Wallace, P.R. (1947) *Phys. Rev.*, **71**, 622.
2. Castro Neto, A.H., Guinea, F., Peres, N.M.R., Novoselov, K.S., and Geim, A.K. (2009) *Rev. Mod. Phys.*, **81**, 109.
3. Goerbig, M.O. (2011) *Rev. Mod. Phys.*, **83**, 1193.
4. Das, A. (2009) Graphene and Carbon nanotubes: Field Induced Doping, Interaction with Nucleobases, Confined Water and Sensors., PhD thesis, Indian Institute of Science, India.
5. Shon, N.H. and Ando, T. (1998) *J. Phys. Soc. Jpn.*, **67**, 2421.
6. Bolotin, K. *et al.* (2008) *Solid State Commun.*, **146**, 351.
7. Morozov, S.V. *et al.* (2008) *Phys. Rev. Lett.*, **100**, 016602.

8. Du, X., Skachko, I., Barker, A., and Andrei, E.Y. (2008) *Nat. Nanotechnol.*, **3**, 491.
9. Nilsson, J., Castro Neto, A.H., Guinea, F., and Peres, N.M.R. (2008) *Phys. Rev. B*, **78**, 045405.
10. McCann, E. (2006) *Phys. Rev. B*, **74**, 161403.
11. McCann, E. and Fal'ko, V.I. (2006) *Phys. Rev. Lett.*, **96**, 086805.
12. Castro, E.V. *et al.* (2010) *J. Phys.: Condens. Matt.*, **22**, 175503.
13. Min, H., Sahu, B., Banerjee, S.K., and MacDonald, A.H. (2007) *Phys. Rev. B*, **75**, 155115.
14. McCann, E., Abergel, D.S., and Falko, V.I. (2007) *Solid State Commun.*, **143**, 110.
15. Castro, E.V. *et al.* (2007) *Phys. Rev. Lett.*, **99**, 216802.
16. Ohta, T., Bostwick, A., Seyller, T., Horn, K., and Rotenberg, E. (2006) *Science*, **313**, 951.
17. Oostinga, J.B., Heersche, H.B., Liu, X., Morpurgo, A.F., and Vandersypen, L.M.K. (2008) *Nat. Mater.*, **7**, 151.
18. Zhang, Y. *et al.* (2009) *Nature*, **459**, 820.
19. Bostwick, A. *et al.* (2007) *New J. Phys.*, **9**, 385.
20. Aoki, M. and Amawashi, H. (2007) *Solid State Commun.*, **142**, 123.
21. Craciun, M.F. *et al.* (2009) *Nat. Nanotechnol.*, **4**, 383.
22. Bao, W. *et al.* (2011) *Nat. Phys.*, **7**, 948.
23. Lui, C.H., Li, Z., Mak, K.F., Cappelluti, E., and Heinz, T.F. (2011) *Nat. Phys.*, **459**, 944.
24. Cong, C. *et al.* (2011) *ACS Nano*, **5**, 8760.
25. Taychatanapat, T., Watanabe, K., Taniguchi, T., and Jarillo-Herrero, P. (2011) *Nat. Phys.*, **7**, 621.
26. Zhang, L., Zhang, Y., Camacho, J., Khodas, M., and Zaliznyak, I. (2011) *Nat. Phys.*, **7**, 953.
27. Fasolino, A., Los, J.H., and Katsnelson, M.I. (2007) *Nat. Mater.*, **6**, 858.
28. Kim, E.-A. and Neto, A.H.C. (2008) *Europhys. Lett.*, **84**, 57007.
29. Chen, J.-H. *et al.* (2008) *Nat. Phys.*, **4**, 377.
30. Adam, S., Hwang, E.H., Galitski, V.M., and Sarma, S.D. (2007) *Proc. Natl. Acad. Sci. U.S.A.*, **104**, 18392.
31. Adam, S. and Sarma, S.D. (2008) *Phys. Rev. B*, **77**, 115436.
32. Sarma, S.D., Adam, S., Hwang, E.H., and Rossi, E. (2011) *Rev. Mod. Phys.*, **83**, 407.
33. Novoselov, K.S. *et al.* (2005) *Nature*, **438**, 197.
34. Novoselov, K.S. *et al.* (2004) *Science*, **306**, 666.
35. Zhu, W., Perebeinos, V., Freitag, M., and Avouris, P. (2009) *Phys. Rev. B*, **80**, 235402.
36. Hwang, E.H., Adam, S., and Sarma, S.D. (2007) *Phys. Rev. Lett.*, **98**, 186806.
37. Fratini, S. and Guinea, F. (2008) *Phys. Rev. B*, **77**, 195415.
38. Rossi, E. and Sarma, S.D. (2008) *Phys. Rev. Lett.*, **101**, 166803.
39. Martin, J. *et al.* (2008) *Nat. Phys.*, **4**, 144.
40. Zhang, Y., Brar, V.W., Girit, C., Zettl, A., and Crommie, M.F. (2009) *Nat. Phys.*, **5**, 722.
41. Deshpande, A., Bao, W., Zhao, Z., Lau, C.N., and LeRoy, B.J. (2009) *Appl. Phys. Lett.*, **95**, 243502.
42. Ghosh, A., Kar, S., Bid, A., and Raychaudhuri A.K. (2004), e-print arXiv:condmat/0402130v1.
43. Jayaraman, R. and Sodini, C.G. (1989) *IEEE Trans. Electron Devices*, **36**, 1773.
44. Lin, Y.M. *et al.* (2006) *Nano Lett.*, **6**, 930.
45. Liu, G. *et al.* (2009) *Appl. Phys. Lett.*, **95**, 033013.
46. Lin, Y.M. and Avouris, P. (2008) *Nano Lett.*, **8**, 2119.
47. Pal, A.N. and Ghosh, A. (2009) *Appl. Phys. Lett.*, **95**, 082105.
48. Xu, G. *et al.* (2010) *Nano Lett.*, **10**, 3312.
49. Heller, I. *et al.* (2010) *Nano Lett.*, **10**, 1563.
50. Pal, A.N. *et al.* (2011) *ACS Nano*, **5**, 2075.
51. Pelz, J. and Clarke, J. (1987) *Phys. Rev. B*, **36**, 4479.
52. Chen, J.H. *et al.* (2008) *Nat. Nanotechnol.*, **3**, 206.
53. Pal, A.N. and Ghosh, A. (2009) *Phys. Rev. Lett.*, **102**, 126805.
54. Jehl, X. *et al.* (1999) *Phys. Rev. Lett.*, **83**, 1660.
55. Saminadayar, L., Glattli, D.C., and Jin, B.E.Y. (1997) *Phys. Rev. Lett.*, **79**, 2526.

56. de Picciotto, R. *et al.* (1997) *Nature*, **389**, 162.
57. Tworzydlo, J. *et al.* (2006) *Phys. Rev. Lett.*, **96**, 246802.
58. Cheianov, V.V. and Fal'ko, V.I. (2006) *Phys. Rev. B*, **74**, 041403(R).
59. Danneau, R. *et al.* (2008) *Phys. Rev. Lett.*, **100**, 196802.
60. DiCarlo, L. *et al.* (2008) *Phys. Rev. Lett.*, **100**, 156801.
61. Suzuura, H. and Ando, T. (2002) *Phys. Rev. Lett.*, **89**, 266603.
62. Tikhonenko, F.V. *et al.* (2009) *Phys. Rev. Lett.*, **103**, 226801.
63. Morpurgo, A.F. and Guinea, F. (2006) *Phys. Rev. Lett.*, **97**, 196804.
64. Morozov, S.V. *et al.* (2006) *Phys. Rev. Lett.*, **97**, 016801.
65. McCann, E. *et al.* (2006) *Phys. Rev. Lett.*, **97**, 146805.
66. Tikhonenko, F.V. *et al.* (2008) *Phys. Rev. Lett.*, **100**, 056802.
67. Yu, Q. *et al.* (2011) *Nat. Mat.*, **10**, 443.
68. Feng, S., Lee, P.A., and Stone, A.D. (1986) *Phys. Rev. Lett.*, **56**, 1960.
69. Kechedzhi, K. *et al.* (2009) *Phys. Rev. Lett.*, **102**, 066801.
70. Kharitonov, M.Y. and Efetov, K.B. (2008) *Phys. Rev. B*, **78**, 033404.
71. Pal, A.N., Kochat, V., and Ghosh, A., e-print arxiv:condmat/1206.3866.
72. Xia, F., Farmer, D.B., Lin, Y.-M., and Avouris, P. (2010) *Nano Lett.*, **10**, 715.
73. Han, M.Y. *et al.* (2007) *Phys. Rev. Lett.*, **98**, 206805.
74. Todd, K. *et al.* (2009) *Nano Lett.*, **9**, 416.
75. Stampfer, C. *et al.* (2009) *Phys. Rev. Lett.*, **102**, 056403.
76. Molitor, F. *et al.* (2009) *Phys. Rev. B*, **79**, 075426.
77. Han, M.Y. *et al.* (2010) *Phys. Rev. Lett.*, **104**, 056801.
78. Bai, J. *et al.* (2009) *Nano Lett.*, **9**, 2083.
79. Bai, J. *et al.* (2010) *Nat. Nanotechnol.*, **5**, 655.
80. Xu, G. *et al.* (2010) *Appl. Phys. Lett.*, **97**, 073107.
81. Xu, G. *et al.* (2010) *Nano Lett.*, **10**, 4590.
82. Danneau, R. *et al.* (2010) *Phys. Rev. B*, **82**, 161405(R).
83. Shklovskii, B.I. (2003) *Phys. Rev. B*, **67**, 045201.
84. Jiao, L. *et al.* (2009) *Nature*, **458**, 877.
85. Kosynkin, D. *et al.* (2009) *Nature*, **458**, 872–876.
86. Jiao, L. *et al.* (2010) *Nat. Nanotechnol.*, **5**, 321.
87. Wang, X. *et al.* (2011) *Nat. Nanotechnol.*, **6**, 563.

6
Suspended Graphene Devices for Nanoelectromechanics and for the Study of Quantum Hall Effect

Vibhor Singh and Mandar M. Deshmukh

6.1
Introduction

Electronic properties of graphene have been studied extensively [1] since the first experiments probing quantum Hall effect (QHE) [2, 3]. In addition to the electronic properties, the remarkable mechanical properties of graphene include a high in-plane Young's modulus of 1TPa probed using nanoindentation of suspended graphene [4] and electromechanical resonators [5–7]. NEMS (nanoelectromechanical system) devices using nanostructures such as carbon nanotubes [8–14], nanowires [15–17], and bulk micromachined structures [18–20] offer promise of new applications and allow us to probe fundamental properties at the nanoscale. NEMS [21]-based devices are ideal platforms to harness the unique mechanical properties of graphene. Electromechanical measurements with graphene resonators [5, 6] suggest that with the improvement of quality factor (Q), graphene-based NEMS devices have the potential to be very sensitive detectors of mass and charge. In addition, the sensitivity of graphene to chemical-specific processes [2, 22] offers the possibility of integrated mass and chemical detection. The large surface-to-mass ratio of graphene offers a distinct advantage over other nanostructures for such applications. As an example of the utility of suspended graphene devices, we show that such devices can be used to probe the strain in these resonators, which can be useful for the study of the coefficient of thermal expansion of graphene ($\alpha_{\text{graphene}}(T)$) as a function of temperature. These measurements indicate that $\alpha_{\text{graphene}}(T) < 0$ for $30\,\text{K} < T < 300\,\text{K}$ [23] and larger in magnitude than in theoretical prediction [24]. These experiments [23] also show that measuring temperature-dependent mechanical properties [25] of suspended structures down to low temperatures gives insight into strain engineering of graphene-based devices [26, 27] and also helps in understanding the role of rippling in degrading carrier mobility at low temperatures. The other advantage that stems from these suspended resonators is the ability to clean them by applying large current. Such devices show very large carrier mobility, which can be used to study the ballistic transport and QHE at low magnetic fields.

Graphene: Synthesis, Properties, and Phenomena, First Edition. Edited by C. N. R. Rao and A. K. Sood.

This chapter is organized as follows: first, we review the basics of QHE for graphene, and then discuss details of the fabrication of suspended graphene and the measurement schemes for the electromechanical actuation. Then, we give one example of the utility of the electromechanical measurements as the measurement of α_{graphene}. After this, we discuss another aspect of the suspended graphene devices, which is the improvement of carrier mobility by current annealing, giving an example of the QHE in these systems.

6.2
Quantum Hall Effect in Graphene

The QHE is a quantum mechanical phenomenon that occurs in two-dimensional electronic systems subjected to magnetic field and is typically observable at low temperatures. In the presence of a magnetic field, the current density (**J**) no longer remains parallel to the applied electric field (**E**), ($\mathbf{J} = \sigma \mathbf{E}$), and hence conductance of the sample is described by 2×2 conductivity tensor with two independent components, namely, longitudinal conductivity σ_{xx} and transverse conductivity σ_{xy}. The localization–delocalization of electrons (for integer QHE) and electron–electron interactions (for fractional QHE) lead to versatile phenomena such as dissipationless transport across the system, and, at the same time, transverse conductance becomes quantized in units of e^2/h irrespective of the material properties and sample geometry. Here, we try to present essential concepts in the case of integer QHE. A complete and detailed description of quantum Hall phenomena is available in the literature [28–30].

Many of the concepts from the previous understanding of integer QHE in 2DEG (two-dimensional electron gas) [28–30] get carried through to describe the QHE in graphene. However, linear dispersion and vanishing density of states at the Dirac point add a different flavor to it (sometimes also referred to as *anomalous QHE* or *relativistic-like QHE*). The quantized tight binding Hamiltonian of graphene in zero magnetic field is given by $\hat{H} = v_F \boldsymbol{\sigma}.\mathbf{p}$, where $\boldsymbol{\sigma}$ are 2×2 Pauli matrices [31]. In the presence of a magnetic field, the Hamiltonian can be obtained by replacing $\mathbf{p}$ by $\mathbf{p} + e\mathbf{A}$. Therefore, the two-dimensional Dirac equation in magnetic field can be written as [31]

$$v_F \boldsymbol{\sigma}.(-i\hbar\nabla + e\mathbf{A})\psi = E\psi \tag{6.1}$$

By taking Landau gauge $\mathbf{A} = (-By, 0)$ and demanding the solution of the form $\psi = e^{ikx}\phi(y)$, the above equation can be simplified as

$$\hbar v_F \begin{pmatrix} 0 & \partial_y - k + \frac{Bey}{\hbar} \\ -\partial_y - k + \frac{Bey}{\hbar} & 0 \end{pmatrix} \phi(y) = E\phi(y) \tag{6.2}$$

We can shift the origin in y-direction to $\xi = \frac{y}{l_B} - l_B k$, where l_B is the magnetic length. By introducing one-dimensional harmonic oscillator operators $O = \frac{1}{\sqrt{2}}(\xi + \partial_\xi)$ and $O^\dagger = \frac{1}{\sqrt{2}}(\xi - \partial_\xi)$, the above equation can be further simplified to

$$(O\sigma^{+} + O^{\dagger}\sigma^{-})\phi(y) = \frac{2E}{\omega_c}\phi(y) \tag{6.3}$$

where $\sigma^{\pm}(= \sigma_x \pm i\sigma_y)$ are raising and lowering operators and ω_c is the cyclotron frequency given by $\sqrt{2}\frac{v_F}{l_B}$. The eigenvalues and eigenfunctions of the above equation were first obtained by McClure [32] while studying the diamagnetism of graphite. The solutions are given by

$$E_{\pm}(n) = \pm\sqrt{2}\frac{\hbar v_F}{l_B}\sqrt{n} = \pm\sqrt{2eB\hbar v_F n} \tag{6.4}$$

and

$$\phi_{n,\pm}(\xi) = \begin{pmatrix} \psi_{n-1}(\xi) \\ \psi_n(\xi) \end{pmatrix} \tag{6.5}$$

where $n = 0, 1, 2, \ldots$ and $\psi_n(\xi)$ are the solution of the one-dimensional harmonic oscillator $\psi_n(\xi) = \frac{1}{\sqrt{2^n n!}} e^{-\frac{\xi^2}{2}} H_n(\xi)$ and $H_n(\xi)$ are Hermite polynomials. The positive sign is taken to describe the electrons in the conduction band and the negative sign is taken to describe the holes in the valence band.

Several key points can be noted in the solution given by Equation (6.4). We note that the linear density of states spectrum splits into nonuniformly spaced LLs (Landau levels), as the position of the LLs is proportional to $\sqrt{n}$ (unlike the case of two-dimensional free electron gas). As one moves to higher index (n) levels, LLs start coming closer to each other (for 2D free electron gas, this gap remains constant $\hbar\omega_c$). Apart from twofold spin degeneracy (g_s), distinct wavefunction with same energies at K′-point gives an additional twofold degeneracy (also called *valley degeneracy* g_v). The fourfold degeneracy ($g_s \times g_v$) provides four independent channels for electron transport. As a result, one would expect that transverse resistance of graphene in quantum Hall limit will follow the sequence $\frac{h}{e^2}\frac{1}{\nu}$, where $\nu = \pm 4, \pm 8, \pm 12, \ldots$ However, the interesting scenario arises because of the $n = 0$ LL, which is equally shared by electrons and holes between K- and K′-points. States below $\mu < 0$ are occupied by the holes, whereas states above $\mu > 0$ are occupied by the electrons. Such a shared filling of zeroth LL shifts the overall quantized sequence to $\pm 2, \pm 6, \pm 10, \ldots$ without a plateau at zero filling. A stronger statement based on the Laughlin's gauge invariance argument can also be made, where it can be shown [31, 33] that because of the presence of zero mode shared between two Dirac points, there are exactly $(4n + 2)$ occupied states that get transferred from one edge to the other when the Fermi level crosses the $n = 0$ LL. Therefore, in the case of monolayer graphene,

$$R_{xy} = \pm\frac{1}{4n+2}\frac{h}{e^2} \tag{6.6}$$

This unique quantization sequence is specific to monolayer graphene [2, 3] and often used to identify devices with monolayer graphene. In Figure 6.1a, we show R_{xx} and R_{xy} with gate voltage at $T = 300$ mK and $B = 9$ T [34]. At fixed field, by changing the gate voltage, the Fermi energy can be moved across the LLs without modifying the energy spectrum. As the gate voltage is swept, vanishing R_{xx} and

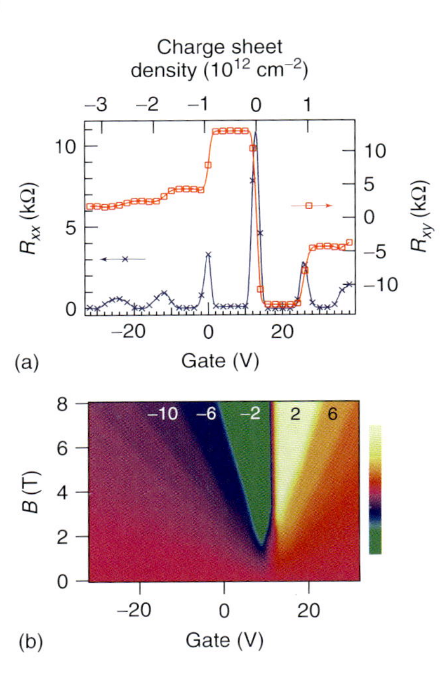

Figure 6.1 Plot of the (a) longitudinal resistance (R_{xx}) and (b) transverse resistance (R_{xy}) for a monolayer graphene device at 300 mK and 9 T.

quantized plateaus in R_{xy} can be seen clearly. These filling factors correspond to monolayer graphene ($\nu = \pm 2, \nu = \pm 6$). In Figure 6.1b, we plot the evolution of R_{xy} as a function of V_g and magnetic field B. As carrier density and magnetic field are changed, the integer filling factors evolve, and it can be seen from the fan diagram shown in Figure 6.1b. The plateaus in R_{xy} corresponding to $\nu = \pm 2, \pm 6$, and -10 are clearly seen.

6.3 Fabrication of Suspended Graphene Devices

To fabricate monolayer graphene electromechanical resonators, we suspended graphene devices using the previously reported [6, 35–37] process that starts with micromechanical exfoliation of graphene from graphite crystals several millimeters thick onto a degenerately doped silicon wafer coated with 300 nm thick thermally grown SiO_2. Following the location of monolayer graphene flakes using optical microscopy, electron beam lithography is used to pattern the resist for fabricating electrodes for electrical contact. The electrodes are fabricated by the evaporation of 10 nm of chromium and 60 nm of gold following the patterning of the resist. To release the graphene from the SiO_2 substrate, a dilute buffered-HF solution is used to selectively etch an area around the graphene device by masking the rest of the substrate using polymer resist. Following an etch for 5 min 30 s, which results in a 170 nm deep trench in SiO_2, the device is rinsed in DI water and isopropanol. Critical point drying, to prevent the collapse of the device due to surface tension, is

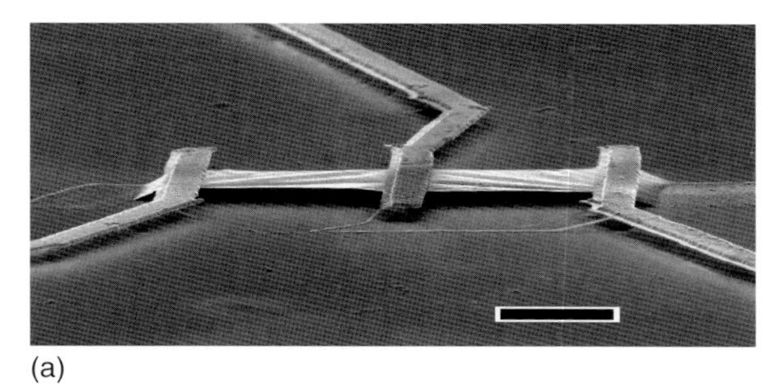

(a)

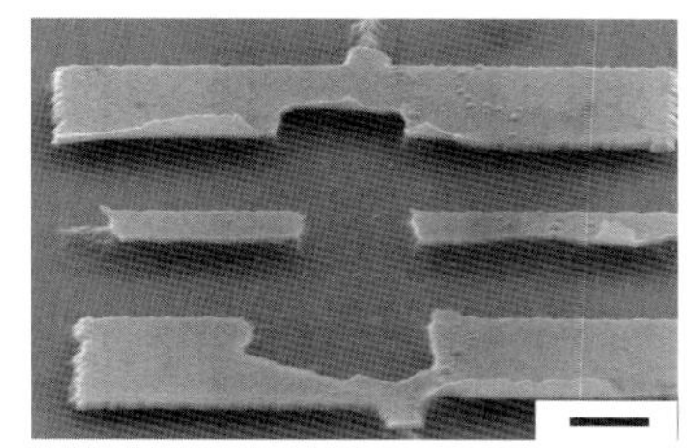

(b)

Figure 6.2 (a) Tilted angle scanning electron microscope (SEM) image of a suspended monolayer graphene device with multiple electrodes. The scalebar represents a length of 2 μm. (b) Tilted SEM image of a suspended graphene device after removal of the graphene flake indicating the etched part of SiO_2 under the gold electrodes. The scale bar corresponds to 1 μm.

the final step in the fabrication of suspended graphene devices. A scanning electron microscopic (SEM) image of a suspended graphene device is shown in Figure 6.2a. An interesting aspect of the suspended graphene devices is that the silicon oxide underneath the graphene buried under metallic electrodes is also etched. In effect, the gold electrodes are holding the freely suspended parts of graphene. This can be clearly seen in Figure 6.2b, where the upper part of the electrodes and the graphene has been removed by sonication. Once the suspended devices are fabricated, they can be used for a variety of experiments; we next focus on the actuation and detection of electromechanical resonators fabricated using graphene.

6.4 Nanoelectromechanics Using Suspended Graphene Devices

Nanoelectromechanical devices, much like stringed instruments, have two important components associated with them: (i) actuation, akin to plucking, and (ii) detection, where one listens to the notes emanating from the stringed instrument. Both these aspects are embedded in our device. Actuation of the device is done by capacitive means, with graphene forming one electrode of a parallel plate capacitor. The detection relies on the crucial fact that as the graphene membrane vibrates in one part of the cycle, it moves closer to the gate electrode, and in the other part, it moves away from the gate. As graphene has a finite transconductance, the motion relative to the gate results in modulation of the graphene membrane conductance. This conductance modulation is largest at the electromechanical resonance, where the amplitude of mechanical motion is maximal. So, measuring the modulation of the conductance as a function of the actuation frequency will allow us to measure the electromechanical resonance. However, the measurement of conductance modulation by simply biasing is challenging, as the parasitic capacitance leads to the generation of displacement current at actuation frequency. As a result, one needs to use a different method to measure the resulting modulation in conductance. We use a technique that implements the device as a heterodyne mixer, and we discuss the details of the technique next.

The electrical actuation and detection is done using the suspended graphene device as a heterodyne mixer [6, 10, 14]. The scheme for electrical actuation and detection is shown in Figure 6.3a superimposed on the SEM image of the device. The electrostatic interaction between the graphene membrane and the back-gate electrode is used to actuate the motion in a plane perpendicular to the substrate. A radio frequency (RF) signal $V_g(\omega)$ and a DC voltage V_g^{DC} are applied at the gate terminal using a bias tee. Another RF signal $V_{SD}(\omega + \Delta\omega)$ is applied to the source electrode (Figure 6.3a). The RF signal applied at the gate $V_g(\omega)$ modulates the gap between graphene and substrate at frequency ω, and V_g^{DC} alters the overall tension and carrier density in the membrane. The amplitude of the current through the graphene membrane at the difference frequency ($\Delta\omega$), also called the *mixing current* $I_{mix}(\Delta\Omega)$, can be written as [6, 9, 10, 12, 14, 38]

$$I_{mix}(\Delta\omega) = \frac{1}{2}\frac{dG}{dq}\left(\frac{dC_g}{dz}z(\omega)V_g^{DC} + C_g V_g(\omega)\right)V_{SD}(\omega + \Delta\omega) \qquad (6.7)$$

where G is the conductance of the graphene device, q is the charge induced by the gate voltage, C_g is the capacitance between the gate electrode and graphene, and $z(\omega)$ is the amplitude of oscillation at the driving frequency ω along the z-axis (perpendicular to the substrate). The difference frequency signal (at $\Delta\omega$) arises from the product of the modulation signals of $V_{SD}(\omega + \Delta\omega)$ and $G(\omega)$. At the mechanical resonance of the membrane, the first term in Equation (6.7) contributes significantly and the second term, which does not depend on the mechanical motion of the graphene membrane, provides a smooth background.

Figure 6.3b shows the result of such a measurement at 7 K for a suspended graphene device while V_g^{DC} is set at −5 V. Using a modified Lorentzian lineshape for the resonance curve [10, 39], we can extract the quality factor $(Q) = 1500$ of the resonator. Figure 6.3c shows the colorscale plot of $I_{mix}(\Delta\Omega)$ as a function of V_g^{DC} and $f = \omega/2\pi$ at 300 K. The resonant frequency increases as the magnitude of V_g^{DC} is increased – the mechanical mode disperses positively with V_g^{DC}. This is well understood in terms of the increase in the tension of the graphene resonator, due to electrostatic attraction between the membrane and the back gate, as the gate voltage is increased.

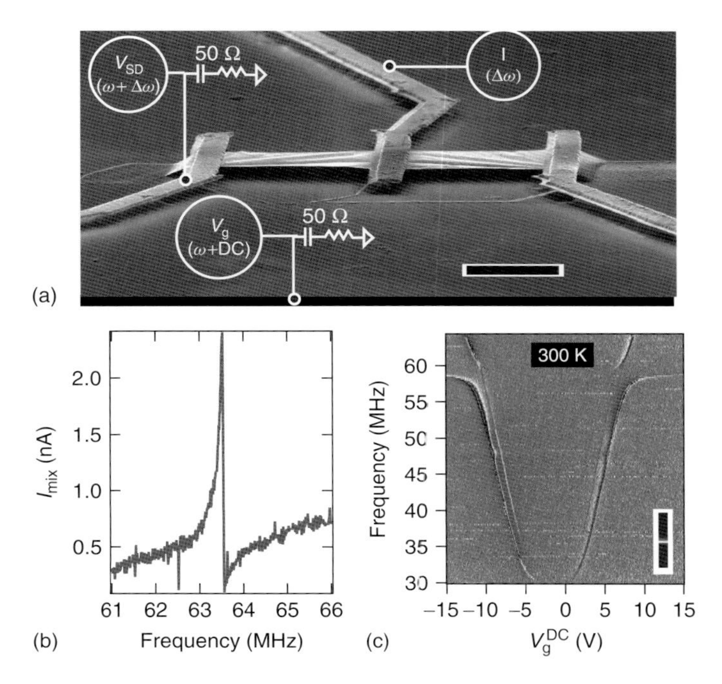

Figure 6.3 (a) Tilted angle scanning electron microscope (SEM) image of a suspended monolayer graphene device and the electrical circuit for actuation and detection of the mechanical motion of the graphene membrane. The scale bar indicates a length of 2 μm. (b) A plot of the mixing current $I_{mix}(\Delta\omega)$ as a function of frequency f. The sharp feature in the mixing current corresponds to the mechanical resonance. (c) Colorscale plot of the mixing current as function of frequency f and DC gate voltage V_g^{DC} at 300 K.

6.5 Using Suspended Graphene NEMS Devices to Measure Thermal Expansion of Graphene

The resonant frequency f_0 of the graphene membrane can be written as

$$f_0\left(V_g^{DC}\right) = \frac{1}{2L}\sqrt{\frac{\Gamma(\Gamma_0(T), V_g^{DC})}{\rho t w}} \tag{6.8}$$

where L is the length of the membrane, w is the width, t is the thickness, ρ is the mass density, $\Gamma_0(T)$ is the in-built tension, and Γ is the tension at a given temperature T and V_g^{DC}. The functional form of $\Gamma(\Gamma_0(T), V_g^{DC})$ is dependent on the details of the model used to take into account the electrostatic and elastic energies.

We now consider how the resonant frequency (f_0) evolves as a function of the temperature. Figure 6.4a shows the result of an evolution of a mode as a function of temperature at $V_g^{DC} = 15$ V. Data acquisition was done during a single sweep over 12 h to allow the resonator to equilibrate, and the acquisition window automatically adjusts to follow the resonance. The resonant frequency increases as the device

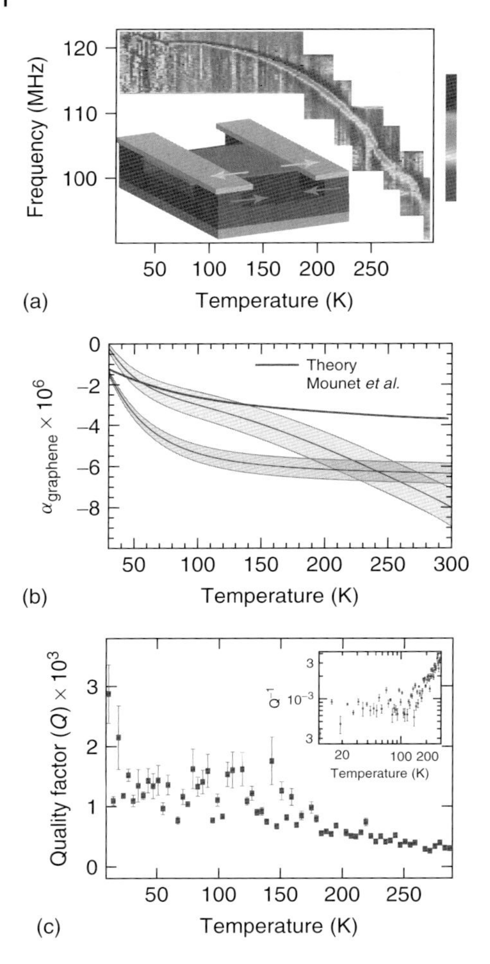

Figure 6.4 (a) Plot showing the evolution of the resonant frequency of a mode for device as a function of temperature for $V_g^{DC} = 15$ V. Inset shows the schematic representation of all the strains external to the suspended graphene membrane as the device is cooled below 300 K. (b) The plot of expansion coefficient of graphene as a function of temperature. Data from two different devices together with the theoretical prediction of Mounet and Marzari [24]. The shaded area represents the errors estimated from the uncertainty of the length of the flake, width of the electrode, and Young's modulus of graphene. (c) The plot shows the variation of the quality factor of resonance with temperature. The inset shows the plot of Q^{-1} with temperature.

is cooled below room temperature. This increase has been seen in all the devices we have studied.

The degree of frequency shift varies from one device to another depending on the device geometry. The origin of this frequency shift with temperature is the increase in tension in graphene because of the expansion/contraction of substrate, gold electrodes, and graphene. The frequency shift can be understood by taking into account the contribution of various strains, as the device is cooled below 300 K. The three main contributions are – first, the thermal strain in unconstrained graphene $\epsilon_{\text{graphene}}(T) = \int_T^{300} \alpha_{\text{graphene}}(t)dt$ due to the coefficient of thermal expansion of graphene $\alpha_{\text{graphene}}(T)$; second, the thermal strain due to the gold electrodes $\epsilon_{\text{gold}}(T) = \int_T^{300} \alpha_{\text{gold}}(t)dt$; and finally, the contribution of the strain induced by the substrate $\epsilon_{\text{substrate}}(T)$. Here, $\alpha_{\text{gold}}(T)$ is the coefficient of thermal expansion for gold [40]. The strain in gold electrodes plays an important role because of the geometry of the device. As mentioned earlier and seen in Figure 6.2b, the underetch that releases the graphene membrane also etches under the graphene covered by the electrodes – resulting in the graphene membrane being suspended off the gold electrodes [6]. The geometry of the resulting device is shown in the inset in Figure 6.4a. The elastic strain in the curved substrate can be calculated using Stoney's equation [41]. Its contribution to the net strain in the graphene membrane is very small, and therefore on cooling, the change of tension in the graphene is due to contraction of gold electrode and expansion/contraction of graphene. We assume that Young's modulus of graphene does not vary significantly over the temperature range [42]. At the interface of gold electrodes supporting the graphene membrane, the net force must balance to zero; however, the stresses are different considering the cross-sectional area of gold electrodes (~500 nm × 60 nm) and graphene (~500 nm × 0.3 nm). The large difference in the cross-sectional area implies that the effective stiffness of gold electrodes is large compared to the stiffness of graphene. As a result, to a very good approximation, the total elastic strain at a given temperature in graphene, that is confined by "rigid" gold electrodes, is $\epsilon_{\text{grapheneclamped}} = \epsilon_{\text{graphene}}(T) + \epsilon_{\text{substrate}}(T) - \epsilon_{\text{gold}}(T)\frac{w_{\text{electrode}}}{L}$, where $w_{\text{electrode}}$ is the average of the width of gold electrodes holding the suspended flake. The change in tension in the membrane can be written as a function of temperature as $\Delta\Gamma_0(T) = wt\epsilon_{\text{grapheneclamped}}(T)E_{\text{graphene}}$, where E_{graphene} is Young's modulus of graphene. Measuring the tension $\Gamma_0(T)$ as a function of temperature offers a way to track the thermal strain in the graphene membrane. Figure 6.4a shows the evolution of resonant frequency as a function of temperature from a device at $V_g^{\text{DC}} = 15$ V, where the increase in frequency is largely due to the contraction of the gold electrodes. However, this rate of increase of resonant frequency is significantly reduced because of the negative α_{graphene} for all $T < 300$ K from the case of frequency change including only the gold's contraction. Using such a measurement of frequency shift while assuming uniform expansion of all the materials and using Equation (6.8), it is possible to calculate the expansion coefficient from the frequency at $V_g^{\text{DC}} = 0$ V as

$$\alpha_{\text{graphene}}(T) = -2f_0(0)\frac{df_0(0)}{dT} \times \frac{(2L)^2\rho}{E_{\text{graphene}}} + \frac{d}{dT}\left(\epsilon_{\text{substrate}}(T) - \epsilon_{\text{gold}}(T)\frac{w_{\text{electrode}}}{L}\right) \tag{6.9}$$

Figure 6.4b shows the result of calculating α_{graphene} for two devices using this analysis and comparison with the theoretical calculation for α_{graphene} by Mounet and Marzari [24]. We find that α_{graphene} is negative and its magnitude decreases with temperature for $T < 300$ K. At room temperature, $\alpha_{\text{graphene}} \sim -7 \times 10^{-6}\ \text{K}^{-1}$, which is similar to the previously reported values measured by others [6, 43]. At 30 K, $\alpha_{graphene} \sim -1 \times 10^{-6}$ K^{-1}. The deviation of $\alpha_{graphene}$ from the theoretically predicted values can possibly be due to the presence of the impurities on the graphene membrane. The knowledge of $\alpha_{graphene}$ is essential for the fabrication of the devices intended for strain engineering applications [26, 27]. Strain engineering of graphene devices at low temperatures using this picture can improve device performance, for example, by enabling temperature compensation [44].

Another effect of the temperature is to modify the quality factor of the graphene resonator. This is clearly seen from Figure 6.4a, where narrowing of the resonance peak with temperature is evident, and by fitting the frequency response at every temperature one can fit the data to extract a Q from the data. The result of such a fit can be seen in Figure 6.4c. On cooling the graphene electromechanical resonators, the resonance frequency as well as the quality factor (Q) increase. Figure 6.4c shows the variation of Q with temperature from device, as it can be seen that Q can be increased by a factor of 4 in this device. All the devices we measured showed increase in Q on cooling. The observation that the increase in resonant frequency is accompanied by an increase in the Q suggests that the loss mechanism may be frequency independent [39]. The inset in the figure shows the plot of Q^{-1}, which is directly proportional to the dissipation in the resonator, with temperature. At lower temperatures, dissipation reduces more slowly when compared to the temperatures greater than 100 K. Similar behavior has been observed by Chen *et al.* [6].

6.6
High-Mobility Suspended Graphene Devices to Study Quantum Hall Effect

We now discuss how these suspended graphene devices can have very high mobility as the scattering due to the dangling bonds and other dopants on the surface of SiO_2 trapped beneath graphene are eliminated. The procedure for fabricating the suspended graphene has been already described.

Figure 6.5a shows the variation of zero bias conductance of a graphene device with gate voltage right after fabrication (top curve). After fabrication, the device has no clear Dirac peak (also the charge neutrality point) and has poor charge carrier mobility. To improve performance, the device is annealed at cryogenic temperatures in cryogenic vacuum by sending gradually larger current through the device (Figure 6.5b) [6, 23, 35–37] and switching the polarity after some changes in conductance are observed. Current annealing increases the local temperature and ablates the resist residue and organic traces due to various lithographic steps. Figure 6.5a shows the result of various repetitions of current annealing procedures on the same device and gradually one observes a sharp Dirac peak as a function of gate voltage. To optimize the cleaning process, the gate response is checked after every cleaning step. At the end of the last annealing step, the mobility of the suspended graphene is in excess of 150,000 cm^2 V^{-1} s^{-1}. Similar mobilities have been observed by other groups [6, 23, 35–37].

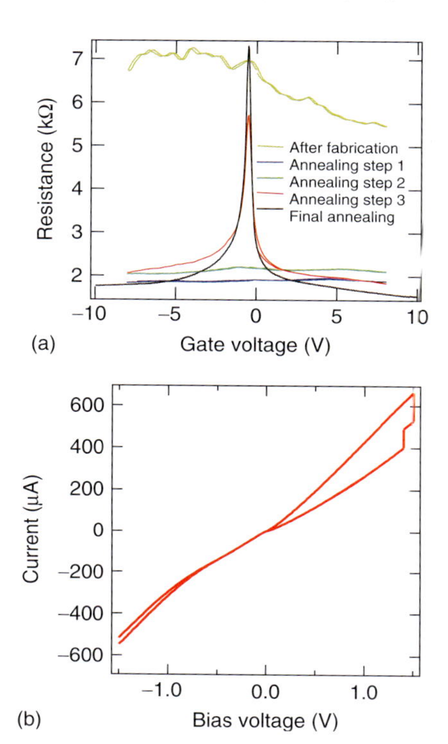

Figure 6.5 (a) Current annealing of the graphene device by sending bipolar current. Initially, the graphene device does not show the Dirac peak and has overall high resistance. (b) Plot of current and voltage as the suspended graphene device is biased. The abrupt jumps at the extreme voltage bias indicate the resistance changes in the device.

The high mobility of the current annealed device can also reflect in the observation of integer quantum Hall effect (IQHE) at very low magnetic fields as the disorder is significantly reduced. The width of the Dirac peak is a measure of the reduction in the charge inhomogeneity. Figure 6.6 shows the colorscale plot of the derivative of two-probe conductance as a function of magnetic field and gate voltage at 4 K. It is clearly seen that for magnetic fields as low as 0.25 T, various LLs can be resolved. Further measurements (not shown) at high magnetic fields (~14 T) show fractional quantum Hall plateaus as seen by other groups [45, 46].

Suspended graphene devices offer a unique system to probe nanoscale electromechanics with application in charge sensing and mass sensing. In addition, suspended graphene devices enable realization of high mobilities and

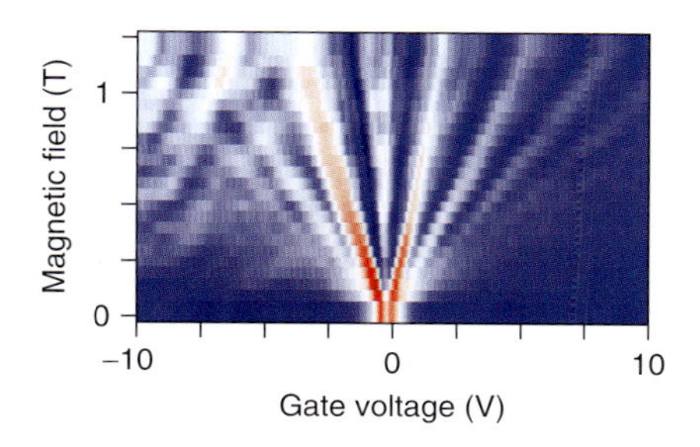

Figure 6.6 Colorscale plot of resistance as a function of the gate voltage and magnetic field after current annealing at cryogenic temperatures. The Landau level spitting can be seen at 1 T at 4 K.

ballistic transport. The reduction in charge disorder also enables observation of IQHE at low fields ($\approx$1 T) and commonly realizable temperature of 4 K.

Acknowledgments

We acknowledge the help from Rohan Dhall, Adrian Allain, Ganesh Subramanian, and Bushra Irfan. Financial support from the Government of India is acknowledged.

References

1. Novoselov, K.S., Geim, A.K., Morozov, S.V., Jiang, D., Zhang, Y., Dubonos, S.V., Grigorieva, I.V., and Firsov, A.A. (2004) *Science*, **306** (5696), 666–669.
2. Novoselov, K.S., Geim, A.K., Morozov, S.V., Jiang, D., Katsnelson, M.I., Grigorieva, I.V., Dubonos, S.V., and Firsov, A.A. (2005) *Nature*, **438** (7065), 197–200.
3. Zhang, Y., Tan, Y.-W., Stormer, H.L., and Kim, P. (2005) *Nature*, **438** (7065), 201–204.
4. Lee, C., Wei, X., Kysar, J.W., and Hone, J. (2008) *Science*, **321** (5887), 385–388.
5. Scott Bunch, J., van der Zande, A.M., Verbridge, S.S., Frank, I.W., Tanenbaum, D.M., Parpia, J.M., Craighead, H.G., and McEuen, P.L. (2007) *Science*, **315** (5811), 490–493.
6. Chen, C., Rosenblatt, S., Bolotin, K.I., Kalb, W., Kim, P., Kymissis, I., Stormer, H.L., Heinz, T.F., and Hone, J. (2009) *Nat. Nano*, **4**, 861–867.
7. Garcia-Sanchez, D., van der Zande, A.M., San Paulo, A., Lassagne, B., McEuen, P.L., and Bachtold, A. (2008) Imaging mechanical vibrations in suspended graphene sheets. *Nano Lett.*, **8** (5), 1399–1403.
8. Huttel, A.K., Steele, G.A., Witkamp, B., Poot, M., Kouwenhoven, L.P., and van der Zant, H.S.J. (2009) *Nano Lett.*, **9** (7), 2547–2552.
9. Lassagne, B., Tarakanov, Y., Kinaret, J., Garcia-Sanchez, D., and Bachtold, A. (2009) *Science*, **325** (5944), 1107–1110.
10. Sazonova, V., Yaish, Y., Ustunel, H., Roundy, D., Arias, T.A., and McEuen, P.L. (2004) *Nature*, **431** (7006), 284–287.
11. Steele, G.A., Huttel, A.K., Witkamp, B., Poot, M., Meerwaldt, H.B., Kouwenhoven, L.P., and van der Zant, H.S.J. (2009) *Science*, **325** (5944), 1103–1107.
12. Chiu, H.-Y., Hung, P., Postma, H.W.Ch., and Bockrath, M. (2008) *Nano Lett.*, **8** (12), 4342–4346.
13. Knobel, R., Yung, C.S., and Cleland, A.N. (2002) *Appl. Phys. Lett.*, **81** (3), 532–534.
14. Lassagne, B., Garcia-Sanchez, D., Aguasca, A., and Bachtold, A. (2008) *Nano Lett.*, **8** (11), 3735–3738.
15. Feng, X.L., He, R., Yang, P., and Roukes, M.L. (2007) *Nano Lett.*, **7** (7), 1953–1959.
16. Solanki, H.S., Sengupta, S., Dhara, S., Singh, V., Patil, S., Dhall, R., Parpia, J., Bhattacharya, A., and Deshmukh, M.M. (2010) *Phys. Rev. B*, **81**, article no. 115459.
17. He, R., Feng, X.L., Roukes, M.L., and Yang, P. (2008) Self-transducing silicon nanowire electromechanical systems at room temperature. *Nano Lett.*, **8** (6), 1756–1761.
18. Carr, D.W., Evoy, S., Sekaric, L., Craighead, H.G., and Parpia, J.M. (1999) *Appl. Phys. Lett.*, **75** (7), 920–922.
19. Naik, A.K., Hanay, M.S., Hiebert, W.K., Feng, X.L., and Roukes, M.L. (2009) *Nat. Nano*, **4** (7), 445–450.
20. Unterreithmeier, Q.P., Weig, E.M., and Kotthaus, J.P. (2009) *Nature*, **458** (7241), 1001–1004.
21. Ekinci, K.L. and Roukes, M.L. (2005) *Rev. Sci. Instrum.*, **76** (6), 061101–061112.

22. Schedin, F., Geim, A.K., Morozov, S.V., Hill, E.W., Blake, P., Katsnelson, M.I., and Novoselov, K.S. (2007) *Nat. Mater.*, **6** (9), 652–655.
23. Singh, V., Sengupta, S., Solanki, H.S., Dhall, R., Allain, A., Dhara, S., Pant, P., and Deshmukh, M.M. (2010) *Nanotechnology*, **21** (16), 165204.
24. Mounet, N. and Marzari, N. (2005) *Phys. Rev. B*, **71** (20), 205214.
25. Bao, W., Miao, F., Chen, Z., Zhang, H., Jang, W., Dames, C., and Ning Lau, C. (2009) *Nat. Nano*, **4** (9), 562–566.
26. Pereira, V.M. and Castro Neto, A.H. (2009) *Phys. Rev. Lett.*, **103** (4), 046801–046804.
27. Guinea, F., Katsnelson, M.I., and Geim, A.K. (2010) *Nat. Phys.*, **6**, 30–33.
28. Prange, R.E. and Girvin, S.M. (1990) *The Quantum Hall Effect*, Spriger-Verlag, New York.
29. Sarma, S.D. and Pinczuk, A. (2004) *Perspectives in Quantum Hall Effects: Novel Quantum Liquids in Low-dimensional Semiconductor Structures*, Wiley-VCH Verlag GmbH, Weinheim.
30. Yennie, D.R. (1987) *Rev. Mod. Phys.*, **59**, 781–824.
31. Castro Neto, A.H., Guinea, F., Peres, N.M.R., Novoselov, K.S., and Geim, A.K. (2009) *Rev. Mod. Phys.*, **81**, 109–162.
32. McClure, J.W. (1956) *Phys. Rev.*, **104**, 666–671.
33. Gusynin, V.P. and Sharapov, S.G. (2005) *Phys. Rev. Lett.*, **95**, 146801.
34. Singh, V. and Deshmukh, M.M. (2009) *Phys. Rev. B*, **80**, 081404.
35. Bolotin, K.I., Sikes, K.J., Hone, J., Stormer, H.L., and Kim, P. (2008) *Phys. Rev. Lett.*, **101**, 096802.
36. Bolotin, K.I., Sikes, K.J., Jiang, Z., Klima, M., Fudenberg, G., Hone, J., Kim, P., and Stormer, H.L. (2008) *Solid State Commun.*, **146** (9–10), 351–355.
37. Du, X., Skachko, I., Barker, A., and Andrei, E.Y. (2008) *Nat. Nano*, **3** (8), 491–495.
38. Jensen, K., Kim, K., and Zettl, A. (2008) *Nat. Nano*, **3** (9), 533–537.
39. Sazonova, V. (2006) A tunable carbon nanotube resonator. PhD thesis. Cornell University.
40. Nix, F.C. and MacNair, D. (1941) *Phys. Rev.*, **60** (8), 597.
41. Freund, L.B. and Suresh, S. (eds) (2003) *Thin Film Materials*, Cambridge University Press.
42. Jiang, J.-W., Wang, J.-S., and Li, B. (2009) *Phys. Rev. B*, **80** (11), 113405–113404.
43. Bao, W., Miao, F., Chen, Z., Zhang, H., Jang, W., Dames, C., and Ning Lau, C. (2009) *Nat. Nanotechnol.*, **4** (9), 562–566.
44. Ho, G.K., Sundaresan, K., Pourkamali, S., and Ayazi, F. (2005) Low-motional-impedance highly-tunable i^2 resonators for temperature-compensated reference oscillators. 18th IEEE International Conference on Micro Electro Mechanical Systems, 2005. MEMS 2005, pp. 116–120.
45. Bolotin, K.I., Ghahari, F., Shulman, M.D., Stormer, H.L., and Kim, P. (2009) *Nature*, **462** (7270), 196–199.
46. Du, X., Skachko, I., Duerr, F., Luican, A., and Andrei, E.Y. (2009) *Nature*, **462** (7270), 192–195.

7
Electronic and Magnetic Properties of Patterned Nanoribbons: A Detailed Computational Study

Biplab Sanyal

7.1
Introduction

The discovery of graphene [1] in 2004 by the Manchester group started a revolution, which is still continuing at full pace. Graphene holds immense promise in future electronics and other technological advancements along with the fundamental understanding of high-energy physics through tabletop low-energy experiments and offers a truly interdisciplinary scientific domain, bringing together physicists, chemists, biologists, and materials scientists. It is astonishing to observe the tremendous potential of such a simple material that consists of carbon atoms in a 2D network. The potential of graphene was greatly recognized by awarding the Nobel Prize in physics in 2010 to A. K. Geim and K. S. Novoselov for synthesizing graphene for the first time. Many excellent review articles have been written on graphene and the readers are referred to those [2]. Not only from the point of view of physics, the chemistry of graphene, especially, the chemical routes for synthesizing graphene have been extremely influential in the field of graphene [3].

Regarding the applications of graphene, one of the most prominent ones is its use in field effect transistors (FETs). A review article by Schwierz [4] discusses the use of graphene in future FET devices. Apart from the usual back-gated applications, a higher doping level via top-gating has been possible, characterized by Raman scattering experiments [5]. Graphene is particularly appealing because of its one-atom thickness, and hence its suitability for high-frequency applications. However, an appropriate band gap is required in graphene to realize the devices for digital logic applications as graphene is a zero band gap semiconductor. In this regard, two major ways are considered to open up a band gap in graphene: (i) chemical functionalization and (ii) graphene nanoribbons (GNRs). Here, we mostly discuss the properties of GNRs where one has the possibility of tuning the band gap by tuning the width and crystallographic orientation of the ribbons. The readers are referred to a review article by Jia *et al.* [6] for fabrication and characterization of graphene edges. Also, the theoretical aspects of edge magnetism in GNRs were discussed in detail in recent review articles [7].

Graphene: Synthesis, Properties, and Phenomena, First Edition. Edited by C. N. R. Rao and A. K. Sood.

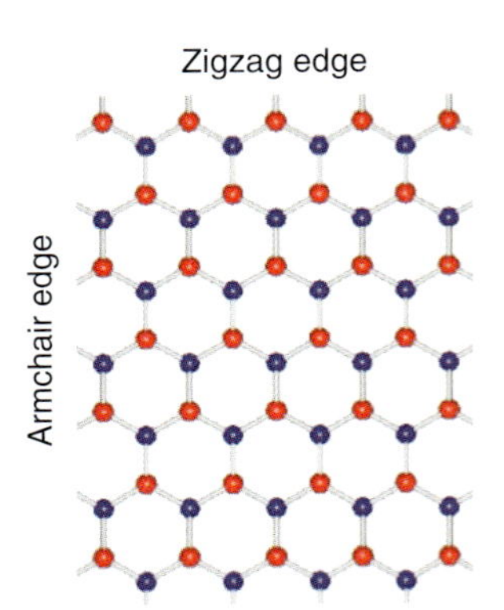

Figure 7.1 Two different types of GNRs are shown. The two colors of the balls indicate two different sublattices of carbon atoms.

GNRs are mainly classified into two types (Figure 7.1): armchair (AC) and zigzag (ZZ) structures. ZZ GNRs are particularly interesting because of the presence of the edge states and many interesting physics associated with them, to be discussed below. It should be noted that in reality, the edge structures of GNRs are quite complicated, with the presence of disorder in the form of vacancies, nonhexagonal carbon structures, and so on. The disordered edges often cause degradation in the carrier mobility, and, therefore, the synthesis of GNRs with atomically sharp edges remains a challenge.

7.2 Experimental Results

Recently, GNRs of various widths have been produced by chemical routes [8]. The GNRs of ZZ and AC edge structures were stabilized in solvents by noncovalent polymer functionalization. It was found from electrical transport measurements that the sub-10 nm GNRs were semiconducting. Also, GNRs had an on–off ratio of about 10^7 at room temperature in relation to FET actions. From the measured high-hole mobilities, it was inferred that the nanoribbons had a high quality and were not too much covalently functionalized to alter the properties.

Electronic transport in lithographically patterned GNRs was studied by Han *et al.* [9]. The presence of an energy gap near the charge neutrality point was observed because of the lateral confinement of charge carriers. It is interesting to note that the temperature-dependent conductance measurements indicated that the energy gap scaled inversely with the ribbon width, in accordance with the theoretical prediction by Son *et al.* [10] (Figure 7.2). This offers a great potential to engineer the band gap of graphene nanostructures by tuning the width of the ribbon by lithographic processes.

The atomic structure and electronic properties of the ribbons have been investigated by scanning tunneling microscopy and tunneling spectroscopy measurements [11]. The authors reported an opening of confinement gaps up to 0.5 eV, thereby providing the possibility of realizing operations of graphene at room

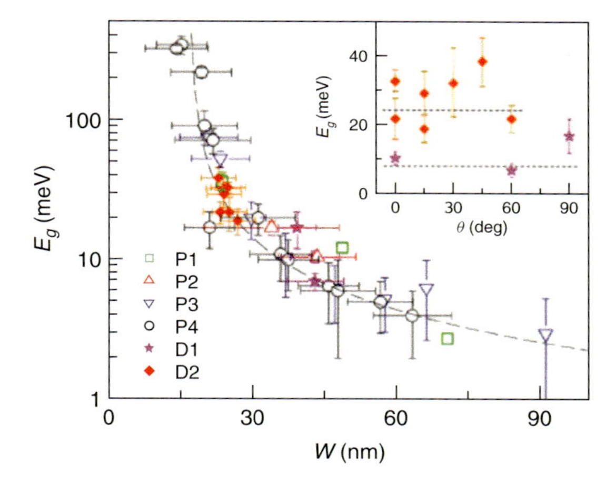

Figure 7.2 Band gap (E_{gap}) versus width (W) for six different device sets considered by Han *et al.* [16]. The inset shows the band gap versus relative angle for the device sets. (Source: Reprinted with permission from Han *et al.* (2007), *Phys. Rev. Lett.*, **98**, 206805. Copyright (2007), the American Physical Society.)

temperature. The structure and dynamics at the edges of holes in the graphene lattice produced by prolonged electron irradiation have been studied [12] by using transmission electron aberration-corrected microscopy. The dynamics of C atoms at the edge of a hole and the rearrangement of bonds were recorded during the growth of the hole. Hence, it was possible to understand the mechanism of edge reconstruction and the formation and stabilization of ZZ edges around the hole.

The transport properties of GNRs of 30 nm width and varying lengths were studied by Gallagher *et al.* [13]. It was found that the transport gap had a weak dependence on the ribbon length. On the contrary, the source–drain gap increased with increasing ribbon length. The ribbons were annealed to reduce the amplitude of the disorder potential and the transport gap shrunk and moved closer to zero gate voltage, whereas the source–drain gap was not affected by the annealing. It was concluded that the presence of charged impurities near the ribbon gave rise to a disorder potential that broke the ribbon into isolated puddles of charge carriers acting as quantum dots. Ribeiro *et al.* [14] performed two-terminal quantum Hall resistance measurements on lithographically patterned GNR devices. The presence of Landau levels was detected in the fragmented electronic spectrum in the presence of a magnetic field, which is a characteristic feature of the confinement of Dirac fermions. The competition between electronic and magnetic confinement at the edges of the nanoribbon gave rise to magnetoelectronic subbands. Also, the disorder-induced mixing between the chiral edges and the disappearance of quantum Hall effect were discussed with the aid of quantum simulations.

Apart from GNRs, nanographene structures have also been studied with regard to their synthesis and characterization [15]. These nanostructures have well-defined AC and ZZ edges. One of the interesting routes to produce nanographenes is through chemical oxidation of the graphene sheet and then cutting the sheet by scanning probe microscopic manipulation technique in the direction along which the epoxide and OH functional groups are attached to carbon atoms [16]. The linear defects with epoxide groups give rise to buckled structures and hence are the precursors of rupturing and formation of well-defined edges of nanographene. Characterization of these structures is done by using noncontact atomic force microscopy and scanning probe microscopy under ultrahigh vacuum conditions. The nanoribbons with ZZ edges give rise to interesting magnetic properties owing to the nonbonding π electrons. Moreover, stacking of a few sheets of nanographenes and the disordered arrangement of nanographene networks yield spin-glass behavior owing to several competing exchange interaction pathways [15].

7.3 Theory of GNRs

The essential features in the electronic structure of graphene were put forward a long time back by Wallace [17] by the single band tight-binding (TB) theory. However, the details of the geometry and electronic structure are better reproduced by materials-specific first principles density functional theory (DFT) [18], especially in connection with the quantitative estimate of materials properties. In DFT calculations, the most common forms of exchange–correlation potentials are within local density approximation (LDA) and generalized gradient approximation (GGA). Some of the popular DFT codes employed in the research on graphene are VASP [19], SIESTA [20], Quantum Espresso [21], and so on. In the following, some works based on both TB and DFT theories are reviewed.

7.3.1 Tight-Binding Method

Nakada *et al.* [22] studied the electronic structure of GNRs using the TB theory to investigate, for example, the characteristics of localized edge states of ZZ GNRs. They have well demonstrated the evolution of band structures of GNRs from bulk graphene. Bulk graphene is a zero band gap semiconductor with a linear dispersion relation. At the Dirac point (denoted as K in the Brillouin zone (BZ)), one finds that the highest occupied and lowest unoccupied bands are degenerate. By projecting the bulk band structure onto the same of GNRs using the zone-folding technique, one finds that the Dirac point of bulk graphene appears at $k = 0$ and $|k| = 2\pi/3$ for AC and ZZ GNR, respectively (Figure 7.3). The width of AC GNRs determines whether the GNRs will be semiconducting or metallic. Metallicity occurs for $N = 3M{-}1$, where M is an integer and N denotes the number of rows in the GNR. For the semiconducting GNRs, the direct gap at $k = 0$ decreases with the increase in the ribbon width. In the asymptotic limit, the projected band structure of graphene is almost reproduced by that of an AC GNR of a very large width. ZZ GNRs show a completely different behavior. In this case, the highest occupied valence band state and the lowest unoccupied conduction band state are always degenerate at $k = \pi$

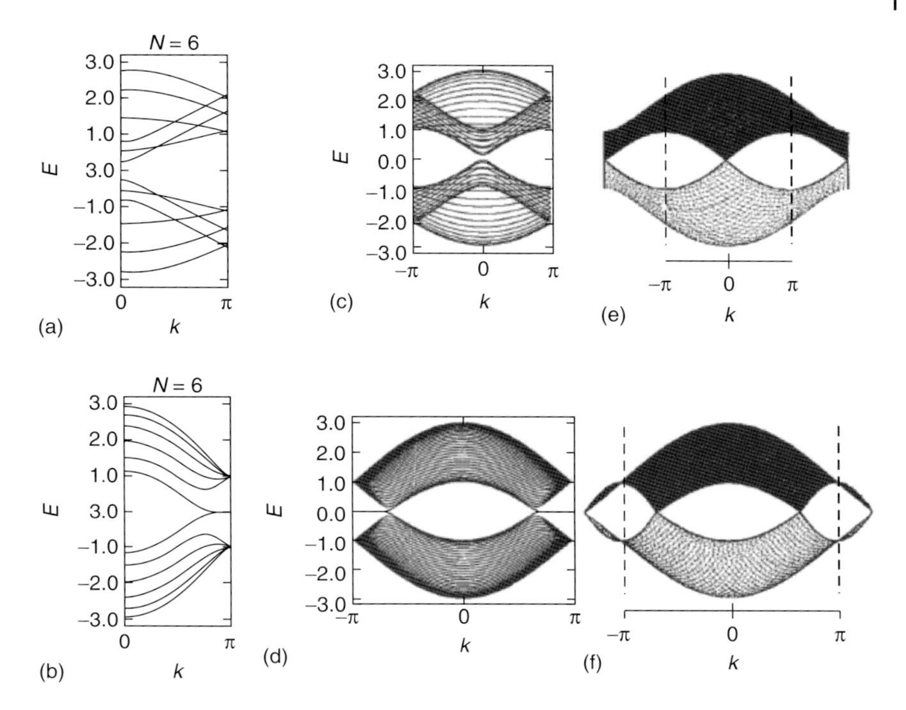

Figure 7.3 (a and b) Calculated band structure of a 6-row-wide armchair and zigzag ribbons, respectively; (c and d) band structure of a 30-row-wide armchair and zigzag ribbons, respectively; (e and f) projected band structure of bulk graphene onto an armchair and zigzag axes, respectively. (Source: Reprinted with permission from Nakada *et al.* (1996), *Phys. Rev. B*, **54**, 17954. Copyright (1996), the American Physical Society.)

although the degeneracy appears already at $|k| = 2\pi/3$ on the basis of the projected band structure of bulk graphene (Figure 7.3). In the 1D BZ, two almost flat bands near the Fermi level are found between $k = 2\pi/3$ and $k = \pi$. The charge density distributions corresponding to these flat bands are located at the GNR edges. The localized state at $k = \pi$ and the delocalized state at $|k| = 2\pi/3$ are connected by the edge state of the semi-infinite graphene sheet. The conditions for the edge state have been analytically derived in the paper by Nakada *et al.* [22] for a semi-infinite graphene sheet.

Another interesting study by Nakada *et al.* is the electronic structure of various edge shapes and sizes of GNRs. They found the presence of an edge state with less-developed ZZ edges with only three or four ZZ sites per sequence. It was concluded that graphene flakes of a nanometer length are the best candidates to exhibit the special edge state. These discussions are quite relevant in the case of porous carbon materials, where various types of edge sites appear. Localization of charge density at the edges dictates the transport properties in these highly inhomogeneous structures. Wakabayashi *et al.* [23] studied the electronic and magnetic properties of AC and ZZ GNRs in the presence of a magnetic field using the TB model. It was found that in a magnetic field, the edge state generates a rational fraction of the magnetic flux in each hexagon of the graphene lattice. Also, the orbital diamagnetic susceptibility is controlled by the structure of the edges. A crossover from high-temperature diamagnetic to low-temperature paramagnetic behavior in the magnetic susceptibility of zigzag graphene nanoribbons (ZGNR)s was reported.

The effects of carrier density on the magnetic properties of ZGNRs within a π orbital Hubbard-model mean field approximation were studied by Jung and MacDonald [24]. In an experimental situation, the carrier doping may be realized by the gate voltage. In this work, the authors used the following Hamiltonian.

$$H = -t \sum_{<i,j>} c_{i\sigma}^{\dagger} c_{j\sigma} + U \sum_{i} n_{i\bar{\sigma}} n_{i\sigma} c_{i\sigma}^{\dagger} c_{i\sigma} + v_{\text{ext}} \sum_{i} c_{i\sigma}^{\dagger} c_{i\sigma} \tag{7.1}$$

t is the nearest-neighbor hopping parameter, whereas U defines the Hubbard parameter for Coulomb interaction. The value of t was chosen to be 2.6 eV, a value adopted by almost all TB theories. $U = 2$ eV was considered to be consistent with LDA-based DFT calculations. v_{ext} is the external potential term describing the interaction with the constant positive background charge. The resulting phase diagram is shown in Figure 7.4. P, F, AF, Fb, AFb, and NC denote paramagnetic, ferromagnetically coupled edges, antiferromagnetically coupled edges, ferromagnetic interaction that allows breaking of symmetry, antiferromagnetic coupling that allows symmetry breaking, and noncollinear configuration. In the AF state of ZGNR, the inversion symmetry across the ribbon is broken in opposite senses in the two spin subsystems. In the low-doping regime, the charge density is distributed asymmetrically across the ribbon, and gives rise to multiferroism in which spin polarization and charge density are coupled. In the weakly doped regime, noncollinear solutions have the ground state with canted spin orientations. Above a certain critical doping density, which is inversely proportional to the ribbon

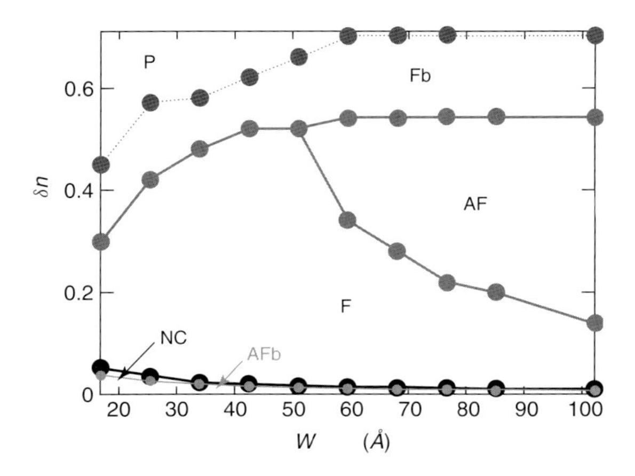

Figure 7.4 Hubbard-model self consistent field (SCF)-theory phase diagram as a function of doping and ribbon width. (Source: Reprinted with permission from Jung and MacDonald (2009) *Phys. Rev. B* **79**, 235433. Copyright (2009), the American Physical Society.)

width, it was found that the system undergoes a transition to an F configuration where the edges are ferromagnetically coupled. When doping is increased further, inversion symmetry breaking occurs across the ribbon. In this high-doping regime the energy differences between different magnetic configurations are small. At sufficiently high doping, only paramagnetic solutions exist. Therefore, the sequence of transition in narrow ribbons is AF - NC - F - Fb - P, whereas for wider ribbons, an additional AF state occurs between the F and Fb regimes.

7.3.2
First Principles Studies

The TB results of Nakada *et al.* [22] were verified by the first principles density functional calculations within local spin density approximation (LSDA) by Son *et al.* [10]. They also predicted a set of doubly degenerate flat edge-state bands, and hence a very large density of states (DOSs) at the Fermi level in a non-spin-polarized calculation for ZZ GNRs. However, the inclusion of spin polarization gives rise to a magnetic ground state with ferromagnetically aligned moments along the edge and antiferromagnetically coupled ones across the width of the GNR. In these calculations, the edge C atoms were passivated by H atoms. This AFM ground state yields a semiconducting electronic structure, although the ferromagnetically coupled edges have a metallic electronic structure. The hierarchy in the total energies from highest to lowest is as follows: nonmagnetic, ferromagnetic, and antiferromagnetic. The total energy difference per edge atom between non-spin-polarized- and spin-polarized edge states increases from 20 meV ($N = 8$) to 24 meV ($N = 16$). The total energy difference between ferromagnetic and antiferromagnetic couplings between edges, however, decreases as the width of the ribbon increases and eventually becomes negligible if the width is significantly larger than the decay length of the spin-polarized edge states. These energy differences per unit cell are 4.0, 1.8, and 0.4 meV for the 8-, 16-, and 32-row ZGNR, respectively. The energy gaps in ZGNRs originate from the staggered sublattice potentials resulting from the magnetic ordering. This happens because the opposite spin states on opposite edges occupy different sublattices. ZGNR is analogous to a single BN sheet because in the former, a band gap opens up owing to the exchange potential difference in the two sublattices, whereas for BN the band gap originates from the ionic potential difference between boron and nitrogen atoms located on different sublattices.

With applied transverse electric fields, Son *et al.* [25] found that the valence and conduction edge-state bands associated with one spin channel close their gap, whereas the opposite happens for the other spin channel. So, under appropriate field strengths, the ZGNRs show a half-metallic electronic structure, resulting in a semiconducting behavior for one spin and metallic behavior for the other. According to the calculations for a 16-row-wide ZGNR, the band gap in one spin channel is completely closed by an external electric field of 0.1 eV Å^{-1}, whereas the gap for the other channel remains very large at 0.30 eV. Another interesting observation is that the energy gap changes to an indirect gap from a direct one as the external field increases, and is closed indirectly. The applied electric-field-induced half-metallicity of the ZGNRs occurs because of the induction of energy-level shifts of opposite signs for the spatially separated spin-ordered edge states. As the ground state of ZGNR reveals antiferromagnetically coupled edges, the effect of the external field on the two edges is opposite. Owing to the application of the field, the electrostatic potential is raised on one side and lowered on the other as the field increases. Son *et al.* also studied the effect of edge defects on half-metallicity. It was found that different defects of concentration 6–12%, such as dangling bonds, vacancies, and Stone–Wales (SW), cannot destroy the half-metallic nature.

In another study, Dutta *et al.* [26] performed DFT calculations (Figure 7.5) to study quasi-one-dimensional H-passivated ZGNRs of various widths with chemical dopants, boron and nitrogen, keeping the whole system isoelectronic with C atoms in graphene. In the case of extreme doping, all C atoms of ZGNRs are replaced by B and N atoms and ZZ BN nanoribbons are formed. The ground state remains antiferromagnetic for all dopings considered. Moreover, the application of an external electric field affects the electronic structure of the nanoribbon giving rise to semiconducting and half-metallic solutions. However, the ZZ BN nanoribbon with a terminating polyacene unit remains half-metallic independent of the width of the ribbon and the strength of the applied electric field.

Barone *et al.* [27] have studied systematically the electronic properties, optical spectra, and relative thermodynamic stability of the semiconducting GNRs by DFT within GGA and hybrid-functional formalism. Ribbons with bare and hydrogen-terminated edges, several crystallographic orientations, and widths up to 3 nm were considered. It was predicted that one needs to have 2–3 nm wide ribbons in order to produce materials with band gaps similar to Ge or InN. To

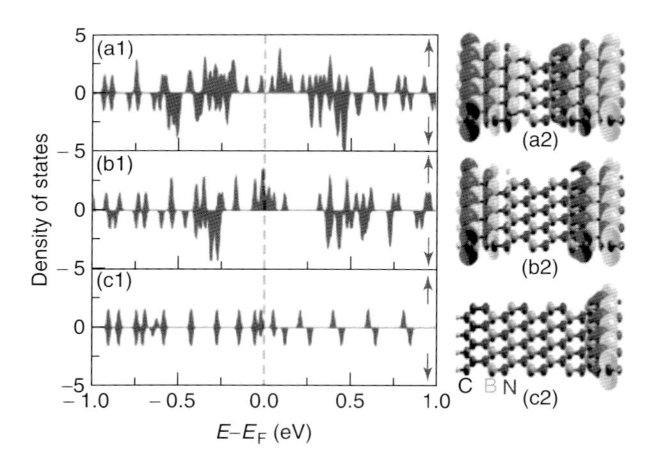

Figure 7.5 Density of states (DOSs) as a function of energy, scaled with respect to the Fermi energy for the 8-row ZGNR with (a1) two, (b1) four, and (c1) six zigzag carbon chains replaced by the zigzag boron–nitrogen chains in the middle of the nanoribbon. The spin density profile of these systems are shown in (a2), (b2), and (c2), respectively. (Source: Reprinted with permission from Dutta *et al.* (2009) *Phys. Rev. Lett.*, **102**, 096601. Copyright (2009), the American Physical Society.)

achieve larger band gaps comparable to those of Si, InP, or GaAs, one should have 1–2 nm wide ribbons. For AC GNRs, the estimated band gap is around 0.3 eV for 8 nm wide ribbons, whereas an 80 nm wide one produces a gap that is 1 order of magnitude less. One can also envisage chiral nanoribbons generated by unfolding carbon nanotubes of indices (n, m) where the chiral angle is defined as $tan\phi = (3^{\frac{1}{2}} m/(2n + m))$. Here, the band gap oscillations rapidly vanish as a function of the chiral angle. The authors have suggested that the optical properties may be used to characterize the nature of GNRs. This is expected to provide a practical way of providing information on the size of the GNRs as well as the nature of the edges.

Recently, Yazyev *et al.* [28] addressed the issue of edge magnetism in chiral GNRs. They used a model Hamiltonian, including electron–electron interactions, to systematically investigate the electronic structure and magnetic properties of chiral GNRs. They suggested a structure–property relationship where the dependence of magnetic moments and edge-state energy splitting on the nanoribbon width and chiral angle was discussed. It was concluded that the magnetic edge states appear as an intrinsic feature of smooth GNRs with chiral edges.

7.4 Hydrogenation at the Edges

7.4.1 Stability of Nanoribbons

Wassmann *et al.* [29] have determined the geometry, stability, and electronic and magnetic structure of hydrogen-terminated GNR edges as a function of the hydrogen content of the environment by means of DFT. Antiferromagnetic ZZ ribbons are stable only at extremely low ultravacuum pressures. Under more standard conditions, the most stable structures are the mono- and dihydrogenated AC edges and a ZZ edge reconstruction with one dihydrogenated and two monohydrogenated sites (Figure 7.6). At high hydrogen concentration, bulk graphene is not stable and spontaneously breaks to form ribbons, in analogy with the spontaneous breaking of graphene into small-width nanoribbons, observed experimentally in solution.

7.4.2 Dihydrogenated Edges

A great body of work exists on single hydrogen-terminated GNRs. Bhandary *et al.* [30] have studied dihydrogenated GNRs using DFT. In a single hydrogenated case, referred to as 1H hereafter, the structure of the nanoribbon remains planar with a slight rearrangement of C atoms to reach a C–C bond distance close to 1.42 Å (which is the bulk graphene bond length) in the middle of the ribbon. The 1H-terminated ZGNR is metallic and spin polarized with an average moment of 0.25 μ_B per C atom. The relevant situation is when two H atoms terminate a C

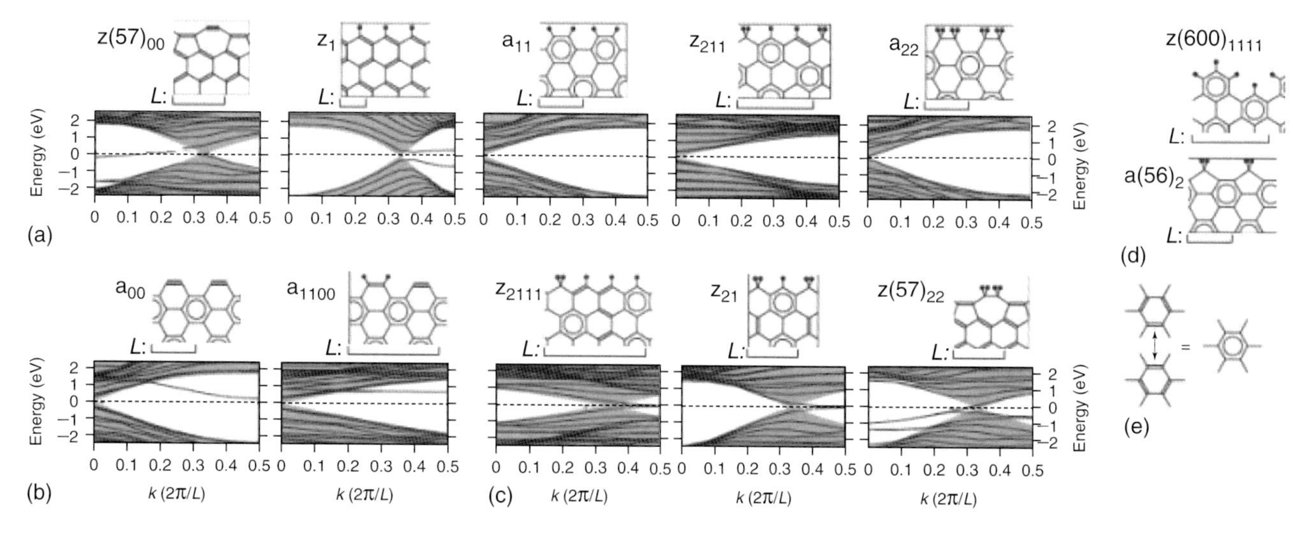

Figure 7.6 (a) Band structures of the five most stable hydrogen-terminated edges of a graphene nanoribbon are shown. Carbon–carbon bonds are represented by the standard notation, while hydrogen atoms are the filled circles. The structures are periodic along the ribbon edge with periodicity *L*. (b and c) Other stable armchair and zigzag terminations. (d) Examples of other types of edges. (e) The benzenoid aromatic carbon ring. (Source: Reprinted with permission from Wassmann *et al.* (2008) *Phys. Rev. Lett.*, **101**, 096402. Copyright (2008), the American Physical Society.)

atom. The chemical binding for this edge C atom is an sp^3 hybrid, with two C atoms and two H atoms as the nearest neighbors.

For the termination of edges with 2H atoms per C, the formation energies are defined in two ways:

$$\begin{aligned} E_f^{(1)} &= E(\text{G2H}) - [E(\text{G1H}) + E(\text{H}_2)] \\ E_f^{(2)} &= E(\text{G2H}) - [E(\text{G1H}) + 2[E(\text{H}) + (\text{BE/H})_{\text{exp}}], \end{aligned} \tag{7.2}$$

where $E(G2\text{H})$ and $E(G1\text{H})$ are the total energies for the ZZ GNRs' edges terminated with 1H and 2H atoms per edge, respectively; $E(\text{H}_2)$ and $E(\text{H})$ in Equation 7.2 are the calculated energies for a H_2 molecule and a H atom, respectively; and $(\text{BE/H})_{\text{exp}}$ is the experimental binding energy per H atom of the H_2 molecule, and it amounts to 2.24 eV. The calculated value is 2.27 eV, which is in very good agreement with the experimental one.

The calculated formation energies increase with the increase in the width of the nanoribbon up to seven rows and then saturate. Moreover, the 3-row ZGNR is spontaneously formed at $T = 0$ K. The calculated formation energies for large widths compare well with the results of Wassmann *et al.* [29]. In order to investigate the influence of finite temperature/elevated gas pressure, Gibbs free energies were calculated as a function of the chemical potential of the hydrogen molecule, according to the following formula given by Wassmann *et al.* [29]:

$$\begin{aligned} G_{\overline{H}} &= E_f^{(1)} - \frac{1}{2}\rho_{\text{H}}\mu_{\text{H}_2} \\ G_{1\text{H}/2\text{H}} &= E_f^{1\text{H}/2\text{H}} - \frac{1}{2}\rho_{\text{H}}\mu_{\text{H}_2} \\ \mu_{\text{H}_2} &= H_{\text{en}}^0(T) - H_{\text{en}}^0(0) - TS^0(T) + k_{\text{B}}T\ln\left(\frac{P}{P^0}\right) \end{aligned} \tag{7.3}$$

In the above equations, $G_{\overline{H}}$ is the Gibbs energy calculated from the zero-temperature formation energy $E_f^{(1)}$ shown before; μ_{H_2} is the chemical potential of H_2 at temperature T and pressure P calculated; H_{en} and S^0 are enthalpy and entropy at temperature T, respectively; P and k_B are the partial pressure and Boltzmann constant, respectively; P^0 is the reference pressure, which is 0.1 bar according to the tabular data; ρ_{H} is the edge hydrogen density expressed as $\frac{N_{\text{H}}}{2L}$, where N_{H} is the number of H atoms in the unit cell; $E(\text{G2H})$, $E(\text{G1H})$, and $E(\text{H}_2)$ have been defined above; and $G_{1H/2H}$ is the Gibbs energy calculated using the formation energy of 1H- or 2H-terminated edges ($E_f^{1\text{H}/2\text{H}}$) with reference to the ZGNR without any H termination, that is, in the presence of the dangling bonds. The results are shown in Figure 7.7 for 300 K (main graph and left inset). The Gibbs free energy is shown after the division by $2L$, where L is the length of the unit cell. One can observe the stability of the nanoribbons terminated with 2H (shown in the main plot) with respect to 1H-terminated edges. The pressures required for stabilizing the 2H-terminated nanoribbons (except for the 3-row ZGNRs) are rather high as seen in the right inset of Figure 7.7. The plot in the right inset shows that the requirement of H_2 pressure for the stabilization of 2H-terminated nanoribbons decreases with the increase in temperature. The left inset shows the Gibbs free energies for

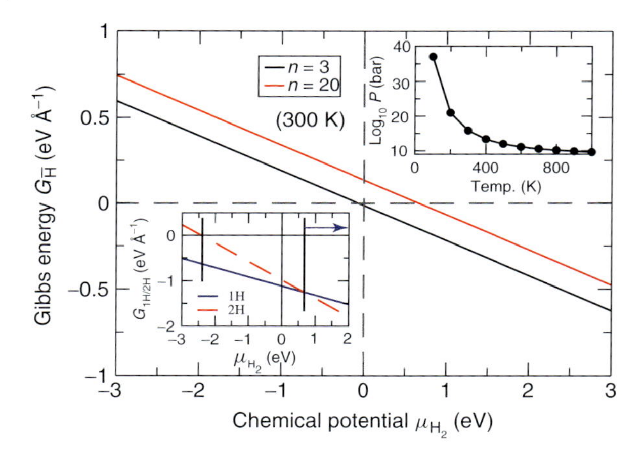

Figure 7.7 Calculated Gibbs formation energies (divided by $2L$) for $n = 3$ and 20-row-thick, 2H-terminated (main plot) and 1H- and 2H-terminated (left inset) ZGNR, at 300 K as a function of chemical potential of the hydrogen molecule according to the equations shown in the text. The left vertical line indicates the chemical potential for which Gibbs energy of the 2H-terminated edge becomes negative, whereas the right vertical line shows the transition point above which the Gibbs energy of 2H is lower than that of the 1H-terminated ZGNR. The blue horizontal arrow indicates that the 2H-terminated ZGNR continues to be stable over the 1H-terminated one after the crossing point. The plot in the right inset shows the logarithmic pressure of H_2 in units of bar for different temperatures at which the Gibbs energy $G_{\overline{H}}$ becomes negative for a 20-row ZGNR. (Source: Reprinted with permission from Bhandary *et al.* (2010) *Phys. Rev. B*, **82**, 165405. Copyright (2010), the American Physical Society.)

1H- and 2H-terminated 20-row ZGNR calculated with reference to the ZGNR with no H termination. It is clear that it is possible to stabilize 1H-terminated ZGNR at very low pressure at 300 K in comparison to 2H-terminated edges. The vertical line shows the chemical potential where a 2H-terminated ZGNR becomes stable. However, the 1H-terminated ZGNR has lower Gibbs energy than the 2H one till the crossing point of the curves marked by another vertical line in the figure is reached. The region after the crossing point of 1H and 2H curves indicates that the 2H-terminated edges can be stabilized over 1H-terminated ones under high pressures.

When two H atoms are attached to an edge C atom, the chemical bonds are completely saturated, giving rise to a vanishing magnetic moment of the edge C atom. Owing to dihydrogenation, the sp^2 planar structure becomes buckled, making an sp^3-like structure with an angle of 102° at the edge carbon. The C–C bond length increases at the edge. However, the inner bond lengths have an oscillatory behavior as one traverses toward the center of the ribbon. The C–C bond length of 1.42 Å in sp^2-bonded graphene is found in the middle of the ribbon, where the distortion is negligible. For the 20-row nanoribbon, the bond length reaches the ideal value in bulk graphene around the 15th bond from the edge. It is obvious that the edge sp^3 structure has a pronounced effect on smaller ribbons by

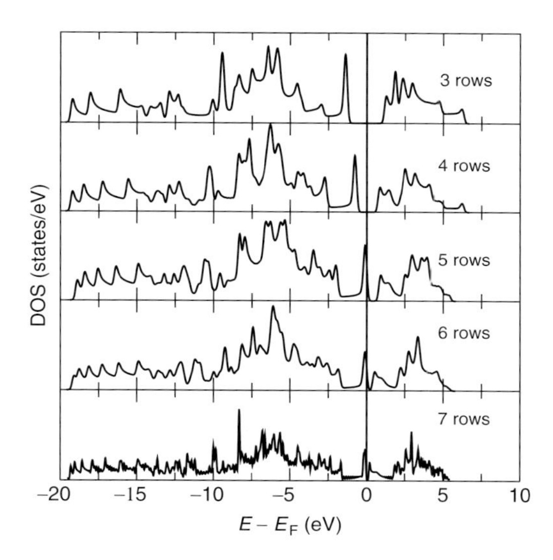

Figure 7.8 Calculated total DOSs of 2H-edge-terminated nanoribbons of various widths. (Source: Reprinted with permission from Bhandary *et al.* (2010) *Phys. Rev. B*, **82**, 165405. Copyright (2010), the American Physical Society.)

perturbing the whole structure, whereas for wider ribbons, this perturbation dies out and one gets a planar valley along with an sp^3 distorted edge.

Dihydrogenation produces an sp^3-like structure at the edge and a nonmagnetic semiconducting electronic structure. The DOS for different widths of ZGNR with 2H edge termination are shown in Figure 7.8. The calculated band gaps for 3- and 4-row ZGNRs are 2.1 and 2.08 eV, respectively. The band gaps decrease with an increase in the ribbon width up to 7-row ZGNR. However, a semiconductor-to-metal transition occurs for an 8-row ZGNR. For this width, the Fermi level cuts through a peak in the nonmagnetic DOS (right-hand side of Figure 7.8). The high value of the DOS at the Fermi level leads to an instability, and hence a spin-polarized solution leads to a lower energy state (Stoner instability). The spin-polarization energies are presented in Table 7.1 where it is clearly seen that the ribbons have a magnetic

Table 7.1 Differences in total energies per unit cell between the nonmagnetic and ferromagnetic states (ΔE) are shown for various widths of the nanoribbons. Also, the total magnetic moments of the unit cells along with the local magnetic moment of a C atom at the edge are shown.

Width (rows)	ΔE (meV)	Total moment (μ_B)	Edge moment (μ_B)
8	1.63	1.04	0.34
10	5.3	1.23	0.38
15	4.9	1.29	0.39
20	3.81	1.3	0.39

ground state, which is also metallic. The observation of a re-entrant magnetism for the dihydrogenated case is an interesting and intriguing phenomenon.

It is found that the second C atom from the edge carries most of the magnetic moment. Also, this second C atom shows a mid-gap state in a nonmagnetic calculation due to a strong perturbation in the potential of the edge C atom. The resulting sharp peak from the mid-gap state drives the magnetic solution of these atoms. The important difference between the 1H- and 2H-terminated ZGNRs is that the geometrical distortions of the network of sp^3-bonded atoms are unlikely for 1H termination, and hence one does not expect the modification of the electronic structure due to structural distortions. Figure 7.9 depicts the projected DOSs of the atoms starting from the edge and going inward for an 8-row ZGNR. Also, the corresponding magnetization density isosurface is shown. The edge C atom shows a gap in the DOS due to sp^3 bonding. In contrast, the DOS of the second

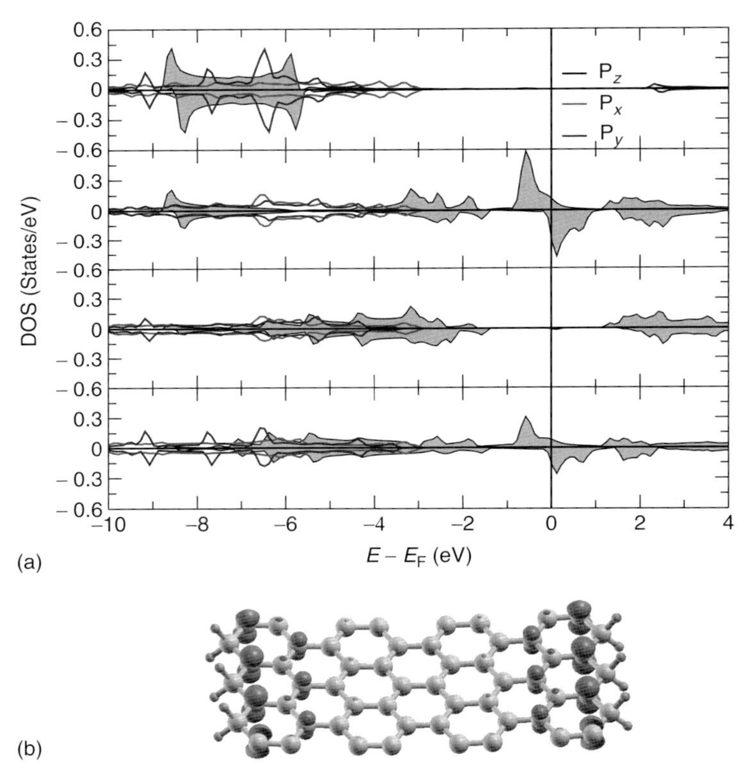

Figure 7.9 (a) Calculated spin-polarized DOS projected on C atoms for a 2H-edge-terminated 8-row nanoribbon. DOSs projected on C–p orbitals of different symmetries are shown. The topmost panel shows the projected DOS of a C atom at the edge, whereas the panels downward are for C atoms toward the center of the nanoribbon. (b) Magnetization density isosurfaces for an 8-row nanoribbon. (Source: Reprinted with permission from Bhandary *et al.* (2010) *Phys. Rev. B*, **82**, 165405. Copyright (2010), the American Physical Society.)

carbon atom from the edge with a p_z orbital character has a sharp peak at the Fermi level, giving rise to a spin-polarized ground state due to Stoner instability. This alternative occurrence of a gap and a spin-polarized state for the neighboring carbon atoms persists for a certain distance from the edge, albeit with a decreasing magnitude as the center of the ribbon is approached. The two different behaviors of carbon atoms situated in different sublattices have been described by the analysis of mid-gap states through a TB model in the paper by Bhandary *et al.* [30].

While graphene is a semimetal, *graphane*, a recently synthesized hydrogenated graphene is an insulator. Chandrachud *et al.* [31] have studied the transformation of graphene on hydrogenation to graphane within the framework of DFT. They have shown that hydrogenation favors clustered configurations, leading to the formation of compact islands. Through the analysis of the charge density and electron localization function, it was argued that, as hydrogen coverage increases, the semimetal turns into a metal, showing a delocalized charge density, then transforms into an insulator. It was also found that a weak ferromagnetic state exists even for a large hydrogen coverage whenever there is a sublattice imbalance in the presence of an odd number of hydrogen atoms. The metallic phase is spatially inhomogeneous as it contains islands of insulating regions formed by hydrogenated carbon atoms and metallic channels formed by contiguous bare carbon atoms. It is interesting to note that the interface between bare and hydrogenated C atoms shows two distinct types: AC and ZZ structures (see Figure 7.10b). The removal of hydrogen atoms along the diagonal of the unit cell, yielding an AC pattern at the edge, gives rise to a band gap of 1.4 eV (Figure 7.10a). This is similar to the usual AC nanoribbons discussed above. On the other hand, the interface parallel to the side of the unit cell gives rise to a ZZ structure, which produces a spin-polarized ground state as shown in Figure 7.10. One may envisage the control of the band gap of the AC interface by controlling the width of the region containing bare C atoms. Therefore, this may be viewed as an alternative way of realizing nanoribbons with controllable properties. It has been shown recently [32] from DFT calculations that Fe impurities deposited in the dehydrogenated channels with a ZZ interface establish a strong magnetic coupling among themselves. For the AC interface, a strong in-plane magnetic anisotropy was predicted.

From a realistic point of view, the graphene edges may have defective structures. Bhowmick and Waghmare [33] have discussed SW defects and their effects on the GNRs regarding the mechanical properties using DFT calculations and continuum theory. The orientation of the SW defect with respect to the ribbon edge dictates the sign of stress. A compressive stress gives rise to local warping in the GNR. The authors demonstrated how warping of GNRs can be nucleated at the SW defects localized at the edges and gives rise to graphene flakes observed in experiments. Also, in ZGNRs, SW defects weaken the magnetic ordering at the edges, yielding a nonmagnetic metallic ground state.

Self-passivating edge-reconstructed edge structures were shown to be more stable than the **zz** or **ac** ones by Koskinen *et al.* [34] using DFT calculations. The various edge structures considered in their study are shown in Figure 7.11. The reconstructed **zz(57)** structure was energetically lowered than the all hexagonal **zz**

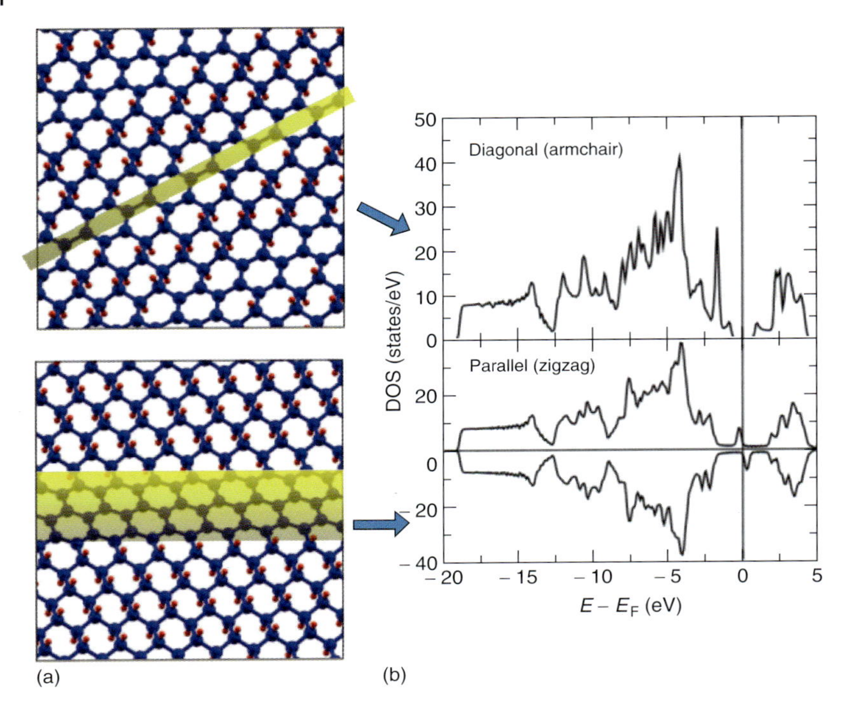

Figure 7.10 (a) Formation of an armchair (upper panel) and a zigzag nanoribbon at the interface between bare and hydrogenated C atoms. The yellow shades denote the region of the nanoribbon formation. (b) Total DOSs for an armchair and a zigzag nanoribbon at the interface correspond to the picture shown in (a). A spin-polarized DOS is shown for the zigzag nanoribbon. (Source: The figure has been taken from Chandarchud *et al.* [31]. Copyright (2011), IOP Publishing Ltd.)

edge by 0.35 eV Å ^{1}. This reconstructed edge has a triple bond of length 1.24 Å with a wider bond angle of 143°. It was also shown that the adsorption of hydrogen at the reconstructed edge becomes weaker because of the weakening of dangling bonds. A recent DFT study [35] has indicated an oscillatory magnetic exchange coupling across the edges of a reconstructed **zz(57)** ribbon decorated by Fe atoms.

7.5 Novel Properties

In organic nanospintronics, one of the objectives is to manipulate the spin states of organic molecules with a d-electron center by suitable external means, for example, temperature, pressure, and light. The central atom with unpaired d-electrons can have several spin states available. If the external agent is able to overcome the energy barrier, one may move from one spin state to another. For example, in the case of an iron porphyrin (FeP) molecule in the gas phase, the ground state has a

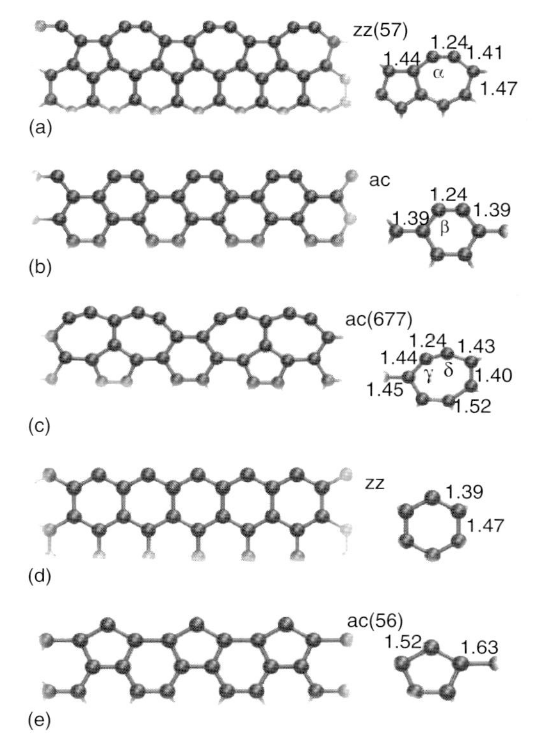

Figure 7.11 Various graphene edges: (a) reconstructed zigzag: **zz(57)**; (b) armchair: **ac**; (c) reconstructed armchair: **ac(677)**; (d) zigzag: **zz**; and (e) pentagonal armchair: **ac(56)**. (Source: Reprinted with permission from Koskinen *et al.* (2008) *Phys. Rev. Lett.*, **101**, 115502. Copyright (2008), the American Physical Society.)

$S = 1$ spin state with another higher energy minimum at $S = 2$ spin state, which is reachable by crossing an energy barrier of around 0.8 eV. Very recently, Bhandary *et al.* [36] demonstrated by first principles density functional calculations that a strain-induced change of the spin state, from $S = 1$ to $S = 2$, takes place for an FeP molecule deposited at a divacancy site in a graphene lattice. The process is reversible in the sense that the application of tensile or compressive strains in the graphene lattice can stabilize FeP in different spin states, each with a unique saturation moment and easy axis orientation. This is demonstrated in Figure 7.12, where the molecular energy diagrams are shown for two spin states along with the magnetization densities for the highest occupied molecular orbital (HOMO). It was well demonstrated that a change in the Fe–N bond length in FeP due to the strain in the graphene lattice gives rise to the change in the molecular-level diagram as well as the interaction between the C atoms of the graphene layer and the molecular orbitals of FeP.

The authors suggested two ways of experimentally detecting the change in the spin state. In X-ray magnetic circular dichroism (XMCD) experiments,

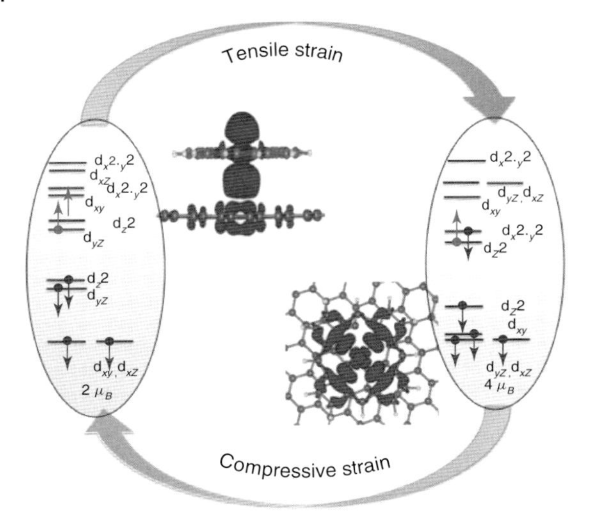

Figure 7.12 Magnetization density isosurfaces for FeP on (left) 0% and (right) 1% strained graphene. The isosurfaces have been plotted for an energy window of 0.4 eV below the Fermi levels in both cases. The spin densities shown in the left up (side view) and right down (top view) parts correspond to spin-up- and spin-down channels, respectively. The energy levels with the d-orbital character of FeP are shown in the extreme left and right for 0% and 1% strained graphene, respectively. (Source: Reprinted with permission from Bhandary *et al.* (2011) *Phys. Rev. Lett.*, **107**, 257202. Copyright (2011), the American Physical Society.)

element-specific magnetic moments can be extracted. The effective spin moment contains a spin-dipole contribution, which is usually very small and negligible. However, for nanostructures, this contribution becomes significant. Bhandary *et al.* [36] predicted that the effective spin moments (sum of spin moment and spin-dipole moment) are very different for $S = 1$ and $S = 2$ spin states, thereby opening a distinct way of detecting the spin change. The other procedure is to monitor the easy axis of magnetization in XMCD. The magnetic anisotropy energy (MAE), and hence the easy axis of magnetization can be calculated by adding the spin-orbit coupling term in the Hamiltonian. By the second-order perturbation theory using a four-level model (see Figure 7.13, where HOMO, HOMO−1, LUMO, and LUMO+1 states are considered), the authors predicted a change in the easy axis of magnetization when the change in the spin state in FeP takes place owing to the strain engineering in graphene.

For some specific technological applications, there is an interest to make graphene metallic. One possible way is to introduce topological disorder in graphene. It has been shown that one may achieve an increase in conductivity by introducing defects in graphene by chemical treatment [37]. This was confirmed both experimentally and theoretically. Recently, Holmström *et al.* [38] predicted the existence of metallic amorphous graphene by first principles calculations. The structures of topologically disordered amorphous graphene were generated by

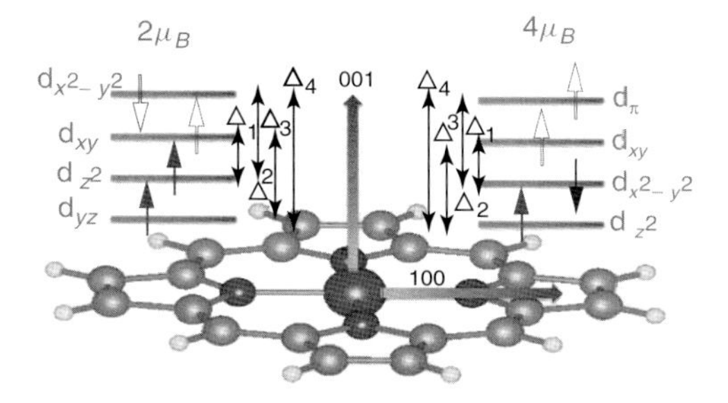

Figure 7.13 Schematic diagram showing the energy levels and their orbital characters for (left) 0% and (right) 1% strained cases, considered for the calculation of MAE. The filled and empty arrows indicate occupied (HOMO and HOMO−1) and unoccupied (LUMO and LUMO+1; lowest unoccupied molecular orbital) d states of Fe, respectively. The big arrows indicate the directions of magnetization. (Source: Reprinted with permission from Bhandary *et al.* (2011) *Phys. Rev. Lett.*, **107**, 257202. Copyright (2011), the American Physical Society.)

the stochastic quenching method. The resulting structures had nanopatches of carbon rings of various coordination, as shown in Figure 7.14. The basic planar configuration that was obtained after the first position-area relaxation cycle is shown in Figure 7.14a. One observes a large number of carbon atoms with four- and fivefold coordination, indicating a high energy state. The further relaxed final planar structure is shown in Figure 7.14b. In this relaxed geometry, all carbon

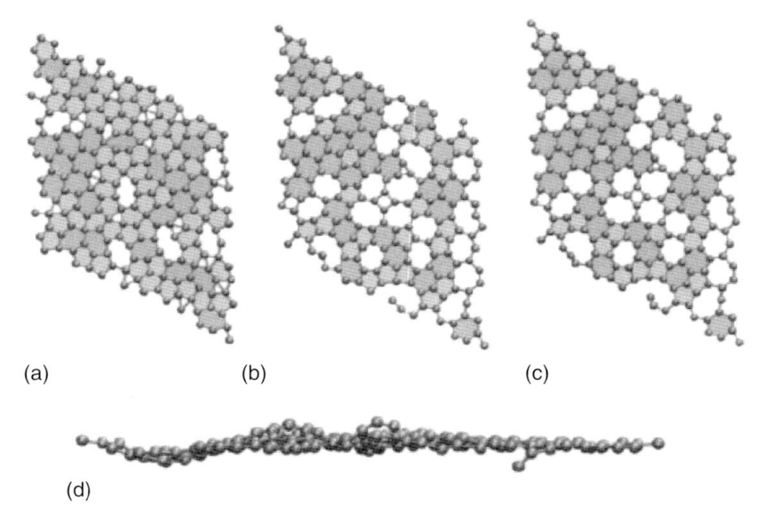

Figure 7.14 (a) Top view of the basic planar structure. Hexagonal rings are marked with dark grey and pentagonal rings are marked with light grey. (b) Top view of the final planar geometry. (c) Top view of the buckled geometry. (d) Side view of the buckled geometry. The distance from left to right in (d) is 41 Å and the buckling amplitude is about 1.7 Å. (Source: Reprinted with permission from Holmström *et al.* (2011) *Phys. Rev. B*, **84**, 205414. Copyright (2011), the American Physical Society.)

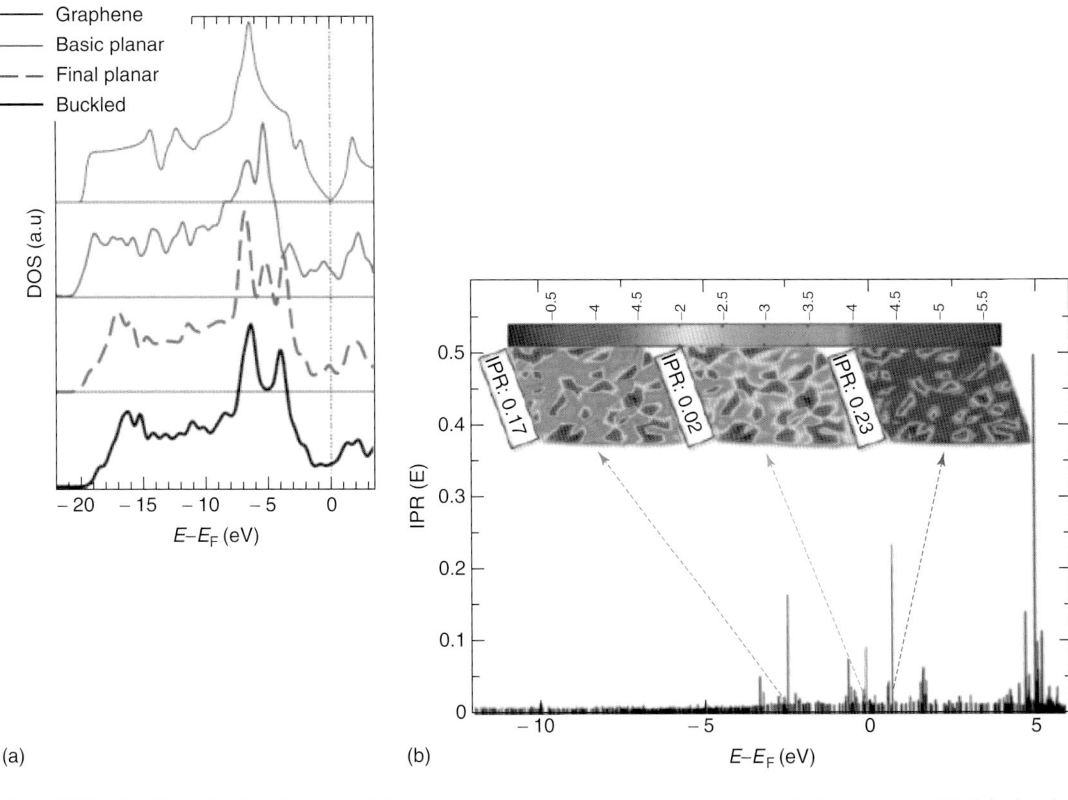

Figure 7.15 (a) Total density of states of the three amorphous geometries compared to bulk graphene. (b) Calculated inverse participation ratio (IPR) of the one-electron states of amorphous graphene for the buckled amorphous geometry. Inset: The imaginary part of the one-electron Green function for states just below E_F, at E_F, and just above E_F. (Source: Reprinted with permission from Holmström *et al.* (2011) *Phys. Rev. B*, **84**, 205414. Copyright (2011), the American Physical Society.)

atoms are sp^2 bonded and threefold coordinated except in two cases with sp-bonded twofold coordination having a 180° bond angle. The fully relaxed buckled structure is shown in Figure 7.14c. The side view in the figure clearly indicates a rippled structure. The resulting DOSs are shown in Figure 7.15 for various configurations. One clearly observes finite DOSs at the Fermi level for the amorphous structures. The authors calculated the inverse participation ratio (IPR) to investigate the nature of the electronic states at various energies. It was found that the electronic states are extended at the Fermi level, signifying increased conductivity due to the topological disorder. Finally, the authors presented the linear response conductance calculation by a model theory as a function of lattice distortions, which results in a similar behavior as the first principles calculation.

7.6
Outlook

It is undoubtedly true that the nanostructures of graphene provide a very rich playground for the practitioners of both experimental and theoretical graphene science. Many issues are still unsettled, especially the occurrence of magnetism at the edges of GNRs at high temperatures. Chemical functionalization by both inorganic and organic species at the edges will provide novel functionalities and, hopefully, will be utilized in devices in future. Sophisticated experimental techniques will provide local information on the atomic scale and will be combined with theoretical investigations to have a proper microscopic understanding of the phenomena exhibited by graphene nanostructures. At the same time, state-of-the-art theoretical and computational studies will continue to provide important understanding and predictions of new properties. Finally, more interdisciplinary research is needed to utilize the full potential of graphene and its nanostructures.

Acknowledgements

The author specifically acknowledges the works of S. Bhandary at Uppsala University. Also, all the collaborators within the KOF program funded by Uppsala University and external collaborators throughout the globe are gratefully acknowledged. The author thanks the national supercomputing facilities allocated by the Swedish National Allocation for Computing (SNAC).

References

1. Novoselov, K.S., Geim, A.K., Morozov, S.V., Jiang, D., Zhang, Y., Dubonos, S.V., Grigorieva, I.V., and Firsov, A.A. (2004) *Science*, **306**, 666.
2. (a) Geim, A.K. (2009) *Science*, **324**, 1530; (b) Geim, A.K. and Novoselov, K.S. (2007) *Nat. Mater.*, **6**, 183; (c) Castro Neto, A.H. et al. (2009) *Rev. Mod. Phys.*, **81**, 109; (d) Abergel, D.S.L.,

Apalkov, V., Berashevich, J., Ziegler, K., and Chakraborty, T. (2010) *Adv. Phys.*, **59**, 261.
3. Rao, C.N.R., Sood, A.K., Subrahmanyam, K.S., and Govindaraj, A. (2009) *Angew. Chem.* **48**, 7752; (b) Rao, C.N.R., Subrahmanyam, K.S., Ramakrishna Matte, H.S.S., and Govindaraj, A. (2011) Graphene: synthesis, functionalization and properties, in *Graphene and its Fascinating Attributes* (eds S.K. Pati, T. Enoki, and C.N.R. Rao), World Scientific publishing co. Pvt. Ltd., Singapore.
4. Schwierz, F. (2010) *Nat. Nanotechnol.*, **5**, 487.
5. Das, A., Pisana, S., Chakraborty, B., Piscanec, S., Saha, S.K., Waghmare, U.V., Novoselov, K.S., Krishnamurthy, H.R., Geim, A.K., Ferrari, A.C., and Sood, A.K. (2008) *Nat. Nanotechnol.*, **3**, 210.
6. Jia, X., Campos-Delgado, J., Terrones, M., Meunier, V., and Dresselhaus, M.S. (2011) *Nanoscale*, **3**, 86.
7. (a) Yazyev, O.V. (2010) *Rep. Prog. Phys.*, **73**, 056501; (b) Dutta, S. and Pati, S.K. (2010) *J. Mater. Chem.*, **20**, 8207.
8. Li, X., Wang, X., Zhang, L., Lee, S., and Dai, H. (2008) *Science*, **319**, 1229.
9. Han, M.Y., Özyilmaz, B., Zhang, Y., and Kim, P. (2007) *Phys. Rev. Lett.*, **98**, 206805.
10. Son, Y.-W., Cohen, M.L., and Louie, S.G. (2006) *Phys. Rev. Lett.*, **97**, 216803.
11. Tapaszto, L., Dobrik, G., Lambin, P., and Biro, L.P. (2008) *Nat. Nanotechnol.*, **3**, 397.
12. Girit, C.O., Meyer, J.C., Erni, R., Rossell, M.D., Kisielowski, C., Yang, L., Park, C.-H., Crommie, M.F., Cohen, M.L., Louie, S.G., and Zettl, A. (2009) *Science*, **323**, 1705.
13. Gallagher, P., Todd, K., and Goldhaber-Gordon, D. (2010) *Phys. Rev. B*, **81**, 115409.
14. Ribeiro, R., Poumirol, J.-M., Cresti, A., Escoffier, W., Goiran, M., Broto, J.-M., Roche, S., and Raquet, B. (2011) *Phys. Rev. Lett.*, **107**, 086601.
15. Enoki, T. (2012) *Phys. Scr.*, **2012**, 014008.
16. Fujii, S. and Enoki, T. (2010) *J. Am. Chem. Soc.*, **132**, 10034.
17. Wallace, P.R. (1947) *Phys. Rev.*, **71**, 622.
18. (a) Hohenberg, P. and Kohn, W. (1964) *Phys. Rev.*, **136**, B864; (b) Kohn, W. and Sham, L.J. (1965) *Phys. Rev.*, **140**, A1133.
19. (a) Kresse, G. and Hafner, J. (1993) *Phys. Rev. B*, **47**, R558; (b) Kresse, G. and Furthmüller, J. (1996) *Phys. Rev. B*, **54**, 11169.
20. Soler, J.M. *et al.* (2002) *J. Phys.: Condens. Matter*, **14**, 2745.
21. Giannozzi, P. *et al.* Quantum Espresso package, *www.quantum-espresso.org.*
22. Nakada, K., Fujita, M., Dresselhaus, G., and Dresselhaus, M.S. (1996) *Phys. Rev. B*, **54**, 17954.
23. Wakabayashi, K., Fujita, M., Ajiki, H., and Sigrist, M. (1999) *Phys. Rev. B*, **59**, 8271.
24. Jung, J. and MacDonald, A.H. (2009) *Phys. Rev. B*, **79**, 235433.
25. Son, Y.-W., Cohen, M.L., and Louie, S.G. (2006) *Nature*, **444**, 347.
26. Dutta, S., Manna, A.K., and Pati, S.K. (2009) *Phys. Rev. Lett.*, **102**, 096601.
27. Barone, V., Hod, O., and Scuresia, G.E. (2006) *Nano Lett.*, **6**, 2748.
28. Yazyev, O.V., Capaz, R.B., and Louie, S.G. (2011) *Phys. Rev. B*, **84**, 115406.
29. Wassmann, T., Seitsonen, A.P., Saitta, A.M., Lazzeri, M., and Mauri, F. (2008) *Phys. Rev. Lett.*, **101**, 096402.
30. Bhandary, S., Eriksson, O., Sanyal, B., and Katsnelson, M.I. (2010) *Phys. Rev. B*, **82**, 165405.
31. Chandrachud, P., Pujari, B.S., Haldar, S., Sanyal, B., and Kanhere, D.G. (2010) *J. Phys.: Condens. Matter*, **22**, 465502.
32. Haldar, S., Kanhere, D.G., and Sanyal, B. (2012) *Phys. Rev. B*, **85**, 55426.
33. Bhowmick, S. and Waghmare, U.V. (2010) *Phys. Rev. B*, **81**, 155416.
34. Koskinen, P., Malola, S., and Häkkinen, H. (2008) *Phys. Rev. Lett.*, **101**, 115502.
35. Haldar, S., Bhandary, S., Bhattacharjee, S., Eriksson, O., Kanhere, D.G., and Sanyal, B. (2012) *Solid State Commun.*, **152**, 1719.
36. Bhandary, S., Ghosh, S., Herper, H., Wende, H., Eriksson, O., and Sanyal, B. (2011) *Phys. Rev. Lett.*, **107**, 257202.
37. (a) Coleman, V.A., Knut, R., Karis, O., Grennberg, H., Jansson, U., Quinlan, R., Holloway, B.C., Sanyal, B., and

Eriksson, O. (2008) *J. Phys. D: Appl. Phys.*, **41**, 062001; (b) Jafri, S.H.M., Carva, K., Widenkvist, E., Blom, T., Sanyal, B., Fransson, J., Eriksson, O., Jansson, U., Grennberg, H., Karis, O., Quinlan, R.A., Holloway, B.C., and Leifer, K. (2010) *J. Phys. D: Appl. Phys.*, **43**, 045404; (c) Carva, K., Sanyal, B., Fransson, J., and Eriksson, O. (2010) *Phys. Rev. B*, **81**, 245405.

38. Holmström, E., Fransson, J., Eriksson, O., Lizarraga, R., Sanyal, B., Bhandary, S., and Katsnelson, M.I. (2011) *Phys. Rev. B*, **84**, 205414.

8
Stone–Wales Defects in Graphene and Related Two-Dimensional Nanomaterials

Sharmila N. Shirodkar and Umesh V. Waghmare

8.1
Introduction

Defects are inevitable in a crystal and form an inherent part of any solid. Defects and impurities are introduced during synthesis and treatment processes in the system. These lattice imperfections determine the electronic, optical, mechanical, and thermal properties of materials. For example, ductility and mechanical strength of metals are determined by defects. These imperfections also govern the electrical and thermal conductance properties in semiconductors [1].

There exist two types of defects: intrinsic and extrinsic [2]. When the crystalline order is not disturbed by foreign atoms, the defects are intrinsic. These foreign atoms are referred to as *impurities*, and they constitute extrinsic defects. Intrinsic defects include point defects (vacancies or interstitial atoms), line defects (dislocations), grain boundaries, stacking faults, and voids in three dimensions. Owing to reduced system dimensions, the types of defects reduce and their effects get pronounced in two-dimensional (2D) materials. As 2D materials are technologically extremely important, we have studied the effects of point defects such as Stone–Wales (SW) defects, bond rotations, and line defects, that is, edges on graphene and related nanomaterials.

SW defects are topological point defects in honeycomb lattices that involve 90° rotation of a C–C bond in graphene. SW defects have been experimentally observed in graphene [3] and are known to lead to out-of-plane deformation of flat 2D structures [4]. This indicates that they may enhance the tendency of formation of nanotubes or fullerenes. In this chapter, we mainly focus on the effects of SW defects and related point defects on 2D materials.

Density functional theory (DFT) is an extremely powerful tool used to determine the electronic ground state and structural properties of atoms, molecules, crystals, and surfaces. A line defect was first theoretically [5] predicted and then experimentally [6] verified in graphene nanotubes. This proves the power of *ab initio* calculations in theoretical prediction as well as in experimental verification of behavior and evolution of defects in crystals. In these calculations, a given type of defect and its interactions can be studied, one at a time. This gives the

Graphene: Synthesis, Properties, and Phenomena, First Edition. Edited by C. N. R. Rao and A. K. Sood.

freedom to analyze any defect configuration and isolate the required interactions that are not possible experimentally. We have therefore used first-principles DFT-based simulations to study the effects of defects on graphene and related nanomaterials.

In this chapter, we present the results of first-principles DFT-based analysis to determine the properties of topological defects in graphene (some of this is being published in Phys. Rev. B. [7]), 2D $MoS_2/MoSe_2$, and $C_{1-x}(BN)_{x/2}$, a 2D solid solution of graphene and hexagonal boron nitride (h-BN).

8.2
Computational Methods

First-principles calculations based on DFT have been proved to be an accurate and realistic approach to calculate the many-electron quantum ground state energy of a material or a molecule or any system of electrons and nuclei. Many properties of materials are determined by differences in (or derivatives of) the total energy, which is the sum of many-electron ground state energy and electrostatic nuclear interaction energy. For example, stress, magnetization, and forces are the first derivatives of energy with respect to strain, magnetic fields, and atom positions, respectively. The second derivatives of energy with respect to strain, magnetic fields, and atomic positions give elastic constants, magnetic susceptibility, and interatomic force constants, respectively.

The three basic methods used in DFT to calculate electronic states are plane wave, linear combination of atomic orbitals (LCAO), and atomic sphere methods. Plane waves are the most popular basis for periodic crystals, as they simplify the algorithms for practical calculations. In crystals, energy eigenfunctions are Bloch functions given by

$$\psi_{\mathbf{k}}(\mathbf{r}) = u_{\mathbf{k}}(\mathbf{r})e^{i\mathbf{k}.\mathbf{r}} = \Sigma_{\mathbf{G}}c(\mathbf{k}-\mathbf{G})e^{i(\mathbf{k}-\mathbf{G}).\mathbf{r}} \tag{8.1}$$

where $\mathbf{k}$ is an allowed wave vector that satisfies the Born–von Karman periodic boundary conditions, $\mathbf{G}$ is a reciprocal lattice vector, $\psi_{\mathbf{k}}(\mathbf{r})$ is a Bloch function, and $u_{\mathbf{k}}(\mathbf{r})$ is its cell periodic part. In a bulk solid, $\mathbf{k}$ and $\mathbf{G}$ are infinitely many. The size of the $\mathbf{G}$ basis is truncated by considering those plane waves whose $(\hbar^2/2m|\mathbf{k}+\mathbf{G}|^2)$ kinetic energy is less than a fixed energy cutoff (Ecut). Brillouin zone (BZ) integration is sampled with a finite number of k-points (i.e., limit $\mathbf{k}$ basis) using the Monkhorst-pack scheme [8].

DFT calculations combine plane waves and periodic boundary conditions (PBCs) to effectively and efficiently calculate the properties of crystalline solids. PBCs simulate an infinite lattice of the system under consideration, with the supercell as the repetitive unit.

We use the Quantum ESPRESSO (QE) [9] package that implements DFT on a plane wave basis. The local density approximated (LDA), Perdew–Zunger (PZ) parametrized form [10] was used for the exchange correlation energy of electrons, and ultrasoft pseudopotentials [11] to represent the interaction between ionic cores and electrons. Kohn–Sham wavefunctions were represented on a plane wave basis with an energy cutoff of 30 Ry for all the systems in our study.

We use a 7×7 hexagonal periodic supercell consisting of 98 atoms with a single SW defect in the graphene study. This generates an infinite spacing of ~ 17 Å between periodic images of the defect. We change the distance between the periodic images of the defects by changing the supercell size and hence control the defect concentration. All BZ integrations were carried out over a $3 \times 3 \times 1$ mesh of k-points.

For graphene and boron nitride ($C_{1-x}(BN)_{x/2}$), a rectangular supercell was constructed with 96 atoms (48 C and 48 BN) in the basis with a single-SW defect. All BZ integrations were carried out over a $3 \times 3 \times 1$ mesh of k-points.

Defects in MoS_2 were simulated with a $4 \times 4 \times 1$ supercell constructed with 48 atoms (16 Mo and 32 S) in the basis with a single-point defect. All BZ integrations were carried out over a $5 \times 5 \times 1$ mesh of k-points.

8.3 Graphene: Stone–Wales (SW) Defects

Graphene is a two-dimensional crystal with a bipartite lattice consisting of two interpenetrating hexagonal lattices. Adatoms, vacancies, point defects, line defects, and edge defects are the different types of defects that occur in graphene [12]. Defects and impurities break the crystalline order and modify structural and electronic properties of graphene and alter the performance of graphene-based devices. SW defects, which are topological defects, have been observed experimentally [3] and are expected to play an important role in the intrinsic rumpling of graphene [4]. It was also observed that SW defects affect the electronic and vibrational properties of graphene [4, 13]. Thus, understanding its effects on the properties of graphene is necessary for the application of this extremely interesting two-dimensional structure.

An SW defect involves an in-plane rotation of a carbon–carbon (C–C) bond with respect to the midpoint of the bond by 90°. As a result of this, a pair of pentagons and heptagons is formed. A disclination in graphene consists of a five- or seven-membered ring in its core with the original threefold coordination of all carbon atoms preserved. An edge-sharing pentagon and a heptagon form a pair of complementary disclinations, also known as a *5/7 dislocation*. An SW defect that consists of two such 5/7 dislocations (i.e., a pair of pentagons and heptagons) forms a dislocation dipole [14] (Figure 8.1). This 5/7 dislocation has a burgers' vector **b** ($b = \sqrt{3}\,1.42$ Å).

We analytically and computationally show that the interaction between SW defects is long ranged. We investigate the effect of SW defects on electronic structure and vibrational properties of graphene, and lattice thermal conductivity.

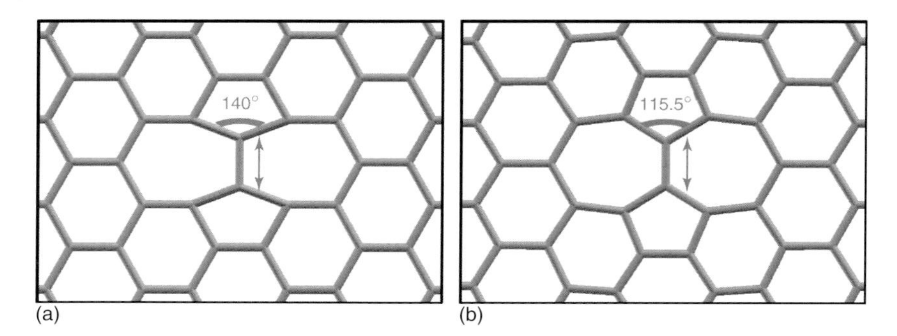

Figure 8.1 Structural change (a) before and (b) after relaxation of the graphene sheet on the introduction of an SW defect.

8.3.1 Structural, Electronic, Magnetic, and Vibrational Properties of Graphene with SW Defect

8.3.1.1 Structural Changes at an SW Defect

We first examine the structural changes in graphene on introduction of an SW defect. On structural relaxation, the rotated C–C bond compresses to 1.31 Å from its original length of 1.41 Å (Table 8.1 and Figure 8.1), which compares well with reported values [4] of compression from 1.42 Å to 1.32 Å. This bond length contraction is also accompanied by the compression of the bond angle at the apex of the pentagon from 140° to 115.5°, which is ∼ −18.5%. These changes in bond length and bond angle increase the pressures in the system from 2.77 kbar before relaxation to 5.95 kbar after relaxation (∼115%). The system experiences compressive stress along the direction of unrotated C–C bonds, and tensile stress along the rotated C–C bond. As studied in previous literature [15], the graphene sheet would reduce this in-plane stress by buckling in the z-direction and therefore helps in forming nonplanar structures. On structural relaxation, the system energy is lowered by ∼6.4 eV, indicating a factor of 2 change in the formation energy of the SW defect. A further lowering in energy is expected from out-of-plane deformation of the sheet.

Table 8.1 Structure before and after relaxation with an SW defect.

Property	Before	After	Difference
Bond length (Å)	1.41	1.31	−7.1%
Bond angle	140°	115.5°	−18.2%
Pressure (kbar)	2.77	5.95	115%

8.3.1.2 Interaction between SW Defects

To study the dependence of interaction between SW defects on distance, we have varied the supercell size. With increasing supercell size, the distance between the periodic images of a defect also increases corresponding to the different planar concentration of the defects. The variation in the formation energy of an SW defect was fitted with x^{-n}, where x is the distance between the defects. The nature of the curve (Figure 8.2) and the fitting parameters (Table 8.2) show that the formation energy falls off $\sim 1/x^2$, showing that the interaction between the defects is long ranged.

8.3.1.3 Electronic Structure of Graphene and Effects of SW Defects

In pristine graphene, each carbon atom undergoes sp^2 hybridization and forms three in-plane σ bonds and one out-of-plane π bond. The in-plane σ bonds are strong covalent bonds that do not contribute to the conductivity of graphene. The out-of-plane π bond is weaker and gives rise to two linear bands crossing the Fermi energy at the K-point, also known as the *Dirac cone*. Our DFT estimate of the electronic band structure of pristine graphene shows the decoupling of σ and π bands (Figure 8.3b). From the band structure, it is evident that the π bands form the Dirac cone. We therefore derive the electronic structure of a single layer of pristine graphene only for π bands, using the tight-binding approach [17, 18].

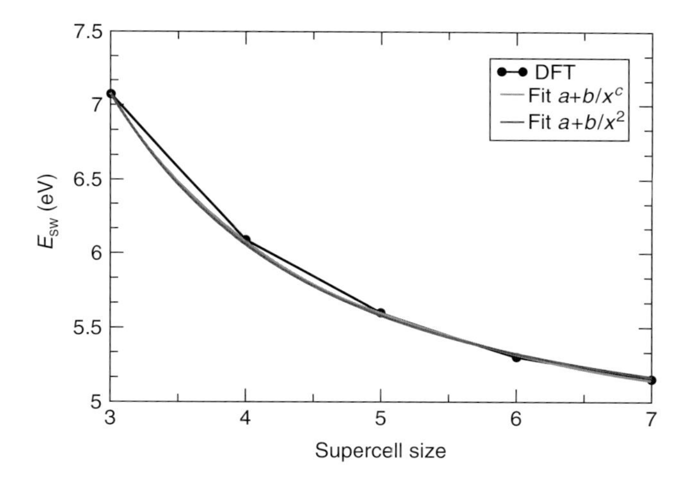

Figure 8.2 Formation energy of an SW defect with varying supercell size.

Table 8.2 Polynomial fit to formation energy.

Fit	*a*	*b*	*c*
$a + b/x^c$	4.59 eV	17.34	1.77
$a + b/x^2$	4.74 eV	21.17	2

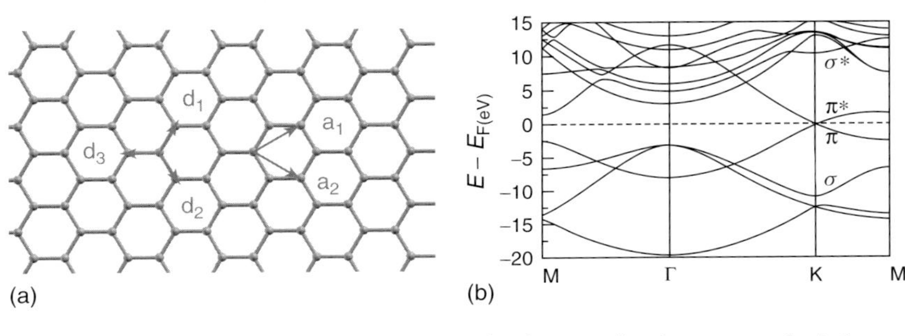

Figure 8.3 Lattice structure and electronic band structure of pristine graphene. (a) Honeycomb lattice of graphene. $\mathbf{a}_1$ and $\mathbf{a}_2$ are the real-space lattice vectors, and $\mathbf{d}_i$, $i = 1, 2, 3$ are the nearest-neighbor vectors. (b) Electronic band structure of pristine graphene along the M-Γ-K-M direction in the Brillouin zone. The contribution to the Dirac cone at K-point is only due to π bands.

The structure of graphene consists of a hexagonal lattice with two atoms per unit cell. The lattice vectors $\{\mathbf{a}_1, \mathbf{a}_2\}$ (Figure 8.3a) of graphene are given by

$$\mathbf{a}_1 = \frac{l}{2}(3\hat{x} + \sqrt{3}\hat{y}), \mathbf{a}_2 = \frac{l}{2}(3\hat{x} - \sqrt{3}\hat{y}) \tag{8.2}$$

where l is the C–C bond length ($l \sim 1.42$ Å). The reciprocal space unit cell vectors, $\{\mathbf{b}_1$ and $\mathbf{b}_2\}$,

$$\mathbf{b}_1 = \frac{2\pi}{3l}(1\hat{x} + \sqrt{3}\hat{y}), \mathbf{b}_2 = \frac{2\pi}{3l}(1\hat{x} - \sqrt{3}\hat{y}) \tag{8.3}$$

In graphene, each C atom has three nearest neighbor C atoms. The nearest neighbor positions $\{\mathbf{d}_1, \mathbf{d}_2, \mathbf{d}_3\}$ (Figure 8.3a) are given by

$$\mathbf{d}_1 = \frac{l}{2}(1\hat{x} + \sqrt{3}\hat{y}), \mathbf{d}_2 = \frac{l}{2}(1\hat{x} - \sqrt{3}\hat{y}), \mathbf{d}_2 = l(-1\hat{x} + 0\hat{y})$$

The tight-binding Hamiltonian with nearest neighbor hopping energy (t) is

$$H = \epsilon_{2p}\Sigma_{i,\sigma}(a_{i,\sigma}^{\dagger}a_{i,\sigma} + b_{i,\sigma}^{\dagger}b_{i,\sigma}) - t\Sigma_{\langle i,j\rangle,\sigma}(a_{i,\sigma}^{\dagger}b_{j,\sigma} + H.C.) \tag{8.4}$$

where i and j are the nearest neighboring sites, σ labels the electron spin, and ϵ_{2p} is the site energy of the $2p_z$ orbital. $a_{i,\sigma}$ ($a_{i,\sigma}^{\dagger}$) and $b_{j,\sigma}$ ($b_{j,\sigma}^{\dagger}$) are the annihilation (creation) operators for electrons of spin σ on sublattice A at site i and B at site j, respectively. As each of the two atoms in the unit cell of graphene contributes only one $2p_z$ orbital toward the electronic band structure in Equation (8.4), the Hamiltonian in Fourier space reduces to a square matrix of order 2 (i.e., 2×2). Hence, the electronic structure of graphene can be obtained by solving the eigenvalue problem for a 2×2 Hamiltonian H and 2×2 overlap integral matrix S. H and S matrices in the Fourier space are

$$H(\mathbf{k}) = \begin{pmatrix} \epsilon_{2p} & -tf(k) \\ -tf^*(k) & \epsilon_{2p} \end{pmatrix} \text{ and } S(\mathbf{k}) = \begin{pmatrix} 1 & 0 \\ 0 & 1 \end{pmatrix}$$

where $\mathbf{k}$ is an allowed Bloch vector, and $\{k_x, k_y\}$ are its Cartesian components. $f(k)$ is given by

$$f(k) = 1 + 2e^{-i3k_x l/2}\cos\left(\frac{\sqrt{3}k_y l}{2}\right)$$

On solving the secular equation $|H(\mathbf{k}) - E(\mathbf{k})S(\mathbf{k})| = 0$, energy bands $E(\mathbf{k})$ are obtained as

$$E(\mathbf{k}) = \epsilon_{2p} \pm tw(\mathbf{k}) \tag{8.5}$$

$$w(\mathbf{k}) = \sqrt{1 + 4\cos^2\left(\frac{\sqrt{3}k_y l}{2}\right) + 4\cos\left(\frac{\sqrt{3}k_y l}{2}\right)\cos\left(\frac{3k_x l}{2}\right)} \tag{8.6}$$

The plus sign in Equation (8.5) is for conduction bands and the negative sign is for valence bands. The two inequivalent points K and K′ in the BZ of graphene are

$$\mathbf{K} = \frac{2\pi}{3l}\left(1\hat{x} + \frac{1}{\sqrt{3}}\hat{y}\right) \text{ and } \mathbf{K}' = \frac{2\pi}{3l}\left(1\hat{x} - \frac{1}{\sqrt{3}}\hat{y}\right)$$

On expanding the electronic energy around the two inequivalent points K and K′, that is, $\mathbf{q} = \mathbf{K} + \mathbf{k}$ and substituting $\epsilon_{2p} = 0$ (i.e., setting Fermi energy to 0), we get

$$w(\mathbf{k}) = \frac{3}{2}|\mathbf{k}|l + \cdots$$

In the linear approximation,

$$E(\mathbf{k}) = \pm t\frac{3}{2}|\mathbf{k}|l \tag{8.7}$$

From Equation (8.7), it is evident that the electronic band structure of graphene is conical near the K and K′-points in the BZ (Figure 8.3b). These are the π and π^* bands formed from the $2p_z$ orbitals of the C atoms. On expanding the Hamiltonian $H(\mathbf{k})$ in the vicinity of K and K′, and substituting $\epsilon_{2p} = 0$, we get

$$H_{\mathrm{K}}(\mathbf{k}) = \hbar v_{\mathrm{F}}\boldsymbol{\sigma}\cdot\mathbf{k} \text{ and } H_{\mathrm{K}'}(\mathbf{k}) = \hbar v_{\mathrm{F}}\boldsymbol{\sigma}^*\cdot\mathbf{k} \tag{8.8}$$

where v_{F} is the Fermi velocity ($v_{\mathrm{F}} \sim 1 \times 10^6$ m s^{-1}), and $\boldsymbol{\sigma} = (\sigma_x, \sigma_y)$ and $\boldsymbol{\sigma}^* = (\sigma_x, -\sigma_y)$ are the Pauli matrices. $H_{\mathrm{K}}(\mathbf{k})$ and $H_{\mathrm{K}'}(\mathbf{k})$ are the Hamiltonians near K and K′-points, respectively, and have the form of the Dirac Hamiltonian for massless fermions Equation (8.8). The linearity in its electronic structure and the mapping of its Hamiltonian to the massless Dirac Hamiltonian are responsible for the exotic thermal and electronic properties of graphene.

To understand the effects of the SW defect in an $n \times n$ supercell, we now analyze the folding of the BZ in graphene while comparing the description with a unit cell (subscript u) to that with a supercell. For a $(3n) \times (3n)$ supercell, high symmetry K and K′-points (K_u and K'_u) of the unit cell fold onto the Γ-point of the BZ of the supercell. In $(3n+1) \times (3n+1)$ and $(3n+2) \times (3n+2)$ supercells, the (K_u, K'_u)-points fold onto the (K, K′)-points and (K′, K)-points in the supercell BZ, respectively.

On the basis of the earlier work, the translational and rotational symmetries of the lattice broken on the introduction of an SW defect lift the degeneracy at the Fermi level and open up a gap in the electronic structure [13, 19]. Our analysis points out the subtlety in this, which has been ignored in former theoretical calculations [13, 19], and we present here a new theory based on electron–phonon coupling in graphene.

Let us define the SW defect potential ($V_{\text{SW}}(\mathbf{r})$) as follows:

$$V_{\text{SW}}(\mathbf{r}) = V_{\text{SW}}^{\text{KS}}(\mathbf{r}) - V_{\text{pristine}}^{\text{KS}}(\mathbf{r})$$

where $V_{\text{SW}}^{\text{KS}}(\mathbf{r})$ and $V_{\text{pristine}}^{\text{KS}}(\mathbf{r})$ are the Kohn–Sham (KS) potentials of graphene with and without an SW defect. With periodic boundary conditions, the SW defect potential acquires the periodicity of the supercell, that is,

$$V_{\text{SW}}(\mathbf{r}) = \Sigma_{\mathbf{G}'} V_{\text{SW}}(\mathbf{G}')\mathrm{e}^{i\mathbf{G}'\cdot\mathbf{r}} \tag{8.9}$$

where $\mathbf{G}'$ is the reciprocal lattice vector of the supercell. This perturbing SW defect potential is nonisotropic and breaks the symmetry of the lattice. The effect of the defect potential on the band energies is estimated through perturbation theory. First-order correction to the energy of degenerate states with wave vectors $\mathbf{k}$ and $\mathbf{k}'$ is given by

$$\Delta E = \langle u_{\mathbf{k}}\mathrm{e}^{i\mathbf{k}\cdot\mathbf{r}}|V_{\text{SW}}(\mathbf{r})|u_{\mathbf{k}'}\mathrm{e}^{i\mathbf{k}'\cdot\mathbf{r}}\rangle$$

$$\Delta E = \Sigma_{\mathbf{G}'}\langle u_{\mathbf{k}}|V_{\text{SW}}(\mathbf{G}')\mathrm{e}^{i(\mathbf{G}'-(\mathbf{k}-\mathbf{k}'))\cdot\mathbf{r}}|u_{\mathbf{k}'}\rangle \tag{8.10}$$

where $u_{\mathbf{k}}(\mathbf{r})$ is the cell periodic part of the Bloch function in the supercell. ΔE is nonzero iff $\mathbf{k} - \mathbf{k}' \in$ reciprocal lattice vectors of the supercell. In case of a $(3n) \times (3n)$ supercell, K_u and K_u' fold onto Γ-point, that is, the Dirac cone of pristine graphene appears at Γ. For $(\mathbf{k}, \mathbf{k}') = \Gamma$, $\mathbf{k} - \mathbf{k}' = \mathbf{0}$, which implies that $\Delta E \neq 0$ and the degeneracy at the Dirac point breaks. The two degenerate bands now have energies $E \pm \Delta E$ with a gap of $\sim$ 0.288 and $\sim$ 0.01 eV in 3×3 and 6×6 supercells (Figures 8.4b and f), respectively, at Γ. In the other two cases of $(3n + 1) \times (3n + 1)$ and $(3n + 2) \times (3n + 2)$ supercells with SW defects, the K_u-point folds onto the K- and K′-points, respectively. As $(\mathbf{K} - \mathbf{K}') \notin$ reciprocal lattice vectors of the supercell, ΔE, the coupling between degenerate states at K and K′ is zero for these two cases. However, the Dirac point in the band structure shifts away from the K and K′-points in the BZ because of the electron–phonon coupling in graphene.

The rotation of the bond involved in the formation of the SW defect can be expressed as a linear combination of in-plane optical phonon modes (o_x and o_y; Figure 8.4b) of graphene. These Γ-point phonon modes, o_x and o_y, couple to the electronic states at the Fermi level and are known to lead to a shift in the Dirac points [20, 21]. However, a gap opens at the shifted Dirac point in case of o_y displacement. These shifts are shown in Figure 8.4b by dark gray and light gray dots. The magnitude of the shift is proportional to the uniform part of the deformation potential of o_x and o_y modes, that is, the SW defect potential $V_{\text{SW}}(\mathbf{G}')$. Although the SW defect breaks the rotational symmetry in the system, the reflection about the x- and y-axis is still preserved with the origin at the center of the

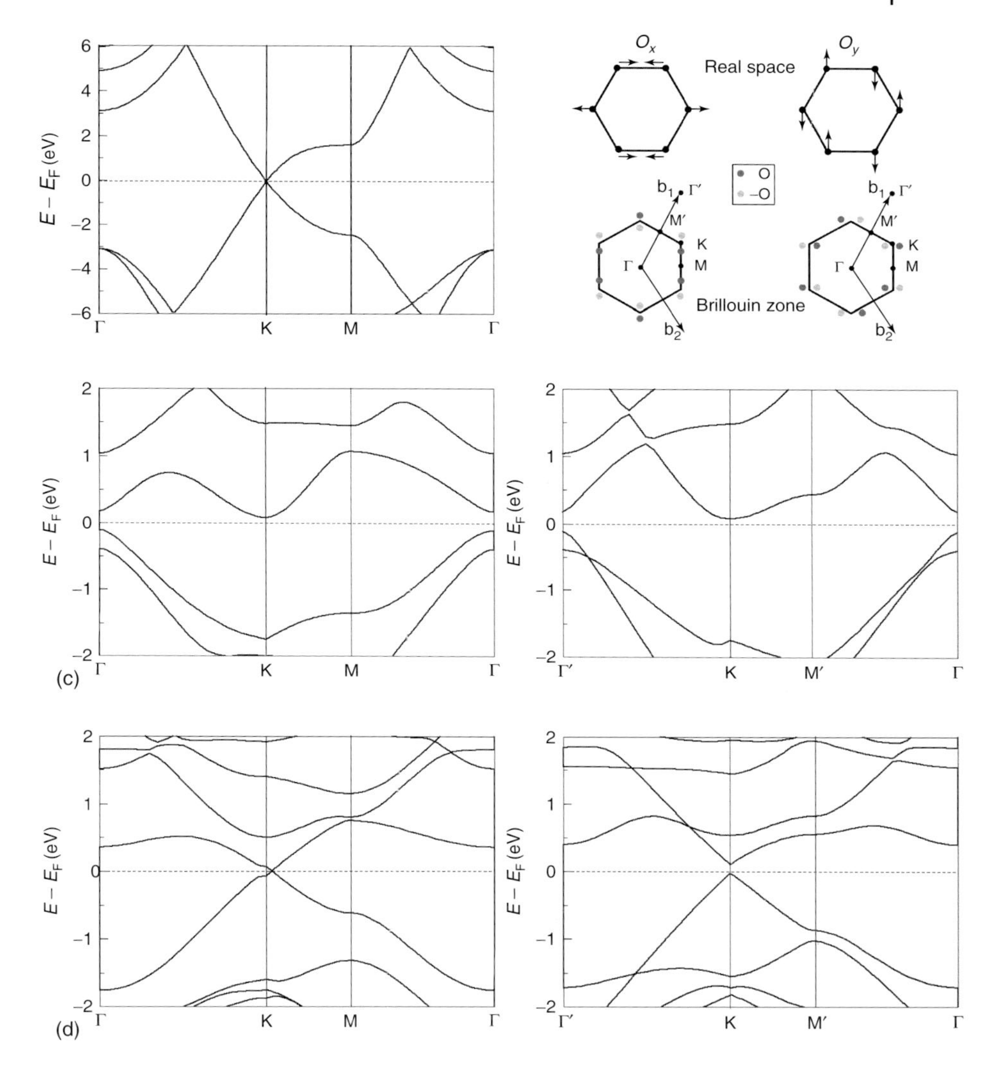

Figure 8.4 Electronic band structure of (a) pristine graphene; a zero bandgap semiconductor. (b) top panel shows displacement of atoms (left, o_x; right, o_y) and bottom panel shows shift of Dirac points from K-point (left, o_x; right, o_y) in the BZ of the supercell. Dark gray and light gray dots correspond to the shift in the Dirac points on reversal of displacement, that is, $\pm o$. Electronic band structure of graphene with an SW defect in (c) 3×3 supercell, (d) 4×4 supercell, (e) 5×5 supercell, (f) 6×6 supercell, (g) 7×7 supercell, and (h) 8×8 supercell along two different directions in reciprocal space. From (c) to (h), the left panel corresponds to Γ-K-M-Γ direction and the right panel corresponds to Γ-K-M′-Γ in the BZ of the supercell. For the 3×3 and 6×6 sized supercells, the K-point of the unit cell folds onto Γ-point of the supercell BZ and a bandgap is observed at Γ. In case of 4×4 and 7×7 supercells, a gap opens along K–M′ and closes along K–M direction. For 5×5 and 8×8 supercells, a gap opens along K–M and closes along K–M′ direction.

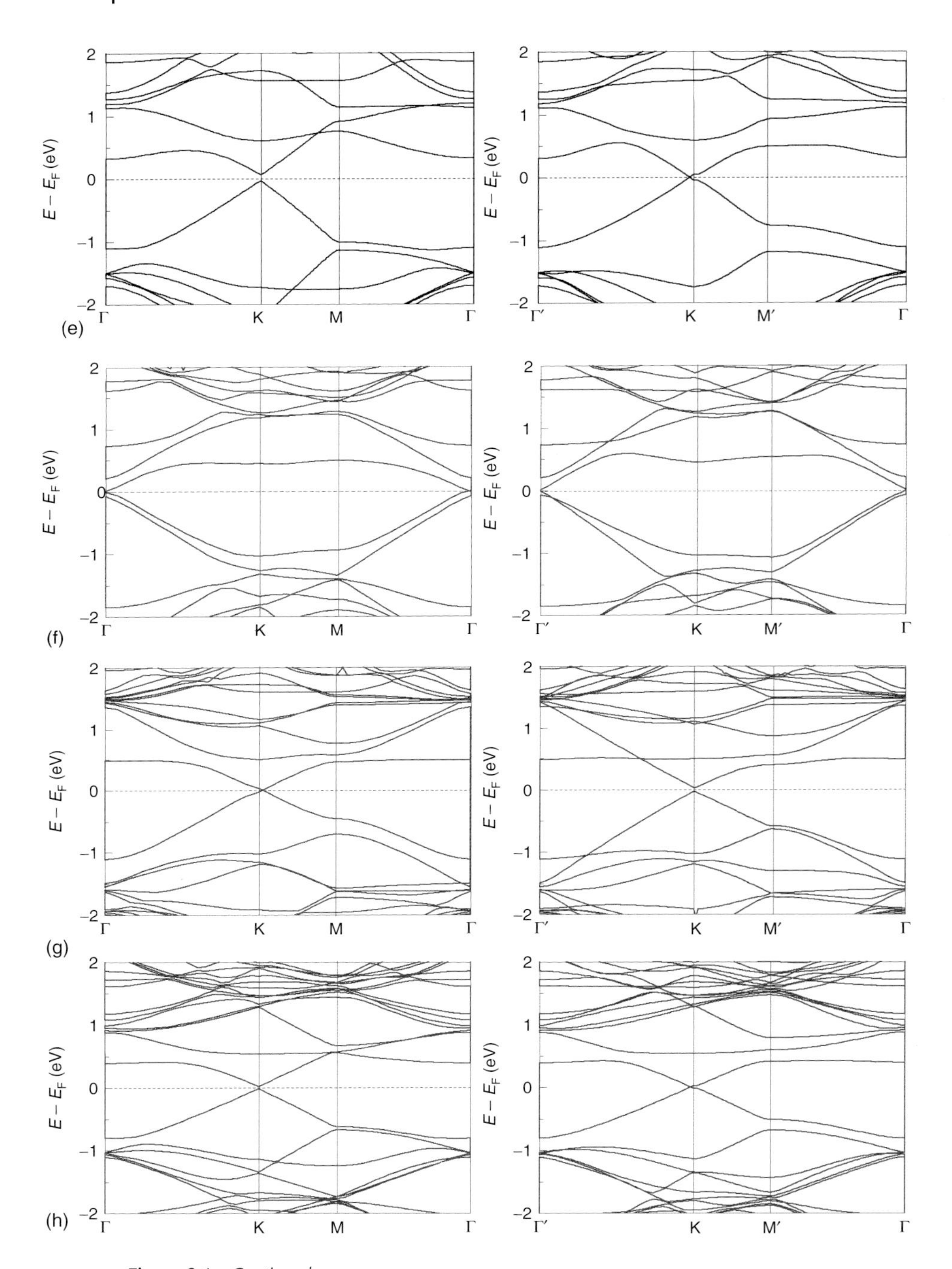

Figure 8.4 *Continued*

rotated bond. The o_x mode retains this symmetry, whereas o_y violates the reflection symmetry (in Figure 8.4b lower panel, refer to only the dark gray or light gray dots). Hence, Dirac cones shift according to o_x displacements, whereas a gap opens up at the points determined by o_y displacements. Because of the difference in BZ folding for $(3n + 1) \times (3n + 1)$ and $(3n + 2) \times (3n + 2)$ supercells, the Dirac-point shifts are out of phase in the two cases. $(3n + 1) \times (3n + 1)$ follows dark gray and $(3n + 2) \times (3n + 2)$ follows light gray, as shown in Figure 8.4b, left panel. Thus, a bandgap does not open in graphene because of the periodic lattice of SW defects but because of those in $(3n) \times (3n)$ supercells.

To further establish this, electronic band structure studies of graphene with varying SW defect concentrations were carried out. The concentration of SW defects was varied by adding one defect per supercell of size $3 \times 3, 4 \times 4, 5 \times 5, 6 \times 6, 7 \times 7$, and 8×8 (Figure 8.4c, d, e, f, g, and h). These supercell dimensions include all the three configurations $(3n) \times (3n)$, $(3n + 1) \times (3n + 1)$, and $(3n + 2) \times (3n + 2)$. In the case of 3×3 and 6×6 supercells, a gap of $\sim$ 0.288 eV and $\sim$ 0.01 eV, respectively, opens up at Γ (Figure 8.4c and 8.4f). With a decreasing SW defect concentration, the electronic bandgap decreases. For $(3n + 1) \times (3n + 1)$ supercells, a gap in the band structure at K-point opens along K to M′ and closes along K to M direction in BZ (Figure 8.4d and g). While in the case of $(3n + 2) \times (3n + 2)$ supercells, a gap in the band structure at K-point opens along K to M and closes along K to M′ direction in BZ (Figure 8.4d and h). Previous studies of graphene with SW defects were carried on in systems with $(3n + 1) \times (3n + 1)$ [19] and $(3n + 2) \times (3n + 2)$ [13] supercells and as the symmetry nonequivalent directions in the BZ were inadequately explored, the presence of a nonzero bandgap was concluded. Our study has analytically and computationally proved that the band structure is highly directional in the vicinity of the Dirac cone (K and K′-points) and supercell size dependent in graphene.

8.3.1.4 Magnetization due to Topological Defects

Graphene is a bipartite lattice, *i.e.*, it is made up of two interpenetrating hexagonal sublattices, A and B. In pristine graphene, the number of carbon atoms in each sublattice is equal and hence the magnetization is zero. On the introduction of defects [22], impurities [23], and boundaries [24], this symmetry may be broken and magnetic moments are likely to arise. Four antiferromagnetic (AFM) and one ferromagnetic (FM) spin orderings were considered. For AFM ordering, sublattice A was initialized with up-spins and sublattice B was initialized with down-spins in configuration (a) (Figure 8.5a), the rotated atoms were initialized with up-spins in configuration (b) (Figure 8.5b), considering the perpendicular bisector of the rotated bond as the dividing line, all atoms above it were initialized with up-spins and below were initialized with down-spins in configuration (c) (Figure 8.5c), rotated atoms at the defect were initialized with opposite spins in configuration (d) (Figure 8.5d). In the FM configuration, spins on all atoms were parallel to each other (Figure 8.5e). On achieving electronic consistency, all the spin configurations converged to a state with zero net magnetic moment.

In the case of vacancy defects and boundaries, carbon atoms have unpaired electrons or dangling bonds that are responsible for the net magnetic moment.

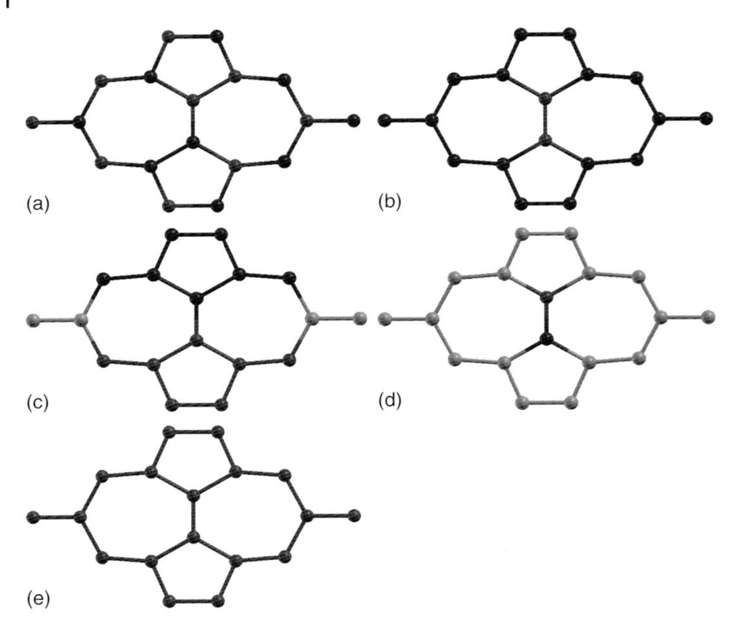

Figure 8.5 Magnetic configurations for an SW defect graphene. (a) A and B sublattices are antiferromagnetically coupled, (b) rotated bond atoms have up-spins and the rest have down-spins, (c) carbon atoms above the rotated bond have down-spins and atoms below the rotated bond have up-spins, (d) one of the rotated atoms has up-spin and other has down-spin, and (e) all are up-spins. Gray indicates up-spins and black indicates down-spins and carbon atoms are in light gray.

Although an SW defect is a topological defect, both the carbon atoms in the defect are bonded with three other C atoms, leaving no possiblity of a dangling bond and hence the SW defect does not induce magnetization in the system.

Similarly, we studied the magnetization induced by electron doping in graphene with SW defect. The doping concentration was varied from 4×10^{12} cm^{-2} to 16×10^{12} cm^{-2}. For both FM states, AFM (a) and AFM (b) states were considered in simulations, and no net magnetization is observed even after doping. Thus, localized unpaired electrons, particularly those at the dangling bonds are essential to magnetization.

8.3.1.5 Effects on Vibrational Properties

With two atoms in its basis, pure graphene has six phonon branches in its dispersion. Three of them consist of acoustic (A) modes and the remaining three consist of optical (O) modes. We denote the phonon modes involving in-plane motion of atoms with the prefix "i" and the modes involving out-of-plane motion of atoms with the prefix "o." The longitudinal modes are denoted by "L" and the transverse modes are denoted by "T." Among the acoustic and optical branches, one mode is an out-of-plane (oT) phonon mode and the other two are in-plane modes, one of which is longitudinal (L) and the other one is transverse (iT). Hence,

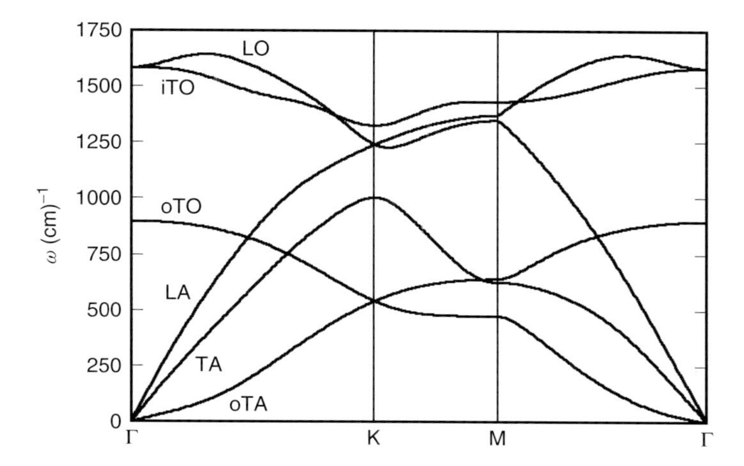

Figure 8.6 Phonon dispersion of perfect graphene [25].

starting from the lowest frequency (i.e., the G band at Γ-point), the six bands in dispersion are named as oTA, iTA, LA, oTO, iTO and LO phonon modes [25] (Figure 8.6). The LO and iTO phonon modes are degenerate at the Γ-point, in pristine graphene. Raman scattering involves excitation of an electron into the conduction band by the absorption of photon, then scattering of this electron by phonon, and finally, de-excitation of the electron back to the valence band accompanied by the emission of photon. The difference in the incident and emitted photon energy is equal to the energy of Raman active phonon scattering of the electron, which corresponds to a peak in the Raman spectrum. A change in the electronic structure changes the excitation spectrum of electrons, which in turn changes the scattering of electrons by phonons. Hence, the Raman spectrum bears signatures of changes in the electronic band structure of graphene with varying number of layers (e.g., monolayer, bilayer, and multilayers) [26] and defects [13] in an easy, nondestructive, and noncontacting way. The most prominent features in the Raman spectra of graphene include two frequencies, namely, G and D bands. The G band is at 1582 cm^{-1}, a mode at Γ-point (Figure 8.7) and the D band (defect mode, as it becomes Raman active because of the presence of defects) is the mode at 1350 cm^{-1} at K-point (Figure 8.8). The integrated Raman intensity ratio of the D and G bands is widely used for characterizing the defect concentration in graphitic materials.

These modes are scattered by an SW defect, and their frequency and eigendisplacements get modified. We analyzed this with frozen phonon calculations for a 7×7 supercell with an SW defect concentration of 2.04%, *i.e.*, 1 defect per 7×7 supercell. On account of zone folding, the D band at K-point is folded back onto the K-point in the BZ of the 7×7 supercell. As the BZ of the 7×7 supercell is very small and its K and Γ-points are close, the eigenvectors of phonons in a given branch for K and Γ-points are approximately the same. Hence, we get the G and D bands from frozen phonon calculations. To identify the G and D bands in graphene

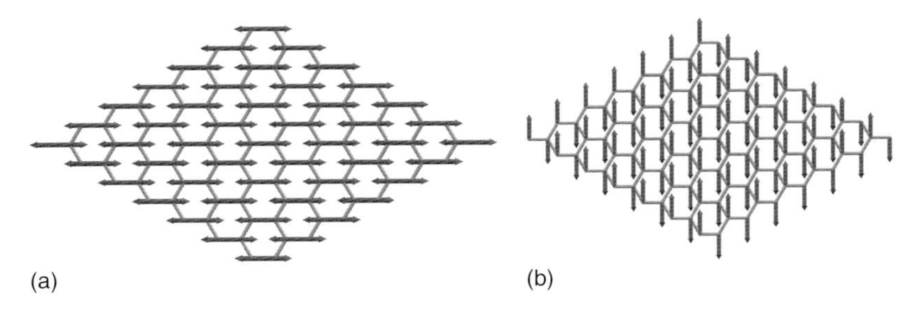

Figure 8.7 G band of pure graphene.

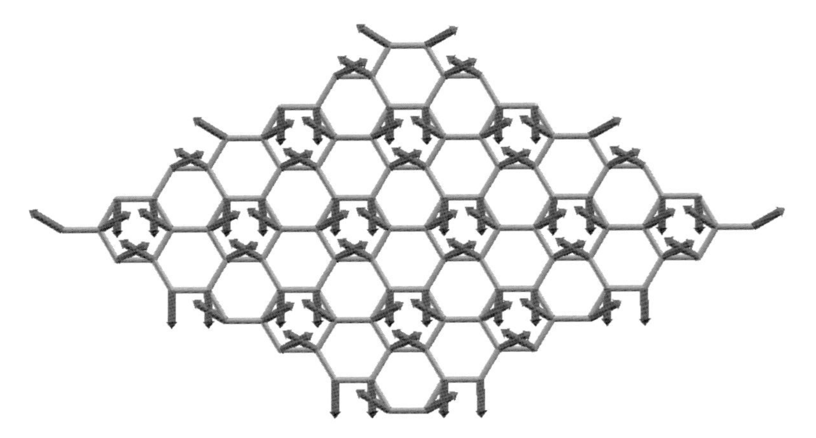

Figure 8.8 D band of pure graphene.

with an SW defect, we projected normal modes of graphene with the SW defect onto those of pristine graphene, by calculating the overlap or correlation matrix:

$$S_{\lambda\lambda'} = \langle e_\lambda | e_{\lambda'} \rangle \tag{8.11}$$

where e_λ is the eigenvector of pure graphene and $e_{\lambda'}$ is the eigenvector of the graphene with an SW defect. The eigenvector of the graphene with the SW defect having strongest correlation with the G and D bands of pristine graphene are identified as the G and D bands respectively, of graphene with the SW defect. See Figures 8.8 and 8.9 for the normal mode displacements of G and D bands of graphene with an SW defect.

Atomic displacements in the G band are excluded by the SW defect, while the D band is localized at the defect (Figures 8.9 and 8.10). As this defect is anisotropic, the symmetry equivalence of the lattice in the x and y directions is broken and the degeneracy of the G band is lifted. The frequency of the G band lowers noticeably from 1582 cm^{-1} to 1556 cm^{-1} for vibrations along the unrotated C–C bonds and from 1582 cm^{-1} to 1544 cm^{-1} for vibrations along the rotated C–C bond. In contrast, the D band hardens from 1324 cm^{-1} to 1337 cm^{-1}. Red shift in the G mode was also reported in the previous literature [13]. These shifts in G and D bands can be understood in terms of the lattice structure at the SW defect.

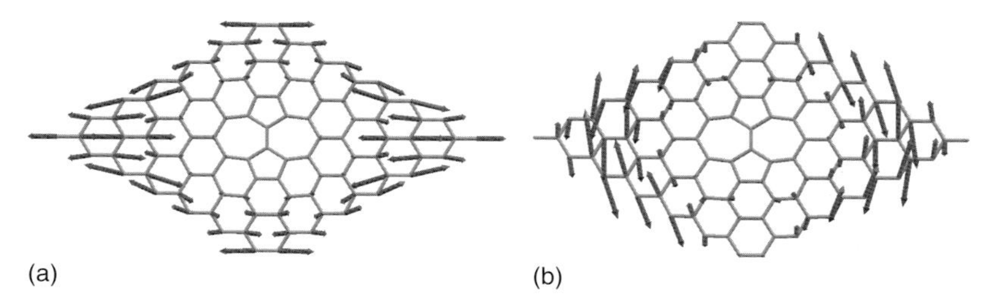

Figure 8.9 G band of graphene with an SW defect.

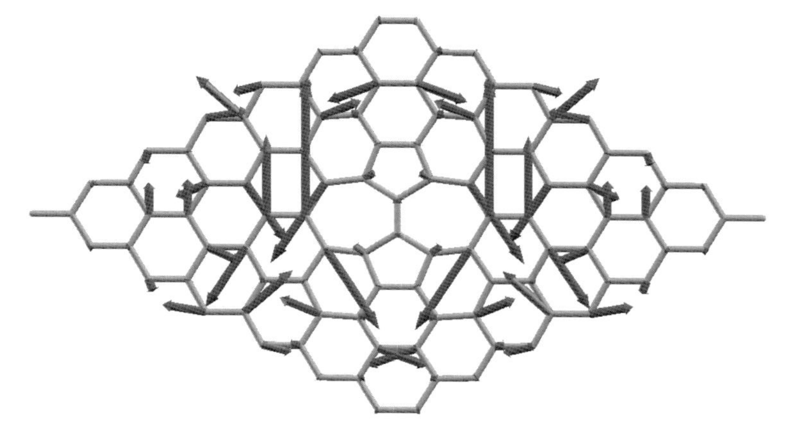

Figure 8.10 D band of graphene with an SW defect.

As discussed previously, the rotated C–C bond at the SW defect is 7% shorter and hence stiffer than other C–C bonds. The stress built up because of bond contraction is released in the bonds surrounding the SW defect. Therefore, the D band that is localized at the SW defect hardens, and the G band that is excluded by the defect softens. There exists a very interesting high-frequency phonon mode that involves in-plane stretching of the C–C bond at the SW defect (implying complete localization of a mode at the defect) also observed in prior theoretical investigations [13]. For visualization of this mode at the frequency of 1862 cm^{-1}, refer to Figure 8.11.

A graphene sheet with an SW defect is not quite stable locally, and this is evident in the two phonon modes with imaginary frequencies. These two modes lead to buckled structures, one with sine wavelike ($\sim$ 65i cm^{-1}) and the other cosine wavelike ($\sim$53i cm^{-1}) deformations, hence referred to as odd and even modes (Figures 8.12 and 8.13). Our results are in agreement with the earlier theoretical investigation [4], in which cosine and sinelike eigen-displacements with imaginary frequencies were obtained. On relaxing the distorted structure obtained by freezing in small eigen-displacements of the odd mode, the graphene sheet with the SW defect buckles. The energy gained on buckling under the LDA approximation is $\approx$ 280 meV (for 7×7 supercell with 1 SW defect) and the difference along z axis

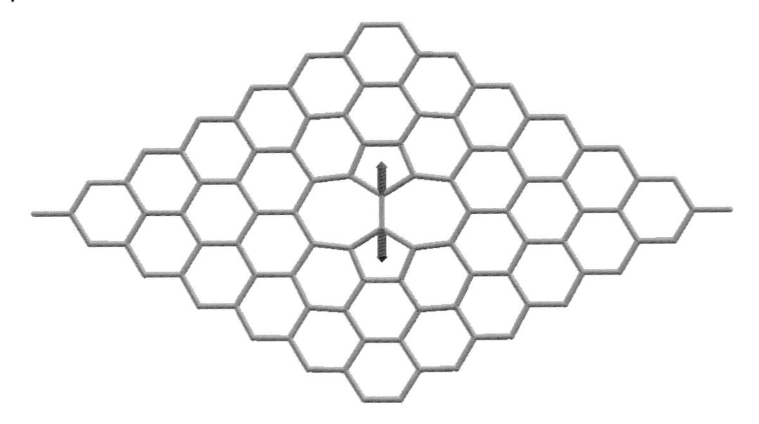

Figure 8.11 High-frequency in-plane stretching mode of graphene with an SW defect.

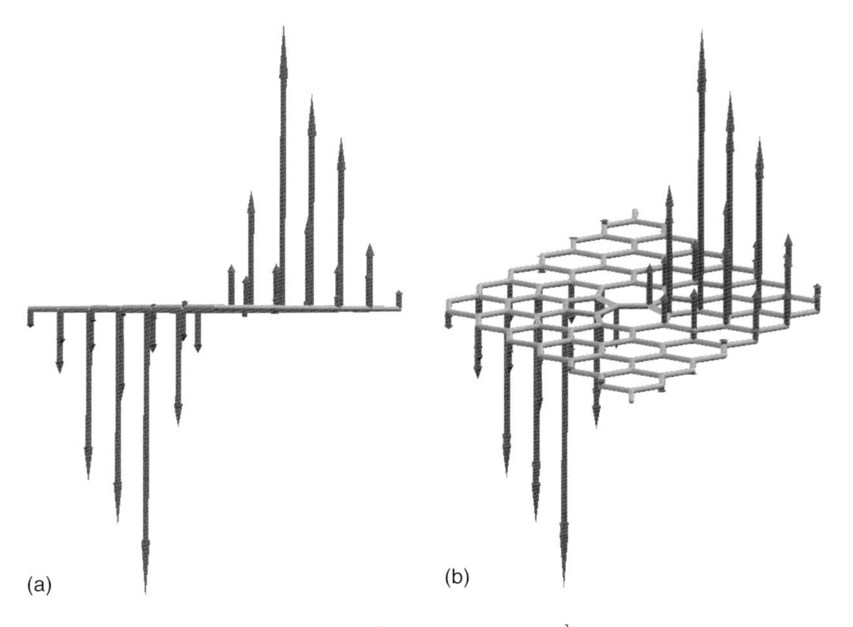

Figure 8.12 Odd mode of imaginary frequency $\sim$ 65i cm^{-1}.

of the highest and lowest carbon atoms is 1.3 Å as compared to the generalized gradient approximation (GGA) of 253 meV (for 8 × 8 supercell with 1 SW defect) and 1.40 Å in earlier work [4].

We have carried out similar studies on the vibrational spectrum of graphene for a higher defect concentration, that is, one defect in a 5 × 5 supercell. The vibrational spectrum shows a similar behavior, *i.e.*, the G band softens and the D band hardens. The extent of softening and hardening is higher in this case because of the higher concentration of SW defects (4%). Here too, two unstable modes are observed with odd and even symmetries. In this case, the even mode is more

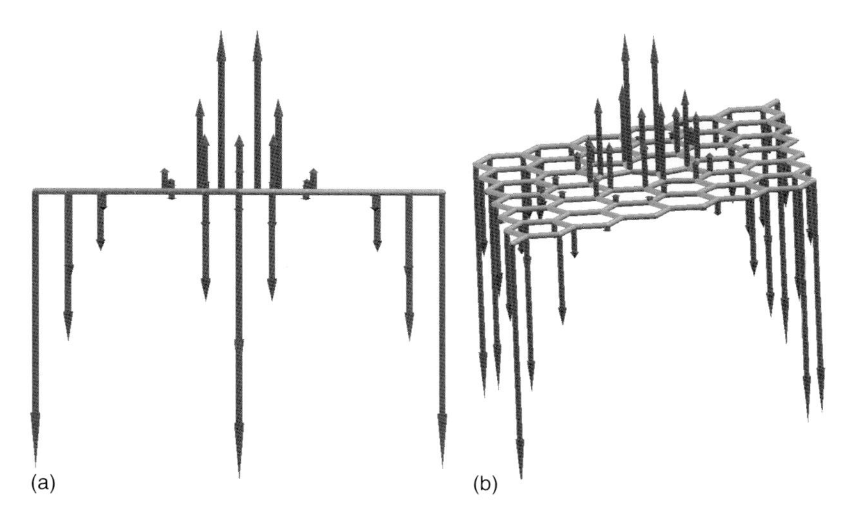

Figure 8.13 Even mode of imaginary frequency $\sim$ 53i cm^{-1}.

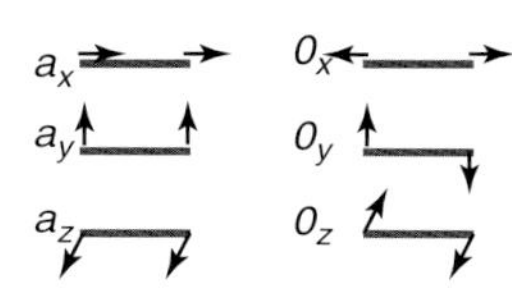

Figure 8.14 Eigendisplacements of acoustic and optical modes in graphene.

unstable than the odd mode. As the two modes differ in frequency by $\sim$ 6i cm^{-1}, we expect, in general that the system gets rippled through a linear combination of both the unstable modes.

We now show how this buckling arises out of the coupling optical modes and strain fields in the continuum limit. Let a_x, a_y, a_z and o_x, o_y, o_z represent the atomic displacements of the acoustic and optical modes in the x, y and z directions respectively (Figure 8.14). An SW defect can be represented as a linear combination of o_x and o_y displacements of carbon atoms (Figure 8.3b). The strain $\epsilon_{\alpha\beta}$ that preserves the invariance of the Hamiltonian under the rigid body rotation of a 2D sheet is given as [27]

$$\epsilon_{\alpha\beta} = \frac{1}{2}\left(\frac{\partial a_\alpha}{\partial x_\beta} + \frac{\partial a_\beta}{\partial x_\alpha} + \frac{\partial a_z(\mathbf{r})}{\partial x_\alpha}\frac{\partial a_z(\mathbf{r})}{\partial x_\beta}\right) \tag{8.12}$$

where $\alpha, \beta = x, y, z$.

The energy of coupling between optical and acoustic modes is given by

$$H(o-\epsilon) = \iint \mathrm{d}x\mathrm{d}y[g_1(o_x^2(\mathbf{r})\epsilon_{xx}(\mathbf{r}) + o_y^2(\mathbf{r})\epsilon_{yy}(\mathbf{r})) + g_2(o_y^2(\mathbf{r})\epsilon_{xx}(\mathbf{r}) + o_x^2(\mathbf{r})\epsilon_{yy}(\mathbf{r})) \\ + 4g_3 o_x(\mathbf{r})o_y(\mathbf{r})\epsilon_{xy} + g_4 o_z^2(\mathbf{r})(\epsilon_{xx}(\mathbf{r}) + \epsilon_{yy}(\mathbf{r})] \tag{8.13}$$

For an SW defect, o_x and o_y are nonzero, and a part of the first term gives

$$g_1 o_x^2 \left(\frac{\partial a_z(\mathbf{r})}{\partial x} \right)^2 \tag{8.14}$$

Atomic displacements are periodic with the supercell and for wave vector "**q**,"

$$a_{\alpha(\mathbf{R})} = e^{i\mathbf{q}\cdot\mathbf{R}} \epsilon_\alpha$$

where **R** is the supercell lattice vector and ϵ is the eigenvector of the acoustic mode. In the long wavelength limit,

$$o_x^2 \left(\frac{\partial a_z(\mathbf{r})}{\partial x} \right)^2 = g_1 o_x^2 q^2 a_z^2 \tag{8.15}$$

$$o_x^2 \left(\frac{\partial a_z(\mathbf{r})}{\partial x} \right)^2 = \frac{1}{\lambda^2} g_1 o_x^2 a_z^2 \tag{8.16}$$

where $q = \frac{2\pi}{\lambda}$ and λ is the wavelength. g_1 corresponds to the change in frequency of the o_x mode on the application of ϵ_{xx} strain in pristine graphene. Our first-principles estimate of g_1 is -0.99 eV Å^{-2}. As $g_1 < 0$, it is clear that the energy of the lattice falls off, as $1/\lambda^2$ (from Equation (8.16)), and in the long wavelength limit it represents the rippling of graphene. The rippling of graphene in the presence of SW defect is thus due to coupling between optical modes and in-plane strain in the system.

8.3.2 Lattice Thermal Conductivity of Graphene with SW Defect

Graphene exhibits exotic electronic and thermal properties. Owing to the unique electronic structure of graphene, electrons have negligible effective mass (*i.e.*, they behave as massless Dirac fermions) around the K-point of the BZ. The large electrical conductivity observed in graphene is due to the large mobility of the carriers. The thermal properties too exhibit a similar behavior since the thermal conductivity of a suspended graphene sheet was measured in the range of 4840–5300 W m^{-1} K^{-1} [28]. In the previous theoretical studies of thermal conductivity [29–31], the effects of Umklapp, boundary, and point defect scattering processes were considered. Here, we use a generic way to estimate the relaxation time and determine the thermal conductivity due to scattering of phonons by point defects or topological defects.

8.3.2.1 Theoretical Model

Phonon thermal conductivity (κ) at temperature T is given by Tamura *et al.* [32]

$$\kappa_{\alpha\beta} = \Sigma_{q,\nu} v_{g\alpha}(q,\nu) v_{g\beta}(q,\nu) C_v(q,\nu) \tau(q,\nu) \tag{8.17}$$

In a cell with N_a number of atoms, ν stands for the number of modes and taking values from 1 to $3N_a$, and q corresponds to the wavevector in the BZ. Here, $v_g(q,\nu)$, $C_v(q,\nu)$, $\tau(q,\nu)$ are the group velocity, specific heat, and relaxation time of a phonon mode, with wavevector q and mode ν, respectively. The specific heat of a given mode is

$$C_v(q,\nu) = \frac{k_B}{V}\left(\frac{\hbar\omega_{q,\nu}}{k_B T}\right)^2 n^o(\omega_{q,\nu})(n^o(\omega_{q,\nu})+1) \tag{8.18}$$

where V is the volume (calculated with height = 3.35 Å i.e., interplanar spacing between planes in graphite), $\omega(q,\nu)$ is the phonon frequency, T is the temperature, n^o is the Bose–Einstein distribution function, and k_B is the Boltzmann constant. v_g and C_v are obtained in a straightforward manner from the knowledge of vibrational spectra at the harmonic level. We use a method to calculate τ for any given phonon mode due to scattering by the SW defect based on the Fermi golden rule for elastic scattering:

$$\tau(q,\nu)^{-1} = \frac{2\pi}{\hbar}\Sigma_{q',\mu}|\langle q,\nu|\Delta H|q',\mu\rangle|^2\delta(\hbar\omega_{q,\nu}-\hbar\omega_{q',\mu}) \tag{8.19}$$

where

$$\langle q,\nu|\Delta H|q',\mu\rangle = \langle q,\nu|H'-H_o|q',\mu\rangle \tag{8.20}$$

$$\langle q,\nu|\Delta H|q',\mu\rangle = \Sigma_{k',m'}\langle q,\nu|k',m'\rangle\langle k',m'|q',\mu\rangle\hbar\omega'_{k',m'} - \hbar\omega_{q,\mu}\delta_{q,q',\mu,\nu} \tag{8.21}$$

where H' is the Hamiltonian of graphene with defect, H_o is the Hamiltonian of pristine graphene and ΔH is the perturbing potential. k', m', and $\omega'_{k',m'}$ represent the phonon wave vector, mode, and frequency of graphene sheet with SW defect. The delta function is the condition for elastic scattering events. Under the harmonic approximation, there exists no scattering between phonons. Hence, the relaxation time $\tau(q,\nu)$ is independent of Bose–Einstein distribution function [33] and the expression for τ is given by

$$\tau(q,\nu)^{-1} = \Sigma_{q',\mu}\frac{2\pi}{\hbar}\delta(\hbar\omega_{q,\nu}-\hbar\omega_{q',\mu})\left|\left(\Sigma_{k',m'}\langle q,\nu|k',m'\rangle\langle k',m'|q',\mu\rangle\hbar\omega'_{k',m'} - \hbar\omega_{q,\mu}\delta_{q,q'\mu,\nu}\right)\right|^2 \tag{8.22}$$

We carried out Γ-point phonon calculations on a 7 × 7 supercell with a single SW defect. This corresponds to a 7 × 7 mesh of k-points in the BZ of a unit cell with two atom basis. Hence, the phonon eigenvectors on 7 × 7 k-mesh in the BZ form a complete orthonormal set. Any phonon eigenvector at Γ-point of a 7 × 7 supercell with defect can thus be expressed as a linear combination of the phonon eigenvectors of pristine graphene. This orthonormal set of phonon eigenvectors is then used to evaluate τ for all the modes in pristine graphene Equation (8.22).

8.3.2.2 κ: Results

Group velocities of longitudinal acoustic (LA) and transverse acoustic (TA) modes from our calculations are 24.5×10^3 m s^{-1} and 15.1×10^3 m s^{-1}, respectively, which are comparable to values from the literature [27]. The scattering potential is independent of temperature, hence the relaxation time is also independent of temperature. Our estimates of relaxation times of the LA and TA modes are 6.3×10^{-10} s and 1.8×10^{-11} s, respectively. The temperature dependence comes through

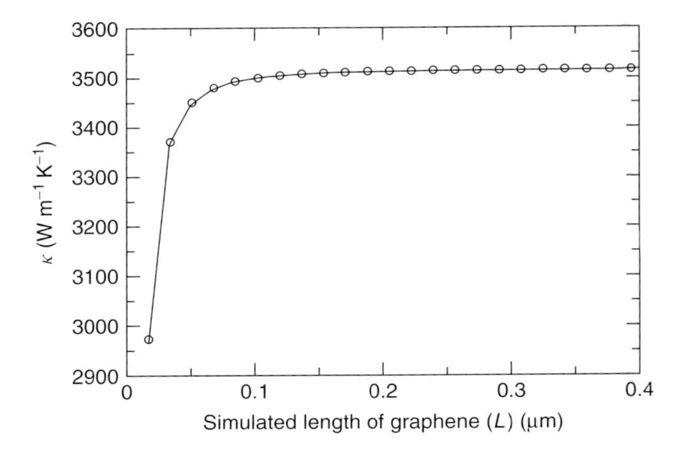

Figure 8.15 Variation of κ (W m^{-1} K^{-1}) with L (μm) at 300 K. κ saturates for $L \geq 0.3$ μm.

the specific heat in this quasiharmonic picture. At Γ-point, the acoustic mode frequency tends to zero, and hence there exists a divergence in the Bose–Einstein distribution function. In two-dimensional systems, the contribution of Γ-point to the displacement autocorrelation function is divergent and is given by Mermin [34]

$$\langle [\mathbf{u}(\mathbf{R}) - \mathbf{u}(\mathbf{R}')]^2 \rangle = ln|\mathbf{R} - \mathbf{R}'|, |\mathbf{R} - \mathbf{R}'| \to \infty$$

where $\mathbf{u}(\mathbf{R})$ is the displacement of the atom at lattice site $\mathbf{R}$. $\langle [\mathbf{u}(\mathbf{R}) - \mathbf{u}(\mathbf{R}')]^2 \rangle$ is the displacement autocorrelation function. This divergence is fixed by taking a graphene sheet of finite size (L). The lowest acoustic frequencies were taken as $\omega = v_g \frac{2\pi}{L}$, which took care of the divergence in the specific heat. The variation of thermal conductivity κ with L at 300 K (Figure 8.15) shows that κ saturates to ~ 3500 W m^{-1} K^{-1} at 0.3 μm. We have fixed the $L = 0.4$ μm in all our further analysis of variation of κ with temperature (Figure 8.16): we observe that thermal conductivity saturates to its maximum value very rapidly at a fairly low temperature. The thermal conductivity saturates to ~ 3500 W^{-1}m^{-1}K at 100 K, indicating that SW defect-based scattering mechanism limits the thermal conductivity mainly at low temperatures. At higher temperatures, Umklapp scattering becomes dominant and the thermal conductivity is expected to decrease [29–31]. Note that this analysis is semiqualitative, carried out to develop understanding the role of SW defect in κ and has to be tested further.

8.3.3
Discussion

We have shown that the energy of interaction between any two SW defects falls as the square of the distance between them. A bandgap opens at the Γ-point in the BZ of $(3n) \times (3n)$ supercells due to an SW defect. The bandgap decreases with decreasing SW defect concentration. The Dirac cones shift from K to K + δ k along K to M in $(3n + 1) \times (3n + 1)$ and along K to M′ in $(3n + 2) \times (3n + 2)$ supercells with SW defect.

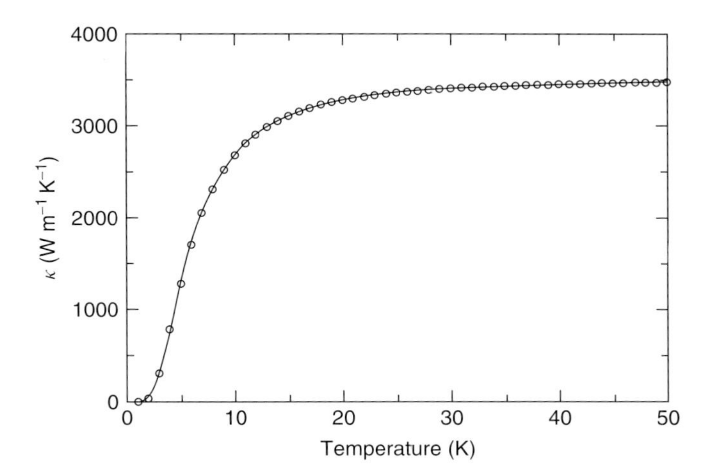

Figure 8.16 Variation of κ ($W\,m^{-1}\,K^{-1}$) with T (K) at $L = 0.4\,\mu m$. κ saturates to its maximum value ($\sim 3500\,W\,m^{-1}\,K^{-1}$) at 100 K. This indicates that the effect of SW defect scattering is mainly at low temperatures.

Graphene with SW defect, with or without electron doping fails to show any net magnetization due to the absence of dangling bonds.

Relaxation with the SW defect causes contraction of the C–C bond at the SW defect, and the release of this stress results in elongation of the bonds in the area around the defect. The G band is scattered by the SW defect and softens, while the D band is attracted/localized at the defect and hardens due to the stiffer C–C bond. This frequency change can be used to identify the presence of defects from the Raman spectra of the graphene sheet.

The SW defect leads to warping instability of the graphene sheet, and it exhibits two imaginary frequencies. One mode shows out-of-plane sine wavelike displacement and the other mode shows cosine wavelike displacement of atoms. These unstable modes make the system buckle and hence may play a role in the intrinsic rumpling of graphene.

8.4
$C_{1-x}(BN)_{x/2}$: C–BN Interfaces

The two-dimensional boron nitride with a honeycomb lattice structure is isoelectronic and isostructural to graphene but exhibits a large bandgap of 5.9 eV [35]. In hexagonal boron nitride (h-BN), every boron atom is bonded to three nitrogen atoms and vice versa. The SW defects are energetically unfeasible in h-BN because a 90° rotation of a B–N bond forms B–B and N–N bonds. These bonds are less favorable over B–N bonds, which makes the formation of SW defects in h-BN highly improbable. A bandgap opens in the electronic structure of graphene when it is mixed with hexagonal boron nitride (h-BN), *i.e.*, $C_{1-x}(BN)_{x/2}$ and can be engineered by varying x [36]. The possibility of an SW defect formation is only at the interfaces of carbon and boron nitride (C–BN), which are present in the solid solution of graphene and h-BN. We have investigated several possible SW defect formations at the armchair (AI) and zigzag interfaces (ZI) of C–BN.

8.4.1
SW Defect at the C–BN Interface

We consider several possible SW defect configurations at the C–BN interface and find that formation energies of these SW defects is in the range of 4–6 eV per defect, which is comparable to the SW defect formation energy in graphene [37]. The stability of such defect structures is determined by various bonds that are present in them. The hierarchy of energetic stability of the bonds is B–N > C–C > C–N > C–B > B–B > N–N [38], which implies that the higher the number of B–N and C–C bonds, the greater is the stability of defects. Using this criterion of bond stability, we describe the stability of SW defects at various C–BN interfaces. The interface energy of AI is $\sim 0.39\,\mathrm{eV}\,\mathrm{Å}^{-1}$ for both the armchair edges in the

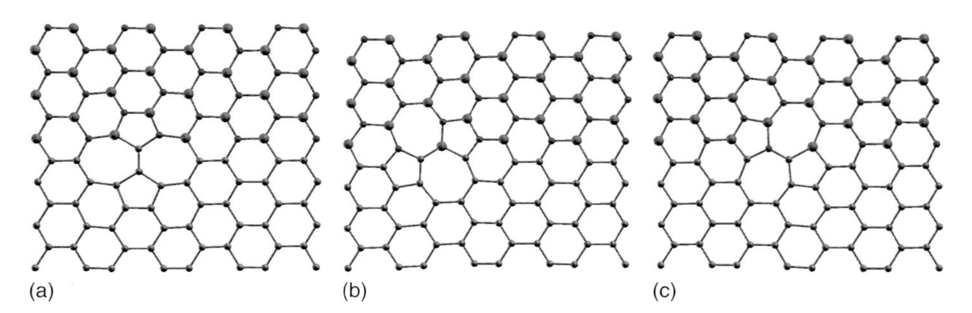

Figure 8.17 SW defect configurations at armchair interface. Top panel: (a) C–C bond rotation, (b) C–B bond rotation, and (c) C–N bond rotation. Bottom panel: (a) C–B bond rotation and (b) C–C bond rotation. C atoms are denoted by light gray, B atoms by gray, and N atoms by black.

Table 8.3 Formation energy of an SW defect at C–BN interface.

Configuration	Formation energy (eV)
AI_1	4.29
AI_2	4.66
AI_3	5.66
ZIN_1	4.48
ZIN_2	4.34
ZIB_1	5.77
ZIB_2	4.38

Table 8.4 Bond lengths at the C–BN interface in Angstroms.

Configuration	C–B	C–N	C–C	B–N
AI_1	1.55	1.39	1.31	–
			1.45	
			1.47	
AI_2	1.42	1.4	1.42	1.45
	1.53			
AI_3	1.54	1.29	1.46	1.55
		1.39		
ZIN_1	1.53	1.31	1.39	1.47
		1.39		
ZIN_2	–	1.40	1.33	–
			1.42	
			1.44	
			1.46	
ZIB_1	1.41	1.42	1.44	1.47
	1.54			
ZIB_2	1.57	–	1.33	–
			1.42	
			1.45	
			1.46	

The equilibrium bond lengths are $d_{C-B} = 1.54$ Å, $d_{C-N} = 1.34$ Å, $d_{C-C} = 1.42$ Å, $d_{B-N} = 1.45$ Å [39].

supercell. For ZI, the interface energy is $\sim 0.58\,\text{eV}\,\text{Å}^{-1}$, which is the sum of both N- and B-terminated interface energies.

We have three possibilities of SW defect formation at AI, which we refer to as AI_1, AI_2, and AI_3 (refer to Figure 8.17a, b, and c, respectively). Formation energies of SW defects for AI_1, AI_2, and AI_3 are 4.29, 4.66, and 5.66 eV, respectively (Table 8.3). In the most stable structure AI_1, there exist three C–C bonds and one of C–N and C–B bonds, but no B–N bonds at the rotated bond of the defect, whereas in AI_2, there is one C–C bond, one C–N bond, two C–B bonds, and one B–N bond. Similarly for AI_3, there is one C–C bond, two C–N bonds, one C–B bond, and one B–N bond at the rotated bond of the defect. As AI_1 has the maximum number of C–C bonds, it shows the highest stability. Although AI_3 has more number of C–N bonds than AI_2, it is less stable because the most energetically stable bond, that is, the BN bond in AI_2 and AI_3 is stretched by 0 and 6.6%, respectively, from its equilibrium bond length of 1.45 Å (refer to Table 8.4 for bond lengths).

Zigzag interfaces can be of two types: N-terminated and B-terminated. We consider two configurations at the N-terminated zigzag interface, ZIN_1 (C–N bond rotated) and ZIN_2 (C–C bond rotated) (refer to Figure 8.18a and b, respectively), and two configurations at the B-terminated interface, ZIB_1 (C–B bond rotated) and ZIB_2

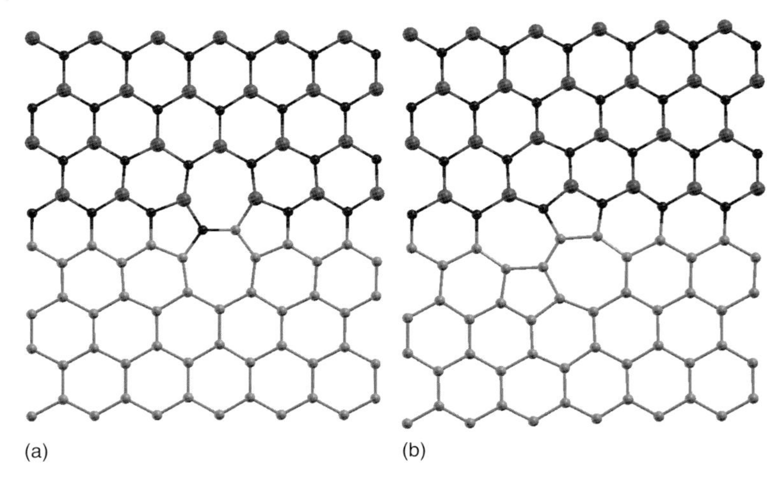

Figure 8.18 SW defect configurations at nitrogen-terminated zigzag interface. (a) C–N bond rotation and (b) C–C bond rotation. C atoms are denoted by light gray, B atoms by gray, and N atoms by black.

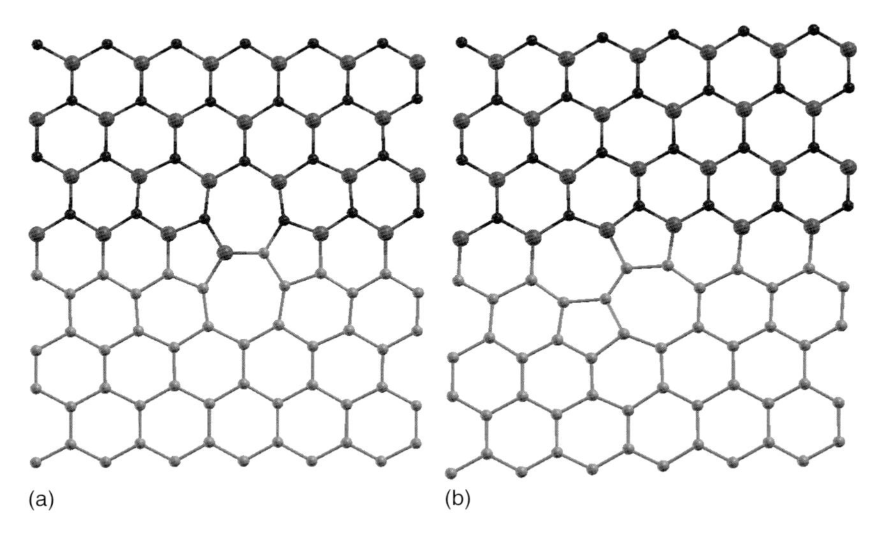

Figure 8.19 SW defect configurations at boron-terminated zigzag interface. (a) C–B bond rotation and (b) C–C bond rotation. C atoms are denoted by light gray, B atoms by gray, and N atoms by black.

(C–C bond rotated) (refer to Figure 8.19a and b, respectively). Formation energies for ZIN_1, ZIN_2, ZIB_1, and ZIB_2 are 4.48, 4.34, 5.77 and 4.37 eV, respectively (Table 8.3). ZIN_2 and ZIB_2 have the maximum number of C–C bonds at the defect, which makes them more stable than ZIN_1 and ZIB_1. Among ZIN_1 and ZIB_1, ZIN_1 is more stable, as the C–N bond in ZIB_1 is stretched by $\sim$ 6%, which leads to increase in formation energy (refer to Table 8.4 for bond lengths).

8.4.2 Discussion

The formation energy of an SW defect at the C–BN interface is in the range of 4–6 eV, and is comparable to the formation energy of an SW defect in graphene. The stability of the defect at the interface is based on the hierarchy in the stability of B–N, C–C, C–N, C–B, B–B, and N–N bonds. We predict that such defects should be present at the C–BN interfaces.

8.5 Two-Dimensional MoS_2 and $MoSe_2$

Bulk MoS_2 or $MoSe_2$ consists of a hexagonal lattice with BAB-BAB-BAB... stacking of Mo and S/Se atoms (Figure 8.20). In the bulk crystalline form, they are indirect bandgap semiconductors [40]. However, MoS_2 and $MoSe_2$ both exhibit direct semiconducting bandgap in their monolayer form [41–43] with energy gaps of 1.80 and 1.29 eV, respectively. We have studied the energetics and stability of various point defects, and the dependence of its electronic properties on the inclusion of planar stacking faults in MoS_2. Although the bandgaps of MoS_2 and $MoSe_2$ in bulk and monolayer forms are much greater than the energy of infrared (IR) radiation, recent experiments [44] report IR absorption in MoS_2 and $MoSe_2$. An explanation for this interesting, yet puzzling, observation can be obtained through first-principles calculations.

8.5.1 Point Defects

Learning from the SW defect in graphene, we generated a point defect by rotating a Mo–S bond in a way such that no two Mo atoms or S atoms come very close to one another. For different angles of rotation of the bond, the system was relaxed and its formation energy was calculated. The bond was rotated through 30°, 60°, 90°, 120°, 150°, and 180° angles about the x-axis (Figure 8.21). For angles less than 90°, the bond reverts to its original position (Figure 8.21a). For angles greater than 90°, the bond does not revert back (Figure 21b–d), and the system settles to a local minimum of energy. For 120° bond rotation, the rotated Mo atom moves down to the plane of S atoms and acquires the A position in the S plane (Figure 8.21b). In doing so, the rotated bond compresses by $\sim$ 8% and system develops uniaxial compressive stresses along both x- and y-directions. This stress would be relieved through in-plane expansion of the MoS_2 monolayer. The formation energy of this defect is 7.82 eV. We henceforth refer to this defect as defect-I. Similarly, on rotating the Mo–S bond through 150° and 180°, the system structurally relaxes to the same state (Figure 8.21c and d), in which the rotated bond compresses by $\sim$ 9%. The system develops compressive stress along x-axis and tensile stress along y-axis. The system will try to reduce these in-plane compressive and tensile

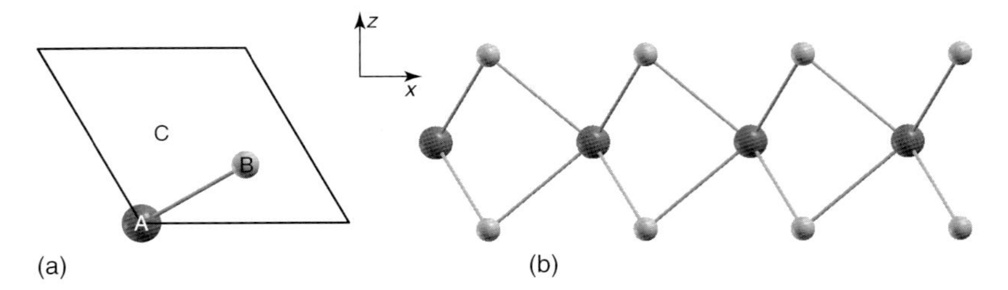

Figure 8.20 (a) Top view and (b) side view. Mo atoms are denoted by dark gray/larger radii and S/Se atoms by light gray/smaller radii.

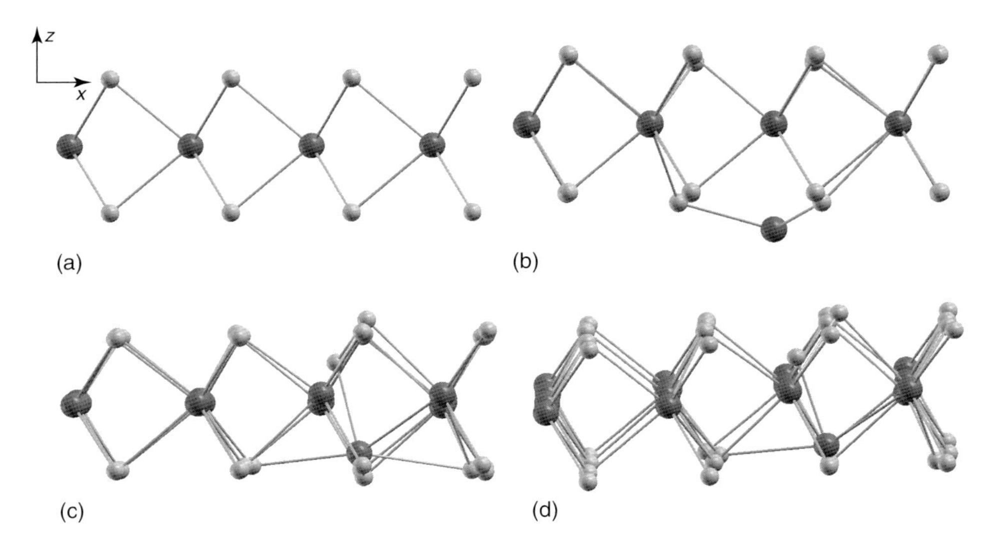

Figure 8.21 Point defect formation after relaxation of bond rotated through (a) angles $\leq 90^\circ$, (b) 120°, (c) 150°, and (d) 180°. Mo atoms are denoted by dark gray/larger radii and S atoms by light gray/smaller radii.

stresses by out-of-plane deformation. The defect formation energy in this case is 8.01 eV, and we henceforth refer to this defect as defect-II. Such large formation energies of defect-I and defect-II imply that these defects would form in very low concentrations.

Furthermore, we have studied the electronic structure of these structures with defect. Both the systems with defect-I and defect-II remain insulating, but their bandgaps reduce. For a concentration of $\sim 6.25\%$ of defect-I (i.e., 1 defect per $4 \times 4 \times 1$ supercell), the bandgap reduces to 0.44 eV at the K-point of system. On the other hand, the system with defect-II turns into an indirect bandgap semiconductor with highest occupied molecular orbital (HOMO) at K-point and lowest occupied molecular orbital (LUMO) at Γ-point, and a bandgap of 0.49 eV. This reduction in bandgap is due to the defect states arising out of the "d" orbitals of the rotated Mo atom and its nearest Mo neighbors. This is confirmed from the isosurface plots of charge density, which depict the HOMO and LUMO states (at Γ-point in the BZ). In the case of defect-I, the "d_{z^2}" orbital of the rotated Mo atom contributes most prominently to the HOMO and LUMO bands (Figure 8.22a and b). For defect-II, the contribution to the HOMO and LUMO bands is mainly from the "d_{yz}" and "d_{zx}" orbitals of the rotated Mo atom (Figure 8.22c and d). As the frontier states are localized on these defects, we expect these defect sites to be highly reactive and favorable for adsorption of foreign atoms and molecules. Even though the formation energies of these defects are high (*i.e.*, low defect concentrations), we expect that the introduction of such point defects in the system during synthesis can be utilized to engineer electronic properties of 2D MoS_2 and $MoSe_2$.

8.5.2 Stacking Faults

Here, we investigate the evolution of electronic properties of MoS_2 with stacking faults. Pure MoS_2 monolayer with B-A-B stacking of S-Mo-S layers is a direct bandgap semiconductor (Figure 8.23a and c). Note that the bands around the bandgap are relatively flat because of the localized nature of "4 d" electronic states of Mo. When one of the S layers is displaced from its B position in hexagonal lattice to the C position (i.e., B-A-C stacking), MoS_2 monolayer turns metallic (refer to Figure 8.23b and d). On projecting out the contributions of individual atoms to the total electronic density of states, we find that the major contribution to bands crossing fermi energy in MoS_2 with stacking fault comes from "4 d" states of Mo. On sliding one of the S layers into the C position, a greater degree of overlap occurs between the p-states of S and d-states of Mo. This overlap leads to higher covalency, and hence, MoS_2 makes a transition from insulating to metallic state due to stacking faults. The stacking fault energy (γ_{isf}) is ~1.49 J m^{-2}, which is much higher than the reported γ_{isf} for C in diamond structure ~0.25 J m^{-2} [45].

8.5.3 IR Radiation Absorption

The bandgap of $MoSe_2$ is smaller than that of MoS_2 because there is a weak overlap between p-states of Se and d-states of Mo as compared to the negligible overlap between p-states of S and d-states of Mo. The bandgap of $MoSe_2$ (1.29 eV) is larger than the energy of incident IR radiation (1.16 eV), and therefore cannot provide an explanation for the observed absorption of IR radiation [44]. We have explored the possibility of this absorption occurring because of edges in the $MoSe_2$ sample by simulating a nanoribbon with an armchair edge. The electronic band structure of the nanoribbon shows that the bandgap reduces to 0.38 eV (Figure 8.24). We see that a dominant contribution to HOMO comes from the "$d_{x^2-y^2}$" and "d_{xy}" orbitals of Mo at the edges of the nanoribbon (Figure 8.25a), whereas the contribution to LUMO comes from "d_{xy}" orbitals of Mo at the edge (Figure 8.25b). As the width of the ribbon is small, the Mo atoms in the interior of the ribbon too contribute to the HOMO state. With increasing width, the contribution of the interior atoms is expected to

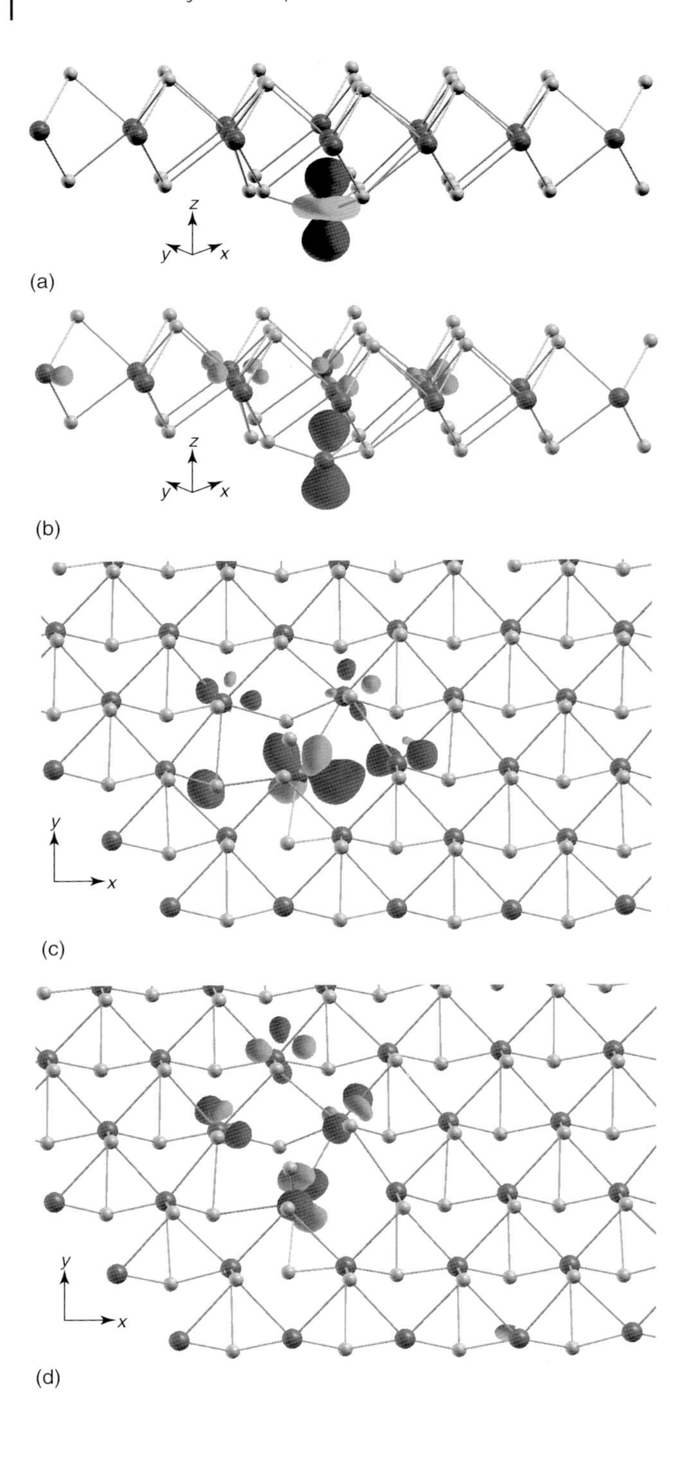
z
y
x
(a)
z
y
x
(b)
y
x
(c)
y
x
(d)

Figure 8.22 Isosurfaces of charge density contributions of the (a) HOMO and (b) LUMO states of system with defect-I (*i.e.*, relaxed structure after 120° bond rotation). The HOMO and LUMO states of the system with defect-II (*i.e.*, relaxed structure after 150° and 180° bond rotation) are shown in (c) and (d), respectively. The main contribution to the defect state is from the "d_{z^2}" orbital at the defect site in case of defect-I and "d_{zx}" and "d_{yz}" orbitals for defect-II. Mo atoms are denoted by dark gray/larger radii and S atoms by light gray/smaller radii.

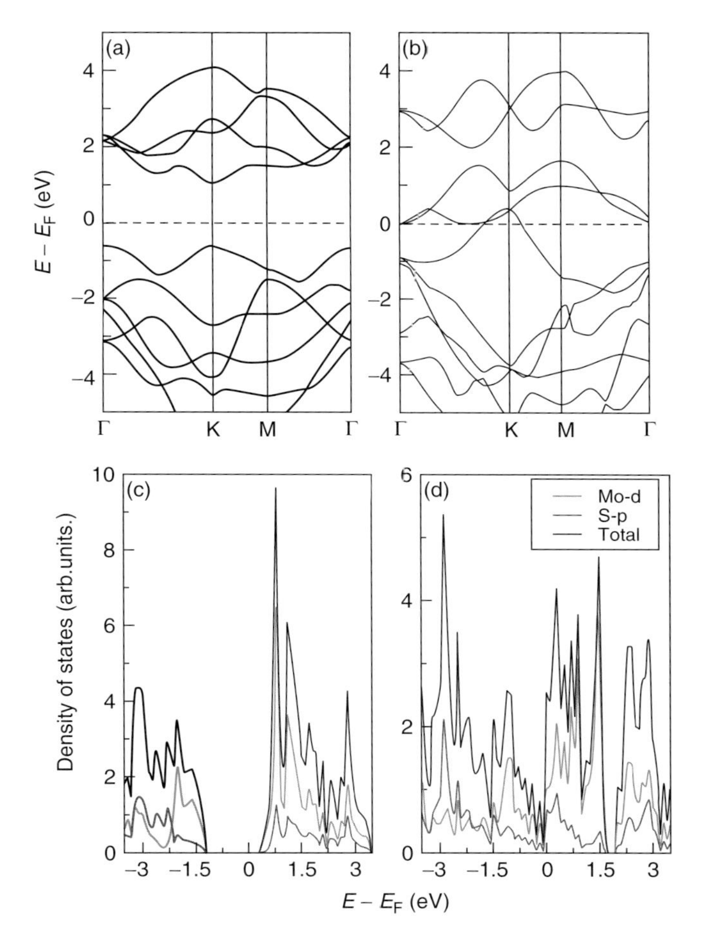

Figure 8.23 Electronic band structure and density of states of MoS_2: Top panel: (a) and (b) are the band structure plots for structure without and with stacking fault, respectively. Bottom panel: (c) and (d) are density of states for structure without and with stacking fault, respectively. Note that the system becomes metallic on the introduction of stacking fault. Increased covalency between d-states of Mo and p-states of S can be clearly seen in (d).

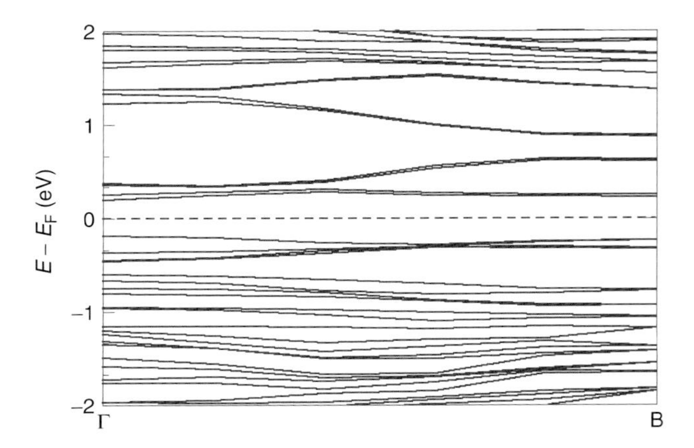

Figure 8.24 Electronic band structure of $MoSe_2$ ribbon with armchair edge. The bandgap reduces to 0.38 eV from 1.29 eV for the nanoribbon.

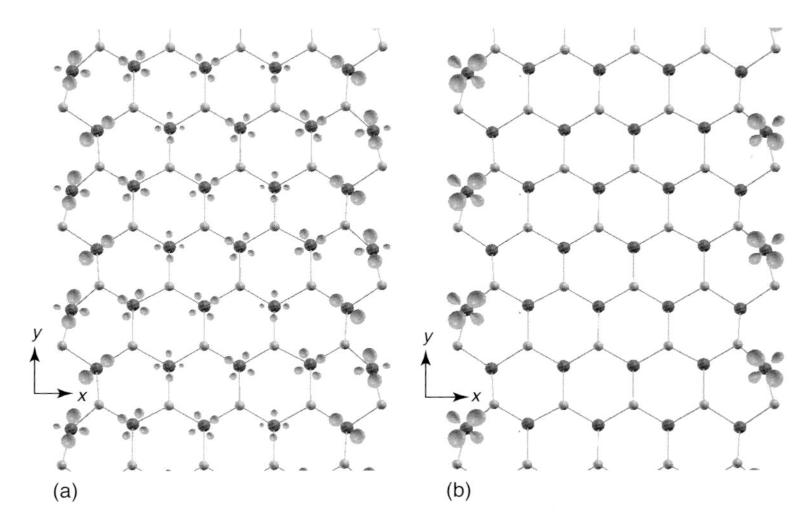

Figure 8.25 Isosurfaces of charge density contributions of the (a) HOMO and (b) LUMO states of nanoribbon of $MoSe_2$. The contribution to the HOMO state is mainly from the "$d_{x^2-y^2}$" and "d_{xy}" orbitals of Mo atoms. The "d_{xy}" orbitals of edge Mo atoms contribute to LUMO state. The "d" orbitals of the edge atoms are the major contributors to the HOMO and LUMO bands. Mo atoms are denoted by dark gray/larger radii and Se atoms by light gray/smaller radii.

fall and hence the contribution to the HOMO and LUMO bands comes mainly from the edges of the ribbon. Also, point defects in MoS_2 reduce the bandgap to ~ 0.5 eV. As $MoSe_2$ and MoS_2 are similar in their electronic character, we conclude that the IR absorption in both the systems occurs mainly due to the defect states introduced by point defects and edge states due to sample edges. The IR active phonon modes in the system further facilitate the absorption of IR radiation, by MoS_2 and $MoSe_2$.

8.5.4 Discussion

Formation energies of point defects in MoS_2 are in the range of 7–8 eV, and hence their concentration is expected to be very small. These point defects introduce defect states near the Fermi energy, which reduce the bandgap to ~0.5 eV. MoS_2, which is an insulator, exhibits an exotic behavior of insulator to metal transition, with the introduction of a planar stacking fault. We propose that by introducing stacking faults and point defects in MoS_2 or $MoSe_2$, their bandgap can be engineered. Absorption of IR radiation occurs largely at the sample edges and better IR absorption is expected to occur in flakes of MoS_2 and $MoSe_2$ samples, possibly with point defects as compared to pristine monolayers.

8.6 Summary

Through an extensive theoretical study, it is clear that topological defects in a two-dimensional system, such as graphene, solid solution of graphene and h-BN, and $MoS_2/MoSe_2$, greatly influence their properties. An SW defect can intrinsically cause rumpling of graphene. Owing to strains arising at the SW defect, the interaction energy between SW defects is long ranged and falls off as an inverse square of the distance between them. Thermal conductivity of the lattice is dominated by defect scattering at low temperatures.

While h-BN is unlikely to exhibit SW topological defects, $C_{1-x}(BN)_{x/2}$ shows a rich variety of SW defects at the C–BN interface. The most likely configuration is the C–C bond rotation at the armchair interface. On account of the structural similarities between graphene and C–BN, it is evident that the C–BN system is expected to buckle because of the presence of the SW defect.

Point defects do occur in MoS_2 monolayers, but probably in very low concentrations because of their high formation energies. On the introduction of a planar stacking fault, the system makes a transition from an insulator to a metal because of the increased covalency between Mo and S; $MoSe_2$ too will exhibit an insulator to metal transition on the inclusion of stacking fault. Our analysis of the interaction of IR radiation with $MoSe_2$ reveals that absorption occurs mainly at the edges of the sample and point defect sites because of the lowering of bandgap by edge and defect states, respectively. We predict that efficiency of IR detectors can be improved with the use of flakes, ribbons, monolayers, and highly defective polycrystalline samples of $MoSe_2$ and MoS_2.

In conclusion, the response of 2D nanomaterials is altered by topological defects, which have relevance to experiments that involve rippling deformation of these materials, Raman spectroscopy, and possible metal to nonmetal transitions in these 2D nanomaterials.

Acknowledgments

SNS is thankful to the Council of Scientific and Industrial Research, India, for Junior Research Fellowship and UVW acknowledges support from AOARD grants (No. FA2386-10-1-4062 and FA2386-10-1-4150).

References

1. Ziman, J.M. (1979) *Models of Disorder: The Theoretical Physics of Homogeneously Disordered Systems*, Cambridge University Press, Cambridge.
2. Dann, S.E. (2000) *Reactions and Characterization of Solids*, The Royal Society of Chemistry.
3. Meyer, J.C., Kisielowski, C., Erni, R., Rossell, M.D., Crommie, M.F., and Zettl, A. (2008) *Nano. Lett.*, **8**, 11.
4. Ma, J., Alfè, D., Michaelides, A., and Wang, E. (2009) *Phys. Rev. B*, **80**, 033407.
5. Upadhyay, M., Singh, S.P., and Waghmare, U.V. (2008) *Small*, **4**, 2209.
6. Lahiri, J., Lin, Y., Bozkurt, P., Oleynik, I.I., and Batzill, M. (2010) *Nat. Nanotechnol.*, **5**, 326.
7. Shirodkar, S.N. and Waghmare, U.V. (2012) *Phys. Rev. B*, under consideration.
8. Monkhorst, H.J. and Pack, J.D. (1976) *Phys. Rev. B*, **13**, 5188.
9. Giannozzi, P., Baroni, S., Bonini, N., Calandra, M., Car, R., Cavazzoni, C., Ceresoli, D., Chiarotti, G.L., Cococcioni, M., Dabo, I., Dal Corso, A., Fabris, S., Fratesi, G., de Gironcoli, S., Gebauer, R., Gerstmann, U., Gougoussis, C., Kokalj, A., Lazzeri, M., Martin-Samos, L., Marzari, N., Mauri, F., Mazzarello, R., Paolini, S., Pasquarello, A., Paulatto, L., Sbraccia, C., Scandolo, S., Sclauzero, G., Seitsonen, A.P., Smogunov, A., Umari, P., and Wentzcovitch, R.M. (2009) *J. Phys.: Condens. Matter*, **21**, 395502. *www.quantum-espresso.org*.
10. Perdew, J.P. and Zunger, A. (1981) *Phys. Rev. B*, **23**, 5048.
11. Vanderbilt, D. (1990) *Phys. Rev. B*, **41**, 7892.
12. Banhart, F., Kotakoski, J., and Krasheninnikov, A.V. (2011) *ACS Nano*, **5**, 26.
13. Popov, V.N., Henrard, L., and Lambin, P. (2007) *Carbon*, **47**, 2448.
14. Yazyev, O.V. and Louie, S.G. (2010) *Phys. Rev. B*, **81**, 195420.
15. Bhowmick, S. and Waghmare, U.V. (2010) *Phys. Rev. B*, **81**, 155416.
16. Arias, T.A. and Joannopoulous, J.D. (1994) *Phys. Rev. Lett.*, **73**, 680.
17. Castro Neto, A.H., Guinea, F., Peres, N.M.R., Novoselov, K.S., and Geim, A.K. (2009) *Rev. Mod. Phys.*, **81**, 109.
18. Saito, R., Dresselhaus, G., and Dresselhaus, M.S. (2000) *Phys. Rev. B*, **61**, 2981.
19. Peng, X. and Ahuja, R. (2008) *Nano. Lett.*, **8**, 12.
20. Dubay, O. and Kresse, G. (2003) *Phys. Rev. B*, **67**, 035401.
21. Pisana, S., Lazzeri, M., Casiraghi, C., Novoselov, K.S., Geim, A.K., Ferrari, A.C., and Mauri, F. (2007) *Nat. Mater.*, **6**, 198.
22. Yazyev, O.V. and Helm, L. (2007) *Phys. Rev. B*, **75**, 125408.
23. Santos, E.J.G., Ayuela, D.S., and Portal, A. (2010) *Phys. Rev. B*, **81**, 125433.
24. Lee, H., Son, Y.-W., Park, N., Han, S., and Yu, J. (2005) *Phys. Rev. B*, **72**, 174431.
25. Pimenta, M.A., Dresselhaus, G., Dresselhaus, M.S., Cancado, L.G., Jorio, A., and Saito, R. (2007) *Phys. Chem. Chem. Phys.*, **9**, 1276.
26. Ferrari, A.C., Meyer, J.C., Scardaci, V., Casiraghi, C., Lazzeri, M., Mauri, F., Piscanec, S., Jiang, D., Novoselov, K.S., Roth, S., and Geim, A.K. (2006) *Phys. Rev. Lett.*, **97**, 187401.

27. Sandeep Kumar, Hembram, K.P.S.S. and Waghmare, U.V., 2010) *Phys. Rev. B*, **82**, 115411.
28. Balandin, A.A., Chosh, S., Bao, W., Calizo, I., Teweldebrhan, D., Miao, F., and Lau, C.N. (2008) *Nano Lett.*, **8**, 902.
29. Singh, D., Murthy, J.Y., and Fisher, T.S. (2011) *J. Appl. Phys.*, **110**, 044317.
30. Kong, B.D., Paul, S., Nardelli, M., Buongiorno, M., and Kim, K.W. (2009) *Phys. Rev. B*, **80**, 033406.
31. Nika, D.L., Pokatilov, E.P., Askerov, A.S., and Balandin, A.A. (2009) *Phys. Rev. B*, **79**, 155413.
32. Tamura, S., Tanaka, Y., and Maris, H.J. (1999) *Phys. Rev. B*, **60**, 2627.
33. Ward, A., Broido, D.A., Stewart, D.A. and Deinzer, G. (2003) *Phys. Rev. B*, **80**, 125203.
34. Mermin, N.D. (1968) *Phys. Rev.*, **176**, 250.
35. Watanabe, K., Taniguchi, T., and Kanda, H. (2004) *Nat. Mater.*, **3**, 404.
36. Xu, B., Lu, Y.H., Feng, Y.P., and Lin, J.Y. (2010) *J. Appl. Phys.*, **108**, 073711.
37. Li, L., Reich, S., and Robertson, J. (2005) *Phys. Rev. B*, **72**, 184109.
38. Nozaki, H. and Itoh, S. (1996) *J. Phys. Chem. Solids*, **57**, 41.
39. Blase, X., Charlier, J.-C., De Vita, A., and Car, R. (1997) *Appl. Phys. Lett.*, **70**, 197.
40. Böker, Th., Severin, R., Müller, A., Janowitz, C., and Manzke, R. (2001) *Phys. Rev. B*, **64**, 235305.
41. Lebègue, S. and Eriksson, O. (2009) *Phys. Rev. B*, **79**, 115409.
42. Mak, K.F., Lee, C., Hone, J., Shan, J., and Heinz, T.F. (2010) *Phys. Rev. Lett.*, **105**, 136805.
43. Ma, Y., Dai, Y., Guo, M., Niu, C., Lu, J., and Huang, B. (2011) *Phys. Chem. Chem. Phys.*, **13**, 15546.
44. Chitara, B., Ramakrishna Matte, H.S.S., Maitra, U., Shirodkar, S.N., Krupanidhi, S.B., Waghmare, U.V., and Rao, C.N.R., a preprint.
45. Thomas, T., Pandey, D., and Waghmare, U.V. (2008) *Phys. Rev. B*, **77**, 121203(R).

9
Graphene and Graphene-Oxide-Based Materials for Electrochemical Energy Systems

Ganganahalli Kotturappa Ramesha and Srinivasan Sampath

9.1
Introduction

Electrochemistry of carbon-based materials dates back to centuries. Various new forms of carbon and their continuous discovery make carbon an interesting material attracting attention of researchers in different areas. Major advantages of carbon-based electrodes include low cost, wide potential window, relatively inert electrochemistry, and electrocatalytic activity for variety of redox reactions. They are very well known in metal production, energy storage in batteries, supercapacitors, and also as catalyst support for various processes [1]. Graphite, a three-dimensional layered material, has been widely used in analytical and industrial electrochemistry since the report of Humphrey Davy to produce alkali metals [1]. Structural polymorphism, chemical stability, rich surface chemistry, and strong carbon–carbon bond along the *xy*-plane are some of the attributes that have placed graphite at an advantageous position in electrochemical studies. The crystallite dimensions, L_a and L_c, in-plane and perpendicular crystallite size are ~1 μm for highly oriented pyrolitic graphite (HOPG), ~10–100 nm for polycrystalline graphite, and ~1–10 nm for carbon black [1]. Graphite consists of atomically ordered hexagonal plane along the "*a*" axis (basal plane) and irregular surface with sp^3-hybridized carbon parallel to the "*c*" axis (edge plane). Anisotropy in electrical conductivity (basal plane resistivity, 4×10^{-5} Ω cm; edge plane resistivity, 0.17 Ω cm) and difference in the availability of functional groups such as carboxyl and hydroxyl along edge planes result in different electrochemical activities for the two surfaces [1]. Electrochemical rate constants at edge planes are reported to be higher ($k \sim 0.02$ cm s^{-1} for diffusing species) than that observed on basal planes ($k < 10^{-9}$ cm s^{-1}) for reversible redox couples such as $Fe^{2+/3+}$ [1].

Other forms of carbon, fullerene, carbon nanotubes, and the recent addition to this family, "graphene and graphene oxide (GO)", have obviously attracted the attention of electrochemists for various applications. Exceptionally attractive and predicted properties such as large specific surface area (2630 m^2 g^{-1}), high intrinsic mobility (200 000 cm^2 V^{-1} s^{-1}), high Young's modulus (~1.0 TPa), thermal conductivity (~5000 W m^{-1} K^{-1}), and optical transmittance (~97.7% in the

Graphene: Synthesis, Properties, and Phenomena, First Edition. Edited by C. N. R. Rao and A. K. Sood.

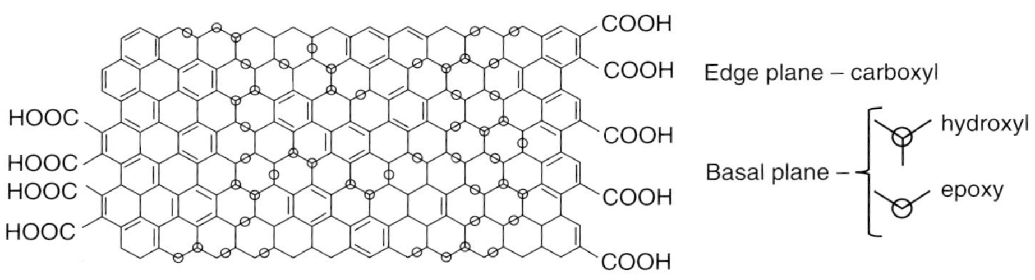

Figure 9.1 Lerf–Klinowski model consisting of carboxylic acid functional groups on the periphery of the basal plane and epoxy and hydroxyl groups along the basal plane. (Adapted from Ref. [13].)

UV–Visible region) along with excellent electrical conductivity have made this a material of choice in recent studies [2–6]. The double layer capacitance values of edge- and basal planes of graphene in the presence of various electrolytes show considerable differences [7, 8] as reported in the literature. Several reviews have appeared on graphene-based electrochemistry in the recent past [6, 7, 9–11]. The present chapter deals with the use of graphene and GO for electrochemical energy systems – batteries, fuel cells, and supercapacitors. As the literature is very extensive, we have decided to concentrate on certain important results to highlight the advantages and disadvantages of this material toward electrochemical energy conversion and storage. A caution note – the term *graphene* used in electrochemical studies is mostly "reduced graphene oxide (rGO)" and not "graphene" possessing long range order.

The oxidized and exfoliated form of graphite [12] having oxygen-containing functionalities such as carboxylic, epoxy, and hydroxyl groups has been studied extensively for several years, although the term *GO* is of recent origin. The widely accepted Lerf-Klinowski model (Figure 9.1) proposes GO to consist of carbon basal plane with randomly distributed epoxy and hydroxyl functional groups and edge planes with carboxyl and phenolic functional groups.

9.2 Graphene-Based Materials for Fuel Cells

Fuel cells are electrochemical energy devices that convert continuously supplied fuel to electricity [14]. Important issues involved in developing efficient fuel cells are related to (i) fuel, (ii) catalyst, (iii) catalyst-support material, and (iv) polymer electrolyte membrane. Polymer electrolyte membrane fuel cells (PEMFCs) operate under ambient conditions to temperatures of about 200 °C. Although research on fuel cells has been carried out for several decades, cost, durability, stability of catalysts, and catalyst supports are still a matter of major concern. The catalyst of choice is platinum (Pt) for several redox reactions. One of the important requirements for the catalyst is its stability under fuel cell operation conditions. The stability is indirectly related to the support material and, very often, carbon used as a support undergoes corrosion even in H_2/O_2 fuel cells [15]. This leads to agglomeration of Pt and subsequently the catalyst peels off from carbon support leading to a decrease in performance. Surface area of the support material influences the amount of catalyst and consequently its dispersibility.

9.2.1 Graphene-Based Catalyst Support for Small Molecule Redox Reactions

GO and rGO have been well studied as support material for various oxidation/reduction reactions [5, 16–20]. Theoretical studies based on DFT calculations have shown that graphene sheets terminated by H atoms bind strongly to Pt clusters [21]. They are shown to have small HOMO–LUMO gap that is retained even after Pt binds to sp^2 carbon to form C–Pt bonds. This should be taken into account while designing supports for catalysts.

Preparation of graphene-based catalysts mostly involves simultaneous reduction of metal ions and GO during which there is every possibility of agglomeration, that is, formation of bi- and multilayer graphene sheets sticking together [5, 17]. This leads to small surface area of the support, which in turn will affect the dispersibility of catalysts and subsequently their performance. In one of the initial studies toward this direction, Kamat and Seger [18] have reported the use of partially reduced GO to support Pt nanoparticles as the cathode material for H_2/O_2 fuel cells. Pt deposition from chloroplatinic acid is affected by using either borohydride or hydrazine as the reducing agent in the presence of GO, which resulted in the formation of (r)GO/Pt composites. The electrochemically active surface area (ECSA) of Pt is found to be large for hydrazine-reduced (r)GO/Pt composites (20 $m^2\,g^{-1}$). However, the ECSA is found to be dependent on the heat treatment of the sample as given in Figure 9.2a. The polarization characteristics of H_2/O_2 fuel cell (Figure 9.2b) using commercial E-TEK/carbon-black/Pt composite as anode and (r)GO/Pt as cathode catalysts show a power density of 161 $mW\,cm^{-2}$ at a current density of 400 $mA\,cm^{-2}$ while unsupported Pt catalyst show 96 $mW\,cm^{-2}$ at $\sim$200 $mA\,cm^{-2}$ under identical conditions. However, the authors point out that the power density decreases drastically with time owing to the instability associated with the catalyst/support. Jafri and coworkers [19] used Pt loaded graphene for H_2/O_2 fuel cell and reported maximum power density values of about 400 $mW\,cm^{-2}$.

Li and coworkers [20] have improved the catalytic activity of graphene/Pt composites by preparing the catalysts in basic conditions (pH 10) and subsequently used them for methanol oxidation (Figure 9.3). The catalyst reveals ECSA of 44.6 $m^2\,g^{-1}$. Xin and coworkers [16] have prepared graphene/Pt composites by reduction of platinic acid using borohydride in the presence of GO and used a lyophilization step instead of normal drying. This procedure is found to reduce irreversible aggregation of graphene sheets resulting in high electrocatalytic activity for methanol oxidation as well as oxygen reduction reaction (ORR).

Although metal–nanoparticle-incorporated graphene structures have been widely reported for redox reactions and fuel cells [5, 16, 16–22], one should be aware that uniform distribution of catalytic particles plays a vital role in the

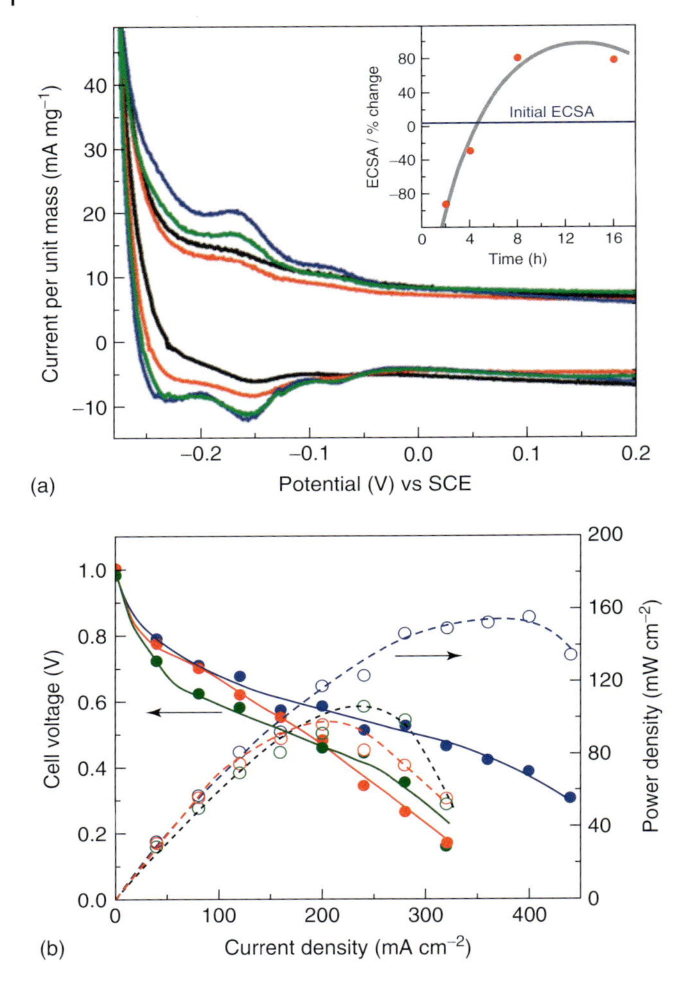

Figure 9.2 (a) Cyclic voltammograms at a scan rate of 20 mV s^{-1} for hydrazine-treated (r)GO-Pt in 0.1 M H_2SO_4 and annealed at 300 °C for different periods of time (black, red, blue, and green correspond to 2, 4, 8, and 16 h, respectively). Inset shows relative increase of ECSA with duration of heating. (b) Galvanostatic fuel cell characteristics at 60 °C. The cathode is composed of (red) Pt, (blue) 1 : 1 GO-Pt, and (green) 1 : 1 (r)GO-Pt (hydrazine, treated at 300 °C) with a loading of 0.2 mg cm^{-2} Pt. (Adapted from Ref. [18].)

catalytic activity. Toward this direction, proteins as templates have been used along with graphene supports to achieve monodispersity. Liu and coworkers [23] have reported the use of protein "bovine serum albumin" (BSA) to control the shape and interaction of nanoparticles with graphene sheet. The forces responsible for stabilization of particles have been proposed to be hydrophobic $\pi-\pi$ stacking interactions and hydrogen bonding. The TEM images (Figure 9.4) show very

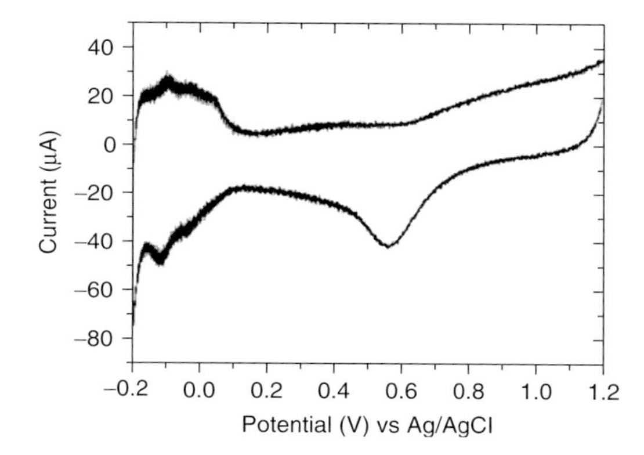

Figure 9.3 Cyclic voltammogram of Pt/graphene electrode in nitrogen saturated aqueous 0.5 M H_2SO_4 at a scan rate of 50 mV s^{-1}. (Adapted from Ref. [20].)

uniform distribution of particles though these systems are yet to be explored for electrochemical and fuel cell activity.

Although platinum is the most effective catalyst used for small-molecule electrochemical oxidation and reduction reactions, several drawbacks such as CO poisoning of Pt sites and instability of catalyst during alcohol oxidation have led to exploration of bimetallic alloys such as Pt-Ru [24], Pt-Au [25], Pt-Pd [26],

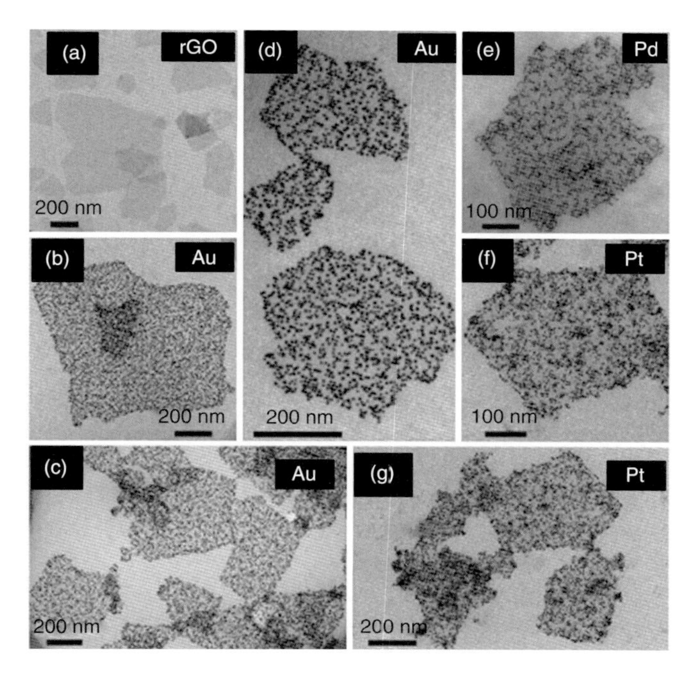

Figure 9.4 TEM images of (a) BSA-rGO, (b–d) Au-BSA-rGO, (e) Pd-BSA-rGO, and (f,g) Pt-BSA-rGO. (Adapted from Ref. [23].)

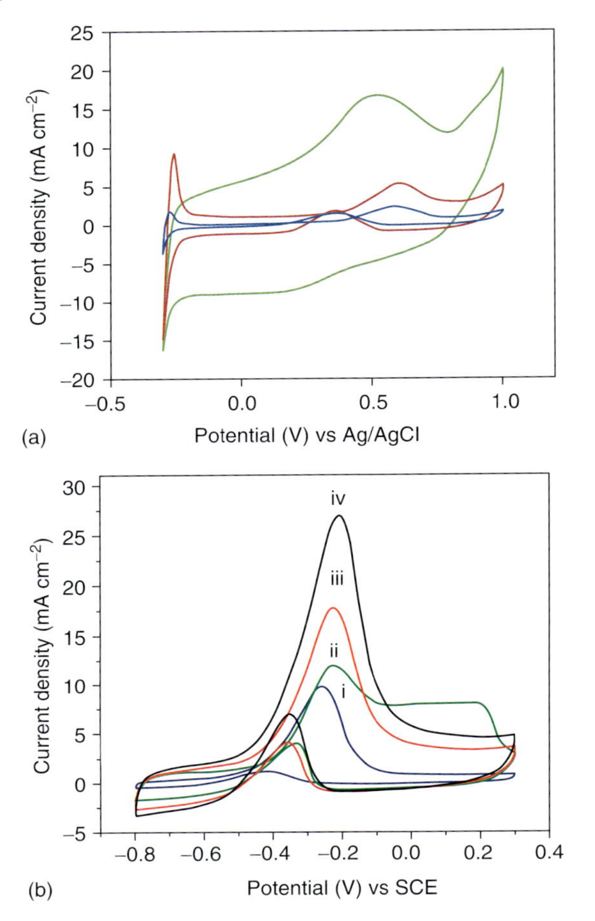

Figure 9.5 (a) Cyclic voltammograms in 1 M CH_3OH/0.5 M H_2SO_4 at a scan rate of 50 mV s^{-1} (blue, red, and green correspond to graphite, Vulcan carbon, and graphene-supported Pt/Ru nanoparticles, respectively) [24]. (b) Cyclic voltammograms of (i) Pt/G, (ii) Pt-Au/G, (iii) Pt-Pd/G, and (iv) Pt-Pd-Au/G catalysts in 1.0 M KOH containing 1.0 M CH_3OH at a scan rate of 50 mV s^{-1}. (Adapted from Ref. [29].)

Pt-Co, Pt-Cr [27], and Pd-Au [28] and trimetallic structures such as Pt-Pd-Au [29] for direct alcohol fuel cells. Efficiency of catalysts can be improved by dispersing them on GO-based supports [24–29]. Toward this direction, Dong and coworkers [24] have reported a comparative study of graphene-supported Pt and Pt-Ru catalyst for methanol and ethanol oxidation and observed forward peak current densities of 19.1 and 9.76 mA cm^{-2} for graphene- and carbon-black-supported Pt nanoparticles, respectively. Reverse peak current, which is attributed to CO poisoning effect, is reported to be suppressed when Pt-Ru supported on graphene catalyst is used (Figure 9.5a). In another related study, Pt-Au electrocatalyst grown on graphene by electrochemical deposition at constant potential has been used by

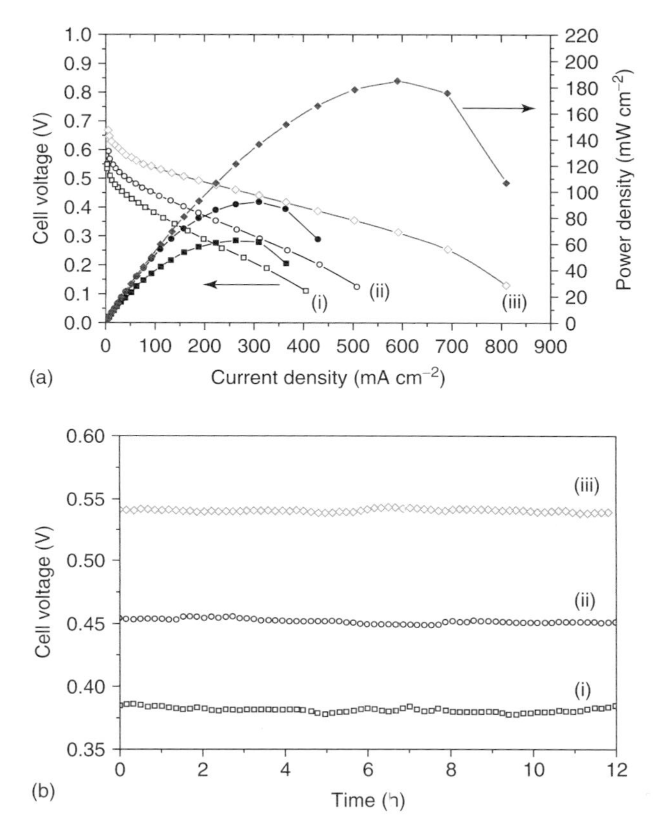

Figure 9.6 (a) $I-V$ polarization curves for (i) graphene/Pt, (ii) graphene/Pt-Au, and (iii) commercial Pt/C anodes for direct formic acid fuel cells at 333 K and at 1 atm. (b) Stability of graphene/Pt, graphene/Pt-Au, and commercial Pt/C anodes at 100 mA cm^{-2}. (Adapted from Ref. [32].)

Hu and coworkers [25] for methanol oxidation and ORR. The catalytic activity is further explored by using graphene composites with trimetallic catalyst, Pt-Pd-Au (Figure 9.5b) [29]. Pt-Pd catalyst is very well known to resist CO adsorption owing to the so-called synergistic effect [30]. Introduction of Au resists CO adsorption, thus decreasing the poisoning effect from adsorbed CO [31].

In addition to methanol oxidation, graphene has been used as a catalyst support for other small-molecule oxidation reactions involving formic acid [32], ethanol [33], hydrazine [34], and glycerol [35]. Graphene/Pt-Au shows 10 times higher catalytic activity for formic acid oxidation with lesser CO poisoning than that observed on graphene/Pt. Power densities of 185, 70, and 53 mW cm^{-2} at 333 K have been observed for graphene/Pt-Au, graphene/Pt, and commercial Pt/C anodes, respectively [32]. The catalysts also showed high cell voltages with good stability as shown in Figure 9.6. In addition to formic acid, graphene-supported Pt-Ru has been explored for glycerol oxidation [35] and Pd-functionalized graphene/SnO_2 for

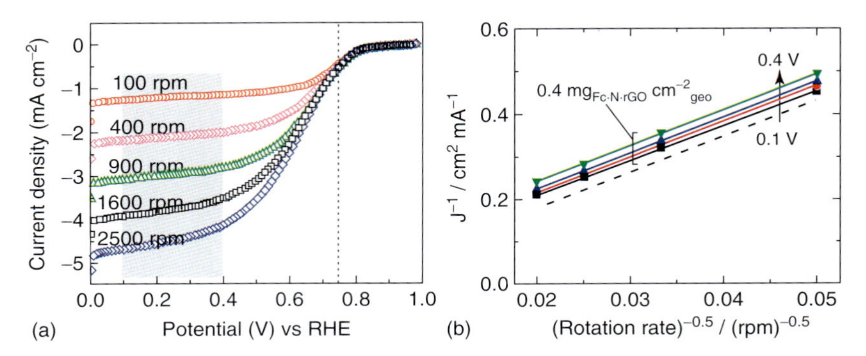

Figure 9.7 ORR activity of Fe-N-rGO in O_2-saturated 0.5 M H_2SO_4 at 10 mV s^{-1} with a catalyst loading of 0.4 mg$_{Fe-N-rGO}$ cm^{-2}. (a) *I–V* curves for Fe-N-rGO at various rotation rates. (b) Koutecky–Levich plots at potential range, 0.1–0.4 V versus reference hydrogen electrode (RHE). Dashed line represents the data for polycrystalline Pt disk at 0.3 V. (Adapted from Ref. [38].)

ethanol oxidation [33]. In a preliminary study, graphene alone without any metallic catalyst is reported to be active toward hydrazine [34] oxidation, although fuel cells have not been constructed.

Nanoparticles such as Co_3O_4 supported on N-doped graphene have recently been reported by Liang and coworkers [36] for ORR and oxygen evolution reactions. The catalyst-supported graphene is prepared by a two-step chemical method by controlled hydrolysis of $Co(OAc)_2$ to obtain 4–8 nm size Co_3O_4 nanoparticles. The composites result in high catalytic activity toward ORR. In another direction, Zhang and coworkers have reported the use of submicron graphite particles of few hundreds of nanometer thickness as the support for Pt nanostructures for ORR [37]. The highly ordered graphitic support is found to be better than commercial Vulcan XC-72 carbon and carbon nanotube supports. Recently, Byon and coworkers [38] have reported graphene-based Fe-N-C catalysts with high oxygen reduction activity. The preparation method involves heating an iron salt, carbon nitride, and rGO together. Rotating disk electrode studies reveal high mass activity for the catalyst and the number of electrons close to four is obtained leading to water as the reduction product (Figure 9.7).

Qu and coworkers [39] have reported N-doped graphene prepared through CVD process (methane as carbon source and ammonia for nitrogen) toward ORR. This metal-free electrode is shown to be more electrocatalytic and more stable for long-term operation than pure platinum. A four-electron pathway is found in alkaline fuel cells based on Figure 9.8.

Recently, Wang and coworkers [40] have showed that a polyelectrolyte (poly(diallyldimethylammonium chloride) PDDA) adsorbed onto graphene yields high ORR activity as compared to commercially available Pt/carbon (Figure 9.9). The stability is found to be high for the composite. The authors have attributed the catalytic activity to be due to intermolecular charge transfer from electron-rich π-conjugated graphene to positively charged PDDA resulting in enhanced

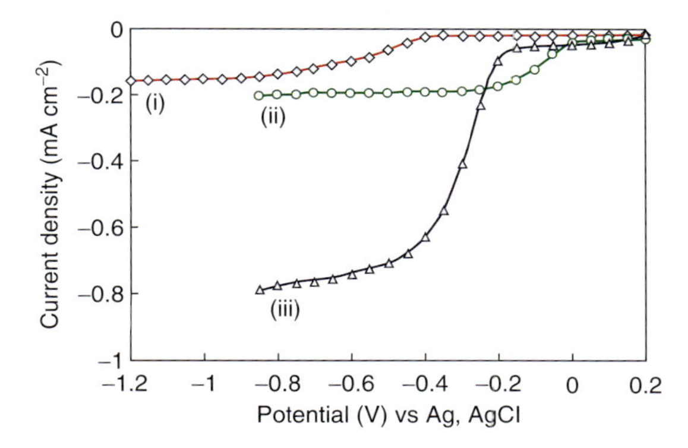

Figure 9.8 Rotating ring disc electrode (RRDE) voltammograms for ORR in air-saturated 0.1 M KOH on (i) graphene, (ii) Pt/C, and (iii) N-graphene electrode. Rotation rate used is 1000 rpm at a scan rate of 0.01 V s^{-1}. Mass of the electrode material used is 7.5 μg. (Adapted from Ref. [39].)

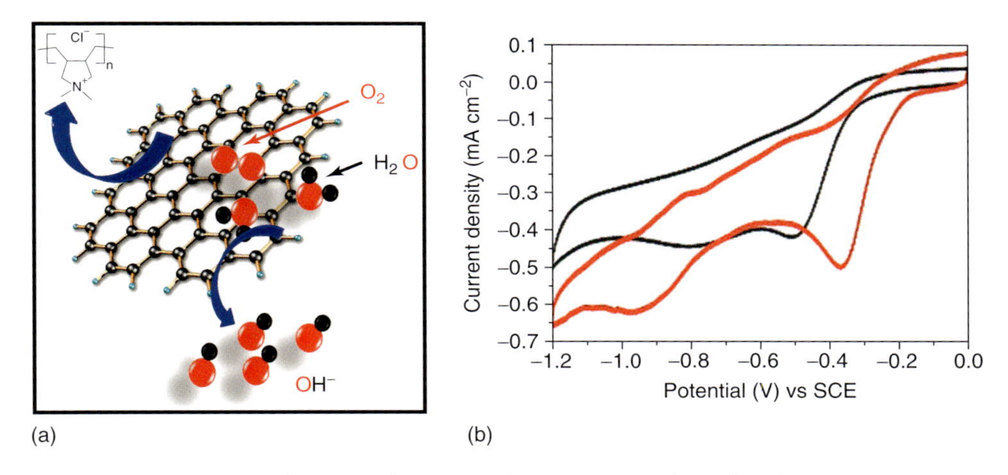

Figure 9.9 (a) Schematic illustration of PDDA/graphene interaction. (b) Cyclic voltammograms of oxygen reduction on graphene (black) and PDDA/graphene (red) electrodes in O_2-saturated aqueous 0.1 M KOH at a scan rate of 50 mV s^{-1}. (Adapted from Ref. [40].)

graphene–oxygen attraction. The interactions of graphene and PDDA have been monitored using IR and Raman spectroscopy.

Jafri and coworkers [41, 42] have reported electrocatalytic activity of Pt-supported graphene/MWCNT composites for ORR and carried out fuel cell studies. The authors have explored the advantages of graphene/MWCNT composite that gives relatively high surface area compared to individual components. Figure 9.10 shows the electrochemical characteristics observed.

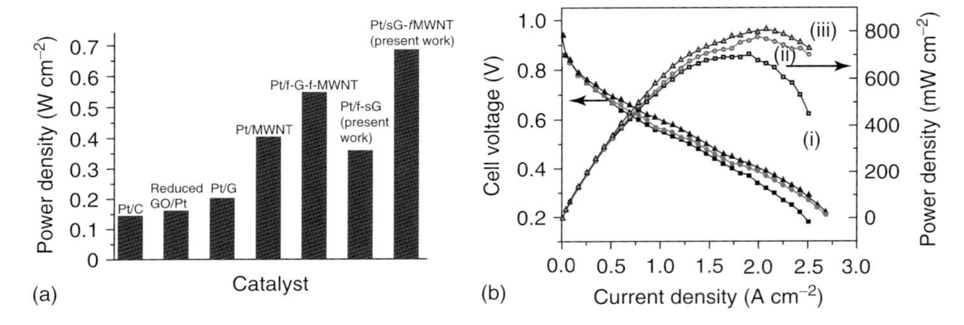

Figure 9.10 (a) Comparison of power densities with various ORR electrocatalysts for H_2/O_2 fuel cells. (b) Polarization curves of Pt/G–fMWNT (cathode catalyst) and Pt/fMWNT (anode catalyst) at temperatures of (i) 40, (ii) 50, and (iii) 60 °C. (Adapted from Ref. [41].)

9.2.2
Graphene-Oxide-Based Proton Conducting Membranes

Apart from catalysts and fuel, the third important component in fuel cells is the ionically conducting membrane. Most of the current fuel cells explore Nafion (sulfonated tetrafluoroethylene-based fluoropolymer-copolymer from Dupont, USA) as the ionically conducting, electronically insulting, polymer electrolyte membrane. Nafion possesses advantages such as the best ratio of hydrophobic (arising from perfluoro carbon chain) to hydrophilic (sulfonic acid groups) regions as well as good ionic conductivity [43]. Although GO is known to be highly hydrophilic, a close examination of its structure shows it to be amphiphilic because of unoxidized (hydrophobic) and oxidized (hydrophilic owing to various functional groups) regions present together. However, only GO as membrane is not stable under liquid and gas flow conditions encountered in fuel cells. GO prepared as composites with other polymer matrices have been explored in this direction. Choi and coworkers [44] have used sulfonated GO/Nafion composite for direct methanol fuel cell (DMFC) and obtained power density of 132 mW cm^{-2} as compared to 101 mW cm^{-2} obtained for Nafion 112 membrane and 120 mW cm^{-2} for GO/Nafion membrane at 60 °C (Figure 9.11a). Zarrin and coworkers [45] have carried out sulfonation of GO after covalent functionalization with 3-mercaptopropyl trimethoxysilane and used the membrane in H_2/O_2 fuel cell. Peak power density of 150 mW cm^{-2} for sulfonated GO/Nafion composite has been observed, which is approximately 3.6 times higher than that observed for recast Nafion (42 mW cm^{-2}) (Figure 9.11b). GO without any functionalization has also been used for H_2/O_2 fuel cell by forming composite with poly(ethylene oxide) (PEO) [46]. Peak power densities obtained are 15 and 60 mW cm^{-2} for PEO/GO and Nafion membranes, respectively at 16 °C. However, when the temperature is increased to 30 and 60 °C, power density increases from 21 to 53 mW cm^{-2} for PEO/GO. Even though performance of PEO/GO membrane is far below that of Nafion, it is cost effective and hence research in this direction is worth pursuing.

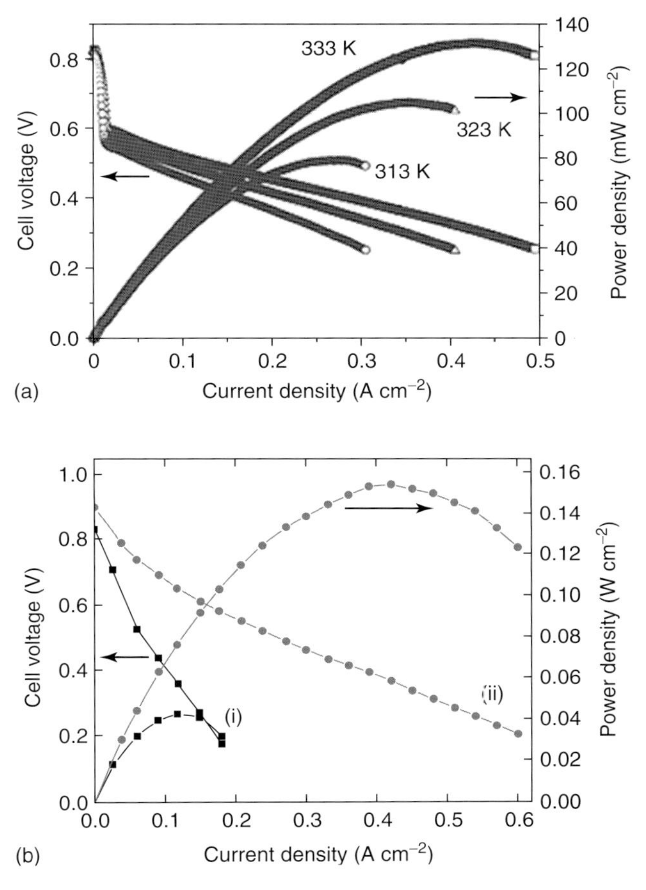

Figure 9.11 Polarization curves of (a) Direct methanol fuel cell using sulfonated GO/Nafion membrane at various temperatures and (b) Sulfonated GO/Nafion (i) and recast Nafion (ii) membranes in H_2/O_2 cell at 120 °C with 25% relative humidity. (Adapted from Refs. [44, 45].)

9.2.3
Graphene-Based Biofuel Cells

Biofuel cells explore catalytic activity of biomolecules to convert biomass into electricity through metabolic activity of microorganisms [47]. Zhang and coworkers [48] explored irontetrasulfophthalocyanine (FeTsPc)-functionalized graphene as catalyst toward ORR. Carbon-paper-supported *Escherichia coli* as anode, FeTsPc–graphene as cathode; phosphate buffer solution (PBS) as catholyte; and PBS-containing 2-hydroxy-1,4-naphthoquinone, glucose, and yeast extract as anolyte are reported to show maximum power density of 817 mW m^{-2}. The same group has explored the possibility of using graphene without functionalization as anode catalyst using carbon-paper-supported *E. coli* as anode material and observed power density

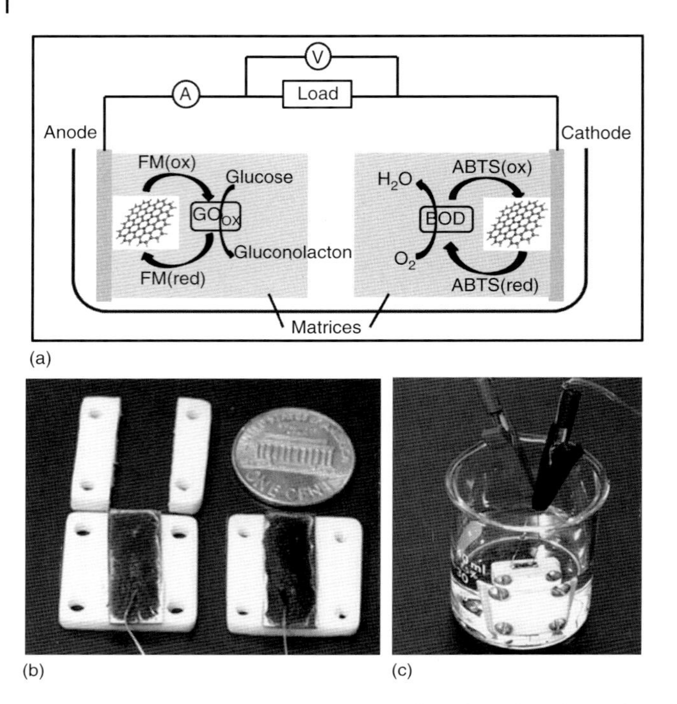

Figure 9.12 (a) Schematic configuration of graphene-based membraneless enzyme fuel cell, (b) graphene-based membraneless EBFC components, and (c) photograph of EBFC. (Adapted from Ref. [50].)

of 2668 mW m^{-2} [49]. Liu and coworkers [50] used graphene–glucose oxidase co-immobilized on gold electrode using silica sol–gel matrix as anode, graphene co-immobilized with bilirubin oxidase as cathode in the configuration shown in Figure 9.12 for enzymatic biofuel cell (EBFC) and obtained maximum power density of about 24.3 ± 4 μW cm^{-2}.

9.3 Graphene-Based Supercapacitors

Supercapacitors or ultracapacitors belong to a class of electrochemical energy storage devices that store electrical energy in the capacitive double layer formed at the electrode–electrolyte interface [14]. The main advantage of electrochemical capacitors is that they possess higher power densities than other electrochemical energy systems. Depending on the charge storage mechanism, supercapacitors are classified into (i) electrochemical double layer (EDL) capacitors (mostly carbon-based materials) and (ii) pseudocapacitors (mostly transition metal oxides and conducting polymers). The third class combines both mechanisms to achieve large capacitance. Three major components of a capacitor are cathode, anode,

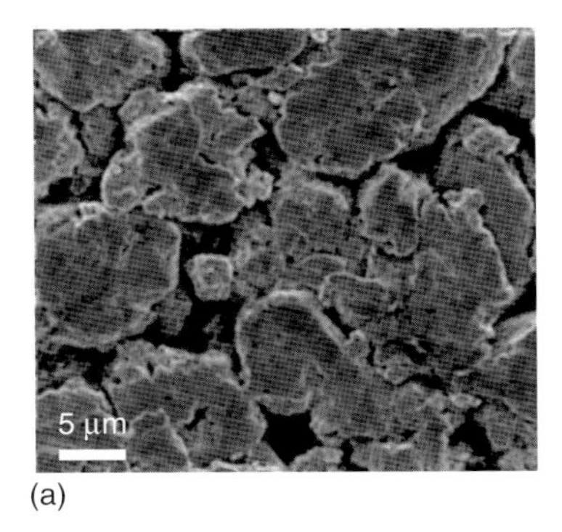

(a)

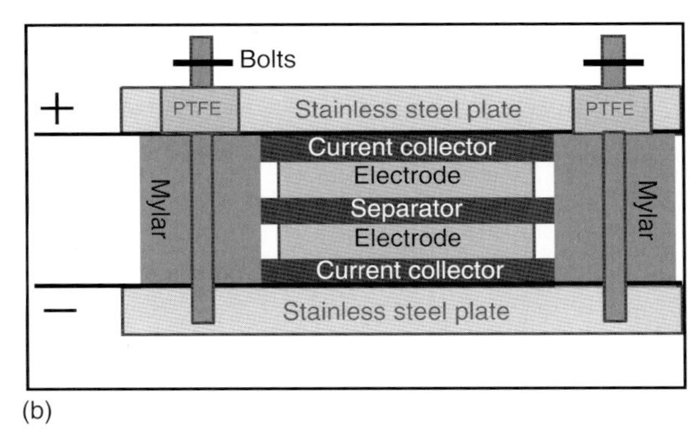

(b)

Figure 9.13 (a) SEM image of chemically modified graphene particles and (b) schematic of test cell assembly used for supercapacitors. (Adapted from Ref. [55].)

and electrolyte [14]. Energy stored in a capacitor is given by $E = \frac{1}{2}CV^2$ where V is the working voltage and C is the capacitance. The present section discusses the application of graphene and graphene-based composites toward capacitor applications.

Specific capacitance and performance characteristics of graphene-based capacitors depend mainly on the synthetic route employed for preparing electrode material. Rao and coworkers [51, 52] synthesized graphene using different procedures: thermal exfoliation of graphite oxide, by heating nanodiamond at 1650 °C in a helium atmosphere, and decomposition of camphor over nickel particles that show surface areas of 925, 520, and 46 $m^2\ g^{-1}$ with specific capacitances of 117, 35, and 6 $F\ g^{-1}$, respectively. The rate capability is reported to be good, with capacity retention of 85% at 1 $V\,s^{-1}$. Thermal reduction in dry state is reported to yield specific capacitances of the order of 150–230 $F\ g^{-1}$ with cycle life of 500 [53, 54].

Stoller and coworkers [55] obtained chemically modified graphene by chemical reduction of GO (Figure 9.13) and realized specific capacitances of 135 and 99 $F\ g^{-1}$ in aqueous and organic electrolytes. However, most of the procedures require binder to fabricate electrodes. Sagar and coworkers [56] have reported the performance characteristics of capacitors based on binderless exfoliated graphite obtained from exfoliation of bisulfate-intercalated graphite at high temperatures of 800 °C. The material, although not referred to as *graphene/graphene oxide*, contains flexible exfoliated sheets that may possibly get classified as few/several layer graphene (GO). Specific capacitances in the range of 0.740–0.980 $mF\,cm^{-2}$ (Figure 9.14) with high operating voltage per cell (3.0 V) and cycle life of 500 are reported in the presence of solid polymer electrolyte containing polyethylene oxide, lithium perchlorate, ethylene carbonate, and propylene carbonate [$(PEO)_9$ $LiClO_4$-40% EC-60% PC] [56]. This study opens up possibilities of using binderless electrodes that can be shaped in any form and size. Absence of binder is expected to result in decreased resistive losses.

Lv and coworkers [57] have performed exfoliation of GO at high temperatures (ambient, atmospheric vacuum) and at low temperatures (high vacuum) resulting

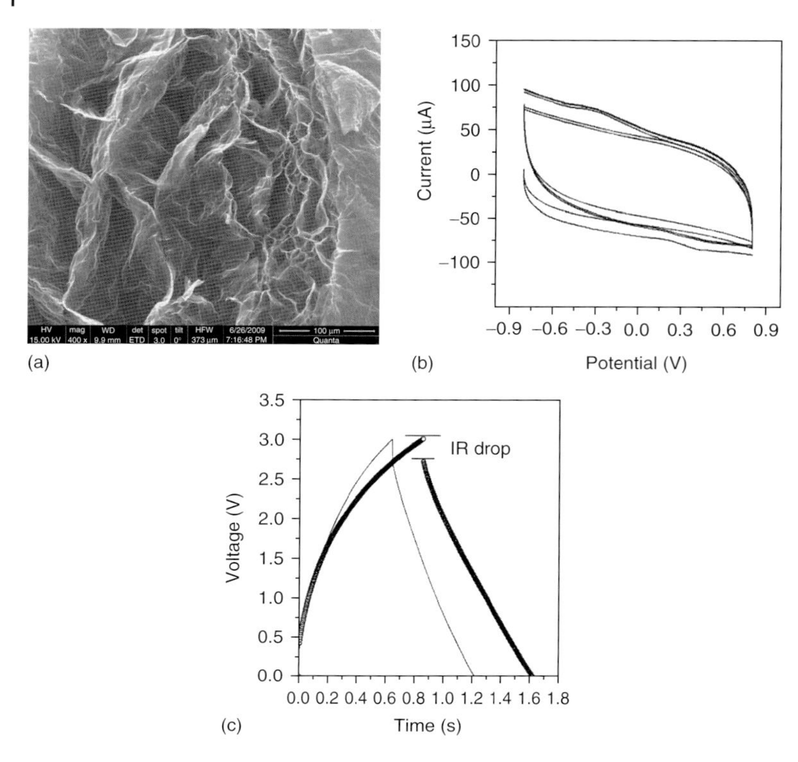

Figure 9.14 (a) SEM image of exfoliated graphite particle prepared at 800 °C (EG800). (b) Successive cyclic voltammograms of EG800 electrode in $(PEO)_9$ $LiClO_4$-40% EC-60% PC electrolyte. (c) Galvanostatic potential-time profiles at 2 mA cm^{-2} for EG800 prepared using graphite from two different sources. (Adapted from Ref. [56].)

in graphene with high specific area of ~400 $m^2\ g^{-1}$, which yields electrochemical capacitance up to 264 F g^{-1}. Murugan and coworkers [58] have reported on the microwave reduction of GO in the presence of organic dispersions to obtain rGO and used it in aqueous media to obtain capacitances of the order of 100 F. Sun and coworkers [59] have reported graphene composite with carbon onion that possesses very high rate capability of 10 A g^{-1} with 54% retention (of 78 F g^{-1}) in capacitance. Other carbon-black/graphene composites also yield high capacitance values with good rate capability [60]. Ultrathin graphene films of thickness 25–100 nm have been used to obtain capacitances of the order of 135 F g^{-1} in aqueous electrolytes with high power density of 7200 W kg^{-1} in 2 M KCl electrolyte [61].

Ionic liquids have also been used in place of aqueous electrolytes. Liu and coworkers [62] proposed materials that possess mesoporous graphene structure accessible to ionic liquid electrolyte and in turn, achieves exceptionally high EDL capacitance. The structure is conducive for permeation, even though ionic liquids contain large-sized ions and high viscosity. Because of the absence of pseudocapacitance contribution, mesoporous graphene electrode is reported to enable fast charging and discharging behavior. Operating voltages as high as 4 V

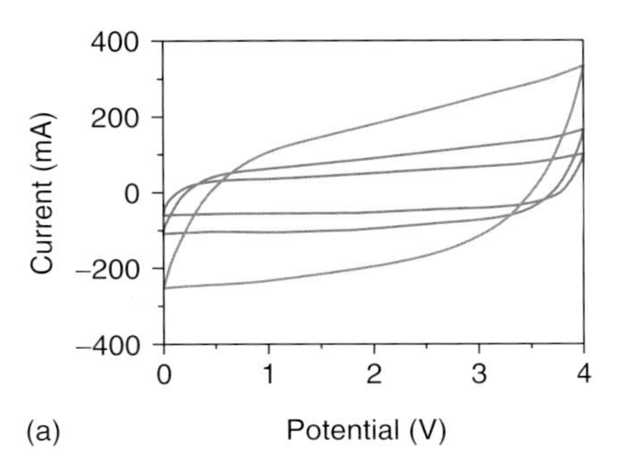

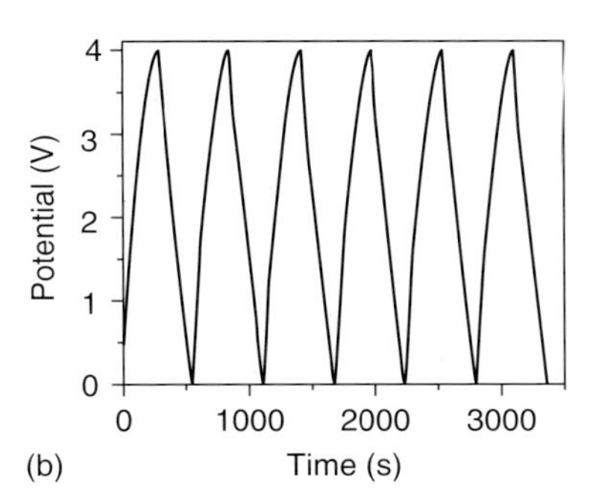

Figure 9.15 (a) Cyclic voltammograms for graphene electrode in EMIM BF_4 ionic liquid electrolyte at different scan rates of 10, 20 and 50 mV/sec (currents increase as the scan rate increases). (b) Charge/discharge curves of at a constant current density of 1 A g^{-1}. (Adapted from Ref. [62].)

(Figure 9.15) have been achieved, which in turn increases the energy density to 90 W h kg^{-1} at 25 °C and 136 W h kg^{-1} at 80 °C, at a current density of 1 A g^{-1} [62].

Specific capacitance of graphene can be enhanced by forming composites of graphene with other materials such as CNT, metal/metal oxide, and conducting polymers. Graphene/CNT composites are generally obtained by mixing individual components under ultrasonication. Lu and coworkers [63] have shown high specific capacitance values for graphene/CNT composites, 265 F g^{-1} with 16 wt% CNTs at a current density of 0.1 A g^{-1}. Graphene seems to be responsible for the increased capacitance observed with the composites. Huang and coworkers [64] have exploited the amphiphilic nature of GO to assemble them on CNTs. Subsequently, the GO/CNT composite film is converted to rGO/CNT by thermal treatment at 220 °C. It is observed that the specific capacitance increases with an increase in CNT in the composite reaching maximum value of 428 and 145 F g^{-1} at current densities of 0.5 and 100 A g^{-1}, respectively. High capacity retention rate of 98% after 10 000 charge/discharge cycles is obtained for the capacitors. Dong and coworkers [65] observed that GO completely wraps CNTs when the two components are sonicated together. This is likely to be due to strong pi–pi interactions resulting in core–shell structures. The specific capacitances for CNT/rGO, rGO, and CNT films are observed to be 194, 155, and 127 F g^{-1} at 0.8 A g^{-1}, respectively. However, the random mixing of GO and CNT is modified by Yu and coworkers [66] to obtain ordered GO/CNT composites using layer-by-layer (L-b-L) technique. Negatively charged GO modified with cationic polymer poly(ethyleneimine) (PEI) through electrostatic interactions is used. The GO/PEI modified substrates are immersed in carboxylic group functionalized CNTs forming films containing GO/PEI and CNT which are further heat treated to obtain [PEI-rGO/CNT-COOH] composite that shows an average capacitance of 120 F/g. Byon and coworkers [67] have explored the L-b-L technique, with negatively charged GO and positively charged CNT (amine functionalized) and obtained high volumetric capacitance of the order of 160 F cm^{-3}. The high specific capacitance values are attributed to various reasons including (i) EDL capacitance and redox reactions between protons and surface

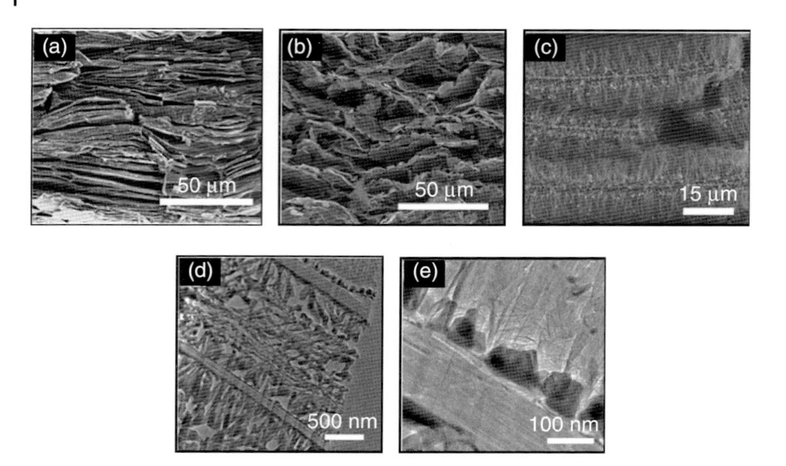

Figure 9.16 SEM images of thermally expanded HOPG (a) without and (b) with SiO_2 coating, (c) 3D pillared vertically aligned carbon nanotubes (VACNT)/graphene architectures, and (d,e) cross-sectional TEM images of 3D pillared VACNT/graphene under different magnifications. (Adapted from Ref. [68].)

oxygen-containing groups on GO; (ii) a high packing density of CNT/GO; and (iii) fast ionic and electronic conduction. When CNTs of controlled length are used to align graphene to stack on one another, by pyrolysis of FePc (Figure 9.16), capacitance increases to 803 F g^{-1} (at 5 mV s^{-1}) [68]. Very often, ECSA decreases owing to agglomeration of graphene nanosheets. To overcome this disadvantage, Yoo and coworkers [69] have recently come up with "in-plane" fabrication approach for ultrathin supercapacitors based on pristine graphene obtained by CVD and multilayer rGO using L-b-L technique. This architecture yields specific capacitances of 80μF cm^{-2}, while rGO multilayers yield high values (394 μF cm^{-2}). Other studies wherein graphene/CNT composites yield good performance in capacitors include that of Fan and coworkers [70] wherein large cycle life of few thousands with little degradation in capacitance has been reported.

Graphene/metal oxide composites have been explored for supercapacitors. Mitra and coworkers [71] have reported the use of exfoliated graphite/ruthenium oxide (RuO_x) composites for electrochemical capacitors and realized few hundreds of farads per gram of the composite. Intercalation/deintercalation process is observed with RuO_2 loading less than 16.5% (confirmed through cyclic voltammetric studies). The capacitance is reported to vary depending on the loading of ruthenium oxide (Figure 9.17). Composite with 16.5 wt% of RuO_x shows specific capacitance of 176 F g^{-1} with energy efficiency of 98.5% at a current density of 3 mA cm^{-2}. Chen and coworkers [72] prepared MnO_2/graphene composites that show capacitances of the order of 200 F g^{-1}. Oxygen-containing functional groups are reported to act as attachment spots for Mn^{2+} that further undergo formation of needle-like MnO_2 nanostructures on graphene during the preparation process. Mn_3O_4 has not been explored in detail for electrochemical applications because of its poor electronic conductivity. The problem of poor conductivity can be overcome

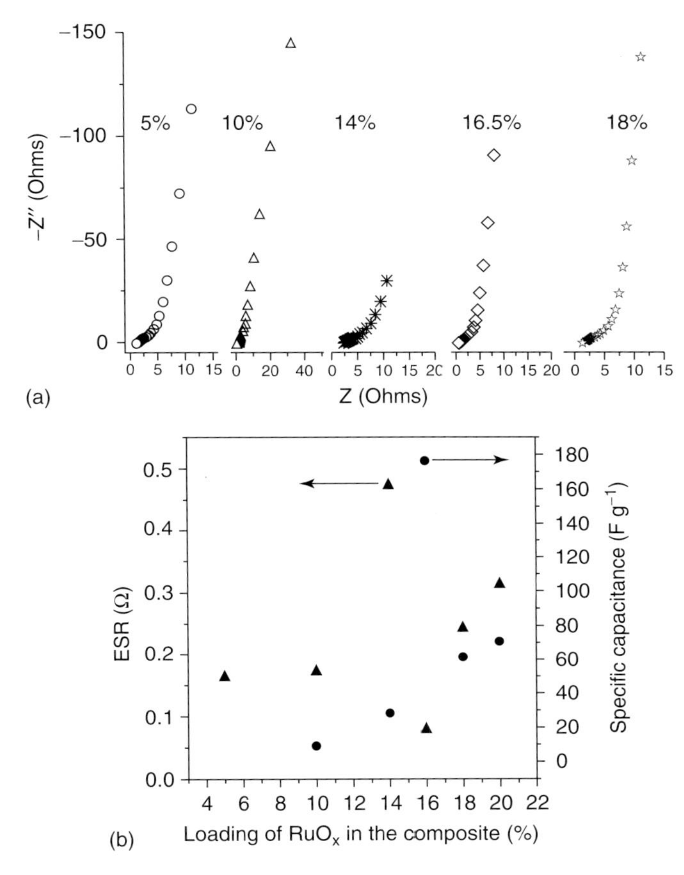

Figure 9.17 (a) Nyquist plots for various composition of EG/RuO_x composite electrodes in 0.5 M H_2SO_4, in the frequency range of 10^5–0.005 Hz. (b) Equivalent series resistance (ESR) and specific capacitance versus ruthenium loading of composite electrodes in 0.5 M H_2SO_4. (Adapted from Ref. [71].)

by forming composites with carbon-based materials. Lee and coworkers [73] have synthesized graphene/Mn_3O_4 composites by hydrothermal reaction and used them as electrodes for supercapacitors. Growth of Mn_3O_4 on exfoliated graphene sheets has been reported to prevent restacking of graphene sheets thus retaining the advantages associated with graphene. While Mn_3O_4 shows the specific capacitance of 25 F g^{-1} at a current density of 0.5 A g^{-1}, graphene/Mn_3O_4 composite shows 121 F g^{-1}. This increase in specific capacitance has been attributed to the higher electronic conductivity of graphene/Mn_3O_4 as compared to Mn_3O_4 alone. Other metal oxide-based graphene composites, such as RuO_2 [74], $Ni(OH)_2$ [75], $Co(OH)_2$ [76], ZnO/SnO_2 [77], and Fe_3O_4 [78], have also been studied in this direction.

Conducting polymers is another class of materials that are known to provide good pseudocapacitive interfaces. Among various conducting polymers explored

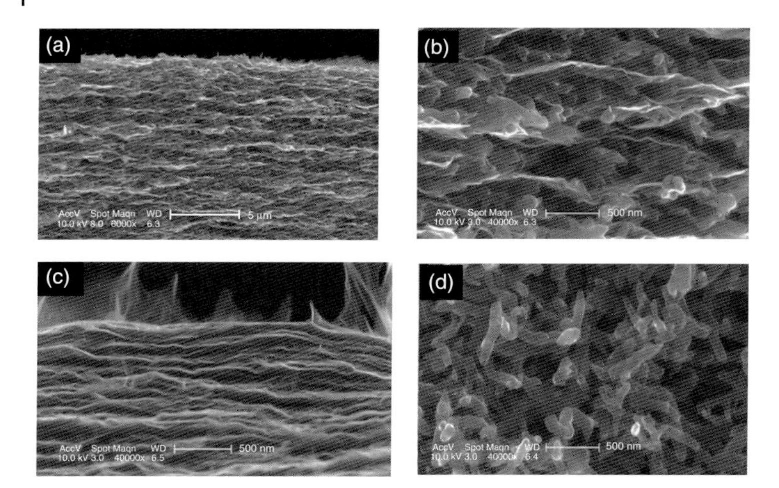

Figure 9.18 Cross section SEM images of (a,b) graphene/PANI nanofibers, (c) pure chemically converted graphene, and (d) PANI nanofiber films prepared by vacuum filtration. The scale bar denotes 500 nm for (b–d) while it is equal to 5 μm for (a). (Adapted from Ref. [79].)

(polyaniline (PANI), polypyrrole and polythiophene etc.), PANI-based composites have attracted considerable attention because of its ease of synthesis, relatively low cost, and environmental stability. Wu and coworkers [79] have reported the formation of G/PANI composites by chemical methods where PANI fibers are sandwiched between graphene layers (Figure 9.18) with interlayer spacing of 10–200 nm. The free-standing, mechanically robust, highly flexible electrodes yield specific capacitance of 210 F g^{-1}.

Recently, Ruoff's research group [80] has reported carbon-based supercapacitors by activation of graphene. Activation of microwave-treated GO (MEGO) using KOH is reported to result in continuous generation of three-dimensional distribution of meso- and micropores (Figure 9.19). This activation process with KOH yields network of pores of small size in the range of ~1 to ~10 nm and restructures MEGO. The MEGO contains small amount of edge-carbon atoms along with O and H atoms. Activated MEGO shows surface area of 3100 m^2 g^{-1}, high electrical conductivity, and low oxygen and hydrogen content with specific capacitance value of 166 F g^{-1} at a current density of 5.7 A g^{-1} in 1-butyl-3-methylimidazolium tetrafluoroborate/acetonitrile electrolyte. Zhang and coworkers [81] have prepared activated graphene oxide (a-GO) by controlled addition of KOH to GO colloid followed by heating to obtain an ink-paste. The ratio of GO/KOH is reported to be crucial as excess KOH results in immediate precipitation of GO. The paste is then deposited on polytetrafluoroethylene membrane and dried under vacuum. The a-GO film revealed specific area of 2400 m^2 g^{-1} and in-plane electrical conductivity of 5880 S m^{-1}. This material (a-GO film) shows a power density of about 500 kW kg^{-1} with energy density of 26 W h kg^{-1} (specific capacitance of 120 F g^{-1}

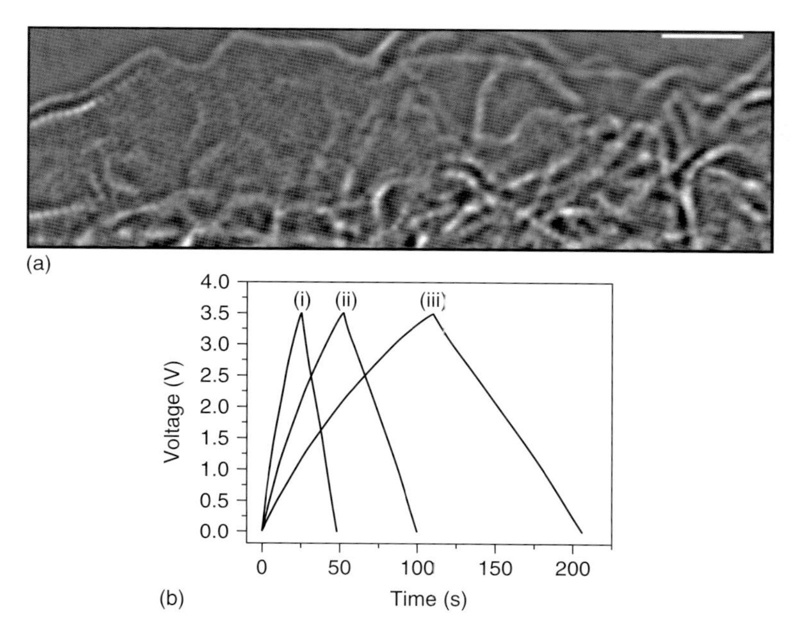

Figure 9.19 (a) Exit wave reconstructed HR-TEM image from the edge of MEGO (scale bar denotes 2 nm) and (b) Galvanostatic charge/discharge curves of MEGO-based supercapacitor under different discharge currents of (i) 5.7, (ii) 2.8, and (iii) 1.4 A g^{-1}, respectively. (Adapted from Ref. [80].)

at the current density of 10 A g^{-1} for an a-GO film) in the presence of organic (1-butyl-3-methylimidazolium (BMIM) BF_4)/AN) electrolyte. Table 9.1 summarizes the electrochemical characteristics for selected graphene and graphene-based systems.

GO has recently been explored as an anode material for hybrid supercapacitors [88]. While supercapacitors provide life time of 100 000 cycles, batteries possess high storage capacity ranging from 25 to 150 W h kg^{-1} [89]. The advantages of the two devices can be combined resulting in hybrid supercapacitors with rapid charge/discharge and long cycle life with high charge storage capacity. Hybrid supercapacitors use a supercapacitor cathode, battery anode, and an electrolyte. Stoller and coworkers [88] have studied the use of high surface area, chemically activated graphene-based material (MEGO) as cathode in Li-ion hybrid supercapacitors. Observed specific capacitances in MEGO/graphite hybrid cell are of the order of 266 F g^{-1} with operating voltage of 4 V and gravimetric energy density of 53.2 W h kg^{-1}.

The performance of supercapacitors depends on the electrolyte and the electrode material used. While in aqueous media, maximum operating voltages can be ~1.2 V (due to decomposition of solvent), the voltage range is high in organic electrolytes [90]. The operating voltage can be enhanced by the use of ionic liquids that have high decomposition potential [91]. Ionic liquids show relatively good ionic

Table 9.1 Electrochemical characteristics of various graphene- based supercapacitor devices [62, 68, 69, 75, 80, 82–87].

Material	Specific surface area ($m^2 g^{-1}$)	Specific capacitance ($F g^{-1}$)	Current density ($A g^{-1}$)	Cycle life (% retention of capacity)	References
Graphene though microwave exfoliation	463	191	0.15	—	[82]
Chemical activation of exfoliated graphene oxide (EGO)	3100	166	5.7	10 000 (97%)	[80]
GO coated on carbon paper	—	238	0.0005	1 000	[75]
Chemically modified graphene	705	135	0.01	—	[83]
Graphene/ionic liquid	—	154.1	1	—	[62]
rGO/polymer multilayer	1310	394 μF/cm^2	—	—	[69]
rGO by solvothermal method	—	276	0.1	—	[84]
rGO/MnO_2	—	188	0.25	1 000 (89%)	[85]
Graphene/CNT	213	1065	22.1	20 000 (96%)	[68]
Graphene/RuO_2	—	220	10	—	[86]
PANI-doped graphene	20.2	480	0.1	—	[87]

conductivity and wide potential window at temperatures above 60 °C [92]. Various ionic liquids such as 1-butyl-3-methylimidazolium hexafluorophosphate (BMIM PF_6) [93], 1-ethyl-3-methylimidazolium tetrafluoroborate (EMIM BF_4) [62], and 1-butyl-3-methylimidazolium tetrafluoroborate (BMIM BF_4) [80] have been explored as electrolytes in supercapacitor applications with graphene (rGO) as the working electrode. Chen and coworkers [93] obtained rGO with electrochemically active oxygen-containing functional groups using weak reducing agent (hydrobromic acid) and studied electrochemical capacitance behavior in BMIM PF_6. A specific capacitance of 158 F g^{-1} at a current density of 0.2 A g^{-1} is obtained. While using ionic liquids as electrolytes, two major factors need to be considered, that is, temperature and concentration. This has been studied by Liu [62] and coworkers using EMIM BF_4 in acetonitrile as the electrolyte. Increase in concentration of EMIM BF_4 results in the increase of ionic conductivity up to 2.0 M (number of ions per unit volume increases) and decreases (owing to weak van der Waals interactions among the ions) thereafter. The specific capacitance increases with an increase in temperature from −20 to 60 °C owing to a decrease in viscosity of the electrolyte. A maximum specific capacitance of 128.2 F g^{-1} with potential window of 2.3 V is obtained for rGO electrode in 2.0 M EMIM BF_4/acetonitrile electrolyte.

9.4 Graphene in Batteries

Beginning with lead-acid battery, which is more than 150 years old, and the latest, lithium ion battery (LIB), the technology related to battery development keeps improving with the advent of new materials. Graphene has been playing a role in this direction for the last few to several years. A battery consists of cathode (positive electrode, an electron acceptor such as lithium cobalt oxide, manganese dioxide, or lead oxide) and anode (negative electrode, lithium, zinc, or lead) physically separated by ionically conducting electrolyte by means of a separator. During discharging/charging process, ions released during oxidation reaction move toward cathode where they are involved in intercalation/deintercalation process.

LIB is quite well studied for various applications especially owing to its high energy density, high voltage, long cycle life, light weight, and good environmental stability. Carbon-based materials are extensively used as anodes owing to their amenability for reversible intercalation/deintercalation process with metal ions and in particular, lithium ions. Other high capacity materials such as Sn [94], Si [95], Mn_3O_4 [96, 97], Co_3O_4 [98], and Ti_5O_{12} [99] have been used in conjunction with carbon (graphite) to improve battery performance. It is recognized that charge storage in graphite-based LIB is limited by ion storage sites within the hexagonal network of sp^2 carbon yielding one Li atom per six carbon atoms leading to LiC_6 [100, 101]. There have been studies to improve the performance of carbon-based electrode materials, by involving double layer adsorption to yield Li_2C_6 and LiC_2 [102]. The functional groups on graphene make it highly electronegative resulting in selective interaction with cationic species. In addition, graphene can be used as binder (eliminating the use of added polymer) along with high electrical conductivity and good mechanical strength.

Wang and coworkers have reported the use of graphene paper obtained from filtration of rGO, as anode material in Li batteries. $LiPF_6$ as the electrolyte, Li as reference and counter electrodes (Figure 9.20) completed the cell [103]. Even though the capacity drops from 680 to 84 $mA\,h\,g^{-1}$, graphene electrode is reported to exhibit consistent flat plateau and a discharge capacity of 528 $mA\,h\,g^{-1}$ with a cut-off voltage of 2.0 V for first discharge. The midpoint of discharge plateau is observed to be 2.20 V (vs Li/Li^+) yielding specific energy density of 1162 $W\,h\,kg^{-1}$. Yoo and coworkers [104] have explored the possibility of increasing interlayer distance of graphene sheets using CNT and fullerene. The specific capacity observed for graphene sheets is 540 $mA\,h\,g^{-1}$, while values of 730 and 784 $mA\,h\,g^{-1}$ are obtained for CNT and fullerene incorporated graphene electrodes, respectively. However, the specific capacity drops from 540 to 290 $mA\,h\,g^{-1}$ after only 20 charge/discharge cycles.

The performance of graphene sheets as electrodes is further enhanced by the use of lithiated graphene obtained by reduction of GO in liquid ammonia and lithium metal, which yields stable reversible capacity of 410 $mA\,h\,g^{-1}$ (Figure 9.21) [105]. Graphite, rGO, and Li-rGO show reversible discharge capacities of 300, 340, and 410 $mA\,h\,g^{-1}$, respectively. This enhancement in the capacity of Li-rGO is

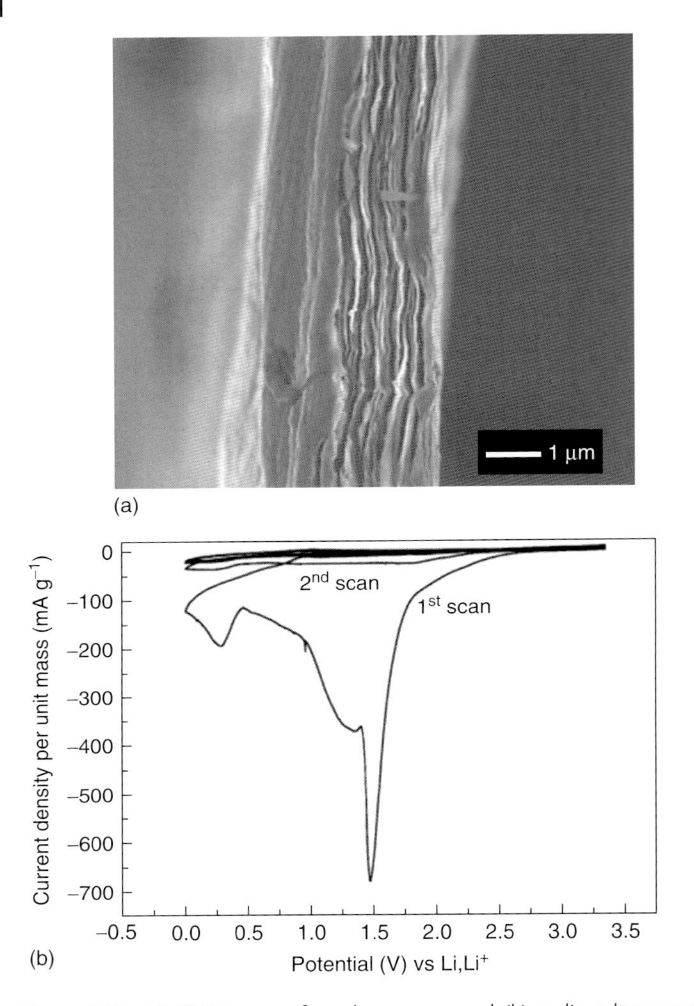

Figure 9.20 (a) SEM image of graphene paper and (b) cyclic voltammograms of graphene paper in $LiPF_6$ with Li as counter and reference electrode at a scan rate of 0.1 mV s^{-1}. (Adapted from Ref. [103].)

attributed to the presence of electrochemically active defects (i.e., edge and basal plane defects) formed during reduction of GO and also to the adsorption of Li on the internal surfaces of disordered graphene sheets.

Tin (Sn) possesses theoretical capacity as high as 994 mA h g^{-1} [94]. However, Sn nanoparticles agglomerate during charging/discharging process resulting in decline in performance [95]. Wang and coworkers [106] tried to overcome the limitation by using graphene along with Sn in 3D nanoarchitectures. The specific capacity of graphene/Sn composite varies from 1250 mA h g^{-1} in the first cycle to 810 mA h g^{-1} in the second cycle and to 508 mA h g^{-1} in the third cycle and stays constant up to 1000 cycles. In most of the graphene-based anode materials,

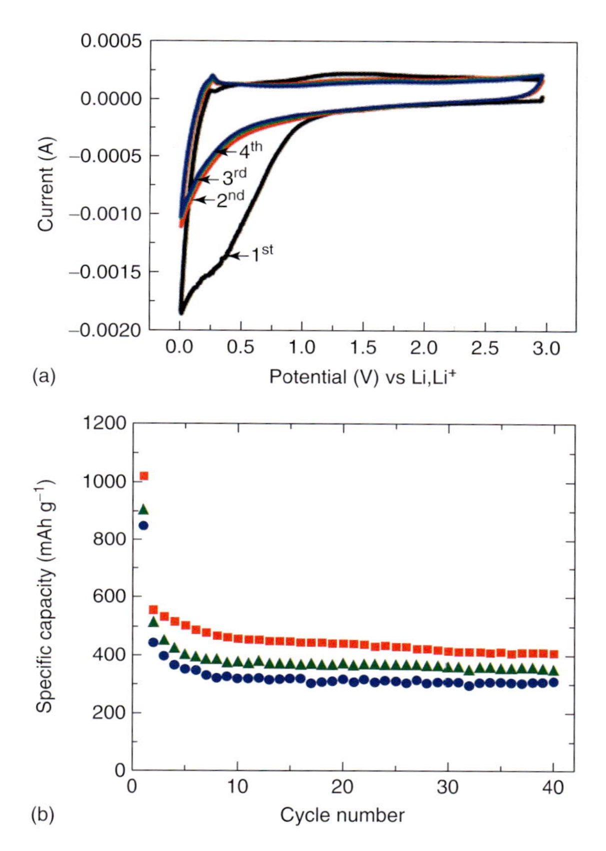

Figure 9.21 (a) Cyclic voltammograms of Li-rGO electrode at a scan rate of 0.1 mV s^{-1} and (b) discharge capacity versus cycle number for graphite (blue), rGO (green), and Li-rGO (red) cycled at a discharge current of 25 mA g^{-1} between 3.0 and 0.02 V versus Li/Li+ in a 1 M solution of $LiPF_6$ in a 1 : 1 (v/v) mixture of ethylene carbonate and dimethyl carbonate as the electrolyte. (Adapted from Ref. [105].)

a large quantity of graphene, ~10–40%, is used [78]. Recently, Rao, Bhattacharya, and coworkers [107] have shown that the presence of very low concentration (<1%) of rGO can be successfully used to improve the performance of mesoporous SnO_2 for LIBs. The authors state that it is sufficient to have low concentrations of rGO to provide necessary conduction pathways during charge/discharge cycles. The first discharge and charge capacities are found to be 1484 and 813 mA h g^{-1} and 1451 and 1163 mA h g^{-1} for SnO_2 and SnO_2-rGO, respectively. Irreversible capacity loss for the composite SnO_2-rGO (288 mA h g^{-1}) is reported to be smaller than that of SnO_2 (671 mA h g^{-1}) with columbic efficiencies of 55 and 80% for SnO_2 and SnO_2-rGO, respectively. Li alloys with group IV materials have attracted attention as anodes when compared to carbon-based ones (e.g., graphite). Among them, silicon and germanium have attracted maximum attention because of high theoretical capacities of 4200 and 1600 mA h g^{-1}, respectively [108, 109]. However, silicon suffers similar problems as that of Sn, in terms of volume changes during charge/discharge process. Chou and coworkers [110] have explored the use of Si/graphene composites and obtained capacities of 1168 $mA\,h\,g^{-1}$ with average columbic efficiency of 93% for about 30 cycles. The enhanced performance is attributed to increased conductivity, specific surface area, and porous nature of the composite. Lee and coworkers [109] improved the performance of graphene/Si composite by additional thermal treatment of graphene/Si composite and observed storage capacity of >2200 $mA\,h\,g^{-1}$ after 50 cycles and >1500 $mA\,h\,g^{-1}$ after 200 cycles. Germanium contributes to two major advantages as compared to silicon: (i) electrical conductivity of Ge is 10^4 times higher than that of Si and (ii) Li diffusion coefficient in Ge is 400 times larger than that in Si at 25 °C [111–113]. However, the poor cyclability of Ge-based electrodes restricts the use of this material. In order to overcome the problems associated with volume changes, Xue and coworkers [114] explored the possibility of using Ge with graphene as composite anode material for LIB. Ge–C core–shell nanoparticles are used in order to avoid direct exposure of Ge nanoparticles to the electrolyte. This protection strategy of active electrode material is a rather new direction to realize high-performance anode materials. The Ge@C/rGO nanocomposites show specific capacity of 380 $mA\,h\,g^{-1}$ at a current density of 3600 $mA\,g^{-1}$ and Ge@C nanoparticles show specific capacity of 100 $mA\,h\,g^{-1}$. These values are higher than the theoretical capacity of graphite (theoretical specific capacity −372 $mA\,h\,g^{-1}$). The improved electrochemical performance of the Ge@C/rGO nanocomposites is attributed to its unique structure. The carbon shell acts as a buffer that plays an important role in minimizing volume changes and also to prevent direct contact between Ge and the electrolyte, facilitating the formation of stable solid electrolyte interphase. Second, rGO nanosheets act as elastic and electronically conductive substrates, resulting in good dispersibility and high electrical conductivity.

Graphene in conjunction with metal oxides is observed to increase the specific capacity of lithium battery systems [115]. For instance, graphene-supported TiO_2 [116] shows enhanced specific capacity of 87 $mA\,h\,g^{-1}$ as compared to 35 $mA\,h\,g^{-1}$ observed for rutile TiO_2 [117]. The composite is prepared by using anionic surfactant, sodium dodecyl sulfate (SDS), that helps in adsorption onto graphene through hydrophobic tails making functionalized graphene sheets (FGSs) highly dispersible and in turn interact with oxide precursors through hydrophilic head groups (Figures 9.22 and 9.23) very effectively. The cooperative interactions between surfactant, graphene, and oxide precursors lead to homogeneous mixing of components, in which hydrophobic graphene most likely resides in the hydrophobic domains of SDS micelles. SnO_2 is proposed to have high theoretical specific capacity of 782 $mA\,h\,g^{-1}$. Large volume changes associated with SnO_2 during charge/discharge process results in poor cyclic performance. Paek and coworkers [118] have explored the possibility of using graphene to reassemble SnO_2 nanoparticles and found that graphene nanosheets help in homogeneously distributing SnO_2 resulting in nanoporous structures with large amount of void space. The obtained capacity of SnO_2/GNS (graphene nanosheet) is stable and remained at 570 $mA\,h\,g^{-1}$ even after 30 cycles (70% retention of reversible capacity), while the

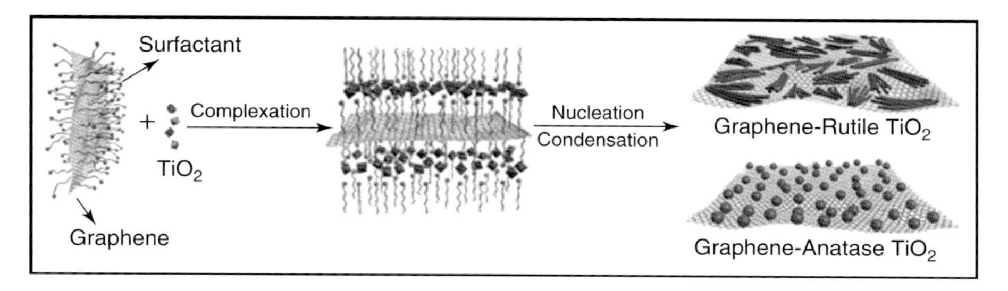

Figure 9.22 Anionic surfactant (sodium dodecyl sulfate)-mediated stabilization of graphene and growth of self-assembled TiO_2/FGS hybrid nanostructures. (Adapted from Ref. [116].)

specific capacity of bare SnO_2 nanoparticles during first charge is 550 mA h g^{-1} and drops rapidly to 60 mA h g^{-1} only after 15 cycles. Yu and coworkers [119] have studied MnO_2/graphene composites as anode materials for LIBs. MnO_2 is well known for its low cost, environmental friendliness, and high theoretical capacity (1232 mA h g^{-1}) [120]. Composites of MnO_2/graphene prepared by L-b-L technique through vacuum filtration method help in obtaining free-standing films with good mechanical stability along with reduced agglomeration of graphene nanoplatelets. The thin free-standing film has been proposed to shorten the diffusion path length and provides fast electron pathways with 2D electron-conducting behavior between the MnO_2 and electrode. The specific capacity of the composite electrode is 581 $mAhg^{-1}$ at a current density of 100 mAg^{-1}, while graphene-free MnO_2 shows only 294 $mAhg^{-1}$.

Most of the reports based on metal oxide/graphene composites attribute the improved performance to the synergistic effect of the components involved because of improved interfacial interaction [123]. Recently, Zhou and coworkers [123] have given possible reasons for the observed performance based on X-ray photoelectron spectroscopy, FT-IR, and Raman spectroscopy. They have studied NiO/graphene composite and found that NiO binds to graphene sheets through oxygen bridges of hydroxyl/epoxy groups. It is reported that the diffusion energy (using DFT calculations) for Ni adatoms (1.37 and 1.84 eV for hydroxyl and epoxy, respectively) on GO is higher than the diffusion barrier (2.23 eV) and hence NiO nanosheets strongly interact with graphene through C-O-Ni bridge resulting in high reversible capacity and excellent rate performance [123].

Recently, Xiao and coworkers [124] have explored the advantage of presence of defects and functional groups on GO for Li-air battery electrode. Exceptionally high capacity of 15 000 mA h g^{-1} is observed. rGO is mixed with microemulsions made of poly(tetrafluoroethylene) (PTFE) along with binder (DuPont Teflon PTFE-TE3859 fluoropolymer resin aqueous dispersion, 60 wt% solids) during which a "broken egg" morphology is observed with interconnected tunnels that pass through the entire electrode length. The tunnels act as arteries that continuously supply oxygen to the interior of the electrode during discharge process. This morphology is achieved during the synthesis procedure where foam-like structure

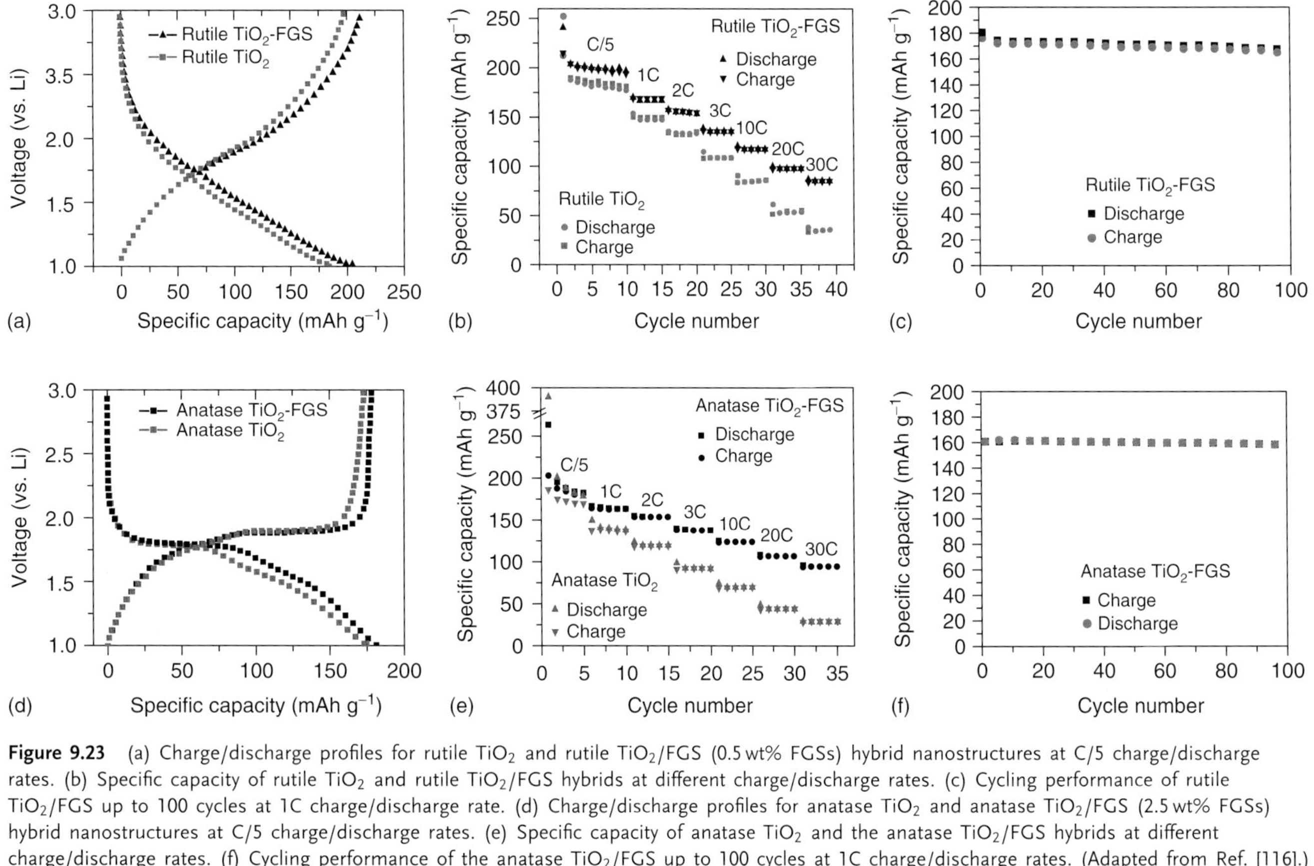

Figure 9.23 (a) Charge/discharge profiles for rutile TiO_2 and rutile TiO_2/FGS (0.5 wt% FGSs) hybrid nanostructures at C/5 charge/discharge rates. (b) Specific capacity of rutile TiO_2 and rutile TiO_2/FGS hybrids at different charge/discharge rates. (c) Cycling performance of rutile TiO_2/FGS up to 100 cycles at 1C charge/discharge rate. (d) Charge/discharge profiles for anatase TiO_2 and anatase TiO_2/FGS (2.5 wt% FGSs) hybrid nanostructures at C/5 charge/discharge rates. (e) Specific capacity of anatase TiO_2 and the anatase TiO_2/FGS hybrids at different charge/discharge rates. (f) Cycling performance of the anatase TiO_2/FGS up to 100 cycles at 1C charge/discharge rates. (Adapted from Ref. [116].)

is formed resulting in bubble formation in which rGO nanosheets reside. It should be noted that the hydrophobicity of rGO plays a large role in this design.

Along with inorganic cathode materials, organic electroactive polymers have recently been explored as new generation of "green" lithium battery electrodes. The advantages include lightness, environmentally benign characteristics, mechanical flexibility, and processing compatibility. Electrochemical activity of organic materials depends on the redox process of functional groups such as quinone [125], anhydride [126], and nitroxide [127] radical, accompanied by the association and disassociation of Li ions or electrolyte anions. The performance of organic electrodes may be enhanced by incorporating them in conductive matrixes. Nakahara and coworkers [128] reported the use of poly(2,2,6,6-tetramethyl-1-piperidinyloxy-4-yl methacrylate) (PTMA) as cathode material with high capacity, high charge/discharge rate and long cycle life. Guo and coworkers [129] have explored the formation of PTMA/graphene composite as cathode material for LIBs. Here, graphene is reported to act as electronic conductivity promoter with PTMA/graphene composite showing two-electron redox reaction, as shown in Figure 9.24. The observed reversible capacity for the composite is 222 mA h g^{-1} at 1 °C with long cycle life of 20 000 cycles at 100 °C. The electrode has 10% of active composite material with 60% graphene. Song and coworkers [130] have used graphene with poly(anthraquinonyl sulfide) (PAQS) and polyimide (PI) as cathode material, to improve the battery performance. Polymer/graphene composite is prepared by *in situ* polymerization method. Figure 9.24b shows the redox reactions of Li ions with PAQS and PI based on quinone and anhydride functional groups [131, 132]. While only graphene shows specific capacity of 32 mA h g^{-1} at 0.1 *C*, graphene/PAQS and graphene/PI composites show 187 and 202 mA h g^{-1} at 0.1 *C*, respectively.

In addition to LIB, graphene has been explored in other types of batteries such as vanadium redox flow battery (VRFB) and zinc-air battery [133, 134]. VRFB proposed by Skyllas-Kazacos and coworkers [135–137] is a stationary battery and uses VO^{2+}/VO_2^+ and V^{2+}/V^{3+} redox couples in sulfuric acid as the positive and the negative half-cell electrolytes, respectively, with open circuit voltage of approximately 1.26 V at 100% state of charge [138]. Observed rate constants for

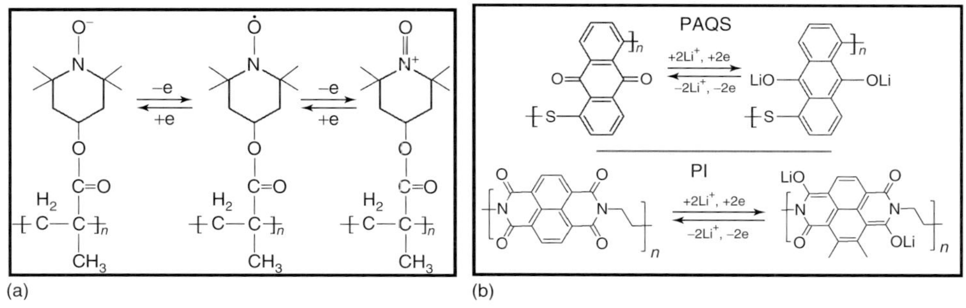

Figure 9.24 (a) Redox reactions of PTMA and (b) redox reactions of Li ions with PAQS and PI. (Adapted from Refs. [129, 130].)

V^{5+}/V^{6+} and V^{2+}/V^{3+} on glassy carbon electrode (GCE) are 7.5×10^{-4} [135] and 1.2×10^{-4} cm s^{-1} [136], respectively. Unlike LIB, where charge/discharge reactions depend on intercalation/deintercalation process, VRFB is based entirely on redox reactions among soluble ionic species [138]. Major electrodes used in VRFB are graphite felt, carbon cloth, and carbon fiber that show poor kinetic reversibility. Han and coworkers [139] have exploited the electrocatalytic activity of hydroxyl and carboxyl acid groups on GO toward redox activity of vanadium species. While graphite shows reversible electrocatalytic activity for VO^{2+}/VO_2^+ with reversible redox peaks at 1.24 and 1.13 V, the redox reaction associated with V^{2+}/V^{3+} shows an irreversible reduction peak at −0.55 V while oxidation peak is absent. The peaks associated with the couple V^{2+}/V^{3+} appear at −0.45 and −0.12 V for GO-50; at −0.44 and −0.18 V for GO-90; at −0.38 and −0.19 V for GO-120 (50, 90, and 120 indicate the temperature (°C) used during the preparation process). The authors attribute large specific surface area and the existence of oxygen-containing functional groups on basal planes and sheet edges of the GO as being responsible for the observed activity. However, poor electrical conductivity of GO nanosheets decreases the rate capability of VRFB resulting in low energy efficiency. In order to overcome this problem, GO is assembled on MWCNTs by electrostatic spray technique on GCE. This results in electrocatalytic hybrids with mixed conducting network which provide rapid ion transport channels along with effective electron transfer to VO^{2+}/VO_2^+ couple. GO and GO/MWCNT show I_{pc} (cathodic peak current) of 0.22 and 0.22 A mg^{-1} with R_{ct} (charge transfer resistance) of 342.70 and 10.79 Ω, respectively. Lee and coworkers have explored ionic liquid modified rGO nanosheets with Mn_3O_4 as composite electrode for ORR in Zn-air battery [103]. The GO is functionalized with ionic liquid (*N*-ethyl-*N*′-(3-dimethylaminopropyl)-carbodiimide methiodide) followed by chemical reduction using hydrazine that is reported to result in conductive rGO functionalized with ionic liquid. Subsequently, Mn_3O_4 nanoparticles are grown using $NaMnO_4$ as precursor. Electrocatalytic activity of the composites toward ORR is studied and the constructed Zn-air batteries show power density of 120 mW cm^{-2} at a current density of 160 mA cm^{-2}.

9.5
Conclusions and Future Perspectives

Graphene and GO have been shown as promising electrode and electrode support for electrochemical energy storage devices. Major advantages include high surface area, improved porosity, controllable electrical conductivity, and high mechanical strength. Even with all the above mentioned efforts, development of graphene-based materials is in its infancy owing to variability in preparation procedures leading to differences in output performance. The major areas of concern are electrode fabrication, ECSA and interactions between graphene and catalyst. One of the possible ways to address the variability issues is to use techniques such as Langmuir–Blodgett method and L-b-L technique for fabricating precise nanostructures with defined dimensions, and self-assembly of graphene-based architectures to improve physicochemical and electrochemical properties. Precise control over number of layers as well as control over extent of reduction of GO may help improve the performance. Additional use of *in situ* spectroscopic techniques will provide fundamental insight into the actual mechanism that, in turn, will help understand and improve the material and its properties.

References

1. McCreery, R.L. (2008) *Chem. Rev.*, **108**, 2646.
2. Rao, C.N.R., Sood, A.K., Voggu, R., and Subrahmanyam, K.S. (2010) *J. Phys. Chem. Lett.*, **1**, 572.
3. Dreyer, D.R., Ruoff, R.S., and Bielawski, C.W. (2010) *Angew. Chem. Int. Ed.*, **49**, 9336.
4. Service, R.F. (2009) *Science*, **324**, 875.
5. Zhu, Y., Murali, S., Cai, W., Li, X., Suk, J.W., Potts, J.R., and Ruoff, R.S. (2010) *Adv. Mater.*, **22**, 3906.
6. Rao, C.N.R., Sood, A.K., Subrahmanyam, K.S., and Govindaraj, A. (2009) *Angew. Chem. Int. Ed.*, **48**, 7752.
7. Ratinac, K.R., Yang, W., Gooding, J., Thordarson, P., and Braet, F. (2011) *Electroanalysis*, **23**, 803.
8. Ambrosi, A., Sasaki, T., and Pumera, M. (2010) *Chem. Asian J.*, **5**, 266.
9. Shao, Y., Wang, J., Wu, H., Liu, J., Aksay, I.A., and Lina, Y. (2010) *Electroanalysis*, **22**, 1027.
10. Guo, S. and Dong, S. (2011) *Chem. Soc. Rev.*, **40**, 2644.
11. Pumera, M. (2010) *Chem. Soc. Rev.*, **39**, 4146.
12. Dreyer, D.R., Park, S., Bielawski, C.W., and Ruoff, R.S. (2010) *Chem. Soc. Rev.*, **39**, 228.
13. Schniepp, H.C., Li, J.L., McAllister, M.J., Sai, H., Herrera-Alonso, M., Adamson, D.H., Prud'homme, R.K., Car, R., Saville, D.A., and Aksay, I.A. (2006) *J. Phys. Chem. B*, **110**, 8535.
14. Winter, M. and Brodd, R.J. (2004) *Chem. Rev.*, **104**, 4245.
15. Borup, R., Meyers, J., Pivovar, B., Kim, Y.S., Mukundan, R., Garland, N., Myers, D., Wilson, M., Garzon, F., Wood, D., Zelenay, P., More, K., Stroh, K., Zawodzinski, T., Boncella, X.J., McGrath, J.E., Inaba, O.M., Miyatake, K., Hori, M., Ota, K., Ogumi, Z., Miyata, S., Nishikata, A., Siroma, Z., Uchimoto, Y., Yasuda, K., Kimijima, K., and Iwashita, N. (2007) *Chem. Rev.*, **107**, 3904.
16. Xin, Y., Liu, J., Zhou, Y., Liu, W., Gao, J., Xie, Y., Yin, Y., and Zou, Z. (2011) *J. Power Sources*, **196**, 1012.
17. Li, D., Müller, M.B., Gilje, S., Kaner, R.B., and Wallace, G.G. (2008) *Nature Nanotechnol.*, **3**, 101.
18. Seger, B. and Kamat, P.V. (2009) *J. Phys. Chem. C*, **113**, 7990.
19. Jafri, R.I., Rajalakshmi, N., and Ramaprabhu, S. (2010) *J. Mater. Chem.*, **20**, 7114.
20. Li, Y., Tang, L., and Li, J. (2009) *Electrochem. Commun.*, **11**, 846.
21. Yumura, T., Kimura, K., Kobayashi, H., Tanaka, R., Okumura, N., and Yamabe, T. (2009) *Phys. Chem. Chem. Phys.*, **11**, 8275.
22. Hou, J., Shao, Y., Ellis, M.W., Moore, R.B., and Yi, B. (2011) *Phys. Chem. Chem. Phys.*, **13**, 15384.
23. Liu, J., Fu, S., Yuan, B., Li, Y., and Deng, Z. (2010) *J. Am. Chem. Soc.*, **132**, 7279.
24. Dong, L., Reddy, R., Gari, S., Li, Z., Craig, M.M., and Hou, S. (2010) *Carbon*, **48**, 781.
25. Hu, Y., Zhang, H., Wu, P., Zhang, H., Zhou, B., and Cai, C. (2011) *Phys. Chem. Chem. Phys.*, **13**, 4083.
26. Guo, S., Dong, S., and Wang, E. (2010) *ACS Nano*, **4**, 547.
27. Rao, C.V., Reddy, A.L.M., Ishikawa, Y., and Ajayan, P.M. (2011) *Carbon*, **49**, 931.
28. Huang, Z., Zhou, H., Li, C., Zeng, F., Fua, C., and Kuang, Y. (2012) *J. Mater. Chem.*, **22**, 1781.

29. Zhang, Y., Gu, Y., Lin, S., Wei, J., Wang, Z., Wang, C., Du, Y., and Ye, Y. (2011) *Electrochim. Acta*, **56**, 8746.
30. Lei, Y., Zhao, G., Tong, X., Liu, M., Li, D., and Geng, R. (2010) *Chem. Phys. Chem.*, **11**, 276.
31. Yang, L., Yang, W., and Cai, Q. (2007) *J. Phys. Chem. C*, **111**, 16613.
32. Rao, C.V., Cabrera, C.R., and Ishikawa, Y. (2011) *J. Phys. Chem. C*, **115**, 21963.
33. Wen, Z., Yang, S., Liang, Y., He, W., Tong, H., Hao, L., Zhang, X., and Song, Q. (2010) *Electrochim. Acta*, **56**, 139.
34. Wang, Y., Wan, Y., and Zhang, D. (2010) *Electrochem. Commun.*, **12**, 187.
35. Kim, H.J., Choi, S.M., Seo, M.H., Green, S., Huber, G.W., and Kim, W.B. (2011) *Electrochem. Commun.*, **13**, 890.
36. Liang, Y., Li, Y., Wang, H., Zhou, J., Wang, J., Regier, T., and Dai, H. (2011) *Nature Mater.*, **10**, 780.
37. Zhang, S., Shao, Y., Li, X., Nie, Z., Wang, Y., Liu, J., Yin, G., and Lin, Y. (2010) *J. Power Sources*, **195**, 457.
38. Byon, H.R., Suntivich, J., and Shao-Horn, Y. (2011) *Chem. Mater.*, **23**, 3421.
39. Qu, L., Liu, Y., Baek, J.B., and Dai, L. (2010) *ACS Nano*, **4**, 1321.
40. Wang, S., Yu, D., Dai, L., Wook Chang, D., and Baek, J.B. (2011) *ACS Nano*, **5**, 6202.
41. Aravind, S.S.J., Jafri, R.I., Rajalakshmi, N., and Ramaprabhu, S. (2011) *J. Mater. Chem.*, **21**, 18199.
42. Jafri, R.I., Arockiados, T., Rajalakshmi, N., and Ramaprabhu, S. (2010) *J. Electrochem. Soc.*, **157**, B874.
43. Mauritz, K.A. and Moore, R.B. (2004) *Chem. Rev.*, **104**, 4535.
44. Choi, B.G., Hong, J., Park, Y.C., Jung, D.H., Hong, W.H., Hammond, P.T., and Park, H.S. (2011) *ACS Nano*, **5**, 5167.
45. Zarrin, H., Higgins, D., Jun, Y., Chen, Z., and Fowler, M. (2011) *J. Phys. Chem. C*, **115**, 20774.
46. Cao, Y., Xu, C., Wu, X., Wang, X., Xing, L., and Scott, K. (2011) *J. Power Sources*, **196**, 8377.
47. Pant, D., Bogaert, G.V., Diels, L., and Vanbroekhoven, K. (2010) *Bioresour. Technol.*, **101**, 1533.
48. Zhang, Y., Mo, G., Li, X., and Ye, J. (2012) *J. Power Sources*, **197**, 93.
49. Zhang, Y., Mo, G., Li, X., Zhang, W., Zhang, J., Ye, J., Huang, X., and Yu, C. (2011) *J. Power Sources*, **196**, 5402.
50. Liu, C., Alwarappan, S., Chen, Z., Kong, X., and Li, C. (2010) *Biosens. Bioelectron.*, **25**, 1829.
51. Subrahmanyam, K.S., Vivekchand, S.R.C., Govindaraj, A., and Rao, C.N.R. (2008) *J. Mater. Chem.*, **18**, 1517.
52. Vivekchand, S.R.C., Rout, C.S., Subrahmanyam, K.S., Govindaraj, A., and Rao, C.N.R. (2008) *J. Chem. Sci.*, **120**, 9.
53. Du, X., Guo, P., Song, H., and Chen, X. (2010) *Electrochim. Acta*, **55**, 4812.
54. Du, Q., Zheng, M., Zhang, L., Wang, Y., Chen, J., Xue, L., Dai, W., Ji, G., and Cao, J. (2010) *Electrochim. Acta*, **55**, 3897.
55. Stoller, M.D., Park, S., Zhu, Y., An, J., and Ruoff, R.S. (2008) *Nano Lett.*, **8**, 3498.
56. Mitra, S. and Sampath, S. (2004) *Electrochem. Solid-State Lett.*, **7**, A264.
57. Lv, W., Tang, D.M., He, Y.B., You, C.H., Shi, Z.Q., Chen, X.C., Chen, C.M., Hou, P.X., Liu, C., and Yang, Q.H. (2009) *ACS Nano*, **3**, 3730.
58. Murugan, A.V., Muraliganth, T., and Manthiram, A. (2009) *Chem. Mater.*, **21**, 5004.
59. Sun, Y., Wu, Q., Xu, Y., Bai, H., Li, C., and Shi, G. (2011) *J. Mater. Chem.*, **21**, 7154.
60. Yan, J., Wei, T., Shao, B., Ma, F., Fan, Z., Zhang, M., Zheng, C., Shang, Y., Qian, W., and Wei, F. (2010) *Carbon*, **48**, 1731.
61. Yu, A., Roes, I., Davies, A., and Chen, Z. (2010) *Appl. Phys. Lett.*, **96**, 253105.
62. Liu, C., Yu, Z., Neff, D., Zhamu, A., and Jang, B.Z. (2010) *Nano Lett.*, **10**, 4863.
63. Lu, X., Dou, H., Gao, B., Yuan, C., Yang, S., Hao, L., Shen, L., and Zhang, X. (2011) *Electrochim. Acta*, **56**, 5115.
64. Huang, Z., Zhang, B., Oh, S., Zheng, Q., Lin, X., Yousefi, N., and Kim, J. (2012) *J. Mater. Chem.*, **22**, 3591.

65. Dong, X., Xing, G., Chan-Park, M.B., Shi, W., Xiao, N., Wang, J., Yan, Q., Sum, T.C., Huang, W., and Chen, P. (2011) *Carbon*, **49**, 5071.
66. Yu, D. and Dai, L. (2010) *J. Phys. Chem. Lett.*, **1**, 467.
67. Byon, H.R., Lee, S.W., Chen, S., Hammond, P.T., and Horn, Y.S. (2011) *Carbon*, **49**, 457.
68. Du, F., Yu, D., Dai, L., Ganguli, S., Varshney, V., and Roy, A.K. (2011) *Chem. Mater.*, **23**, 4810.
69. Yoo, J.J., Balakrishnan, K., Huang, J., Meunier, V., Sumpter, B.G., Srivastava, A., Conway, M., Reddy, A.L.M., Yu, J., Vajtai, R., and Ajayan, P.M. (2011) *Nano Lett.*, **11**, 1423.
70. Fan, Z., Yan, J., Zhi, L., Zhang, Q., Wei, T., Feng, J., Zhang, M., Qian, W., and Wei, F. (2010) *Adv. Mater.*, **22**, 3723.
71. Mitra, S., Lokesh, K.S., and Sampath, S. (2008) *J. Power Sources*, **185**, 1544.
72. Chen, S., Zhu, J., Wu, X., Han, Q., and Wang, X. (2010) *ACS Nano*, **4**, 2822.
73. Lee, J.W., Hall, A.S., Kim, J., and Mallouk, T.E. (2012) *Chem. Mater.*, **24**, 1158.
74. Wang, H., Liang, Y., Mirfakhrai, T., Chen, Z., Casalongue, H.S., and Dai, H. (2011) *Nano Res.*, **4**, 729.
75. Wang, H.L., Casalongue, H.S., Liang, Y.Y., and Dai, H.J. (2010) *J. Am. Chem. Soc.*, **132**, 7472.
76. Chen, S., Zhu, J.W., and Wang, X. (2010) *J. Phys. Chem. C*, **114**, 11829.
77. Lu, T., Zhang, Y., Li, H., Pana, L., Li, Y., and Sun, Z. (2010) *Electrochim. Acta*, **55**, 4170.
78. Shi, W., Zhu, J., Sim, D.H., Tay, Y.Y., Lu, Z., Zhang, X., Sharma, Y., Srinivasan, M., Zhang, H., Hng, H.H., and Yan, Q. (2011) *J. Mater. Chem.*, **21**, 3422.
79. Wu, Q., Xu, Y., Yao, Z., Liu, A., and Shi, G. (2010) *ACS Nano*, **4**, 1963.
80. Zhu, Y., Murali, S., Stoller, M.D., Ganesh, K.J., Cai, W., Ferreira, P.J., Pirkle, A., Wallace, R.M., Cychosz, K.A., Thommes, M., Su, D., Stach, E.A., and Ruoff, R.S. (2011) *Science*, **332**, 1537.
81. Zhang, L.L., Zhao, X., Stoller, M.D., Zhu, Y., Ji, H., Murali, S., Wu, Y., Perales, S., Clevenger, B., and Ruoff, R.S. (2012) *Nano Lett.*, **12**, 1806.
82. Zhu, Y., Murali, S., Stoller, M.D., Velamakanni, A., Piner, R.D., and Ruoff, R.S. (2010) *Carbon*, **48**, 2118.
83. Hsieh, C., Hsu, S., Lin, J., and Teng, H. (2011) *J. Phys. Chem. C*, **115**, 12367.
84. Lin, Z., Liu, Y., Yao, Y., Hildreth, O.J., Li, Z., Moon, K., and Wong, C. (2011) *J. Phys. Chem. C*, **115**, 7120.
85. Zhang, J., Jiang, J., and Zhao, X.S. (2011) *J. Phys. Chem. C*, **115**, 6448.
86. Mishra, A.K. and Ramaprabhu, S. (2011) *J. Phys. Chem. C*, **115**, 14006.
87. Zhang, K., Zhang, L.L., Zhao, X.S., and Wu, J. (2010) *Chem. Mater.*, **22**, 1392.
88. Stoller, M.D., Murali, S., Quarles, N., Zhu, Y., Potts, J.R., Zhu, X., Ha, H., and Ruoff, R.S. (2012) *Phys. Chem. Chem. Phys.*, **14**, 3388.
89. Burke, A. (2009) *Ultracapacitor Technologies and Application in Hybrid and Electric Vehicles*, Research Report, Institute of Transportation Studies, University of CA, Davis.
90. Frackowiak, E. (2007) *Phys. Chem. Chem. Phys.*, **9**, 1774.
91. Kurig, H., Vestli, M., Janes, A., and Lust, E. (2011) *Electrochem. Solid-State Lett.*, **14**, A120.
92. Hapiot, P. and Lagrost, C. (2008) *Chem. Rev.*, **108**, 2238.
93. Chen, Y., Zhang, X., Zhang, D.C., Yu, P., and Ma, Y.W. (2011) *Carbon*, **49**, 573.
94. Idota, Y., Kubota, T., Matsufuji, A., Maekawa, Y., and Miyasaka, T. (1997) *Science*, **276**, 1395.
95. Larcher, D., Beattie, S., Morcrette, M., Edstrom, K., Jumas, J., and Tarascon, J. (2007) *J. Mater. Chem.*, **17**, 3759.
96. Pasero, D., Reeves, N., and West, A.R. (2005) *J. Power Sources*, **141**, 156.
97. Thackeray, M.M., David, W.I.F., Bruce, P.G., and Goodenough, J.B. (1983) *Mater. Res. Bull.*, **18**, 461.
98. Li, W.Y., Xu, L.N., and Chen, J. (2005) *Adv. Funct. Mater.*, **15**, 851.
99. Radich, J.G., McGinn, P.J., and Kamat, P.V. (2011) *Electrochem. Soc. Interface*, 63–66.
100. Whittingham, M.S. (2004) *Chem. Rev.*, **104**, 4271.

101. Bruce, P.G., Scrosati, B., and Tarascon, J.-M. (2008) *Angew. Chem., Int. Ed.*, **47**, 2930.

102. Dahn, J.R., Zheng, T., Liu, Y., and Xue, J.S. (1995) *Science*, **270**, 590.

103. Wang, C., Li, D., Too, C.O., and Wallace, G.G. (2009) *Chem. Mater.*, **21**, 2604.

104. Yoo, E.J., Kim, J., Hosono, E., Zhou, H., Kudo, T., and Honma, I. (2008) *Nano Lett.*, **8**, 2277.

105. Kumar, A., Reddy, A.L.M., Mukherjee, A., Dubey, M., Zhan, X., Singh, N., Ci, L., Billups, W.E., Nagurny, J., Mital, G., and Ajayan, P.M. (2011) *ACS Nano*, **5**, 4345.

106. Wang, G., Wang, B., Wang, X., Park, J., Dou, S., Ahn, H., and Kim, K. (2009) *J. Mater. Chem.*, **19**, 8378.

107. Shiva, K., Rajendra, H.B., Subrahmanyam, K.S., Bhattacharyya, A.J., and Rao, C.N.R. (2012) *Chem. Eur. J.*, **18**, 4489.

108. Winter, M., Besenhard, J.O., Spahr, M.E., and Novak, P. (1998) *Adv. Mater.*, **10**, 725.

109. Lee, J.K., Smith, K.B., Hayner, C.M., and Kung, H.H. (2010) *Chem. Commun.*, **46**, 2025.

110. Chou, S., Wang, J., Choucair, M., Liua, H., Stride, J.A., and Dou, S. (2010) *Electrochem. Commun.*, **12**, 303.

111. Wang, D., Chang, Y.-L., Wang, Q., Cao, J., Farmer, D.B., Gordon, R.G., and Dai, H.J. (2004) *J. Am. Chem. Soc.*, **126**, 11602.

112. Fuller, C.S. and Severiens, J.C. (1954) *Phys. Rev.*, **96**, 21.

113. Cui, G., Gu, L., Zhi, L., Kaskhedikar, N., van Aken, P.A., Mullen, K., and Maier, J. (2008) *Adv. Mater.*, **20**, 3079.

114. Xue, D.-J., Xin, S., Yan, Y., Jiang, K.-C., Yin, Y.-X., Guo, Y.-G., and Wan, L.-J. (2012) *J. Am. Chem. Soc.*, **134**, 2512.

115. Poizot, P., Laruelle, S., Grugeon, S., Dupont, L., and Tarascon, J.M. (2000) *Nature*, **407**, 496.

116. Wang, D., Choi, D., Li, J., Yang, Z., Nie, Z., Kou, R., Hu, D., Wang, C., Saraf, L.V., Zhang, J., Aksay, I.A., and Liu, J. (2009) *ACS Nano*, **3**, 907.

117. Hu, Y.S., Kienle, L., Guo, Y.G., and Maier, J. (2006) *Adv. Mater.*, **18**, 1421.

118. Paek, S., Yoo, E.J., and Honma, I. (2009) *Nano Lett.*, **9**, 72.

119. Yu, A., Park, H.W., Davies, A., Higgins, D.C., Chen, Z., and Xiao, X. (2011) *J. Phys. Chem. Lett.*, **2**, 1855.

120. Wu, M.S., Chiang, P.C.J., Lee, J.T., and Lin, J.C. (2005) *J. Phys. Chem. B*, **109**, 23279.

121. Wu, Z., Ren, W., Wen, L., Gao, L., Zhao, J., Chen, Z., Zhou, G., Li, F., and Cheng, H. (2010) *ACS Nano*, **4**, 3187.

122. Sun, Y., Hu, X., Luo, W., and Huang, Y. (2011) *ACS Nano*, **5**, 7100.

123. Zhou, G., Wang, D.-W., Yin, L.-C., Li, N., Li, F., and Cheng, H.-M. (2012) *ACS Nano*, Article ASAP, **6**, 3214–3223. doi: 10.1021/nn300098m.

124. Xiao, J., Mei, D., Li, X., Xu, W., Wang, D., Graff, G.L., Bennett, W.D., Nie, Z., Saraf, L.V., Aksay, I.A., Liu, J., and Zhang, J. (2011) *Nano Lett.*, **11**, 5071.

125. Chen, H., Armand, M., Demailly, G., Dolhem, F., Poizot, P., and Tarascon, J.-M. (2008) *Chem. Sus. Chem.*, **1**, 348.

126. Walker, W., Grugeon, S., Mentre, O., Laruelle, S., Tarascon, J.-M., and Wudl, F. (2010) *J. Am. Chem. Soc.*, **132**, 6517.

127. Oyaizu, K. and Nishide, H. (2009) *Adv. Mater.*, **21**, 2339.

128. Nakahara, K., Iwasa, S., Satoh, M., Morioka, Y., Iriyama, J., Suguro, M., and Hasegawa, E. (2002) *Chem. Phys. Lett.*, **359**, 351.

129. Guo, W., Yin, Y.-X., Xin, S., Guo, Y.-G., and Wan, L.-J. (2012) *Energy Environ. Sci.*, **5**, 5221.

130. Song, Z., Xu, T., Gordin, M.L., Jiang, Y.-B., Bae, I.-T., Xiao, Q., Zhan, H., Liu, J., and Wang, D. (2012) *Nano Lett.*, Article ASAP, **12**, 2205–2211. doi: 10.1021/nl2039666.

131. Song, Z., Zhan, H., and Zhou, Y. (2009) *Chem. Commun.*, 448.

132. Song, Z., Zhan, H., and Zhou, Y. (2010) *Angew. Chem., Int. Ed.*, **49**, 8444.

133. Han, P., Yue, Y., Liu, Z., Xu, W., Zhang, L., Xu, H., Dong, S., and Cui, G. (2011) *Energy Environ. Sci.*, **4**, 4710.

134. Lee, J., Lee, T., Song, H., Cho, J., and Kim, B. (2011) *Energy Environ. Sci.*, **4**, 4148.

135. Sum, E., Rychcik, M., and Skyllas-Kazacos, M. (1985) *J. Power Sources*, **16**, 85.

136. Sum, E. and Skyllas-Kazacos, M. (1985) *J. Power Sources*, **15**, 179.

137. Skyllas-Kazacos, M., Rychcik, M., Robins, R.G., Fane, A.G., and Green, M.A. (1986) *J. Electrochem. Soc.*, **133**, 1057.

138. Gattrell, M., Qian, J., Stewart, C., Graham, P., and MacDougall, B. (2005) *Electrochim. Acta*, **51**, 395.

139. Han, P.X., Wang, H.B., Liu, Z.H., Chen, X., Ma, W., Yao, J.H., Zhu, Y.W., and Cui, G.L. (2011) *Carbon*, **49**, 693.

10
Heterogeneous Catalysis by Metal Nanoparticles Supported on Graphene

M. Samy El-Shall

10.1
Introduction

Graphene consists of a hexagonal monolayer network of sp^2-hybridized carbon atoms [1–7]. The unique properties of this 2D material include the highest intrinsic carrier mobility at room temperature of all known materials and very high mechanical strength and thermal stability [1–7]. Electrons in graphene have zero effective masses and can travel for micrometers without scattering at room temperature allowing them to behave ballistically, similar to massless particles [1–7]. Furthermore, graphene also conducts heat ballistically [1]. Phonons move like electrons across the solid giving graphene unusual thermal capabilities [1–7]. Most of the outstanding electronic properties of graphene can only be observed in samples with high perfection of the atomic lattice because structural defects that may appear during growth or processing deteriorate the performance of graphene electronic devices [1–7]. These properties inspire many new applications in a wide range of areas including nanoelectronics, supercapacitors, batteries, photovoltaics, solar cells, fuel cells, transparent conducting films, sensors, and many others [1–8]. The recognition of this important new area of research and its tremendous potential applications have been highlighted by awarding the 2010 Nobel Prize in physics to Andre Geim and Konstantin Novoselov for *Groundbreaking experiments regarding the two-dimensional material graphene* [9].

In addition to the unique electronic properties of graphene, other properties such as high thermal, chemical, and mechanical stability as well as high surface area also represent desirable characteristics as 2D support layers for metallic and bimetallic nanoparticles in heterogeneous catalysis [10–23]. The large surface area ($2600\ m^2\ g^{-1}$, theoretical value) of graphene and its high thermal and chemical stability provide an excellent catalyst support [10–23]. Supporting metal nanoparticles on graphene sheets could provide a large surface area and thermally stable system for potential applications in catalysis, fuel cells, chemical sensors, and hydrogen storage. In catalysis applications, unlike electronic applications, structural defects in the graphene lattice can be useful, as they make it possible to tailor the localized

Graphene: Synthesis, Properties, and Phenomena, First Edition. Edited by C. N. R. Rao and A. K. Sood.

properties of graphene to achieve new surface functionalities, which could enhance the interactions with the anchored metal nanoparticles [10–23].

This chapter is focused on the application of graphene as a unique high surface area catalyst support for a variety of important chemical transformations. The organization of the chapter is as follows. First, Section 10.2 describes different methods for the synthesis of graphene and metal nanoparticles supported on graphene using microwave-assisted chemical reduction and laser irradiation of graphene oxide (GO) in different solvent environments. Section 10.3 deals with the application of Pd nanoparticles supported on graphene as efficient heterogeneous catalysts for the carbon–carbon cross-coupling reactions. The activity of the Pd/graphene catalyst prepared by the microwave-assisted chemical reduction method is discussed compared to the laser synthesis of Pd nanoparticles supported on partially reduced graphene oxide (PRGO). Section 10.4 presents the application of metal/metal-oxide nanoparticles supported on PRGO for CO oxidation. Finally, in Section 10.5, we present a summary and outlook for unique properties and versatility of graphene as a catalyst support in heterogeneous catalysis.

10.2 Synthesis of Graphene and Metal Nanoparticles Supported on Graphene

10.2.1 Chemically Converted Graphene by Microwave-Assisted Chemical Reduction of Graphene Oxide

Chemical reduction methods using microwave heating have been employed for the production of metal/metal-oxide nanoparticles supported on the defect sites of reduced graphene oxide (GO) nanosheets [24–26]. The main advantage of microwave irradiation (MWI) over other conventional heating methods is the rapid and uniform heating of the reaction mixture [24–31]. Owing to the difference in the solvent and reactant dielectric constants, selective dielectric heating can provide significant enhancement in the transfer of energy directly to the reactants, which causes an instantaneous internal temperature rise [27–31]. This temperature rise in the presence of a reducing agent allows the simultaneous reduction of the metal ions and GO and the dispersion of the metallic nanoparticles on the large surface area of the resulting graphene nanosheets. Unlike conventional thermal heating, MWI allows a better control of the extent of GO reduction by hydrazine hydrates (HHs) as both the MWI power and time can be adjusted to yield a nearly complete reduction of GO. The reduction of GO proceeds by rapid deoxygenation to create C–C and C=C bonds [24–26, 32–36]. The occurrence of nonequilibrium local heating of graphene oxide is likely to result in the generation of hot spots (electric discharge) in the reduced graphene oxide, which could lead to the development of structural defects in the graphene lattice. These defect sites act as nucleation centers for the formation of the metal nanoparticles, which can be anchored to the graphene sheets.

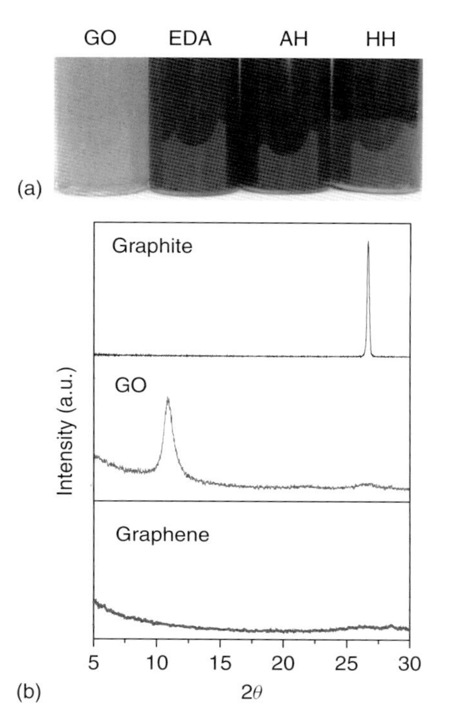

Figure 10.1 (a) Images of the graphene oxide (GO) suspension in water before and after MWI in the presence of the reducing agents ethylenediamine (EDA), ammonium hydroxide (AH), and hydrazine hydrate (HH). (b) XRD patterns of graphite, GO, and graphene prepared by MWI of GO using hydrazine hydrate as a reducing agent [24].

Figure 10.1a displays a digital photograph of the exfoliated GO dispersed in water ($5\,\mathrm{mg\,ml^{-1}}$) [37] before and after MWI in the presence of ethylenediamine (EDA), ammonium hydroxide (AH), and HH [24]. Following MWI for 30–60 s, the yellow-brown color of the dispersed GO in water changed to black indicating the reduction of the GO. This reduction is supposed to lead to deoxygenation of the GO (removal of oxygen functionalities such as hydroxyl and epoxide groups) and to significant restoration of the sp^2 carbon sites and re-establishment of the conjugated graphene network [32, 33]. It has been shown that HH is one of the best reducing agents for the chemical reduction of exfoliated GO to produce very thin graphene sheets, although the chemical pathways for the GO reduction remain unclear [32–36]. In addition to HH, we found that MWI of exfoliated GO in the presence of EDA and AH could also lead to fast deoxygenation, consistent with a recent report where graphene suspensions were prepared by heating exfoliated GO in NaOH or KOH at moderate temperatures (50–90 °C) [36].

The XRD patterns displayed in Figure 10.1b confirm the chemical reduction of the GO and the formation of chemically converted graphene (CCG) sheets using HH as the reducing agent [24, 35, 36]. The initial graphite powder shows the typical sharp diffraction peak at $2\theta = 26.658^\circ$ with the corresponding d-spacing of 3.34 Å. The oxidation process results in the insertion of hydroxyl and epoxy groups between the carbon sheets mainly on the centers while the carboxyl groups are typically inserted on the terminal and lateral sides of the sheets [37, 38]. The insertion of these groups leads to a decrease in the van der Waals forces between the graphite sheets in the exfoliated GO. As shown in Figure 10.1b, the XRD pattern of the exfoliated GO shows no diffraction peaks from the parental graphite material and only a new broad peak at $2\theta = 10.89^\circ$ with a d-spacing 8.14 Å is observed. This indicates that the distance between the carbon sheets has increased because of the insertion of inter planer groups. The GO sheets are expected to be thicker than the graphite sheets because of the presence of covalently bounded oxygen atoms and the displacement of the sp^3-hybridized carbon atoms slightly above and below the original graphene sheets. After MWI of the GO in the presence of HH as the reducing agent, the XRD pattern of the resulting CCG sheets shows the disappearance of the 10.89° peak confirming the complete reduction of the GO sheets [24, 38].

Raman spectroscopy is one of the most widely used techniques to characterize the structural and electronic properties of graphene including disorder and defect structures, defect density, and doping levels [39, 40]. The Raman spectrum of graphene is characterized by three main features, (i) the G-mode arising from emission of zone-center optical phonons (usually observed at $\sim$1575 cm^{-1}), (ii) the D-mode arising from the doubly resonant disorder-induced mode ($\sim$1350 cm^{-1}), and (iii) the symmetry-allowed 2D overtone mode ($\sim$2700 cm^{-1}) [39, 40]. The shift and line shape associated with these modes have been used to distinguish single free-standing graphene sheets from bilayer and few-layer graphene (FLG) [39, 40]. To characterize the properties of the CCG, we measured the Raman spectra of the original graphite, GO, and the as-prepared CCG sheets (by MWI in the presence of HH) using the 457.9 nm radiation from an Ar-ion laser and the results are shown in Figure 10.2. The spectrum of the graphite powder (Figure 10.2a) shows the in-phase vibration of the graphite lattice at 1575 cm^{-1} (G-band). The spectrum of the exfoliated GO (Figure 10.2a) shows broadened and blue shifted

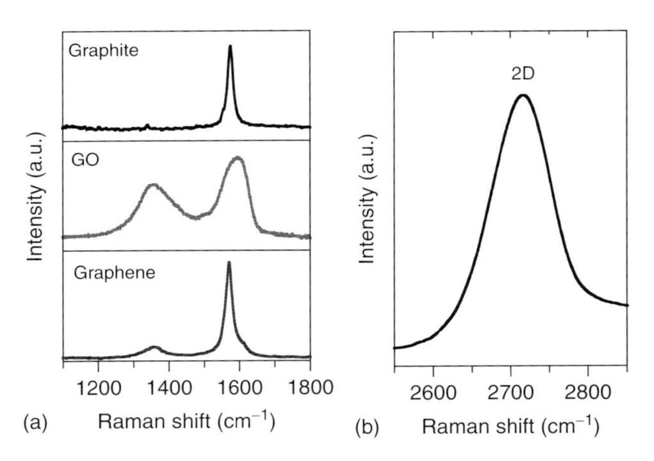

Figure 10.2 (a) The Raman spectra of the original graphite sample, the exfoliated graphene oxide (GO), and the chemically converted graphene using MWI of GO in the presence of HH. (b) The Raman spectrum of the chemically converted graphene in the region of the 2D-band showing a strong broad peak around 2700 cm^{-1} [24].

G-band (1594 cm^{-1}) and the D-band with small intensity at 1354 cm^{-1}. The blue shift of the G-band in GO is usually attributed to the presence of isolated double bonds, which resonate at higher frequencies than the G-band of graphite [41]. The spectrum of the CCG (Figure 10.2a) shows a strong G-band around 1571 cm^{-1} almost at the same frequency as that of graphite with a small shoulder, identified as the D′-band around 1616 cm^{-1}, and a weak D-band around 1357 cm^{-1}. The D-band at 1357 cm^{-1} and the D′-shoulder at 1616 cm^{-1} have been attributed to structural disorder at defect sites and finite size effects, respectively [39, 42–44]. The intensity ratio of the D-band to the G-band is used as a measure of quality of the graphitic structures because for highly ordered pyrolitic graphite, this ratio approaches zero [42, 45]. As shown in Figure 10.2a, the Raman spectrum of the CCG sheets exhibits a weak disorder-induced D-band with the D–G intensity ratio of only 0.10–0.12, thus indicating the high quality of the synthesized graphene sheets. It is also interesting to note that the frequency of the G-mode in the synthesized CCG sheets (1571 cm^{-1}) is very similar to that observed in the graphite powder (1575 cm^{-1}) and significantly different from that of the GO (1594 cm^{-1}). The same trend has been found in the frequency of the G-mode of the free-standing graphene layer, which is appreciably downshifted as compared with that of the supported layer [40].

The high energy second-order 2D-band of the synthesized graphene sheets around 2700 cm^{-1} is shown in Figure 10.2b. The position and shape of the 2D-peak depend on the number of graphene layers, and therefore the 2D-peak can be used to distinguish between single-layer graphene, bilayer graphene, and FLG [39, 43]. For example, it has been shown that a single-layer graphene exhibits a single sharp 2D-band located below 2700 cm^{-1}, while bilayer sheets have a broader 2D peak at $\sim$2700 cm^{-1} [39]. Sheets with more than five layers have broad 2D-peaks significantly shifted to positions greater than 2700 cm^{-1} [39]. Since the synthesized graphene sample using HH under MWI shows a single 2D-band around 2720 cm^{-1}, it can be concluded that the analyzed region of the sample consisted of FLG probably with five or more layers.

Figure 10.3 displays typical SEM and TEM images of the synthesized graphene sheets by MWI in the presence of HH. The SEM micrographs clearly show extended sheets of lateral dimensions ranging from a few micrometers to tens of micrometers in length with layered structures. The TEM images show a few stacked layers (two to three layers) and a lateral size up to a few micrometers. Also, the TEM images show that some of the graphene layers are folded on one edge with isolated small fragments on the surfaces.

10.2.1.1 Metal Nanoparticles Supported on Graphene by Microwave Synthesis

Supporting the metal nanoparticles on the graphene sheets could prevent the formation of stacked graphitic structures because the metal nanoparticles can act as spacers to increase the distance between the sheets. This could lead to an increase in the surface area of the nanoparticle–graphene composites [10–23]. These materials may have promising potential applications in catalysis, fuel cells, chemical sensors, and hydrogen storage.

Figure 10.3 (a) SEM and (b) TEM images of the chemically converted graphene sheets using MWI of GO in the presence of HH [24].

Good dispersion of the metal nanocrystals on the graphene sheets can be achieved by the simultaneous reduction of the metal salts and GO during the MWI irradiation process [24]. Figure 10.4 compares the TEM images of the Pd/graphene sample (Figure 10.4a–c) prepared by mixing separately prepared Pd nanoparticles (using 100 μl HH and 100 μl Pd nitrate under MWI for 10–15 s) and CCG sheets (using 100 μl HH and 50 mg GO and 25 ml water under MWI for 20–25 s) with the sample (Figure 10.4d–f) prepared by the simultaneous reduction of GO (50 mg in 25 ml water) and 100 μl Pd nitrate using 100 μl HH (under MWI for 20–25 s). It is clear that the simple physical mixing of the nanoparticles and graphene sheets results in significant aggregation of the metal nanoparticles with very poor dispersion on the graphene sheets. On the other hand, the simultaneous reduction of the metal salt with GO results in well-dispersed nanoparticles on the graphene sheets thus suggesting that specific interaction between the metal nanocrystals and the graphene sheets may be responsible for dispersion of the nanoparticles.

10.2.2
Laser-Converted Graphene by Laser Reduction of Graphene Oxide

Another method for the reduction of GO is laser irradiation in solution to produce laser-converted graphene (LCG) sheets [46–49]. This method provides a solution

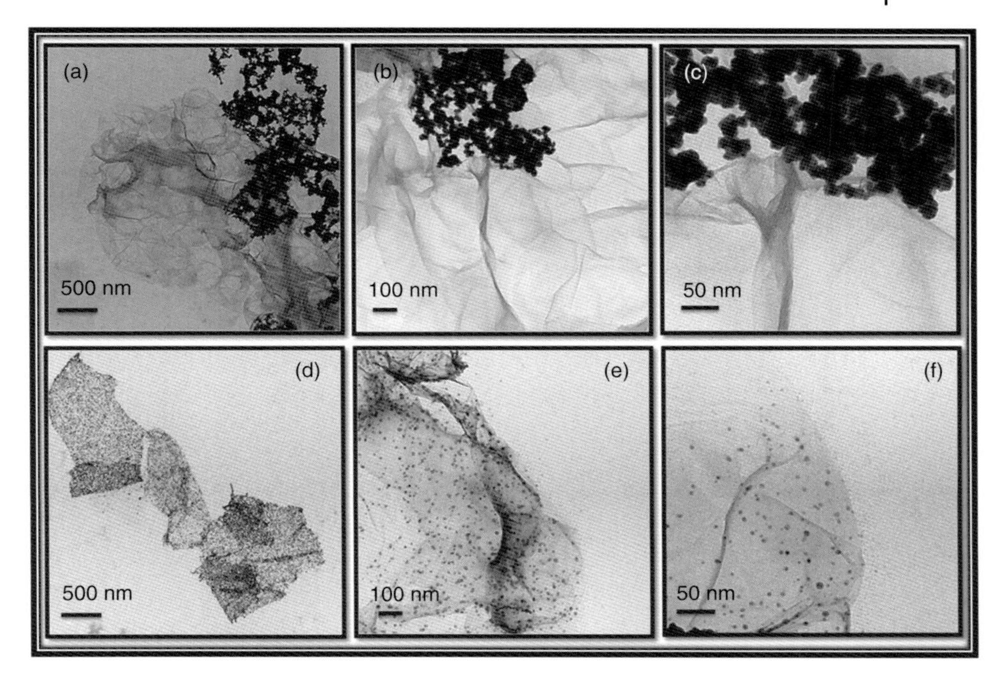

Figure 10.4 TEM images of the chemically converted graphene sheets containing Pd nanoparticles prepared by (a–c) mixing separately prepared Pd nanoparticles and CCG sheets (a–c) and simultaneous reduction of GO and Pd nitrate in water using hydrazine hydrate under MWI (d–f) [24].

processable synthesis of individual graphene sheets in water at ambient conditions without the use of any chemical reducing agent. The UV–vis spectrum of GO is characterized by significant absorption below 400 nm with a characteristic shoulder at 305 nm attributed to $n \rightarrow \pi^*$ transitions of C=O bonds [46, 50]. With the 532 nm laser irradiation, two-photon absorption is expected to contribute significantly to the absorption of the laser energy by GO because of excited state nonlinearities, which enhance the two-photon absorption of GO at 532 nm in the nanosecond regime [50, 51].

The irradiation time required for the deoxygenation of GO using the 532 or the 355 nm lasers varies from a few to several minutes depending on the laser power, the concentration of GO and the volume of the solution. Experiments using the fundamental of the YAG laser (1064 nm, 30 Hz, 5 W) resulted in a rapid partial deoxygenation of the GO but with no change in the color of the solution suggesting no conversion to graphene [46]. At higher laser power (7 W, 30 Hz), the solution was bleached after 5–6 min with the evolution of CO_2 gas and H_2O vapor and the formation of a graphitic powder indicating the decomposition of GO [46]. With the 355 nm irradiation, bleaching of the black solution was observed at longer irradiation times (after 6–7 min using an average power of 5 W, 30 Hz).

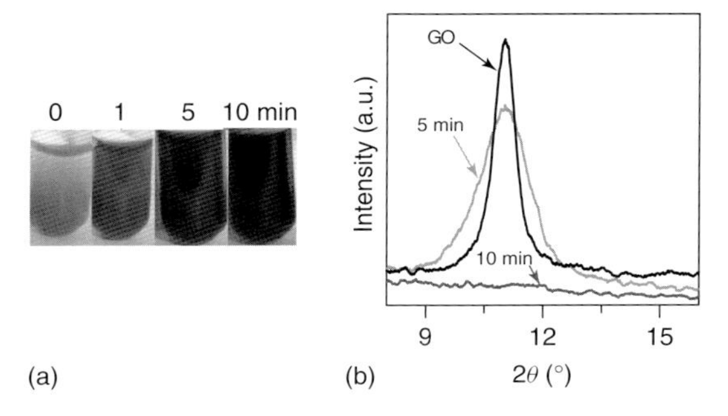

Figure 10.5 (a) Digital photographs of GO solutions and (b) XRD of the solid samples as a function of the 532 nm laser irradiation time (5 W, 30 Hz) [46].

However, with the 532 nm irradiation, no bleaching was observed even after 10 min of irradiation using an average power of 7 W at a repetition rate of 30 Hz [46].

Figure 10.5a displays digital photographs of GO solutions in water at different irradiation times using the 532 nm laser (5 W, 30 Hz). It is clear that after 10 min of laser irradiation, the yellow-golden color of the GO changes to black indicating the reduction of GO. As shown in Figure 10.5b, following the 532 nm laser irradiation for a few minutes, the $2\theta = 10.9^{\circ}$ peak decreases as the yellow-golden color of the GO solution changes gradually to brown and finally to black with the complete disappearance of the 10.9° peak thus confirming the deoxygenation of the GO sheets and the restoration of the sp^2 carbon sites in the LCG [32, 33]. A similar result was obtained following irradiation using the third harmonic of the YAG laser (355 nm, 5 W, 30 Hz) (Figure 10.5) [46].

To understand the nature of the species absorbing the laser energy and elucidate the mechanism of the laser deoxygenation of GO, the UV–vis spectrum of GO was measured in water and compared with the spectra of the LCG produced following the 532 and 355 nm laser irradiation of GO as shown in Figure 10.6. It is important to note that the shoulder at 305 nm disappears after the 532 or the 355 nm irradiation of GO (Figure 10.6), and the absorption peak of GO at 230 nm red shifts to about 270 nm because of the $\pi \rightarrow \pi^*$ transitions of extended aromatic C-C bonds thus suggesting that the electronic conjugation within the graphene sheets is restored in the LCG [52, 53]. It is interesting that the gradual redshift of the 230 nm peak of GO and the increase in absorption in the whole spectral region (230 nm) can be observed as a function of the laser irradiation time as shown in Figure 10.6. The redshift and the increase in absorption do not show much change after 10 min of laser irradiation, indicating complete deoxygenation of GO and formation of LCG within 10 min. A similar result was obtained using the 355 nm laser irradiation [46].

The observed trend suggests that the laser irradiation method can provide a way to tune the degree of deoxygenation of GO because the extent of the redshift is related to the degree or aromaticity in the LCG. A similar trend was observed for the HH reduction of GO where gradual redshift of the 231 nm peak to 270 nm and an increase in the absorption in this region were observed as a function of reaction time [53]. However, in the CCG using HH reduction, the reaction was completed

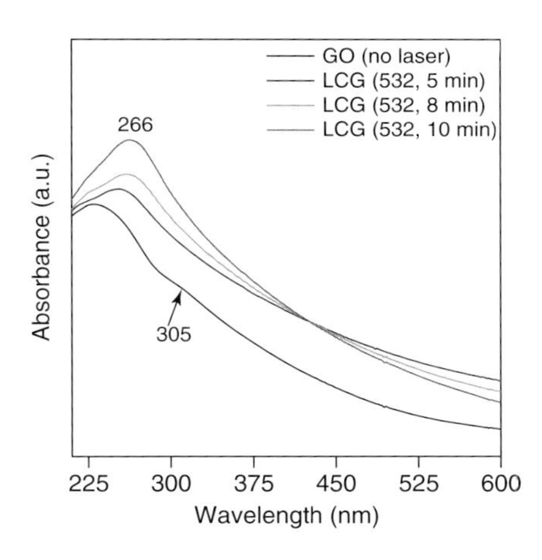

Figure 10.6 The UV–vis spectra showing the change of GO solution in water as a function of laser irradiation time (532 nm, 5 W, 30 Hz) [46].

after 1 h at 95 °C as compared to 10 min at room temperature for the current LCG method.

In addition to the XRD and UV–vis data, the successful conversion of GO into LCG using the 532 and 355 nm irradiation can be further verified by FT-IR, X-ray photoelectron spectroscopy (XPS), and Raman spectroscopy as shown in Figure 10.7. Figure 10.7a compares the FT-IR spectra of GO and the LCG. The GO spectrum shows strong bands corresponding to the C=O stretching vibrations of the COOH groups at 1740 cm^{-1}, the O–H deformations of the C–OH groups at 1350–1390 cm^{-1}, the C–O stretching vibrations at 1060–1100 cm^{-1}, and the epoxide groups (1230 cm^{-1}) [54, 55]. These bands were completely absent from the spectrum of the LCG thus confirming the conversion of GO into graphene following the 532 nm laser irradiation of GO in water.

The XPS C1s spectrum of GO shows peaks corresponding to oxygen-containing groups between 285.5 and 289 eV in addition to the sp^2-bonded carbon C=C at 284.5 eV (Figure 10.7b). Typically, peaks at 285.6, 286.7, 287.7, and 289 eV are assigned to the C1s of the C–OH, C–O, C=O, and HO–C=O groups, respectively [23, 32]. The XPS data of the LCG clearly indicates that most of the oxygen-containing groups in GO are removed after the 532 nm laser irradiation of GO in water (Figure 10.7b).

The Raman spectra of the prepared GO and LCG are shown in Figure 10.7c. The spectrum of the exfoliated GO shows a broadened and blue shifted G-band (1594 cm^{-1}) and the D-band with small intensity at 1354 cm^{-1} (as compared to graphite) [39]. The spectrum of the LCG shows a strong G-band around 1572 cm^{-1}, almost at the same frequency as that of graphite with a small shoulder, identified as the D′-band around 1612 cm^{-1}, and a weak D-band around 1345 cm^{-1}. The D-band and the D′-shoulder have been attributed to structural disorder at defect sites and finite size effects, respectively [39, 44, 45]. The intensity ratio of the D-band to the G-band is used as a measure of the quality of the graphitic structures because for highly ordered pyrolitic graphite, this ratio approaches zero [44, 45]. As shown in Figure 10.7c, the Raman spectrum of the LCG sheets exhibits a

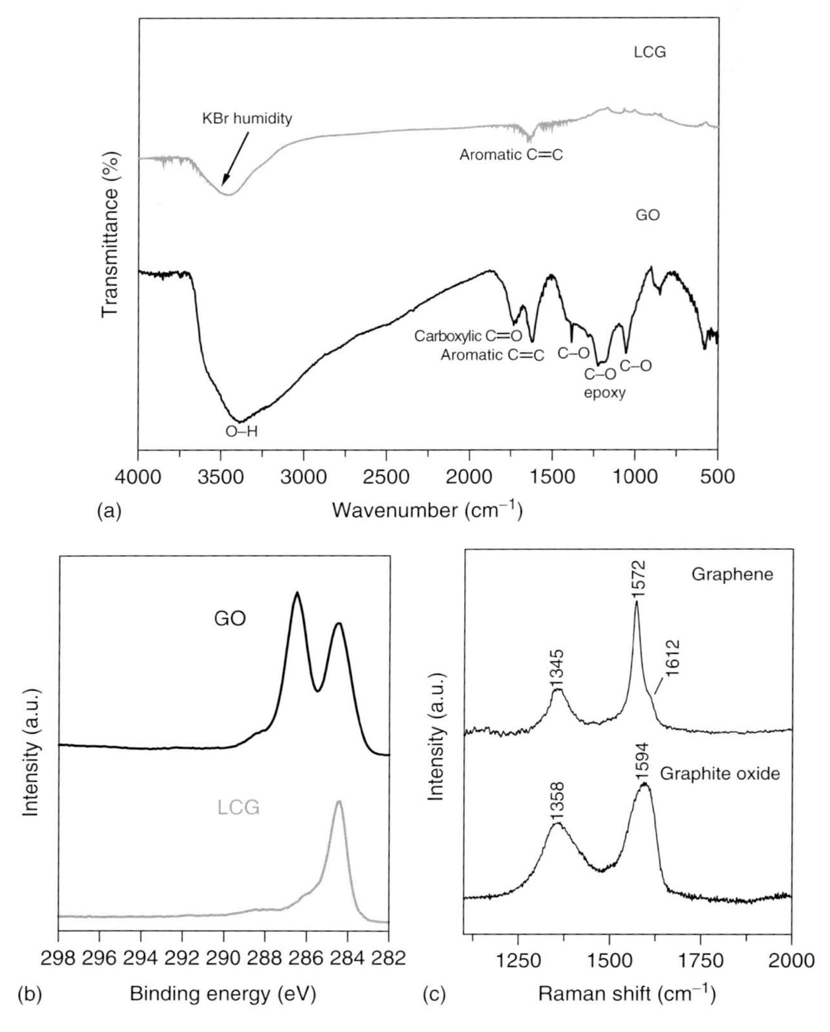

Figure 10.7 (a) FT-IR spectra of graphene oxide (GO) and laser-converted graphene (LCG), (b) XPS C1s spectra of graphene oxide (GO) and laser-converted graphene (LCG), and (c) Raman spectra of graphene oxide and graphene formed after laser irradiation of GO [46].

weak disorder-induced D-band with the D/G intensity ratio of about 0.29 as compared to 0.67 in GO thus indicating a significant reduction of the degree of disorder and defect sites following the laser deoxygenation of GO [44, 45]. However, whilecomparing the LCG to the CCG, the CCG exhibits less defects as evident by the low value of the intensity ratio of the D/G-bands (0.10–0.12, as shown in Figure 10.2) as compared to 0.29 in the LCG. This could be due to incomplete reduction of GO by laser irradiation in water in the absence of a reducing environment such as alcohols or glycols.

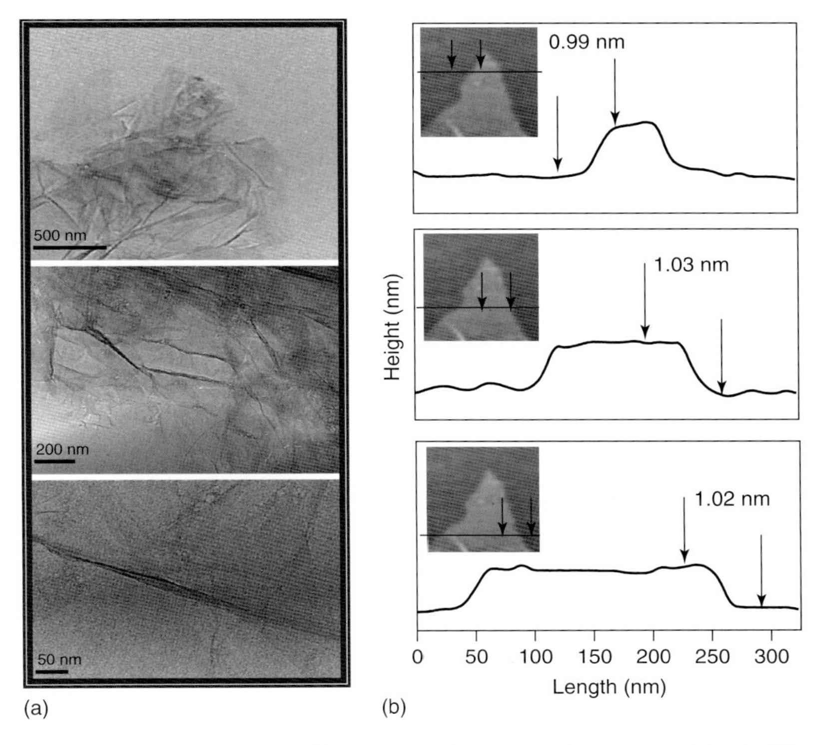

Figure 10.8 (a) TEM images of laser-converted graphene sheets prepared by the 532 nm irradiation of graphene oxide solution in water. (b) AFM images and cross-sectional analysis of the laser-converted graphene [46].

TEM images of the LCG sheets (Figure 10.8a) show wrinkled and partially folded sheets with a lateral dimension of up to a few micrometers in length [46]. AFM images (Figure 10.8b) with cross-section analysis show that most of the flakes consist of a single graphene sheet. The vertical distances measured across the sheet at different lateral locations were determined to be 0.99, 1.03, and 1.02 nm as shown in Figure 10.8b [46]. This is consistent with the reported AFM results on graphene, where the single-layer graphene is ~1 nm [8, 32–35].

10.2.2.1 **Laser-Assisted Photoreduction of Graphene Oxide in Different Solvents**

Figure 10.9a compares the UV–vis spectra of GO following the 532 nm laser irradiation in 50% ethanol–water and 2% PEG–water mixtures, and in pure water. In all cases, the characteristic shoulder of GO at 305 nm disappears after laser irradiation, and the absorption peak of GO at 230 nm red shifts to about 270 nm because of the $\pi \rightarrow \pi^*$ transitions of extended aromatic C–C bonds as the electronic conjugation within graphene is restored in the LCG [46, 47, 52, 55]. The XPS data of the LCG in 50% ethanol–water, 2% PEG–water, and pure water clearly (Figure 10.9b) indicate that most of the oxygen-containing groups in GO are removed after the laser irradiation.

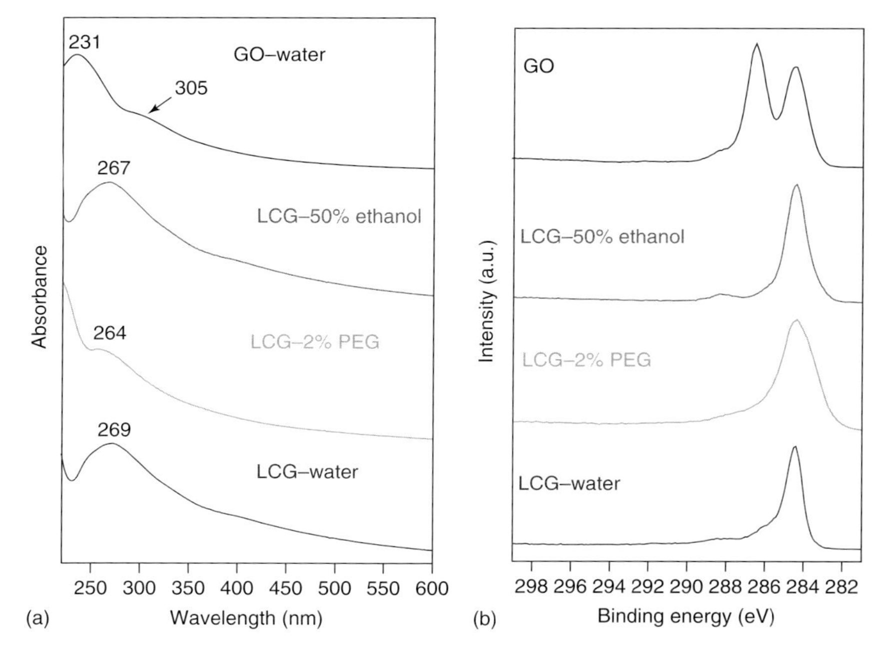

Figure 10.9 (a) UV–vis and (b) XPS C1s spectra of graphene oxide (GO) and laser-converted graphene (LCG) prepared by 532 nm laser irradiation (4 W, 30 Hz) of GO for 10 min in different solvents as indicated [47].

The reduction of GO in the presence of ethanol or PEG is much faster than in pure water under identical solution volume, concentration, and laser power conditions. For example, 532 nm irradiation with 1 W laser power (30 Hz) converts GO in pure water into graphene in about 40 min (7.2×10^3 laser pulses), while in the presence of 50 vol% ethanol or 2 vol% PEG, the same concentration of GO can be reduced in about 22 or 32 min, respectively. Obviously, the ethanol and PEG solutions exhibit higher reduction efficiencies than pure water [56–58]. This is attributed to the role of ethanol or PEG in scavenging the holes generated by the laser irradiation of GO. The evolution of gas bubbles (most likely H_2) is also observed during the laser irradiation of GO in different solvent environments [47]. It is interesting to note that the rate of gas evolution from the laser irradiation of GO in aqueous methanol solution is higher than that in ethanol–water solution and much higher than the rate measured in pure water as shown in Figure 10.10. These results are in full agreement with the recently published work of Yeh *et al.* 59, which showed that GO under irradiation with UV or visible light photocatalyzes H_2 production from 20 vol% aqueous methanol solution and pure water. Also, the amount of H_2 produced from the mercury-lamp irradiation of GO in aqueous methanol solution was much more than that produced in pure water. However, the time course of the production of H_2 using visible light irradiation was several hours (up to 12 h) in comparison with the few minutes in the laser irradiation experiments.

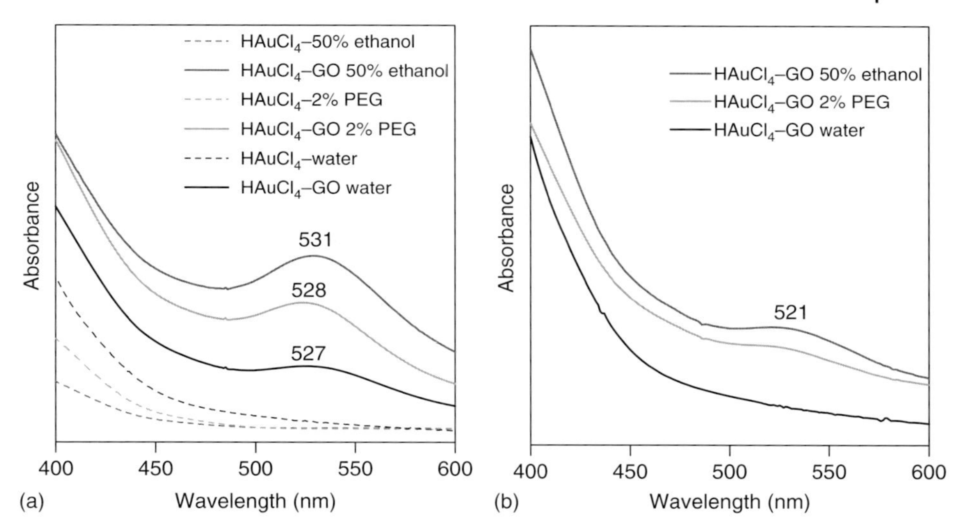

Figure 10.10 (a) Absorption spectra of 25 μl $HAuCl_4$ + GOin 50 vol% ethanol–water, 2 vol% PEG–water, and pure water recorded after 2 min of laser irradiation (532 nm, 4 W, 30 Hz). Dotted lines represent data of blank solutions containing the same amount of $HAuCl_4$ but no GO under identical laser irradiation conditions. (b) Absorption spectra of the same solutions in (a) irradiated with lower laser power (532 nm, 1 W, 30 Hz) showing no formation of gold nanoparticles in the pure water solution (black) [47].

Following the absorption of two photons by GO in the laser irradiation experiments and the generation of electron–hole (e-h) pairs, the holes are scavenged by methanol, ethanol, or PEG thus leaving the photogenerated electrons for the injection into the solution phase for the H_2 production and for the reduction of GO. It is likely that mutual photocatalytic reduction between different GO sheets occurs at the early stages of the photocatalytic reaction. The increased efficiency of H_2 production by irradiation of GO in alcohol–water mixtures as compared to pure water is attributed to the role of alcohol as a sacrificial hole scavenger, which interacts effectively with the photogenerated holes in the VB of GO to prevent electron–hole recombination. However, in pure water, in the absence of hole scavengers, the holes generated in the valence band of GO may not be positive enough to oxidize water to O_2 and therefore, the process is expected to be very slow, consistent with the experimental observations, and it may also result in the formation of other oxidation products such as hydrogen peroxides.

10.2.3 Photochemical Reduction of Metal Ions and Graphene Oxide

10.2.3.1 Photoreduction of Gold Ions and GO in Different Solvents

Figure 10.10a compares the absorption spectra of GO solutions containing the same amount of $HAuCl_4$ in 50 vol% ethanol–water and 2 vol% PEG–water mixtures and in pure water before and after the 532 nm irradiation (4 W, 30 Hz) for 2 min. The reduction of the gold ions and formation of Au nanoparticles is clearly evident by the observation of the surface plasmon resonance (SPR) band of Au nanoparticles (527–531 nm) [60, 61] as shown in Figure 10.10a.

It is clear that laser excitation of GO is involved in the reduction of the Au ions because irradiation of the $HAuCl_4$ solutions in the absence of GO under identical conditions does not result in the formation of Au nanoparticles as shown in Figure 10.10a.

The increase in the SPR band intensity of the Au nanoparticles for the 50 vol% ethanol and the 2 vol% PEG solutions as compared to pure water (Figure 10.10a) is attributed to increase in the concentration of the Au nanoparticles in the presence of ethanol or PEG consistent with the increased reduction efficiencies of these solutions over pure water. At low laser power (1 W, 30 Hz), the reduction of the Au ions is not observed in pure water as indicated by the absence of the SPR band of Au nanoparticles as shown in Figure 10.10b. This suggests that the laser reduction mechanism is different in water than in the presence of alcohol or PEG.

In the three solvent systems, an increase in the SPR band intensity of the Au nanoparticles by increasing the laser irradiation time from 2 to 10 min is observed thus providing direct evidence for increasing the nanoparticles' concentration with the number of photons absorbed by GO. This confirms that the reduction of the Au ions is directly coupled with the absorption of the 532 nm photons by GO. Furthermore, the SPR band of Au nanoparticles shows a systematic red shift with irradiation time indicating the formation of larger or aggregated Au nanoparticles at longer irradiation times [60].

As both the Au ions and the sp^2 regions of GO compete for the photogenerated electrons with the Au ions being more efficient in capturing the electrons, the resulting Au nanoparticles are dispersed on PRGO. The partial reduction of GO in the presence of the Au ions is attributed to the efficiency of the Au ions in capturing the photogenerated electrons. It is also possible that the strong absorption of the 532 nm photons by the Au nanoparticles (because of the SPR $\sim$530 nm) [60, 61] decreases the probability of the two-photon absorption by GO. This would result in decreasing both the number of the photogenerated electrons needed for the reduction of GO as well as the photothermal energy conversion.

Figure 10.11 displays TEM images of the Au nanoparticles dispersed on the surface of the LCG graphene sheets formed by the 2 min laser irradiation of the GO solutions containing the same amount of $HAuCl_4$ in 50 vol% ethanol–water and 2 vol% PEG–water mixtures and in pure water. It is clear that the concentration of the Au nanoparticles formed in the GO-water solution is significantly lower than the concentration formed in the presence of ethanol or PEG. Again, this is a consequence of the favorable reducing environment created by ethanol or PEG as compared to water.

10.2.3.2 **Photoreduction of Silver Ions and GO in Different Solvents**

Similar to the formation of Au nanoparticles, irradiation of GO in water by the 532 nm laser in the presence of $AgNO_3$ results in the formation of Ag nanoparticles.

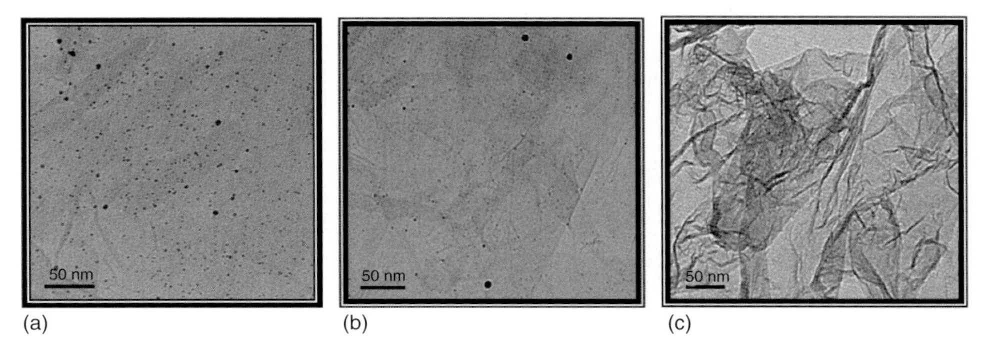

Figure 10.11 TEM images of Au nanoparticles deposited on partially reduced GO prepared by the 532 nm laser irradiation (4 W, 30 Hz) of the $HAuCl_4$–GO mixtures for 2 min in (a) 50% ethanol–water, (b) 2% PEG–water, and (c) pure water solvents [47].

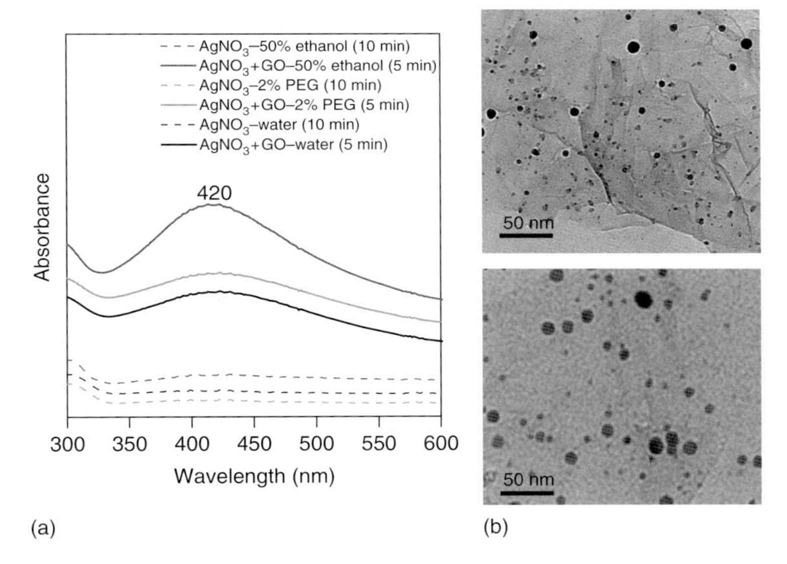

Figure 10.12 (a) Absorption spectra of $AgNO_3$–GO solutions in 50 vol% ethanol–water, 2 vol% PEG–water, and pure water recorded after 5 min laser irradiation (532 nm, 4 W, 30 Hz). Dotted lines represent data of blank solutions containing the same amount of $AgNO_3$ but no GO afte 10 min laser irradiation (532 nm, 4 W, 30 Hz). (b) TEM images of Ag nanoparticles incorporated within graphene sheets prepared by laser irradiation of $AgNO_3$–GO solutions in 50 vol% ethanol–water (top) and pure water (bottom) [47].

This is evident from the observation of the SPR band of Ag nanoparticles ~420 nm shown in Figure 10.12a [61] and the TEM images of the reduced graphene oxide-containing Ag nanoparticles shown in Figure 10.12b.

The XRD patterns of GO before and after laser irradiation in the presence of $AgNO_3$ in the 50 vol% ethanol–water mixture as well as in pure water show the disappearance of the GO peak at $2\theta = 10.9^\circ$. This result is different from the Au^{3+} experiments where only partial reduction of GO was observed as indicated by the incomplete disappearance of the GO peak following the irradiation of the $HAuCl_4$-GO solution.

10.2.3.3 Mechanism of Photocatalytic Reduction

Although fully oxidized GO is an insulator, partially oxidized GO has semiconductor properties with a band gap that is determined by the extent of oxygenation of graphite [24, 62–64]. This is due to the local changes in the carbon hybridization from sp^2 to sp^3 as oxidation of graphite sheets takes place. Band gap energies between 2.5 and 4.0 eV have been calculated for an O/C ratio >0.5 [63]. Absorption of a single 355 nm photon (3.5 eV) or two 532 nm photons by GO excites valence-to-conduction transitions (2.3 eV), which generates electron–hole pairs within the semiconductor GO [46, 65]. Since electron-lattice temperature equilibrium occurs on the picosecond timescale [66, 67], the resulting thermal energy is sufficient for the rapid removal of O_2, CO, CO_2, and H_2O from GO thus forming deoxygenated GO (or reduced GO) [46, 65, 68]. In addition to the photothermal effects, the generation of electron–hole pairs in GO in the presence of a hole scavenger molecule such as alcohol or glycol will be highly efficient in activating the reduction of GO. Furthermore, in the presence of metal ions, reduction of the metal ions could take place thus producing metal nanoparticles supported on the reduced or PRGO.

The reduction mechanism of the metal ions (M^+) appears to involve photogenerated electrons from GO. In the presence of a reducing environment provided by ethanol or PEG, the holes are scavenged to produce protons and reducing organic radicals. The electrons are used for the reduction of the metal ions, and as the alcohol radicals (C_2H_4OH) are strong reducing agents, they undergo oxidation to CH_3CHO and therefore reduce GO [56, 58, 59]. A possible mechanism can be summarized as follows [47, 49]:

$$GO + h\nu \rightarrow GO\ (h^+ + e^-) \tag{10.1}$$

$$h^+ + C_2H_5OH \rightarrow {}^\bullet C_2H_4OH + H^+ \tag{10.2}$$

$$M^+ + GO(e^-) \rightarrow GO + M \tag{10.3}$$

$$n{}^\bullet C_2H_4OH + GO \rightarrow RGO + n\,CH_3CHO + \frac{n}{2}\,H_2O \tag{10.4}$$

In the absence of a reducing environment as in pure water, the mechanism may be summarized as follows:

$$GO + h\nu \rightarrow GO\ (h^+ + e^-) \tag{10.5}$$

$$4\,h^+ + 2H_2O \rightarrow O_2 + 4H^+ \tag{10.6}$$

$$M^+ + GO\ (e^-) \rightarrow GO + M \tag{10.7}$$

$$GO + 4e^- + 4H^+ \rightarrow RGO + 2H_2O \tag{10.8}$$

The partial reduction of GO observed in the presence of Au ions suggests that the reduction of the Au ions takes place very rapidly (reaction shown in Equation 10.7

is more efficient than Equation 10.8), and as the resulting Au nanoparticles absorb the 532 nm photons much more efficiently than GO, the photothermal energy conversion by GO is decreased and complete reduction of GO is not possible.

The photocatalytic approach leads to the formation of metal nanocrystals dispersed on the reduced or PRGO surfaces without the use of chemical reducing or capping agents, which tend to significantly reduce the catalytic activity and poison the nanoparticle catalysts.

10.3 Pd/Graphene Heterogeneous Catalysts for Carbon–Carbon Cross-Coupling Reactions

Metal-catalyzed carbon-carbon bond forming reactions have rapidly become one of the most effective tools in organic synthesis for the assembly of highly functionalized molecules [69–73]. The development of palladium-catalyzed cross-coupling reactions represents one of the most significant advancements in contemporary organic synthesis. This area of chemistry has increased the accessibility to molecules of greater chemical complexity, particularly in the area of pharmaceutical drug discovery and development [74, 75]. In 2010, Richard Heck, Ei-ichi Negishi, and Akira Suzuki received the Nobel Prize in Chemistry for their groundbreaking work in cross-coupling catalysis [76]. These reactions have typically been carried out under homogeneous reaction conditions, which require the use of ligands to solubilize the catalyst and broaden its window of reactivity [75, 77]. However, the use of these catalysts under homogeneous conditions has limited their commercial viability because of product contamination as a direct result of inability to effectively separate the catalyst from the reaction product [74, 78–80]. The issue of product contamination is of particular importance in pharmaceutical applications where this chemistry is practiced extensively. Ligand-free heterogeneous palladium catalysis presents a promising option to address this problem as evidenced by the significant increase in research activity in this area [10, 26, 49, 80–85].

10.3.1 Pd/Graphene Catalysts Prepared by Microwave-Assisted Chemical Reduction of GO

The synthesis of CCG sheets from exfoliated graphene oxide using HH as a reducing agent and microwave heating in an aqueous medium allows for concurrent metal reduction and deposition onto the graphene surface [24, 26]. This procedure generates highly uniform metal nanoparticles supported on the CCG sheets [26]. This is particularly true when palladium is used as the metal in the deposition process as illustrated in Figure 10.13. This palladium/graphene catalyst has been shown to possess remarkable catalytic activity in Suzuki cross-coupling reactions (Reaction 1) with reported turnover frequencies (TOFs) of 108 000 h^{-1} [26]. Furthermore, these reactions are complete in approximately 20 min at room

$$\text{PhBr} + \text{PhB(OH)}_2 \xrightarrow[\text{K}_2\text{CO}_3,\ \text{EtOH/–H}_2\text{O (1:1)},\ \text{Room temp, 30 min.}]{\text{Pd/graphene (0.3 mole \%)}} \text{Ph–Ph} \quad (1)$$

94% Isolated yield

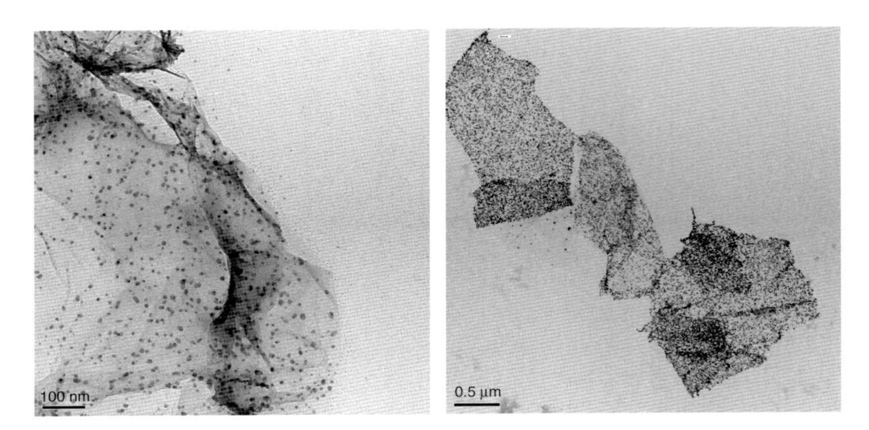

Figure 10.13 TEM images of palladium/graphene prepared by microwave irradiation of a mixture of graphene oxide and palladium nitrate in the presence of hydrazine hydrate. Notice the homogeneous distribution of the Pd nanoparticles on very large (several microns) graphene sheets [26].

temperature with quantitative conversions [26]. In contrast, when the same reaction was carried out under homogeneous conditions with palladium acetate and triphenylphosphine in refluxing acetonitrile, the reaction was required to run for a minimum of 8 h in order to obtain an 82% isolated yield [26]. The Pd/graphene catalysts could also be easily recovered and recycled under batch reaction conditions [26]. In contrast, conventional palladium-catalyzed homogeneous cross-coupling reactions must be heated to elevated temperatures for several hours in order to achieve comparable results [26].

10.3.1.1 Catalytic Activity and Range of Utility

The Pd/G catalyst system was also evaluated to determine the breadth of catalytic activity by screening a series of Suzuki cross-coupling reactions containing a broad range of chemical functionality as illustrated in Table 10.1 [26]. In this context, a broad range of aryl bromide containing electron-donating (**1d**, **e**) and electron-withdrawing groups (**1a–c**) were effectively incorporated in the coupling products. In addition, phenyl boronic acids bearing useful functionality such as 4-dimethylamino (**1c**, **g**), 4-amino carbonyl (**1d**, **f**), or 4-thiomethyl (**1b**) all led to high yield of Suzuki products. Even a more difficult substrate such as 4-nitro-1-chlorobenzene was able to undergo the Suzuki coupling with phenyl boronic acid in good yield (**1h**).

Table 10.1 Evaluation of the Pd/graphene catalyst for Suzuki reaction[a] [26].

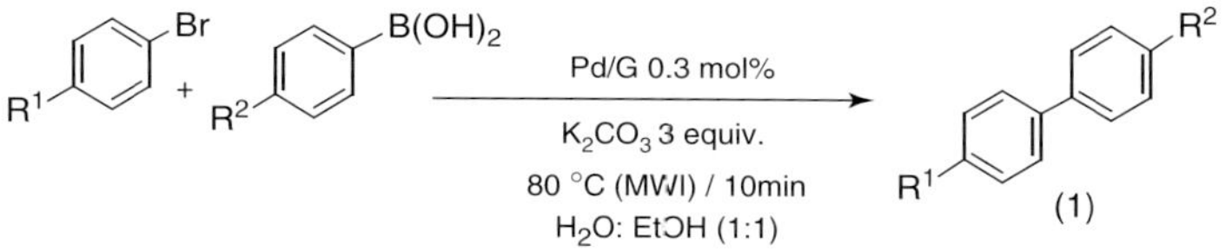

R^1–Br + R^2–$B(OH)_2$ → (Pd/G 0.3 mol%; K_2CO_3 3 equiv.; 80 °C (MWI) / 10min; H_2O: EtOH (1:1)) → R^1–R^2 (1)

cpd	Aryl halide	Boronic acid	I (%)[c]
a[b]	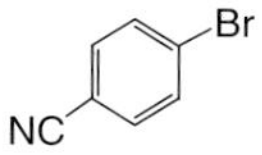 NC, Br	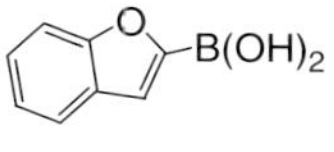 O, $B(OH)_2$	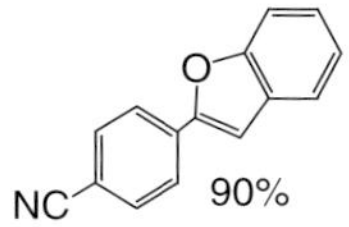 NC, O; 90%
b	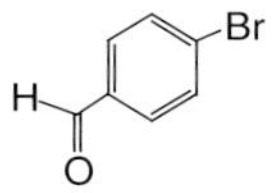 H, O, Br	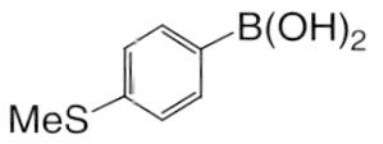 MeS, $B(OH)_2$	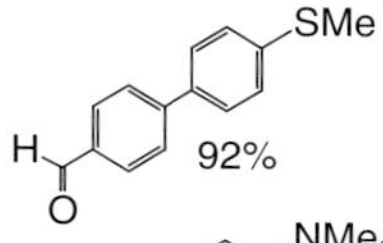 H, O, SMe; 92%
c	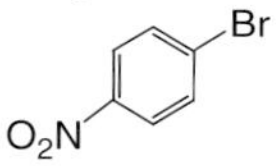 O_2N, Br	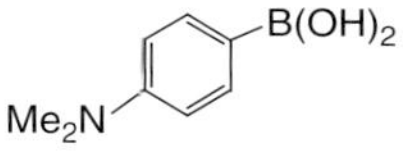 Me_2N, $B(OH)_2$	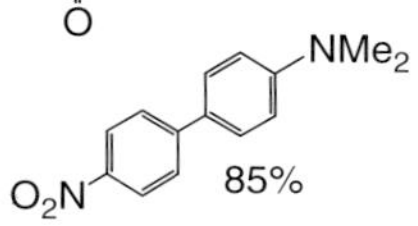 O_2N, NMe_2; 85%
d	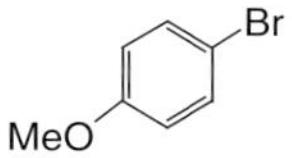 MeO, Br	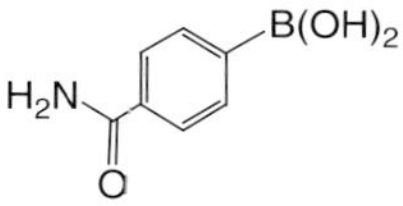 H_2N, O, $B(OH)_2$	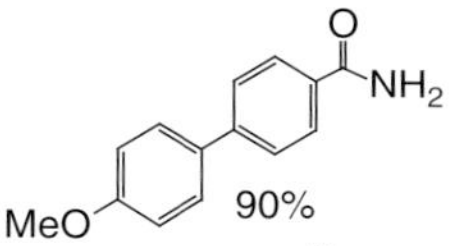 MeO, O, NH_2; 90%
e[b]	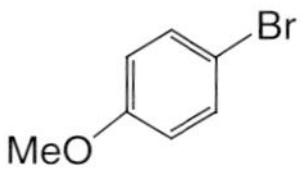 MeO, Br	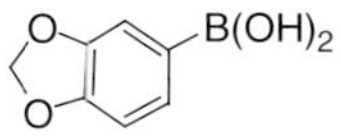 O, O, $B(OH)_2$	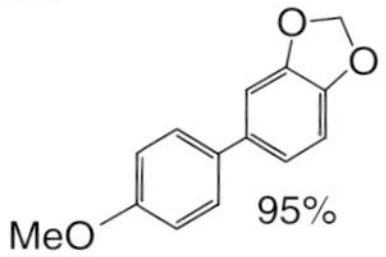 MeO, O, O; 95%
f	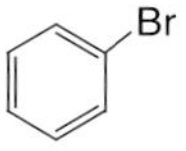 Br	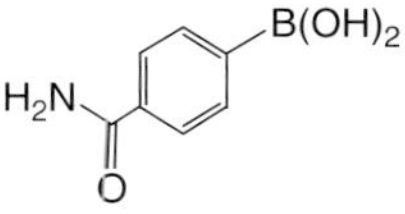 H_2N, O, $B(OH)_2$	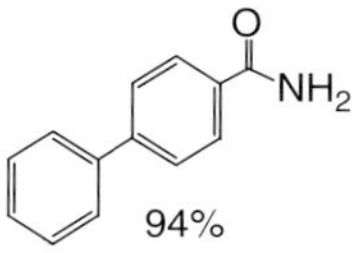 O, NH_2; 94%
g	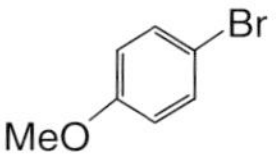 MeO, Br	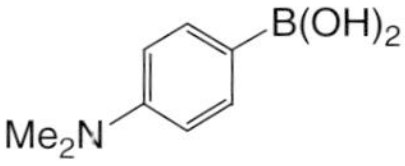 Me_2N, $B(OH)_2$	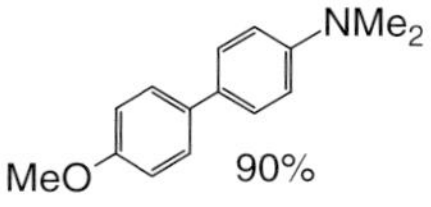 MeO, NMe_2; 90%
h	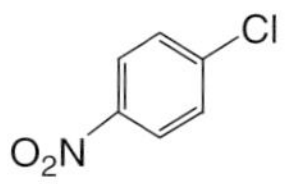 O_2N, Cl	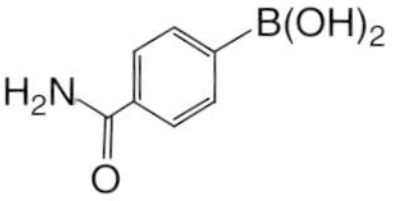 H_2N, O, $B(OH)_2$	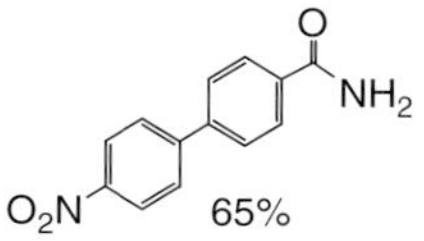 O_2N, O, NH_2; 65%

[a] Aryl halide (0.51 mmol), boronic acid (0.61 mmol, 1.2 equiv), potassium carbonate (212 mg, 1.53 mmol, 3 equiv), and Pd/G (2.1 mg, 1.53 μmol, 0.3 mol%) in 8 ml H_2O/EtOH (1 : 1) was heated at 80 °C (MWI) for 10 min.

[b] Reactions were completed at room temperature after 30 min.

[c] Isolated yields.

10.3.1.2 Catalyst Recyclability

An important element of the development of a new catalyst is to demonstrate the ability to recycle the catalyst system. The Pd/graphene nanoparticles could be easily recycled for up to eight times while achieving a quantitative conversion to the product [26]. The activity of the catalyst dropped in run nine showing only a 62% conversion. Evidence for deactivation of the Pd/G catalyst was obtained from the TEM image after the 10th run, which clearly demonstrates agglomeration of the Pd nanoparticles on the surface of graphene [26]. This result indicates that the mechanism of deactivation of the catalyst is likely to involve the formation of aggregated Pd nanoparticles, which leads to the decrease of the surface area and saturation of the coordination sites. Although there is evidence for a leaching/redeposition mechanism with other solid-supported palladium catalysts, it is evident that the solid-state mechanisms in this area are not yet fully understood [86, 87]. However, it is apparent that graphene is an excellent support system for palladium in cross-coupling applications [26, 49]. The unique structure and the presence of C=C bonds along with the defect sites within the graphene matrix may play an important role in the enhanced catalytic performance. The electron-rich labile support system appears to stabilize the metal center, impede catalytic deactivation by agglomeration, and increase the cross-coupling catalytic activity [26, 49].

The extraordinary catalytic activity of this palladium/graphene system is believed to be derived from a combination of the preparative method, which yields uniform palladium nanoparticles of high surface area along with the unique structural properties of the graphene support system [26, 49]. In addition to the large surface area and thermal stability of graphene, the structural defects formed during the synthesis of the Pd/graphene catalyst could lead to a stronger catalyst–support interaction, which may impede particles' migration and thus contribute to the enhanced catalytic performance.

The presence of native as well as physically and chemically induced defects in graphene and metal-graphene systems have been predicted theoretically and observed by high-resolution transmission electron microscopy [88–92]. The simple intrinsic defect types in graphene include single and double vacancies due to the absence of one and two lattice atoms, respectively [89]. However, the removal of a large number of atoms leads to the formation of holes with unsaturated bonds around their circumferences as observed in electron microscopy experiments [92]. Although one of the unique properties of graphene is its ability to reconstruct its lattice around intrinsic defects, the presence of metal atoms and nanoparticles prevents this occurrence [89, 92].

One of the most efficient ways of controlling graphene defects is by controlling the reduction of graphene oxide (GO) because GO is essentially a highly defective graphene sheet functionalized with oxygen atoms [24, 26, 49, 91, 93]. Through the control of the GO reduction process using different chemical and physical methods, the sp^2 domains of graphene can be partially restored and the remaining structural defects can be used for the development of strong interaction sites with the metal nanoparticle catalysts. *The defect sites on the surface of the graphene provide an excellent electronic environment for metal deposition and as a result, play a major role in imparting these exceptional catalytic properties.*

10.3.2 Pd/PRGO Catalysts Prepared by Laser Partial Reduction of GO

Photochemical and photothermal reduction methods to prepare metal nanoparticles supported on graphene have been developed without a need for chemical reducing agents thus providing a green approach for the synthesis and processing of metal-graphene nanocomposites [47–49, 56, 65, 68, 94, 95]. Other advantages of photochemical reactions include the possibility of creating structural defects on the graphene nanosheets which could lead to a stronger catalyst–support interaction by impeding particles' migration and thus contributing to an enhanced catalytic performance. Through the control of the GO photoreduction process, the sp^2 domains of graphene can be partially restored and the remaining structural defects can be used for the development of strong interaction sites with the metal nanoparticle catalysts [46–49].

10.3.2.1 Laser Synthesis of Pd Nanoparticles on Structural Defects in Graphene

As in any other catalyst support system, defects can be introduced into graphene deliberately, for example, by chemical treatments as shown in Section 10.3.1. Another approach to create structural defects is through the laser-induced photochemical reduction of graphene oxide as demonstrated recently [46–49]. This method provides a solution-based synthesis of individual graphene sheets in water at ambient conditions without the use of any chemical reducing agent [46].

Using the 532 and 355 nm YAG laser irradiation of mixtures of GO and palladium nitrate in water, Pd nanoparticles supported on the defects generated in the reduced or PRGO can be produced [47–49]. The defects in reduced GO are formed through the reactions of the photogenerated holes (h^+) and electrons (e^-) with the oxygen functional groups in the GO lattice as, for example, with the epoxy group according to

$$\text{C–O–C (GO)} + H_2O + 2h^+ \rightarrow \text{C (defect carbon)} + CO_2 + 2H^+ \qquad (10.9)$$

$$\text{C–O–C (GO)} + 2H^+ + 2e^- \rightarrow \text{C–C (defect carbon)} + H_2O \qquad (10.10)$$

Three Pd nanoparticle catalysts supported on PRGO nanosheets (Pd/PRGO) were prepared by laser irradiation of GO in different solvent environments [49]. Although the Pd contents in the initial solutions ($Pd(NO_3)_2$ + GO in water, 50% ethanol–water, or 50% methanol–water) are all similar (40 wt%), the amount of Pd deposited in each catalyst following the laser irradiation process is found to be different depending on the solvent environment [49]. The Pd/PRGO catalyst prepared in pure water (catalyst A) has the lowest Pd content (18.2 wt%) followed by catalyst B prepared in 50% ethanol–water (24.3 wt%) and then catalyst C prepared in 50% methanol–water (29.4 wt%). The Pd content in the Pd/PRGO catalysts appears to increase with increasing the reducing environment of the solvent used. This also correlates with the laser irradiation times (under similar conditions: 20 μl

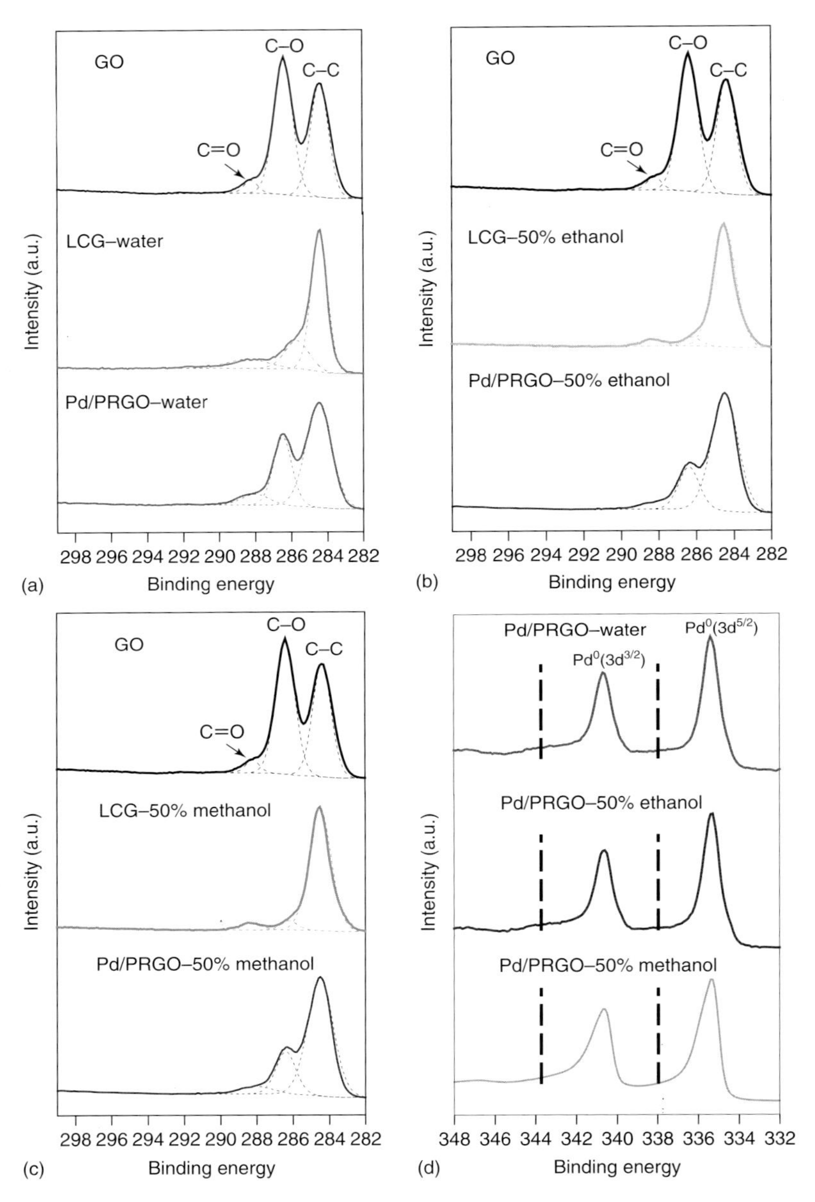

Figure 10.14 Comparisons of the XPS (C1s) binding energies of GO and that of the laser-converted graphene (LCG) and Pd/PRGO prepared by the 532 nm laser irradiation (5 W, 30 Hz) of GO (0.2 mg ml^{-1}) or GO + $Pd(NO_3)_2$ in (a) water, (b) 50% ethanol–water, and (c) 50% methanol–water. (d) XPS (Pd-3d) binding energies of the Pd/PRGO catalysts prepared in different solvents as indicated. Dashed lines show the locations of the 3d electron binding energies of Pd^{2+} [49].

$Pd(NO_3)_2$ in 6 ml of GO solution (2 mg GO/10 ml water), 532 nm laser, 5 W, 30 Hz) required for the reduction of the Pd ions and the partial reduction of GO (9, 5, and 3.5 min in pure water, 50% ethanol–water and 50% methanol–water, respectively).

Figure 10.14 displays the XPS spectra of the Pd-3d electron in the Pd/PRGO catalysts prepared in water, 50% ethanol–water, and 50% methanol–water solvent systems. The data show that most of the Pd nanoparticles are present as Pd(0) consistent with the observed binding energies of 335 eV (Pd(0) $3d^{5/2}$) and 341.1 eV (Pd(0) $3d^{3/2}$) [96]. As shown in Figure 10.14, the reference binding energies of the Pd(II)-3d electrons are significantly higher than the observed Pd(0) values [96].

Figure 10.15 displays representative TEM images of the Pd/PRGO catalysts (A–C) prepared in pure water, 50% ethanol–water, and 50% methanol–water, respectively, as described earlier. From the measured particle size distributions, catalyst A prepared in pure water has the smallest Pd nanoparticles with an average size of 7.8 ± 3.5 nm followed by catalyst B (prepared in 50% ethanol–water mixture) with an average Pd particle size of 10.8 ± 4.1 nm, and then catalyst C (prepared in 50% methanol–water mixture) where the average particle size is 14.8 ± 4.2 nm [49]. It is interesting to note that catalyst A contains a significant number of very small Pd nanoparticles in the size range of 1–2 nm as shown in the TEM image of catalyst A. The TEM results are consistent with both the inductively coupled plasma mass spectrometry (ICP-MS) data and the laser irradiation times, which together confirm the general trend that increasing the reducing environment of the solvent during the laser irradiation of the GO-$Pd(NO_3)_2$ mixture results in a rapid reduction of the Pd ions and an increase in the growth rate of the resulting Pd nanoparticles, and consequently an increase in the average size and the degree of agglomeration of the final Pd nanoparticles supported on the PRGO nanosheets [49].

10.3.2.2 Mechanism of Partial Reduction of GO and Defect Generation

The above results provide direct evidence for the formation of Pd nanoparticles on PRGO following the 532 nm laser irradiation of the GO-$Pd(NO_3)_2$ mixture in

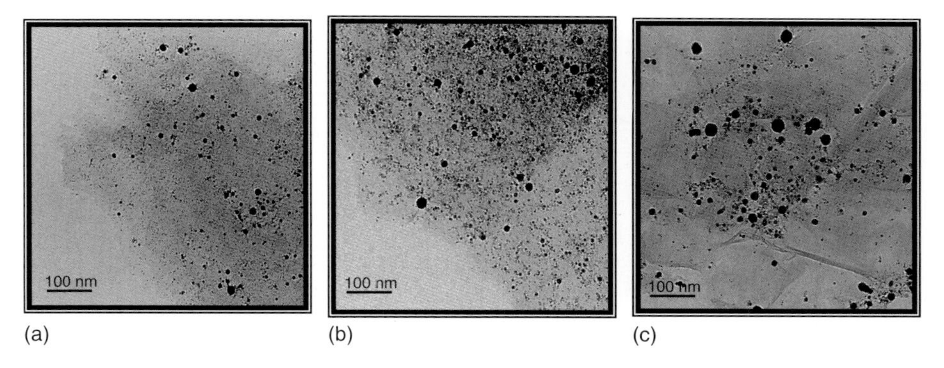

Figure 10.15 TEM images and particle size distributions of the Pd/PRGO catalysts (a–c) prepared in pure water, 50% ethanol–water, and 50% methanol–water, respectively [49].

different solvent environments. As indicated earlier, partially oxidized GO contains a mixture of hydrophobic π-conjugated sp^2 and hydrophilic oxygen-containing sp^3 domains, which lead to semiconductor properties with a band gap that depends on the extent of oxygenation of graphite (2.5–4.0 eV depending on the O/C ratio) [62–64]. Absorption of two 532 nm photons generates an electron–hole pair ($e^- - h^+$) within the semiconductor GO (Equation 10.1) [46, 91]. The photogenerated electrons in GO are used for the reduction of the Pd^{2+} ions (Equation 10.3) simultaneously with partial reduction of GO depending on the solvent environment. In the presence of a reducing environment provided by ethanol, the holes are scavenged to produce protons and reducing radicals (Equation 10.2). As the $^\bullet C_2H_4OH$ radicals are strong reducing agents, they undergo oxidation to CH_3CHO and therefore reduce GO (Equation 10.4) [56, 59, 65]. This is consistent with increasing the degree of reduction of GO in the presence of alcohols. In the presence of methanol, CO_2 and H_2 gases can be produced according to Equations (10.11) and (10.12); respectively [59].

$$6H^+ + CH_3OH + 6OH^- \rightarrow CO_2 + 5H_2O \quad (10.11)$$

$$6H^+ + 6e^- \rightarrow 3H_2 \quad (10.12)$$

In the absence of a reducing environment as in pure water, the photogenerated holes can react with the oxygen-containing functional groups of GO and water to produce CO_2 and protons and result in the formation of carbon defect sites in GO (Equations 10.9 and 10.13) [91]. The electrons and protons remove OH or oxygen groups from the GO to form H_2O (Equations 10.10 and 10.14). These reactions result in the significant number of carbon defect sites in GO that are considered important active sites for the production of H_2 [91]. Equations 10.13 and 10.14 describe the formation of carbon defect sites through the photoreaction of water with the hydroxyl groups of GO [91].

For the hydroxyl group in GO,

$$\text{C–C–OH (GO)} + H_2O + 3H^+ \rightarrow \text{C (defect carbon)} + CO_2 + 3H^+ \quad (10.13)$$

$$\text{C–C–OH (GO)} + H^+ + e^- \rightarrow \text{C–C (defect carbon)} + H_2O \quad (10.14)$$

For the epoxy group in GO,

$$\text{C–O–C (GO)} + H_2O + 2h^+ \rightarrow \text{C (defect carbon)} + CO_2 + 2H^+ \quad (10.15)$$

$$\text{C–O–C (GO)} + 2H^+ + 2e^- \rightarrow \text{C–C (defect carbon)} + H_2O \quad (10.16)$$

10.3.2.3 **Application of Pd/PRGO Nanocatalysts to Suzuki Reaction**

The catalytic activities of the Pd/PRGO nanocatalysts A–C produced by the 532 nm laser irradiation of $Pd(NO_3)_2$–GO mixtures in water, 50% ethanol, and 50% methanol as solvents, respectively, were evaluated for cross-coupling reactions [49]. Figure 10.16 illustrates the catalytic activity of the three catalysts at different catalyst loadings using the Suzuki cross-coupling reaction of bromobenzene and phenyl boronic acid in a mixture of H_2O/EtOH (1 : 1) at room temperature (Reaction 1). As shown in Figure 10.16, at 0.5 mol% concentration, catalysts A and B show

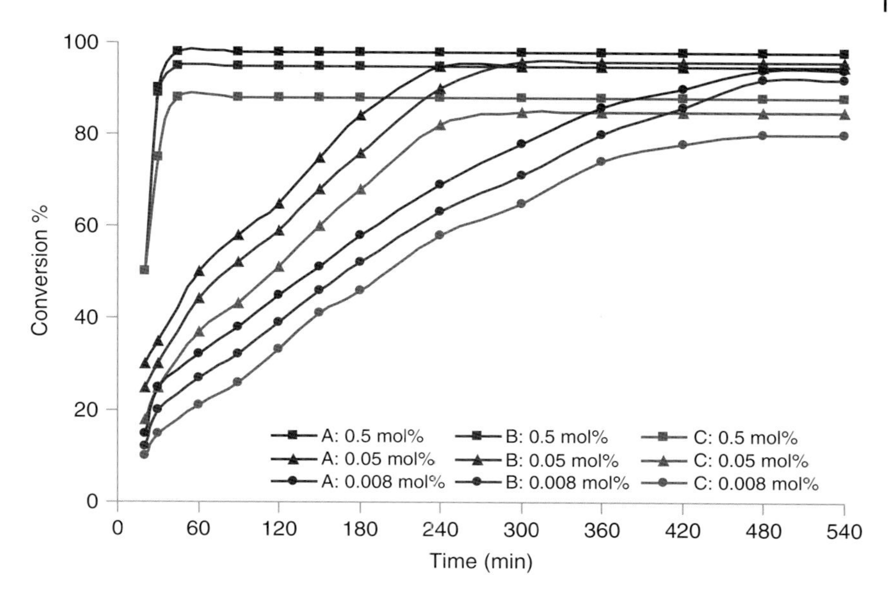

Figure 10.16 Effect of concentration of the Pd/PRGO catalysts A–C on the conversion of the Suzuki reaction [49].

similar reactivity and result in a high conversion of 100 and 95%, respectively, after 45 min reaction time at room temperature. Catalyst C shows only 88% conversion within 45 min under these conditions. Similarly, at lower catalyst loading of 0.05 mol%, catalysts A and B work effectively, giving a conversion of 95 and 90%, respectively, after 4 h at room temperature. Catalyst C gives only 82% yield under these conditions. A similar trend is observed when the amount of catalyst loading is decreased to 0.008 mol% in which catalysts A–C result in 95, 92, and 80% conversions, respectively, after 8 h at room temperature. Interestingly, when the same Pd loading of 0.008 mol% is applied at 120 °C under MWI, the reaction is surprisingly fast for catalyst A, converting 62% of bromobenzene after 2 min. Continuing the reaction under the same conditions (120 °C, MWI) leads to complete formation of the biphenyl (100%) product after 5 min. It should be noted that at this very low catalyst loading (0.008 mol%), higher microwave heating of 120 °C was required to afford a full conversion to the Suzuki product. The 100% conversion of the 0.008 mol% catalyst A indicates a remarkable turnover number (TON) of 7800 and a TOF of 230 000 h^{-1} for a microwave-assisted Suzuki cross-coupling reaction at 120 °C. However, with 0.008 mol% of catalyst B, the Suzuki reaction of bromobenzene and phenyl boronic acid results in a 45% formation of the biphenyl product after 2 min at 120 °C under MWI. This reaction is also completed after 5 min under MWI at 120 °C. This provides a TON of 5600 and a TOF of 169 000 h^{-1} for the Pd/PRGO catalyst B for a microwave-assisted Suzuki cross-coupling reaction at 120 °C.

It is interesting to note that when a similar Suzuki cross-coupling reaction (bromobenzene and phenyl boronic acid with potassium carbonate in a mixture of H_2O/EtOH (1 : 1) at 120 °C) was performed under conventional thermal heating using 0.008 mol% of catalyst A, only a small conversion of 5% was observed after 6 h refluxing conditions at 120 °C. This result clearly demonstrates the effect of MWI in increasing reaction rates by providing a direct and rapid heating source for the cross-coupling reactions [26]. This observation is consistent with recent reports on the selective heating of the surface of the heterogeneous catalysts and the formation of hot spots by MWI, which can lead to nonequilibrium local heating at the surface of the metal nanoparticle catalysts [97–99].

The activity and recyclability of catalyst A are truly remarkable considering that the ICP-MS analysis indicates that this catalyst (prepared in pure water) has the lowest Pd content (18.2 wt%) as compared to catalyst B (prepared in 50% ethanol–water, 24.3 wt% Pd) and catalyst C (prepared in 50% methanol–water, 29.4 wt% Pd). However, the high activity of catalyst A correlates nicely with the smallest average size of the Pd nanoparticles (7.8 ± 3.5 nm) as compared to catalyst B (10.8 ± 4.1 nm) and catalyst C (14.8 ± 4.2 nm) as shown in Figure 10.15. Catalyst A also exhibits well-dispersed Pd nanoparticles on the PRGO nanosheets with no evidence of significant particles' agglomeration as shown in the TEM images of Figure 10.15. As discussed earlier, increasing the reducing environment of the solvent during the laser irradiation of the GO–$Pd(NO_3)_2$ mixture results in a rapid reduction of the Pd ions and an increase in the growth rate of the resulting Pd nanoparticles, and consequently an increase in the average size and the degree of agglomeration of the final Pd nanoparticles supported on the PRGO nanosheets as clearly demonstrated in catalyst C. We also believe that the reduction mechanism discussed earlier plays an important role in the remarkable activity and reusability of catalyst A. Specifically, the reduction of the oxygen-containing functional groups of GO into carbon defect sites as a result of the reactions of water with the photoexcited GO appears to play a major role in imparting the exceptional catalytic properties to catalyst A. These defect sites can provide favorable structural environment and possible electronic interaction for anchoring the palladium nanoparticles on the surface of the PRGO nanosheets.

10.3.2.4 **Recyclability of the Pd/PRGO Nanocatalysts in Suzuki Reaction**

A significant practical application of heterogeneous catalysis is in the ability to easily remove the catalyst from the reaction mixture and reuse it for subsequent reactions until the catalyst is sufficiently deactivated. The recyclability of the Pd/PRGO nanocatalysts A and B was examined using the Suzuki cross-coupling reaction of bromobenzene and phenyl boronic acid at 80 °C under microwave heating using 0.5 mol% of catalyst loading. For catalyst A, the completion of reaction was accomplished after 5 min for the first four runs. The catalytic activity was dropped to 60% in fifth and sixth runs, 50% in run seven, and then significantly dropped below 50% for the next recycling reactions. For catalyst B, full conversions were obtained in the first two runs. The activity slightly dropped in the third run, yielding 86%

conversion. The catalytic activity was further decreased in runs four and five–63 and 50% conversions, respectively. Lower conversions of below 50% were achieved after the fifth run. In contrast to catalysts A and B, no effective recyclability was found for catalyst C. This catalyst gave a 100% conversion only for the first run, and activity was then diminished to 65 and 54% and even lower for the subsequent runs [49].

It is instructive to compare the performance of the current Pd/PRGO catalysts with other Pd nanoparticle catalysts developed for cross-coupling reactions. There are many examples in the literature of Pd nanoparticles stabilized by capped polymeric materials for cross-coupling reactions, and in particular the Suzuki reactions [81–85, 100–103]. These include Pd nanoparticles stabilized by polyvinyl pyrrolidone (PVP), and some examples of using polystyrene-polyethylene oxide (PS-PEO), and poly(amido-amine) (PAMAM) dendrimers as well as other polymers [103]. Although these nanocatalysts demonstrated reactivity toward the Suzuki reactions, most catalysts have shown some limited catalytic activity such as lower product yields using low catalyst loadings, lack of efficient recyclability, and low TON and TOF [103]. For example, the yields of the biphenyl products using the Pd-PVP and Pd-MAMAM nanoparticles were found to be 39 and 34%, respectively, for the first cycle of the reaction and significantly lower yields for the second cycle [84]. In addition, it has been found that in case of Pd supported on PVP, increasing the amount of the PVP stabilizer resulted in decreasing the catalyst activity in the Suzuki reactions because of the capping effect of the PVP which reduces the access to the active coordination sites on the metal centers for the catalysis [83]. Another example also includes the Pd/PVP catalysts with very low catalyst loadings (about 0.005–0.01 mol%) under microwave reaction conditions, which resulted in 75–79 and 43–87% yields for aryl iodides and bromides, respectively [101]. Recently, microporous polymers have been used as both stabilizer and support for the Pd nanoparticles and resulted in highly recyclable catalysts for the Suzuki coupling of bromoacetophenone and phenyl boronic acid (up to nine times using 0.3 mol% catalyst loading) [102]. On the basis of these and other examples discussed in Ref. [103], it can be concluded that the current Pd/PRGO nanoparticles provide highly stable and efficient catalysts for cross-coupling reactions as compared to other Pd nanoparticles stabilized by polymeric molecules [103].

10.3.2.5 **Applications of the Pd/PRGO Catalyst A to Heck and Sonogashira Reactions**

The utility of laser-synthesized Pd/PRGO catalyst A was further evaluated for other carbon-carbon bond forming processes such as the Heck and Sonogashira reactions [73, 77]. While these reactions are typically carried out in the presence of a homogeneous palladium catalyst and appropriate ligands, the Pd/PRGO catalysts were examined under ligandless MWI conditions [49]. Thus, the reactions of iodobenzene with styrene or phenyl acetylene, and potassium carbonate in H_2O/EtOH solvents were investigated in the presence of 0.5 mol% catalyst A, under microwave heating at 180 °C for 10 min. As shown in Scheme 10.1, both these reactions can be successfully performed under these conditions, providing a

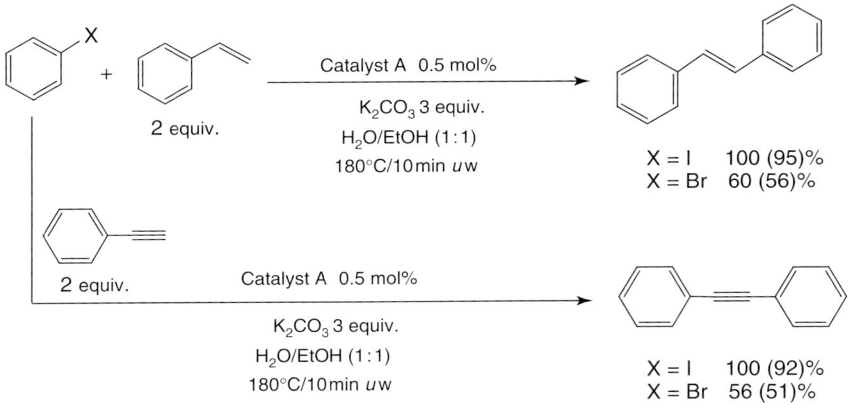

X
+
2 equiv.
Catalyst A 0.5 mol%
K_2CO_3 3 equiv.
H_2O/EtOH (1 : 1)
180°C/10 min uw
X = I 100 (95)%
X = Br 60 (56)%

2 equiv.
Catalyst A 0.5 mol%
K_2CO_3 3 equiv.
H_2O/EtOH (1 : 1)
180°C/10 min uw
X = I 100 (92)%
X = Br 56 (51)%

Scheme 10.1 The Heck and Sonogashira coupling reaction using the Pd/PRGO catalyst A.

complete conversions of 100% and high isolated yields of 95 and 92% of the corresponding Heck and Sonoagashira products, respectively. Utilizing bromobenzene as the starting materials afforded lower conversions of 60 and 56% for the Heck and Sonogashira reactions, respectively. With a more difficult substrate such as 4-nitro-1-chlorobenzene, however, there were only 10–15% conversions for both the Heck and Sonogashira reactions.

10.4 CO Oxidation by Transition-Metal/Metal-Oxide Nanoparticles Supported on Graphene

Using 532 or 355 nm laser irradiation of a mixture of graphene oxide and metal ion precursors in water, we reported the synthesis of metal nanoparticle catalysts well dispersed and supported on reduced GO [47, 48]. The Pd/CoO composite nanoparticles exhibit significantly enhanced activity for CO oxidation over both the Pd and the CoO catalysts [48]. The simple synthesis method without chemical reducing agents, the formation of alloy nanoparticles with variable catalyst loading and composition, and the low cost of GO could enhance the viability of these supported catalysts.

The reduction of the metal ions and the formation of metal nanoparticles are confirmed by the XRD patterns of the resulting metal-graphene nanocomposites as shown in Figure 10.17 for the Pt/graphene and Pd/graphene nanocomposites. Both of the Pt and Pd samples showed a typical face-centered-cubic (fcc) pattern. It should be noted that laser irradiation of the same metal ion precursors in water under identical solution concentration and laser irradiation conditions but in the absence of GO does not result in the formation of metal nanoparticles. This indicates that the reduction of the metal ions is directly coupled to the absorption

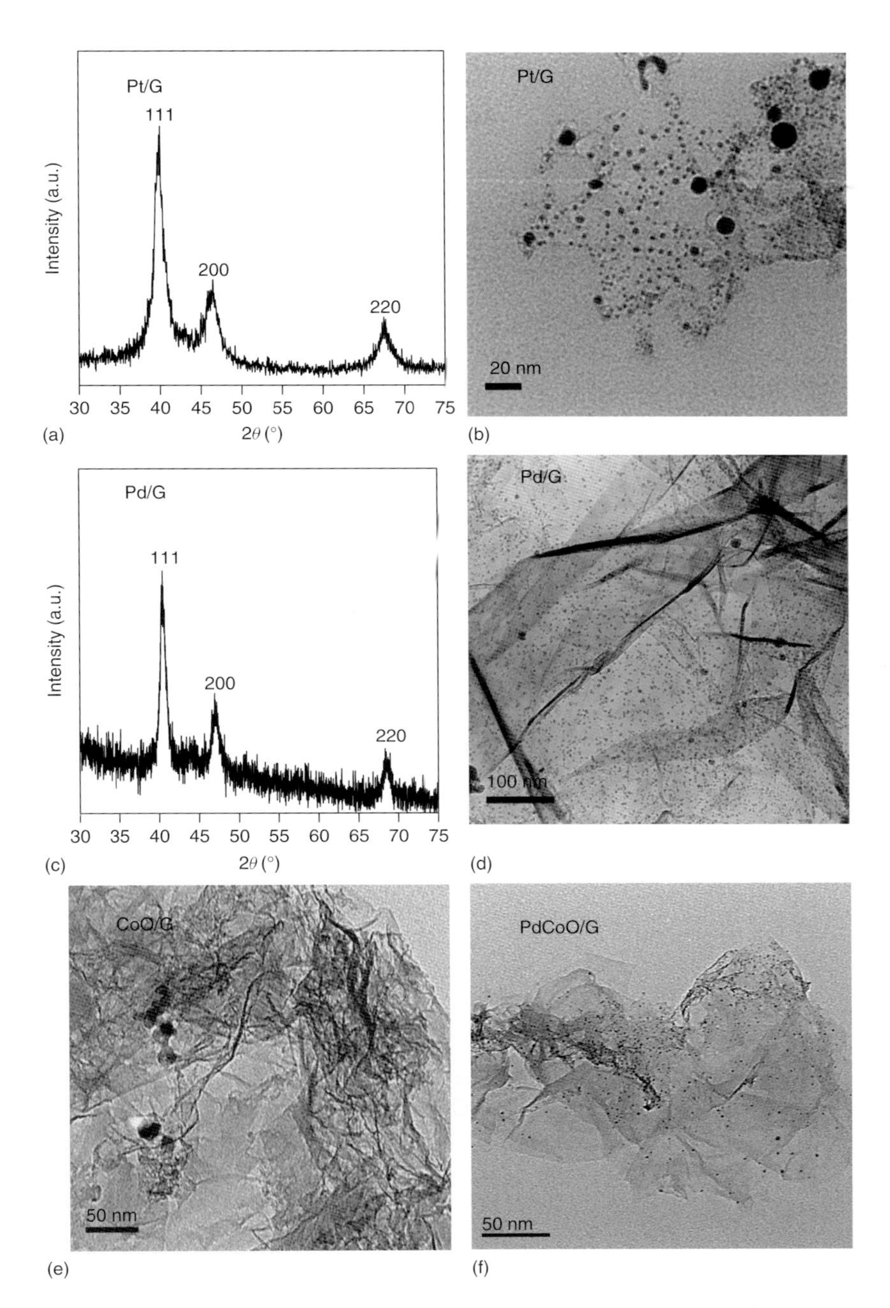

Figure 10.17 (a) XRD data and (b) TEM image of Pt/G. (c) XRD data and (d) TEM image of Pd/G. (e,f) TEM images of CoO/G and Pd-CoO/G, respectively [48].

of the 532 nm (two photons) or the 355 nm (one photon) light by GO, and thus laser excitation of GO must be involved in the reduction of these ions.

The TEM images of the Pt, Pd, CoO, and Pd-CoO nanoparticles dispersed on the surface of the reduced GO sheets show that most of the nanoparticles are in the size range of 5–7 nm, with only a few particles with diameters >10–12 nm. The percentage loading of the nanoparticles is determined by the concentration of the metal precursor in the laser irradiated GO solution. For the TEM images shown in Figure 10.17, the weight percentage of the metal loadings in the Pt/G, Pd/G, CoO/G, and Pd-CoO/G nanocomposites with respect to GO are 5, 2.5, 10, and 2.5–10, respectively.

The formation of Pd/CoO composite nanoparticles is confirmed by the high-resolution TEM and NANO-EDX data shown in Figure 10.18. Two samples containing 1.25 wt% Pd–10 wt% CoO and 2.5 wt% Pd–10 wt% CoO supported on reduced GO were examined by high-resolution TEM and EDX analysis with a probe beam of 0.5 nm and the results are shown in Figure 10.18. In the NANO-EDX measurements, the electron beam was focused on a number of particles. It was found that all particles consisted of Pd and Co and that more Pd was detected in the sample containing 2.5 wt% Pd–10 wt% CoO as evident from the corresponding EDX spectra shown in Figure 10.18.

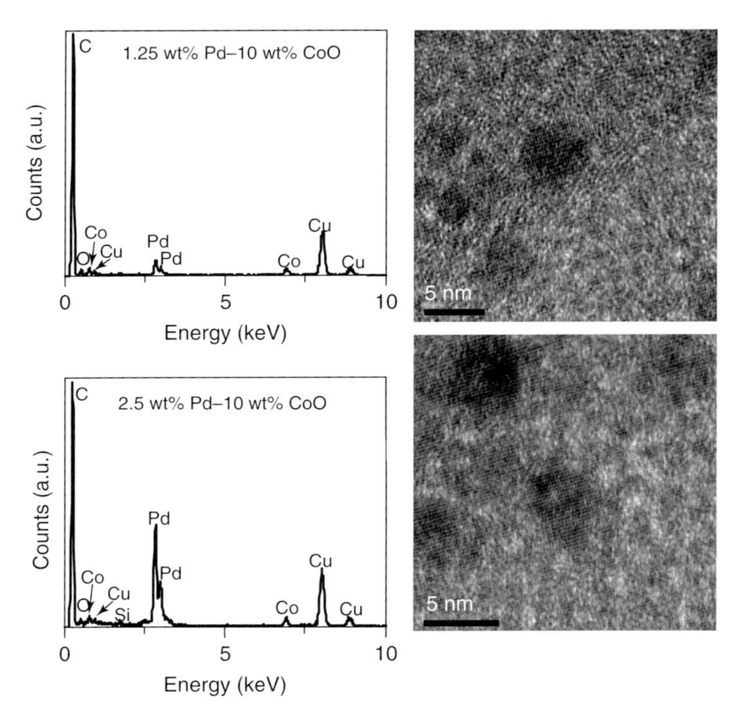

Figure 10.18 High-resolution TEM images and NANO-EDX spectra of the nanocomposites containing 1.25 wt% Pd–10 wt% CoO and 2.5 wt% Pd–10 wt% CoO supported on reduced GO [48].

It should be noted that the formation of PdCo alloy nanoparticles is thermodynamically favorable, as these metals have the same fcc crystal structure with similar lattice constants. The alloy nanoparticles are likely to be formed by atom substitution of one metal for the other through diffusion processes in the solution of the reduced binary metal salts in the presence of the GO solution.

The Pd/G, CoO/G, and Pd-CoO/G nanocomposites were tested for the catalytic oxidation of CO using a flow reactor. Figure 10.19 compares the catalytic oxidation of CO over the 2.5 wt% Pd/G, 10 wt%CoO/G, 1.25 wt% Pd–10 wt%CoO/G, and 2.5 wt% Pd–10 wt%CoO/G nanocomposites and the LCG reference sample. It is clear that the LCG sample does not show any significant activity for CO oxidation, and therefore, it can be regarded as an inert support. It is also clear that the activity of the Pd/G catalyst is significantly higher than that of the

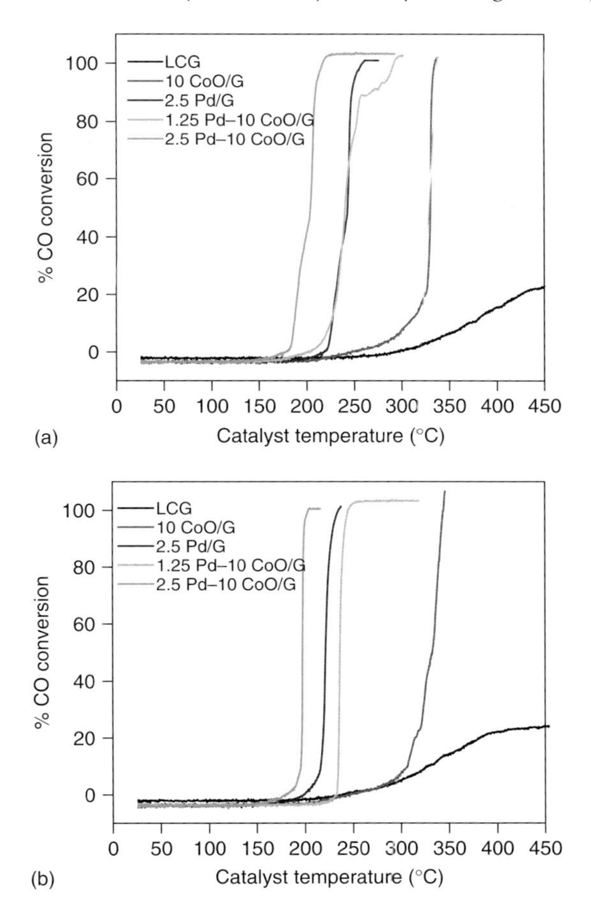

Figure 10.19 CO oxidation as a function of temperature for the reactant gas mixture containing 4.0 wt% CO and 20.0 wt% O_2 in helium (20 mg catalyst, flow rate = 100 cc min^{-1}) for catalysts prepared using (a) 355 nm and (b) 532 nm laser irradiation of GO aqueous solutions containing metal ions as described in the experimental section of Ref. [48].

CoO/G catalyst consistent with the well-established high activity of supported Pd nanoparticles for CO oxidation [30, 104–108]. Although the performance of the 2.5 wt% Pd/G nanocatalyst is better than that of the 10 wt%CoO/G catalyst, the nanocomposite containing 2.5 wt% Pd–10 wt%CoO/G exhibits significantly higher activity as compared to both the 2.5 wt% Pd/G and the 10 wt% CoO/G catalysts. This is an interesting result, which could be attributed to the formation of alloy nanoparticles containing Pd and CoO where CO and O_2 molecules are adsorbed on the Pd and CoO sites, respectively, thus providing a favorable catalytic pathway for the formation of CO_2. This trend has been observed in alloy nanoparticles; and electronic and surface compositions have been suggested as the origin of the catalytic enhancement in alloy nanoparticles [30, 104–106]. In the present system, the formation of CoO results in a stronger interaction with the Pd atoms as compared to the interaction of Pd nanoparticles with GO. It is well known that metal oxide surfaces provide strong metal–support interaction for improving the catalytic oxidation of CO over metal nanoparticles [30, 104–108]. For example, a 5 wt% Pd/CeO_2 nanoparticle catalyst prepared in an aqueous medium shows a 100% conversion of CO to CO_2 at 151 °C [107], in comparison with the current 2.5 wt% Pd–10 wt%CoO/G catalyst (prepared by the 532 nm laser irradiation), which exhibits a 100% CO oxidation at 195 °C. However, the Pd-CoO/G catalyst exhibits higher activity than the Pd nanoparticle catalyst supported over CeO_2–TiO_2 mixed oxide prepared by two steps involving sol–gel and incipient impregnation methods (100% conversion at ~220 °C) [108]. Similarly, the Pd-CoO/G catalyst has better performance than a PdAu alloy nanoparticle catalyst supported on TiO_2 prepared by a dendrimer-encapsulated method (100% conversion at ~250 °C) [105]. Clearly, the method or preparation has a great effect on the activity of the supported metal nanoparticle catalysts [30, 104–108].

The catalytic activity of the catalysts supported on reduced GO can be further enhanced by optimizing the size, shape, and composition of the nanoparticles through the laser synthesis of selected promising nanoalloy catalysts such as PdAu, PdCu, and PdRh in order to enhance their activities.

10.5 Conclusions and Outlook

Graphene is a unique high surface area support for metallic and bimetallic nanoparticle catalysts for a variety of important chemical transformations. In addition to the large surface area, the high thermal, chemical, and mechanical stability make graphene an excellent catalyst support. Furthermore, structural defects in the graphene lattice can be useful in anchoring the metal nanoparticles to the graphene surface thus achieving new surface functionalities with tunable metal–support interaction. Several studies have demonstrated unusual catalytic activity for metallic and bimetallic nanoparticles supported on the defect sites of the graphene nanosheets. Specifically, the palladium/graphene catalyst has been shown to possess remarkable catalytic activity in carbon-carbon bond forming reactions such as the Suzuki, Heck, and Sonogashira cross-coupling reactions. Furthermore, these catalysts could also be easily recovered and recycled under batch reaction conditions, thus providing high economic viability. Other catalysts such as transition-metal/metal-oxide nanoparticles supported on PRGO demonstrate good activity for CO oxidation.

Microwave-assisted chemical reduction and laser irradiation of graphene oxide in solution offer possibilities for simultaneous reduction of graphene oxide and formation of metal nanoparticles supported on the defect sites created in the reduced graphene oxide. With these methods, catalyst–support interaction can be tailored to produce highly efficient, selective, and recyclable heterogeneous catalysts for a variety of chemical transformations including the production of biofuels and valuable organic compounds from biomass feedstock, which could have a significant impact on addressing global energy problems and advancing chemical industry.

Acknowledgment

We thank the National Science Foundation (**CHE-0911146**) and NASA (**NNX08AI46G**) for support.

References

1. Geim Andre Konstantin (2009) *Science*, **324**, 1530–1534.
2. Geim, A.K. and Novoselov, K.S. (2007) *Nat. Mater.*, **6**, 183–191.
3. Novoselov, K.S., Geim, A.K., Morozov, S.V., Jiang, D., Katsnelson, M.L., Grigorieva, I.V., Dubonos, S.V., and Firsov, A.A. (2005) *Nature*, **438**, 197–200.
4. Allen, M.J., Tung, V.C., and Kaner, R.B. (2009) *Chem. Rev.*, **110**, 132–145.
5. Rao, C.N.R., Sood, A.K., Subrahmanyam, K.S., and Govindaraj, A. (2009) *Angew. Chem. Int. Ed.*, **48**, 7752–7777.
6. Wu, J., Pisula, W., and Mullen, K. (2007) *Chem. Rev.*, **107**, 718–747.
7. Novoselov, K.S., Jiang, Z., Zhang, Y., Morozov, S.V., Stormer., H.L., Zeitler, U., Maan, J.C., Boebinger, G.S., Kim, P., and Geim, A.K. (2007) *Science*, **315**, 1379.
8. Eda, G., Fanchini, G., and Chhowalla, M. (2008) *Nat. Nanotechnol.*, **3**, 270–274.
9. The Nobel Prize in Physics 2010 was awarded jointly to Geim, A. and Novoselov, K. "for Ground-breaking Experiments Regarding the Two-Dimensional Material Graphene", *http://nobelprize.org/nobel_prizes/physics/laureates/2010/press.html* (accessed 2011).
10. Scheuermann, G.M., Rumi, L., Steurer, P., Bannwarth, W., and Mulhaupt, R. (2009) *J. Am. Chem. Soc.*, **131**, 8262–8270.
11. Kamat, P.V. (2010) *J. Phys. Chem. Lett.*, **1**, 520–527.
12. Seger, B. and Kamat, P.V. (2009) *J. Phys. Chem. C*, **113**, 7990–7995.
13. Williams, G., Seger, B., and Kamat, P.V. (2008) *ACS Nano*, **2**, 1487–1491.
14. Si, Y.C. and Samulski, E.T. (2008) *Chem. Mater.*, **20**, 6792–6797.
15. Xu, C., Wang, X., and Zhu, J.W. (2008) *J. Phys. Chem. C*, **112**, 19841–19845.
16. Muszynski, R., Seger, B., and Kamat, P.V. (2008) *J. Phys. Chem. C.*, **112**, 5263.

17. Goncalves, G., Marques, P.A.A.P., Granadeiro, C.M., Nogueira, H.I.S., Singh, M.K., and Gracio, J. (2009) *Chem. Mater.*, **21**, 4796–4802.
18. Zhou, X., Huang, X., Qi, X., Wu, S., Xue, C., Boey, F.Y.C., Yan, Q., Chen, P., and Zhang, H. (2009) *J. Phys. Chem. C*, **113**, 10842–10846.
19. Lu, Y.H., Zhou, M., Zhang, C., and Feng, Y.P. (2009) *J. Phys. Chem. C*, **113**, 20156–20160.
20. Jasuja, K. and Berry, V. (2009) *ACS Nano*, **3**, 2358–2366.
21. Jasuja, K., Linn, J., Melton, S., and Berry, V. (2010) *J. Phys. Chem. Lett.*, **1**, 1853–1860.
22. Kong, B.S., Geng, J., and Jung, H.T. (2009,) *Chem. Commun.*, 2174–2176.
23. Jin, Z., Nackashi, D., Lu, W., Kittrell, C., and Tour, J.M. (2010) *Chem. Mater.*, **22**, 5695–5699.
24. Hassan, H.M.A., Abdelsayed, V., Khder, A.E.R., AbouZeid, K.M., Terner, J., El-Shall, M.S., Al-Resayes, S.I., and El-Azhary, A.A. (2009) *J. Mater. Chem.*, **19**, 3832–3837.
25. Zedan, A.F., Sappal, S., Moussa, S., and El-Shall, M.S. (2010) *J. Phys. Chem. C*, **114**, 19920–19927.
26. Siamaki, A.R., Khder, A.E.R.S., Abdelsayed, V., El-Shall, M.S., and Gupton, B.F. (2011) *J. Catalysis*, **279**, 1–11.
27. Panda, A.B., Glaspell, G.P., and El-Shall, M.S. (2006) *J. Am. Chem. Soc.*, **128**, 2790.
28. Panda, A.B., Glaspell, G.P., and El-Shall, M.S. (2007) *J. Phys. Chem. C*, **111**, 1861.
29. Abdelsayed, V., panda, A. B., Glaspell, G. P., and El-Shall, M. S. (2008) in *Nanoparticles: Synthesis, Stabilization, Passivation, and Functionalization*, ACS Symposium Series 996, Chapter 17 (eds R. Nagarajan and T. Alan Hatton), American Chemical Society, Washington, DC, pp. 225–247.
30. Abdelsayed, V., Aljarash, A., El-Shall, M.S., Al Othman, Z.A., and Alghamdi, A.H. (2009) *Chem. Mater.*, **21**, 2825–2834.
31. Herring, N.P., AbouZeid, K., Mohamed, M.B., Pinsk, J., and El-Shall, M.S. (2011) *Langmuir*, **27**, 15146–15154.
32. Stankovich, S., Piner, R.D., Chen, X., Wu, N., Nguyen, S.T., and Ruoff, R.S. (2006) *J. Mater. Chem.*, **16**, 155–158.
33. Stankovich, S., Dikin, D.A., Piner, R.D., Kohlhaas, K.M., Kleinhammes, A., Jia, Y., Wu, Y., Nguyen, S.T., and Ruoff, R.S. (2007) *Carbon*, **45**, 1558–1565.
34. Park, S. and Ruoff, R.S. (2009) *Nat. Nanotechnol.*, **4**, 217–224.
35. Gilije, S., Han, S., Wang, M., Wang, K.L., and Kaner, R.B. (2007) *Nano Lett.*, **7**, 3394.
36. Fan, B.X., Peng, W., Li, Y., Li, X., Wang, S., Zhang, G., and Zhang, F. (2008) *Adv. Mater.*, **20**, 4490.
37. Hummers, W.S. Jr. and Offeman, R.E. (1958) *J. Am. Chem. Soc.*, **80**, 1339–1339.
38. McAllister, M.J., Li, J.L., Adamson, D.H., Schniepp, H.C., Abdala, A.A., Liu, J., Herrara-Alonso, M., Milius, D.L., Car, R., Prud'homme, R.K., and Aksay, I.A. (2007) *Chem. Mater.*, **19**, 4396.
39. Ferrari, A.C. (2007) *Solid State Commun.*, **446**, 60.
40. Berciaud, S., Ryu, S., Brus, L.E., and Heinz, T.F. (2008) *Nano Lett.*, doi: 10.1021/nl8031444
41. Kudin, K.N., Ozbas, B., Schniepp, H.C., Prud'home, R.K., Aksay, I.A., and Car, R. (2008) *Nano Lett.*, **8**, 36.
42. Malesevic, A., Vitchev, R., Schouteden, K., Volodin, A., Zhang, L., Van Tendeloo, G., Vanhulsel, A., and Van Haesendonck, C. (2008) *Nanotechnology*, **19**, 305604.
43. Dato, A., Radmilovic, V., Lee, Z., Phillips, J., and Frenklach, M. (2008) *Nano Lett.*, **8**, 2012.
44. Dresselhaus, M.S., Dresselhaus, G., Satio, R., and Jorio, A. (2005) *Phys. Rep.*, **409**, 47.

45. Graf, D., Molitor, F., Ensslin, K., Stampfer, C., Jungen, A., Hierold, C., and Wirtz, L. (2007) *Nano Lett.*, **7**, 238.
46. Abdelsayed, V., Moussa, S., Hassan, H.M., Aluri, H.S., Collinson, M.M., and El-Shall, M.S. (2010) *J. Phys. Chem. Lett.*, **1**, 2804–2809.
47. Moussa, S., Atkinson, G., El-Shall, M.S., Shehata, A., AbouZeid, K.M., and Mohamed, M.B. (2011) *J. Mater. Chem.*, **21**, 9608–9619.
48. Moussa, S., Abdelsayed, V., and El-Shall, M.S. (2011) *Chem. Phys. Lett.*, **510**, 179–184.
49. Moussa, S., Siamaki, A.R., Gupton, B.F., and El-Shall, M.S. (2012) *ACS Catal.*, **2**, 145–154.
50. Bagri, A., Mattevi, C., Acik, M., Chabal, Y.J., Chhowalla, M., and Shenoy, V.B. (2010) *Nat. Chem.*, **2**, 581–587.
51. Acik, M., Mattevi, C., Gong, C., Lee, G., Cho, K., Chhowalla, M., and Chabal, Y.J. (2010) *ACS Nano*, **4**, 5861–5858.
52. Paredes, J.I., Villar-Rodil, S., Martinez-Alonso, A., and Tascon, J.M.D. (2008) *Langmuir*, **24**, 10560–10564.
53. Li, D., Muller, M.B., Gilje, S., Kaner, R.K., and Wallace, G.G. (2008) *Nat. Nanotechnol.*, **3**, 101–105.
54. Park, S., An, J., Jung, I., Piner, R.D., An, S.J., Li, X., Velamakanni, A., and Ruoff, R.S. (2009) *Nano Lett.*, **9**, 1593–1597.
55. Bourlinos, A.B., Gourins, D., Petridis, D., Szabo, T., Szeri, A., and Dekany, I. (2003) *Langmuir*, **19**, 6050–6055.
56. Vinodgopal, K., Neppolian, B., Lightcap, I.V., Grieser, F., Ashokkumar, M., and Kamat, P.V. (2010) *J. Phys. Chem. Lett.*, **1**, 1987–1993.
57. Zhou, Y., Bao, Q., Tang, L.A.L., Zhong, Y., and Loh, K.P. (2009) *Chem. Mater.*, **21**, 2950–2956.
58. Su, C.Y., Xu, Y., Zhang, W., Zhao, J., Liu, A., Tang, Z., Tsai, C.H., Huang Y., and Li, L.J. (2010) *ACS Nano*, **4**, 5285–5292.
59. Yeh, T.F., Syu, J.M., Cheng, C., Chang, T.H., and Teng, H. (2010) *Adv. Funct. Mater.*, **20**, 2255–2262.
60. Link, S. and El-Sayed, M.A. (2000) *Int. Rev. Phys. Chem.*, **19**, 409–453.
61. Burda, C., Chen, X., Narayanan, R., and El-Sayed, M.A. (2005) *Chem. Rev.*, **105**, 1025–1102.
62. Ito, J., Nakamura, J., and Natori, A. (2008) *J. Appl. Phys.*, **103**, 113712.
63. Yan, J.A., Xian, L., and Chou, M.Y. (2009) *Phys. Rev. Lett.*, **103**, 086802.
64. Lahaye, R.J.W.E., Jeong, H.K., Park, C.Y., and Lee, Y.H. (2009) *Phys. Rev. B*, **79**, 125435.
65. Sokolov, D.A., Shepperd, K.R., and Orlando, T.M. (2010) *J. Phys. Chem. Lett.*, **1**, 2633–2636.
66. Carbone, F., Baum, P., Rudolf, P., and Zewail, A.H. (2008) *Phys. Rev. Lett.*, **100**, 035501.
67. Lenner, M., Kaplan, A., Huchon, C., and Palmer, R.E. (2009) *Phys. Rev. B*, **79**, 184105–184116.
68. Cote, L.J., Cruz-Silva, R., and Haung, J. (2009) *J. Am. Chem. Soc.*, **131**, 11027–11032.
69. (a) Buchwald, S.L. (2008) *Acc. Chem. Res.*, **41**, 1439–1564; (b) Martin, R. and Buchwald, S.L. (2008) *Acc. Chem. Res.*, **41**, 1461; (c) Rouhi, M. (2004) *C&E News*, **82** (36), 49–58; (d) Negishi, E. and de Mrijere, A. (eds) (2002) *Handbook of Organopalladium Chemistry for Organic Synthesis*, Wiley-Interscience, New York; (e) Diederich, F. and Stang, P.J. (eds) (1998) *Metal-Catalyzed Cross-Coupling Reactions*, Wiley-VCH Verlag GmbH, Weinheim.
70. Yin, L. and Liebscher, J. (2007) *Chem. Rev.*, **107**, 133.
71. Heck, R.F. (1979) *Acc. Chem. Res.*, **12**, 146.
72. Miyaura, N. and Suzuki, A. (1995) *Chem. Rev.*, **95**, 2457.
73. Beletskaya, I.P. and Cheprakov, A.V. (2000) *Chem. Rev.*, **100**, 3009.
74. de Vries, J.G. (2001) *Can. J. Chem*, **79**, 1086.
75. Zapf, A. and Beller, M. (2002) *Top. Catal.*, **19**, 101.
76. The Nobel Prize in Chemistry 2010 was awarded jointly to Heck, R.F., Negishi, E., and Suzuki, A. "for Palladium-Catalyzed Cross Couplings in Organic Synthesis", *http://nobelprize.org/nobel_prizes/*

chemistry/laureates/2010/press.html (accessed 2011).

77. Chinchilla, R. and Najera, C. (2007) *Chem. Rev.*, **107**, 874–922.
78. Garrett, C. and Prasad, K. (2004) *Adv. Synth. Catal.*, **346**, 889.
79. (a) Cole-Hamilton, D.J. (2003) *Science*, **299**, 1702; (b) Widegren, J.A. (2003) *J. Mol. Catal. A*, **198**, 317; (c) Phan, T.S.N., Van der Sluys, M., and Jones, C.W. (2006) *Adv. Synth. Catal.*, **348**, 609–678; (d) Tucker, C.E. and de Vries, J.G. (2002) *Top. Catal.*, **19**, 111.
80. Bhanage, B.M. and Arai, M. (2001) *Catal. Rev. Sci. Eng.*, **43**, 315.
81. Köhler, K., Heidenreich, R.G., Soomro, S.S., and Pröckl, S.S. (2008) *Adv. Synth. Catal.*, **350**, 2930.
82. Djakovitch, L. and Koehler, K. (2001) *J. Am. Chem. Soc.*, **123**, 5990.
83. Narayanan, R. and El-Sayed, M.A. (2003) *J. Am. Chem. Soc.*, **125**, 8340.
84. Narayanan, R. and El-Sayed, M.A. (2004) *J. Phys. Chem. B*, **108**, 8572.
85. Ellis, P.J., Fairlamb, I.J.S., Hackett, S.F.J., Wilson, K., and Lee, A.F. (2010) *Angew. Chem. Int. Ed.*, **49**, 1820.
86. Davies, I.W., Matty, L., Hughes, D.L., and Reider, P.J. (2001) *J. Am. Chem. Soc.*, **123**, 10139.
87. Leadbeater, N.E., Williams, V.A., Barnard, T.M., and Collins, M.J. (2006) *Org. Process Res. Dev.*, **10**, 833.
88. Banhart, F., Kotakoski, J., and Krashennnikov, A.V. (2011) *ACS Nano*, **5**, 26–41.
89. Kim, G. and Jhi, S.H. (2011) *ACS Nano*, **5**, 805.
90. Bagri, A., Grantab, R., Medhekar, N.V., and Shenoy, V.B. (2010) *J. Phys. Chem. C*, **114**, 12053–12061.
91. Matsumoto, Y., Koinuma, M., Ida, S., Hayami, S., Taniguchi, T., Hatakeyama, K., Tateishi, H., Watanabe, Y., and Amano, S. (2011) *J. Phys. Chem. C*, doi: 10.1021/jp206348s
92. Loh, K.P., Bao, Q., Eda, G., and Chhowalla, M. (2010) *Nat. Chem.*, **2**, 1015–1024.
93. Eda, G., Lin, Y.Y., Mattevi, C., Yamaguchi, H., Chen, H.A., Chen, I.S., Chen., C.W., and Chhowalla, M. (2010) *Adv. Mater.*, **22**, 505–509.
94. Ng, Y.H., Iwase, A., Kudo, A., and Amal, R. (2010) *J. Phys. Chem. Lett.*, **1**, 2607–2612.
95. Li, H., Pang, S., Feng, X., Mullen, K., and Bubeck, C. (2010) *Chem. Commun.*, **46**, 6243–6245.
96. NIST X-ray Photoelectron Spectroscopy Database, NIST Standard Reference Database 20, Version 3.5 (2007) *http://srdata.nist.gov/xps/*. (assessed 2011)
97. Tsukahara, Y., Higashi, A., Yamauchi, T., Nakamura, T., Yasuda, M., Baba, A., and Wada, Y. (2010) *J. Phys. Chem. C*, **114**, 8965–8970.
98. Gutmann, B., Schwan, A.M., Reichart, B., Gspan, C., Hofer, F., and Kappe, C.O. (2011) *Angew. Chem. Int. Ed.*, **50**, 4636–4640.
99. Horikoshi, S., Osawa, A., and Serpone, N. (2011) *J. Phys. Chem. C*, **115**, 23030–23035.
100. Jansat, S., Durand, J., Favier, I., Malbosc, F., Pradel, C., Teuma, E., and Gomez, M. (2009) *ChemCatChem*, **1**, 224.
101. Martins, D.L., Alvarez, H.M., and Aguiar, L.C.S. (2010) *Tetrahedron Lett.*, **51**, 6814.
102. Ogasawara, S. and Kato, S. (2010) *J. Am. Chem. Soc.*, **132**, 4608.
103. Balanta, A., Godard, C., and Claver, C. (2011) *Chem. Soc. Rev.*, **40**, 4973, and references therein.
104. Yang, Y., Saoud, K.M., Abdelsayed, V., Glaspell, G., Deevi, S., and El-Shall, M.S. (2006) *Catal. Commun.*, **7**, 281–284.
105. Scott, R.W.J., Sivadinarayana, C., Wilson, O.M., Yan, Z., Goodman, D.W., and Crooks, R.M. (2005) *J. Am. Chem. Soc.*, **127**, 1380–1381.
106. Wang, D., Xin, H.L., Yu, Y., Hongsen Wang, H., Eric Rus, E., Muller, D.A., and Abruna, H.D. (2010) *J. Am. Chem. Soc.*, **132**, 17664–17666.
107. Glaspell, G., Fuoco, L., and El-Shall, M.S. (2005) *J. Phys. Chem. B*, **109**, 17350–17355.
108. Zhu, H., Qin, Z., Shan, W., Shen, W., and Wang, J. (2007) *Catal. Today*, **126**, 382–386.

11
Graphenes in Supramolecular Gels and in Biological Systems

Santanu Bhattacharya and Suman K. Samanta

11.1
Introduction

11.1.1
Overview of 2D-Nanomaterials

Graphene is a two-dimensional planar carbon allotrope [1] with outstanding materials properties and tremendous applications compared to its other counterparts such as zero-dimensional fullerenes [2] and one-dimensional nanotubes [3]. The family of graphene includes single- and multilayered graphene, graphene oxide (GO), and reduced graphene oxide (RGO), and they exhibit highly conducting [4–6], optoelectronic [7, 8], and interesting mechanical [9, 10] properties that enable them to be useful in a number of areas. Graphenes exhibit high Young's modulus [11], impressive thermal conductivity [12], and extremely high charge-carrier mobilities at room temperature [13]. Graphene is also a zero-band gap semiconductor with long-range ballistic transport [13], possesses high surface area, and holds great promise for potential applications in fabricating electronic devices, such as field-effect transistors [14], ultrasensitive sensors [15], lithium-ion batteries [16], fuel cells [17], and transparent conductive electrodes used in touch screens and displays [18]. Several of the above-mentioned properties in three-dimensional (3D) matrices derived from graphene-based assemblies are, however, highly limited till date.

11.1.2
Overview of Physical Gels

Supramolecular gels are a class of soft materials that could act as a medium or a host to develop various 3D architectures and hold tremendous potential [19, 20]. Low-molecular-mass gels (LMMGs) are described as an immobilized mass composed of certain organic molecules and a specific solvent or a mixture [21, 22]. Depending on the choice of the solvent, either organic or aqueous, the immobilized mass is denoted as "organogel" or "hydrogel," respectively. The intermolecular

Graphene: Synthesis, Properties, and Phenomena, First Edition. Edited by C. N. R. Rao and A. K. Sood.

weak interactions such as hydrogen bonding, van der Waals, π-stacking, and electrostatic and dipolar interactions lead to the entangled fiber formation to hold the solvent molecules in its solvent pocket in the microenvironment. Therefore, supramolecular gels could be a host for the incorporation of various dopants such as nanoparticles [23], nanotubes [19, 20], and biological substances such as drugs or even DNA [24, 25], which make such systems superior materials involving applications in drug delivery, tissue engineering, and stimuli responsiveness and as pH sensors [26–28].

11.1.3
Different Types of Graphenes, Their Preparation, Functionalization, and Gelation

Graphene comprises a family of 2D nanomaterials including single-layer and multilayer graphene, GO, RGO, and functionalized graphene. Depending on their chemical composition, their solubility and respective properties are different. Graphenes are generally obtained from the exfoliation of graphite, where the individual layers are peeled off from the surface of graphite. Single-layer graphene has been prepared through simple mechanical exfoliation of graphite using Scotch tape [1]. In another process, thermal exfoliation of GO is used, where GO acts as an intermediate in the preparation of graphene [29]. GO has been prepared using the Hummers method, wherein the oxidation of graphite followed by reduction at high temperatures produces few-layer graphene [30]. Graphene looks like wrinkled flakes and is devoid of any functional groups. Thus, it is totally hydrophobic in nature and not soluble in an aqueous medium. GO contains hydroxyl, –COOH, and –CO– functionalities (Figure 11.1) and is therefore soluble in water, showing a brown aqueous solution. RGO is prepared with mild reduction of GO using hydrazine (Hz), H_2O_2, $NaBH_4$, and so on [31]. This reduction step removes the carbonyl functionalities from the backbone of graphene, although the hydroxyl group still remains, which makes RGO hydrophilic in nature.

Functionalization is undertaken in order to increase the "solubility" or dispersibility of graphene in a particular solvent. For example, covalent functionalization with long aliphatic hydrocarbon chains increases the solubility of graphene

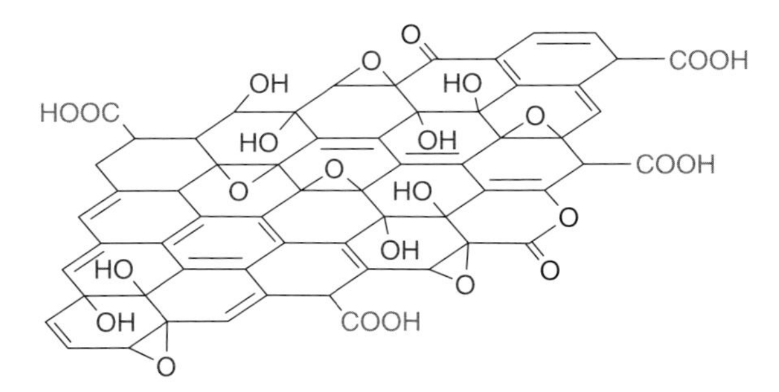

Figure 11.1 Schematic molecular structure of GO [32].

in various organic solvents. However, covalent functionalization often destroys the native framework of graphene and hence their corresponding properties also show certain changes [33]. For example, their electrical conductivity decreases owing to the decrease in their relative π-conjugations. To increase the solubility, noncovalent strategies are thus better where several noncovalent interactions exist between the graphene and the host material. Supramolecular gels are one such class of hosts that incorporate graphenes to obtain corresponding composite materials. Most importantly, functionalization through noncovalent means leaves the integrity of the graphene frameworks. However, chemical functionalization such as oxidation (to produce GO) introduces functionality such as carboxylic acid, epoxide, and hydroxide, which increases the polarity at the expense of their structural integrity. This, in turn, makes GO soluble in water and in special situations hydrogels are obtained via supramolecular interactions. Thus, excellent mechanical, thermal, electrical, and various other physical properties of this class of materials, as described earlier, lead them toward the development of various composite materials with small molecules, metal ions, or even with polymers through various noncovalent interactions. The properties are summarized in the following sections.

11.2 Toward the Gelation of GO

GO holds a great potential because of its unique structure and outstanding physical and chemical properties [34, 35]. As described, GO manifests amphiphilicity through the hydrophilic edges and a more hydrophobic basal plane [36]. This makes GO an attractive 2D-building block for the construction of various graphene-based supramolecular architectures [37]. Recently, a complete study on the gelation of GO was carried out by Shi *et al.* [32], which includes the supramolecular interaction of GO with small molecules, surfactants, polymers, metal ions, various cross-linkers, and even the effect of pH (Figure 11.2).

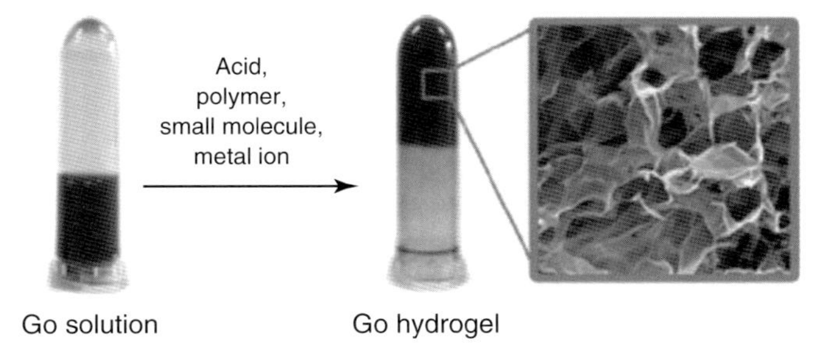

Figure 11.2 Possible way of hydrogel formation by GO and the 3D architecture as revealed under SEM image. (Source: Adapted with permission from Ref. [32]. Copyright 2011, American Chemical Society.)

11.2.1 Effect of pH on the Gelation of GO

GO prepared via the exfoliation of graphite through the peeling-off process using Hummer's method produces single-layer graphite that is functionalized with various hydrophilic oxygenated functional groups such as carboxylic acid, hydroxyl groups, epoxy group, and carbonyl moieties. These hydrophilic groups induce amphiphilicity in the graphene sheets, and, as a result, the GO forms a stable colloidal dispersion in aqueous medium. However, the electrostatic repulsion between the ionized carboxyl groups in the GO sheets prevents their aggregation in aqueous medium. Therefore, acidification of a GO dispersion weakens the electrostatic repulsion and leads to the formation of an aggregated state. With decreasing pH of a GO dispersion, the zeta potential (ζ) increases, indicating the protonation of the COO^- groups (Figure 11.3). Therefore, GO sheets adhere to each other and tend to phase separate in a strong acidic medium because of insufficient mutual repulsion in aqueous solution. However, at a higher concentration of GO (> 4mg ml^{-1}), a hydrogel, instead of an amorphous precipitation, results on acidification, as confirmed by the tube inversion method. Owing to this low critical gel concentration (CGC), GO can be included in the class of a "super gelator" [32]. Clearly, a 3D GO network is formed through the self-assembly of the acidified GO sheets. Acidification weakens the electrostatic repulsion between GO sheets, and their hydrogen-bonding force is enhanced because of the protonation of carboxyl groups. However, further lowering the pH value (<0.6) led to the precipitation of GO sheets [38].

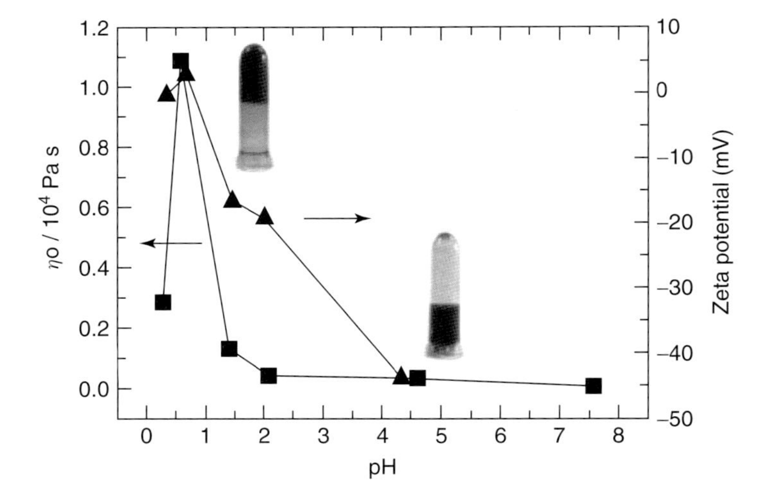

Figure 11.3 Zero-shear viscosities and zeta potentials of a GO (5 mg ml^{-1}) aqueous solution at different pH values. The insets show the photographs of a GO solution at pH = 4.6 and a hydrogel at pH = 0.6. (Source: Adapted with permission from Ref. [32]. Copyright 2011, American Chemical Society.)

11.2.2
Effect of the Dimension of GO toward Gelation

The observed hydrogel is dependent on the lateral dimension of the GO sheets. A stable hydrogel formation on acidification takes place for the GO sheets having the lateral dimensions of several micrometers. No gelation occurred at any pH if the lateral dimensions of most GO sheets were smaller than 1 μm even if the concentration of GO was as high as 9 mg ml^{-1}. In this case, the small GO sheets turned cloudy in several minutes and finally precipitated from the acidified solution.

Owing to the amphiphilic nature of GO, they make contact with each other even in a dilute solution. Therefore, a weak dynamic GO network is possible in the dispersion of GO sheets depending on the balance between the electrostatic repulsion and binding interactions (hydrogen bonding, π-stacking, hydrophobic effect, etc.). Thus, gelation occurs through a greater bonding force and a weaker repulsion force among the GO network in solution. Both gelation and precipitation of GO sheets may be induced by the optimization of these two opposing forces and primarily depend on the stacking states of the GO sheets. In a hydrogel, the GO sheets are randomly orientated, whereas they adopt a parallel arrangement on their precipitation. The parallel arrangement is energetically more favorable due to the larger area of contact between the GO sheets. However, due to a poor mobility of the larger GO sheets in solution, their parallel orientations are highly restricted. On the other hand, the precipitation is a slower process compared to the gelation. In gel, the large conjugated basal planes induce restriction in orientation of the GO sheets which leads to the formation of a stable network. However, in a very slow process the gel transforms into a precipitate due to the conversion into a thermodynamically more stable state. The precipitation process accelerated with a smaller size of GO sheets (smaller than 1 μm). Therefore, the stability of the hydrogel depends on the ratio of repulsion and bonding force between the GO sheets.

11.2.3
Cross-Linker (Small Molecule/Polymer)-Induced GO Gels

Addition of a cross-linker (Figure 11.4) can promote gelation by increasing the attraction force between the GO sheets [32]. Thus, the introduction of a small amount of a polymer component such as poly(vinyl alcohol) (PVA) can induce the gelation of GO due to the additional hydrogen-bonding interactions between the GO sheets and PVA chains. Hydrogels are also obtained by mixing small amounts of poly(ethylene oxide) (PEO), hydroxypropylcellulose (HPC), and poly(vinyl pyrrolidone) (PVP) into GO dispersions. Among them, the polymers, e.g., PEO and PVP have intrinsic hydrogen bond acceptor properties while HPC has both hydrogen bond donor and acceptor properties. Thus, the gelation is promoted by the additional bonding force through the hydrogen bonds among these polymers with the adjacent GO sheets.

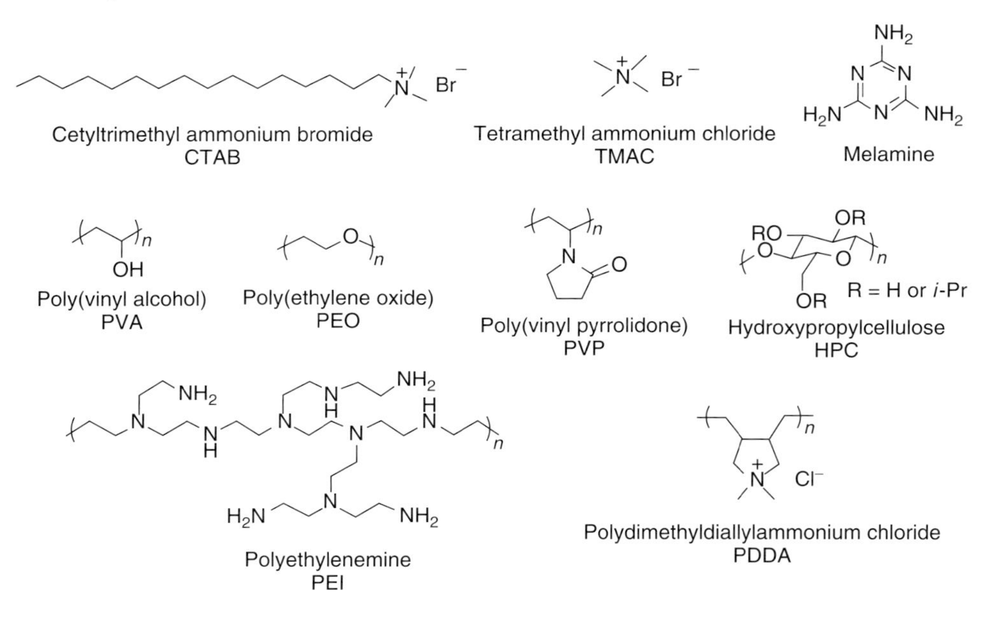

Figure 11.4 Chemical structures of some cross-linkers used to prepare GO composite hydrogels [32].

The formation of GO/polymer composite hydrogels depends on various factors. These include concentration of the GO and the polymer, the pH of the medium as well as the molar masses of the polymers. Among these parameters, the concentration of GO is a primary factor for the hydrogel formation. A lower concentration of GO than the CGC of the GO/polymer hydrogel can not impart sufficient interactions between the GO sheets. Thus, it fails to form a hydrogel even in the presence of a cross-linker.

Depending on the bonding forces caused by different cross-linkers, GO hydrogels show a sol–gel transitions at various pHs. Also, with increasing molar masses of the polymer cross-linker, the viscoelasticity of hydrogels increases. More number of hydrogen bonds are possible for a longer polymer chain with the GO sheets. This also increases the possibility of interaction between two or more GO sheets. Thus, HPC is an effective cross-linker for the formation of a GO hydrogel, whereas monosaccharide glucose fails to become a cross-linker because of its smaller molecular size.

Electrostatic interaction is another driving force for the supramolecular self-assembly of the graphene sheets. Polydimethyldiallylammonium chloride (PDDA) is a model cross-linking agent for excluding the possibility of forming hydrogen bonds. In this case, the gelation takes place through the dominating electrostatic interaction between the quaternary ammonium groups of PDDA and the carboxylate residues of the GO sheets. Thus, PDDA was found to be an efficient cross-linker with a CGC as low as 0.1 mg ml^{-1}.

Effective electrostatic interaction was observed for polyethylenimine (PEI) as the cross-linking reagent toward the aggregation of GO sheets into hydrogels. PEI chains contain primary, secondary, and tertiary amines, and most of these amino groups are protonated under the gelation condition. Thus, both hydrogen-bonding and electrostatic interactions are operational between GO sheets and PEI chains. Therefore, PEI is a good cross-linker with a CGC as low as 0.2 mg ml^{-1}. However, a higher concentration of PEI (0.5 mg ml^{-1}) leads to the GO precipitation. Therefore, polycationic polymers emerged as strong cross-linkers below a certain concentration above which the GO precipitates. GO hydrogels are also observed in the presence of a small amount of quaternary ammonium salts, such as cetyltrimethyl ammonium bromide (CTAB) and tetramethylammonium chloride (TMAC). The gelation takes place mainly due to the long-range electrostatic force, and it is stronger than the hydrogen-bond interaction. Furthermore, the hydrophobic interactions of the long aliphatic chain of CTAB with GO sheets led to a lowering of the CGC value of CTAB (0.3 mg/mL) than TMAC (1.9 mg/mL). Also, a small molecule such as melamine which possesses strong hydrogen bond acceptors and bears multiple charges upon protonation, shows excellent cross-linking ability (CGC 0.3 mg/mL).

11.2.4 Cation-Induced GO Gels

Similar to the cross-linkers, a variety of metal ions can induce the gelation of GO solutions. Divalent and trivalent ions (e.g., Ca^{2+}, Mg^{2+}, Cu^{2+}, Pb^{2+}, Cr^{3+}, and Fe^{3+}) promote the formation of GO hydrogels while monovalent ions (e.g., K^{+}, Li^{+}, and Ag^{+}) fail to induce gelation [32]. The driving force for the GO gelation is the coordination of metal ions with the hydroxyl and carboxyl groups on GO sheets. As the coordination stability of the multivalent transition metal ions is greater than the alkali and alkaline earth metal ions, the former has greater cross-linking abilities and show smaller CGC values. Thus, in a concentrated solution, the GO sheets associate with each other through the coordination of the metal ions to form 3D networks which leads to hydrogelation.

Gel-like three-dimensional (3D) architectures were also achieved from RGO sheets on connecting with divalent cations (Ca^{2+}, Ni^{2+}, or Co^{2+}) and water molecules via hydrothermal treatment [39]. In the assembly, the RGO sheets, water molecules, and divalent ions acts as a skeleton, filler, and linker, respectively, through chemical bonds and hydrogen bonds. RGO is prepared through partial reduction of GO (obtained by Hummers method) with H_2O_2. In the 3D architecture, RGO sheets remain as single or double layer through the isolation of interlamellar water molecules, as shown in the model (Figure 11.5). Large extent of water content (~99 wt%) in the assembly facilitates the introduction of a polymeric strengthening agent such as PVA. The self-assembly route for the generation of 3D-RGO structure is useful for preparing versatile host–guest composites for the applications to energy storage or even in biology.

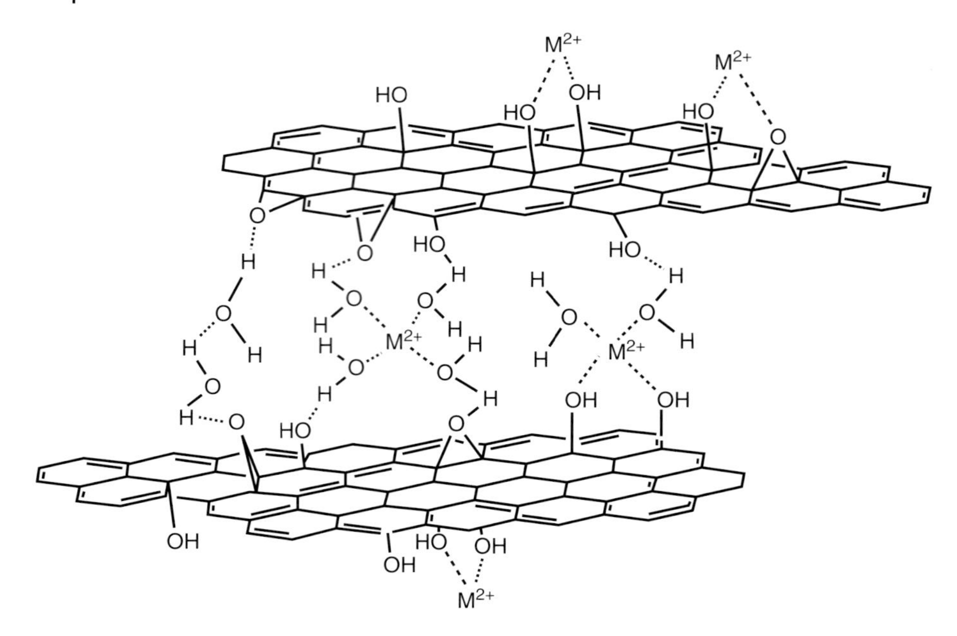

Figure 11.5 Schematic illustration of the formation of a gel-like RGO cylinder with divalent ion linkage. M^{2+} represents the divalent ion (Ca^{2+}, Ni^{2+}, or Co^{2+}). (Source: Adapted with permission from Ref. [39]. Copyright 2010, American Chemical Society.)

11.2.5 Surfactant-Induced GO Gels

Nonionic surfactants such as Brij 700 and P-123 can exfoliate and disperse a large extent of pristine graphene from graphite in water at significant concentrations (1 mg ml^{-1}) with some specific surfactants [40]. The surfactants (Brij 700, HTAB, P-123, PSS, SDBS, TDOC, and Tween 80) lead to increases in the amount of the suspended material in water with sonication time. The dispersions are observed to be stable for at least several weeks and are constituted by single- and few-layer graphene sheets whose basal planes are essentially free of structural defects. These dispersions can be processed into potentially useful macroscopic materials, such as paper-like films by vacuum filtration through polycarbonate membranes that show significant values of electrical conductivity. Supramolecular hydrogels were prepared from the P-123-stabilized graphene dispersions through the formation of inclusion complexes between α-cyclodextrin and the PEO chains. The electrical contacts between the neighboring graphene platelets and corresponding electrical transport in the films are hindered by the presence of surfactant molecules. Thus, owing to the larger fraction of surfactant in the films, it showed lower resulting electrical conductivity.

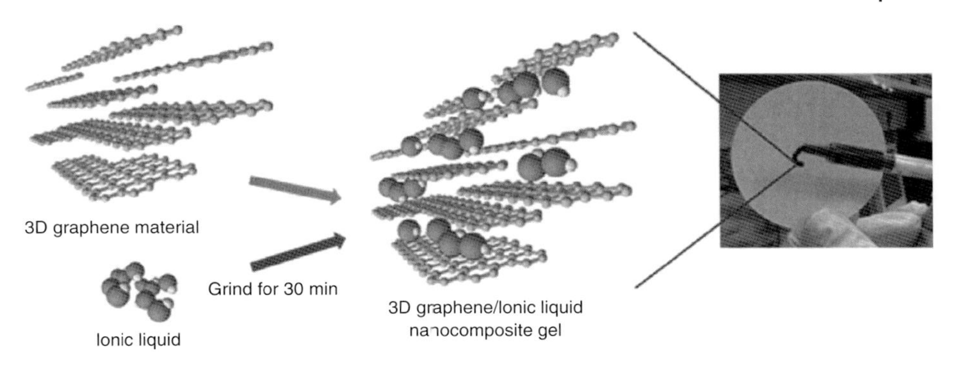

Figure 11.6 Schematic representation of the fabrication of the 3D graphene/IL nanocomposite as well as its photograph. (Source: Adapted with permission from Ref. [41]. Copyright 2011, John Wiley and Sons.)

11.2.6 Ionic-Liquid-Induced GO Gels

Graphene/ionic liquid (IL) nanocomposite gel with uniform porous structure was prepared by simple mixing of the individual components [41]. The nanocomposite gel was prepared using 16 mg of graphene mixed with 1.6 ml of IL 1-butyl-3-methylimidazolium hexafluorophosphate (BMIMPF6) in an agate mortar on grinding for 30 min. The uniform nanocomposite was formed possibly by the good interaction of graphene with IL through cation–π and/or π–π interactions as shown in Figure 11.6. Thus, these nanomaterials with homogeneous porous structures can be a stoichiometric electron acceptor and act as a host for electron-donating guest species such as nitric oxide (NO). The nanocomposite gel has a large electroactive surface area and thus, it behaves as a combination of a planar disk electrode and a thin-layer electrochemical cell demonstrating a fast response, high sensitivity, and low detection limit for NO. The high performance of the gel composite is due to the porous structure of the graphene material with a high specific surface area and superior conductivity. The nanocomposite-based NO sensor shows a large dynamic response range and excellent stability suggesting that the nanocomposite is useful for the sensitive detection of NO and can be applied in various electrochemical devices.

11.2.7 Gelation of Hemoglobin by GO and Sensing

Two-dimensional GO sheets are capable of forming stable composite hydrogels in presence of hemoglobin (Hb) [42]. The GO/Hb composite hydrogels were prepared by adding 4 mg of Hb dissolved in 0.1 ml deionized water and 0.8 ml of GO solution (9 mg/ml). The mixture was shaken vigorously for ~15 s to obtain a self-standing hydrogel having high water content (98.5%). The pH of the GO aqueous solution was 1.87, which is much lower than the isoelectric point of Hb (pH 6). A strong electrostatic attraction between the negatively charged GO sheets and the positively charged Hb molecules is the driving force for the gel formation.

The hydrogels were used for the catalyzing oxidation of pyrogallol by H_2O_2 in organic solvents. The activity of the catalyzing process is greater in the hydrogel compared to the free Hb or GO. The hydrogel provides an aqueous microenvironment for the protection of enzyme from deactivation as well as it serves as a platform to attract the substrate and exclude the product.

The Hb- and graphene-based hydrogel containing chitosan-dimethylformamide (CS-DMF) was developed for an amperometric nitrite (NO_2^-) sensor [43]. The surface morphologies of the modified electrode (Hb-CS-DMF/graphene film) exhibited a three-dimensional network porous structure. The electrochemistry of the composite indicated that Hb immobilized on the surface of the graphene-modified electrode could keep its bioactivity, as it exhibited a surface-controlled electrochemical process and had a fast heterogeneous electron-transfer rate. Thus, the introduction of graphene facilitates the electron transfer between Hb molecules and the underlying electrode owing to the excellent electrical conductivity of graphene and the biocompatibility of chitosan. As a result, the composite showed a good sensitivity and stability for the amperometric determination of nitrite.

11.2.8 Gelation of DNA by GO with Dye-Absorption and Self-Healing Properties

A novel and facile 3D self-assembly method has been developed for the preparation of GO/DNA composite hydrogels with high mechanical strength, excellent environmental stability, high dye-adsorption capacity, and self-healing function [44]. The GO/DNA hydrogel has been prepared by mixing equal volumes of the aqueous dispersion of GO (6 mg ml^{-1}) and the double-stranded DNA (ds-DNA) (10 mg ml^{-1}) followed by heating the homogeneous mixture at 90 °C for 5 min. Upon heating, the ds-DNA unwound to single-stranded DNA (ss-DNA) which bridged the adjacent GO sheets through strong noncovalent interactions to furnish self-standing hydrogels. However, a direct interaction or π-stacking between the GO sheets are absent in the hydrogels. Thus, the ss-DNA acts a 'glue' for holding the GO sheets together to form the hydrogel.

The GO/DNA hydrogel is stable in a variety of harsh conditions such as strong acidic (pH 2), basic (pH 13), or salty (1 M NaCl) aqueous solution or even in organic solvents such as tetrahydrofuran (THF). The stability of the gel arises due to the strong interaction between GO and DNA through $\pi-\pi$ stacking and hydrophobic interactions of the DNA bases and the 2D GO sheets. Additional electrostatic and hydrogen bonding interactions among the amine groups of the DNA bases and the oxygen containing functional groups of GO increases the stability of the gels.

The GO/DNA gel show impressive dye adsorption properties. Dye adsorption was observed for a 0.2 ml volume of the gel in presence of 0.6 ml of an aqueous solution of a model dye safranine O (0.1 mg/ml). The dye adsorption reached more than 80% after 12 h and nearly 100% after 24 h (Figure 11.7a,b). The maximum dye

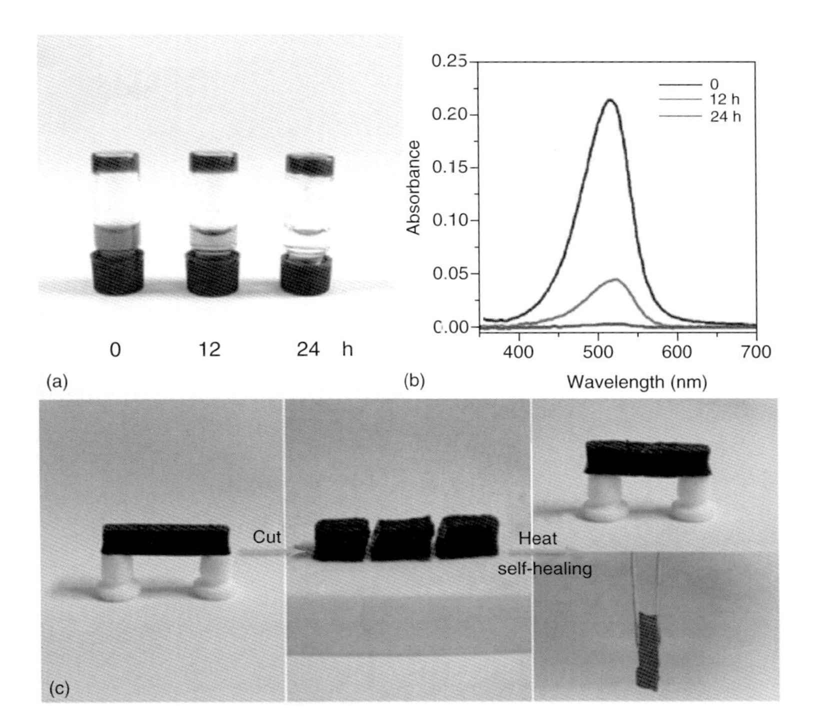

Figure 11.7 (a) Photographs and (b) absorption spectra of an aqueous solution of safranine O after adsorption by GO/DNA gel at different times; 0.6 ml of dye solution (0.1 mg ml^{-1}) was added onto 0.2 ml of the gel. (c) Self-healing process: the as-prepared free-standing gel was cut with a razor into three small blocks and the blocks could adhere to each other by pushing the freshly formed surfaces to contact together followed by heating at 90 °C in air for 3 min. (Source: Adapted with permission from Ref. [44]. Copyright 2010, American Chemical Society.)

loading capacity of the gel to adsorb safranine O was 960 mg/g of GO. The high dye-loading ability of the gel is due to the strong electrostatic interactions between positively charged safranine O and negatively charged GO as well as DNA. Also, the large specific surface area within the porous gel network increases the affinity of the dye molecules towards the GO sheets.

The GO/DNA gel possesses interesting self-healing property. A 2.5 cm long self-standing gel of GO/DNA was cut into three pieces to expose fresh surfaces (Figure 11.7c). The small pieces were then held together and heated at 90 °C for 3 min. Upon heat treatment the pieces were joined together to produce a 'self-healed' gel which is as strong as to bridge the two posts horizontally. The excess of ss-DNA present in the gel as unbound or weakly bound are converted into ds-DNA during the cooling process. On heating, the ds-DNA present in the surface of the two adjacent blocks unwind again to form ss-DNA. These *in situ* ss-DNA linked the GO sheets to join the block and thus 'repairs' the damage.

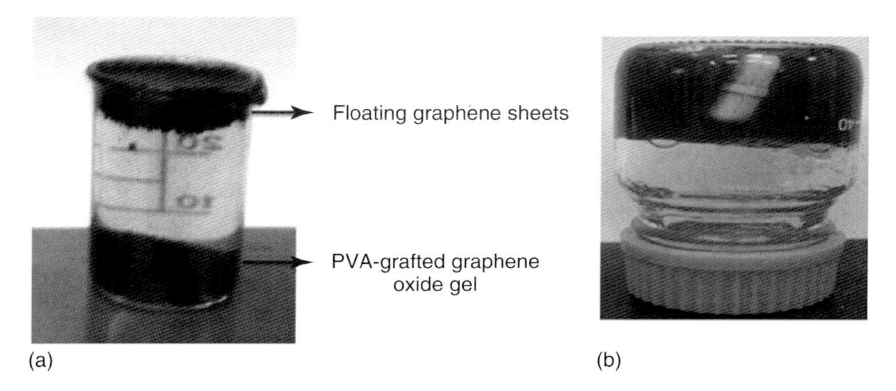

Figure 11.8 (a) PVA-mediated separation of graphene. (b) GO-reinforced PVA Hydrogels. (Source: Adapted with permission from Ref. [45]. Copyright 2010, Elsevier.)

11.2.9 Gelation-Assisted Isolation of Graphene from Graphene-GO Mixture

A simple and rapid coagulation technique to separate graphene sheets from GO and other impurities has been developed through the formation of GO-based hydrogels [45]. GO was prepared through the Hummers method and was reduced with sodium borohydride followed by Hz. However, the reduced GO so obtained was not fully reduced but contained some remnants of oxidized graphene moieties including nitrogen impurities. The partially reduced GO obtained in this way was dispersed in water and a PVA solution was added. When the temperature of the mixture was raised to 45–50 °C, graphene flakes appeared on the surface of the solution while GO precipitated rapidly (Figure 11.8a). The floating graphene layer existed as fluffy flakes, which were collected on a substrate and dried. The precipitate was a gelatinous sticky mass of PVA-functionalized GO, which forms graphene-reinforced PVA hydrogels as shown in Figure 11.8b. The hydrophilic functionalities render the GO layers hydrophilic in nature, so that water molecules can readily intercalate into the interlayer galleries. The hydroxyl groups in PVA are hydrogen-bonding acceptors, which increase its interaction with the water molecules making PVA soluble in water. Therefore, the hydroxyl groups of PVA form strong hydrogen bonds with the hydroxyl and carboxylic acid groups present on GO, thereby forming hydrogen-bonded cross-links between GO and PVA molecules leading them to the hydrogel formation.

11.3 Polymer-Assisted Formation of Multifunctional Graphene Gels

Gelation of graphene (mostly GO) was observed in several instances with the assistance of different polymeric systems mostly in aqueous medium. The effect of external stimuli on the gelation was observed to be multifaceted. In each case, the polymer–graphene composite emerges as new nanomaterials with novel properties.

11.3.1 Thermal and pH Regulated GO-Polymer Hydrogels

A facile one-step strategy for GO interpenetrating poly(*N*-isopropylacrylamide) (PNIPAM) hydrogel networks have been developed in water through a highly efficient cross-linking reaction [46]. The resulting hydrogel network showed better mechanical strength and also exhibited dual thermal and pH response with good reversibility. GO sheets are uniformly dispersed in PNIPAM hydrogel networks with a disordered alignment. The incorporation of GO sheets into PNIPAM microgel networks raises the sol-to-gel transition temperature during cooling. The GO/PNIPAM hydrogel and PNIPAM microgel networks are pH sensitive. As the pH increases, the sol–gel transition temperature increases. This is due to the fact that, an increase in the pH value induces ionization of the carboxylic acids which in turn reduces the number of hydrogen bonds. This leads to a swollen structure or a higher transition temperature of the gel.

In another instance, stimuli-responsive PNIPAM hydrogels containing partially exfoliated graphite have been prepared by frontal polymerization [47]. With increasing the concentration of graphene, the maximum temperature of the front (T_{max}) increases. Thus, when the amount of graphene exceeds ~0.07 wt%, it showed a sharp increase of T_{max}. The graphite-containing samples exhibit lower values of lower critical solution temperature (LCST) than that of the neat PNIPAM hydrogel (28 °C). However, with increasing contents of graphite filler, LCST increased.

11.3.2 Gelation-Assisted Polymer Nanocomposites

Commercially available nonionic pluronic copolymers (poly-(ethylene oxide)-*block*-poly(propylene oxide)-*block*-poly(ethyleneoxide), PEO-*b*-PPO-*b*-PEO) were employed as the solubilizing agent for chemically exfoliated GO and graphene formed through *in situ* reduction by Hz [48]. The stable aqueous copolymer-coated graphene solution (Figure 11.9a) thus obtained was found to be macroscopically homogeneous, but microscopically heterogeneous. In the composite the hydrophobic PPO segments were bound to the hydrophobic surface of graphene via hydrophobic effect, while the hydrophilic PEO chains were exposed into the water (Figure 11.9c). Thus, owing to this dual role, pluronic copolymer can form a supramolecular hydrogel with α-CDs through the penetration of PEO chains into the cyclodextrin cavities (Figure 11.9b). Utilizing this phenomenon, a facile and effective method to hybridize the well-dispersed graphene into a supramolecular hydrogel was developed. The hybrid hydrogel showed similar rheological behavior such as the shear thinning property. However, the viscoelasticity of the hybrid hydrogel decreased significantly compared to the native hydrogel.

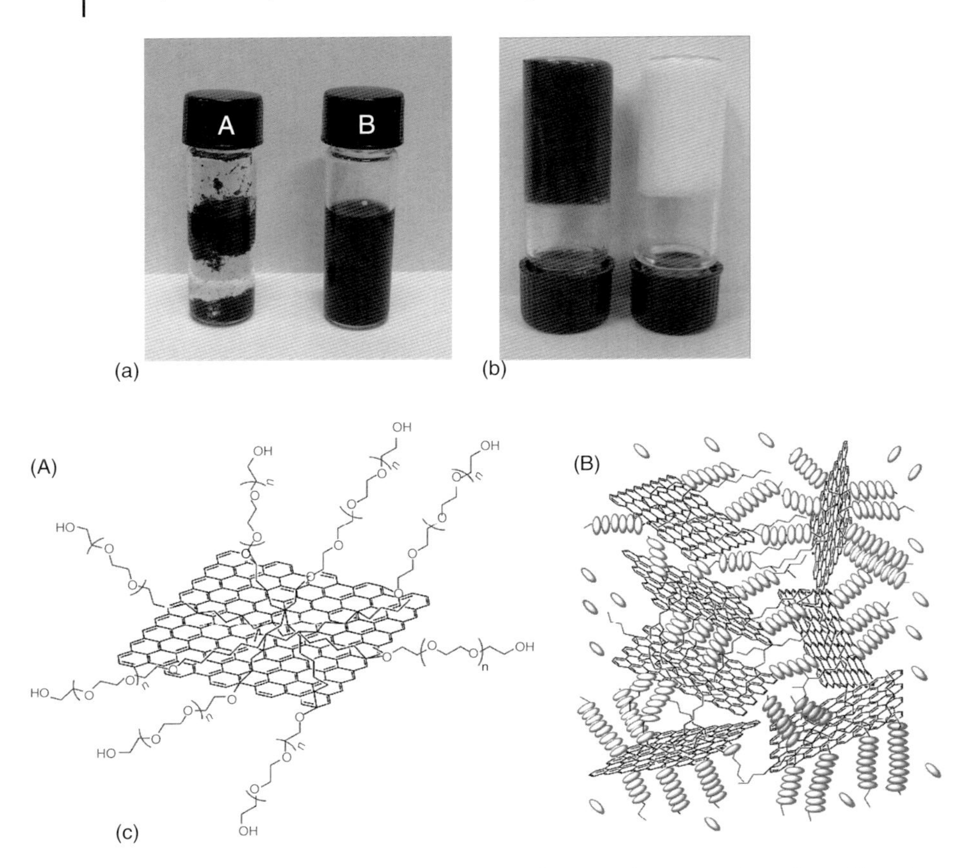

Figure 11.9 (a) Photograph of reduced graphite oxide (1.0 mg ml^{-1}) (A) and copolymer-coated graphene in aqueous solution (B). (b) Hybrid supramolecular hydrogel (on left) compared with the native polymer hydrogel. (c) Proposed structure of the copolymer-coated graphene (A) and supramolecular well-dispersed graphene sheet containing hybrid hydrogel (B). (Source: Adapted with permission from Ref. [48]. Copyright 2009, American Chemical Society.)

Spontaneous formation of uniform microspheres were achieved via poly(vinylidene fluoride hexafluoro propylene) (PVDF-HFP)/graphene composite gels [49]. Gelation occurs through the adsorption of water vapor by a mixture of PVDF-HFP/graphene in DMF (Figure 11.10A, B). The hybrid gel after solvent exchange and freeze drying emerged into the porous microspheres with average diameter of 8–10 μm (Figure 11.10). However, the native gel in absence of graphene shows cellular morphology upon liquid–liquid demixing process. The native gels contained dense and partly connected droplets of diameter ~40 nm. However, the size of the droplets decreases (~20 nm) in the graphene hybrid gels along with the accumulation of long-branched fibrillar microstructures. This indicates the presence of scrolled and entangled graphene sheets which thus helps the formation of spherical microstructures. The driving force is

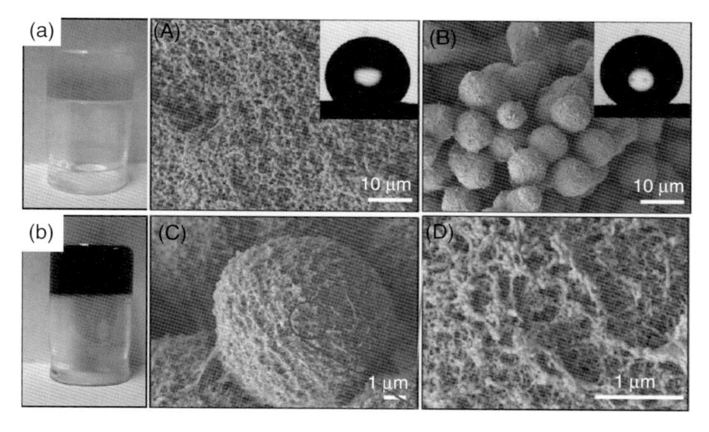

Figure 11.10 Photographs of (a) a PVDF-HFP gel and (b) a PVDF-HFP/graphene gel. SEM images of (A) PVDF-HFP and (B) PVDF-HFP/graphene dried gels. The inset optical images show a water drop sitting on these porous dried gels. The contact angles of the PVDF-HFP and PVDF-HFP/graphene hybrid composites are $132.7 \pm 2.7^{\circ}$ and $151.6 \pm 1.4^{\circ}$, respectively. (C,D) Magnified surface morphology of PVDF-HFP/graphene microspheres. (Source: Adapted with permission from Ref. [49]. Copyright 2011, American Chemical Society.)

the affinity of PVDF-HFP towards graphene which induces the aggregation of PVDF-HFP chains around the entangled graphene sheets. As the hybrid gel is composed of several microspheres having the surface roughness, the gels manifest superhydrophobic property. Due to this property, the gel can be used as catalyst supporting materials or to generate superhydrophobic polymer coatings.

11.3.3
Mechanical Properties of GO-Polymer Hydrogels

GO nanosheets have been doped into various polymers matrices for the enhancement of mechanical properties in composite materials [50]. An *in situ* radical polymerization reaction of acrylamide (AM) in presence of GO nanosheets leads to a GO/polyacrylamide (PAM) hydrogels. During the polymerization reaction, the PAM molecules were well grafted onto the GO nanosheets. Therefore, the compatibility between PAM and GO in the GO-g-PAM composite was revived. Thus, the composite exhibited improved mechanical properties in both dry and wet states. However, the increase in the mechanical strength highly depends on the amount of GO present in the composite. Thus, a six-fold increase in the compressive strength of the GO/PAM hydrogels were observed compared to that of the pure PAM.

GO/PVA composite hydrogels were also obtained by a freeze/thaw method [51]. The composite hydrogel succeed to show improved properties such as tensile strength, breaking elongation and compressive strength as compared to the pure PVA hydrogels. Thus, upon incorporation of 0.8 wt% of GO into the PVA, the composite hydrogel showed 132% increase in the tensile strength and 36% increase in the comressive strength.

11.3.4
Electrical Properties of GO-Polymer Hydrogels

GO-polymer hydrogels composed of 2-aminoanthraquinone (AAQ) grafted onto chemically modified graphene have been prepared for the applications in super-capacitor electrodes [52]. High specific capacitance value of 258 F/g at a discharge current density of 0.3 A/g was obtained for the composite electrode while pure graphene hydrogel shows much lesser value (193 F/g). Also, the composite based electrode showed high rate capability and stability, mostly due to the additional redox capacitance of the covalently bonded AAQ moieties. In the composite electrode, graphene provides large specific surface area for the formation of electrical double layers which induce charge transfer and electrolyte diffusion and hence improves the performance of the electrode.

Electrical conductivities of the graphene based hydrogels were improved by the hydrothermal reduction of GO with Hz or hydroiodic acid [53]. These reduced GO hydrogels show high conductivity (1.3–3.2 S/m) as well as large specific capacitance (220 F/g at 1 A/g discharge current density). This GO hydrogel based material showed high power density and long cycle life.

Macroassembled 3D graphene sheets exhibit ultralow density, large internal surface area and high electrical conductivities [54]. Using sol–gel techniques, the individual GO sheets were cross-linked covalently to obtain graphene aerogels as monolithic solids. These gels showed ultralow density ($\sim$10 mg/cm^3) as well as improved electrical conductivity ($\sim 1 \times 10^2$ S/m) which is about 2 orders of magnitude greater than that of noncovalently cross-linked graphene sheets.

11.3.5
Multifunctional GO Hydrogels

Towards the end of functional GO based hydrogels, multifunctional composites were also accomplished. For instance, self-assembled graphene hydrogels were prepared by chemical reduction of GO with sodium ascorbate which exhibit porous structure, electrical conductivity (1 S/m), mechanical elasticity as well as electrochemical performance. Moreover, the method of preparation is mild and environmental friendly.

Multifunctional chemically converted graphene (CCG) based gels were prepared at the solution-filter membrane interface upon filtration process [56]. The CCG sheets were assembled themselves in a sheet-by-sheet arrangements in which a layer of water molecules are adsorbed in between the sheets due to the hydrophilic groups on the CCG surface (Figure 11.11). Thus, both attractive hydration forces and repulsive electrostatic forces between the hydrated CCG sheets held them to a loosely bound state. Through partial $\pi-\pi$ stacking interactions the CCG sheets are arranged in a parallel orientation. The directionally oriented CCG films showed

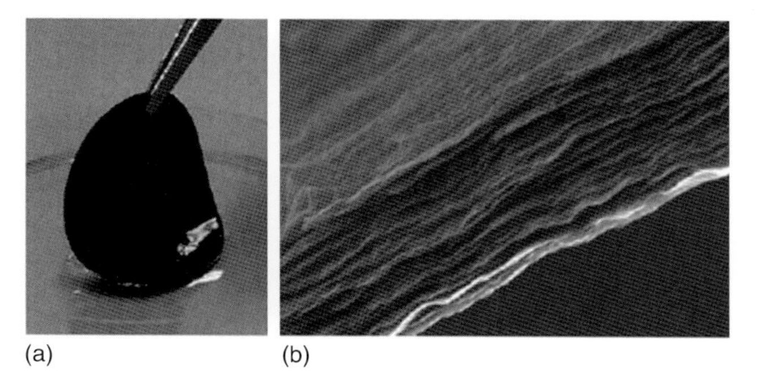

Figure 11.11 (a) Photograph of the as-formed oriented graphene hydrogel films peeled off from the filter membrane and (b) SEM image of the cross section of a freeze-dried graphene hydrogel film. Scale bar: 1 mm. (Source: Adapted with permission from Ref. [56]. Copyright 2011, John Wiley and Sons.)

high electrical conductivity (0.58 S/cm), remarkable chemical and thermal stability which makes it useful for flexible electronic devices.

Hydrothermal reduction of an aqueous dispersion of GO also leads to the multifunctional hydrogels [57]. Although the gel contains only 2.6% of graphene and a large extent of water (97.4%), it shows an electrical conductivity as high as 5×10^{-3} S/cm. The gel shows high thermal stability in the temperature range of 25–100 °C, high storage modulus (450–490 kPa) as well as high specific capacitance (175 F/g). Therefore, the multifunctional nature of these GO hydrogels is useful for various applications including device applications, supercapacitors, high performance nanocomposites as well as in biology.

11.3.6 Stimuli-Responsive Hydrogels and Their Applications

A light-responsive PNIPAM hydrogel nanocomposite incorporating glycidyl-methacrylate-functionalized graphene oxide (GO–GMA) has been described [58]. This nanocomposite hydrogel can undergo a large volumetric change in response to infrared (IR) light illumination, owing to the highly efficient photothermal conversion of GO–GMA. The GO–GMA hydrogel also possesses a large swelling ratio by water uptake. The transition temperature of the gel is lowered by ~10 °C owing to the incorporation of GO. It has been realized that a microvalve in a microfluidic channel implementing such a hydrogel nanocomposite and the microvalve can control the fluidic flow within the microchannel through remote IR light actuation.

In another instance, graphene/Fe_3O_4 hybrids were synthesized using a one-step solvothermal method [59]. In the hybrid, the graphene sheets are decorated by Fe_3O_4 particles distributed randomly throughout the nanosheet. Stimuli-responsive PNIPAM/graphene/Fe_3O_4 (P/G) and PNIPAM/hectorite (P/H) gels were prepared through solution polymerization using graphene/Fe_3O_4 and hectorite as cross-linkers respectively. The P/G and P/H gels exhibited

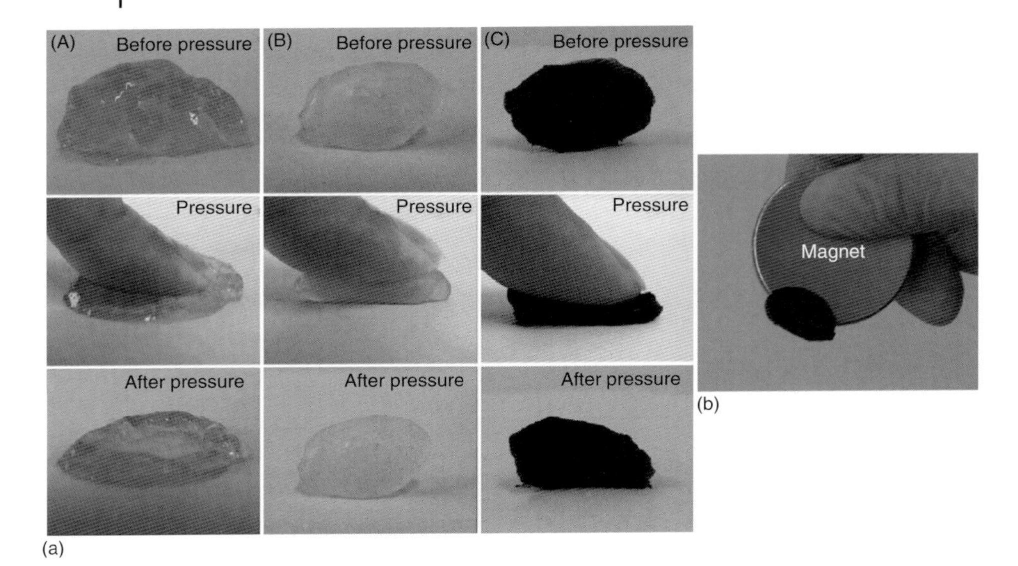

Figure 11.12 (a) Photographic images of the as-prepared PNIPAM gels. (A) Using no cross-linker, (B) using hectorite as a cross-linker, and (C) using graphene/Fe_3O_4 hybrids as a cross-linker. (b) Photographic image of the response of the PNIPAAm/graphene/Fe_3O_4 gel to a magnet. (Source: Adapted with permission from Ref. [59]. Copyright 2011, Elsevier.)

high strength and excellent resilience while the gel without any cross-linkers was broken into pieces under external pressure (Figure 11.12a). These indicate that the polymer chains can form effective networks around the graphene/Fe_3O_4 cross-linker. Furthermore, due to the intrinsic magnetic property of Fe_3O_4 the P/G gel showed response towards external magnet (Figure 11.12b).

11.4 Graphene Aerogels

GO and graphene-based carbon aerogels were demonstrated for the first time by freeze-drying of GO/water dispersions [60]. The preparation of graphene aerogels was accomplished by first exfoliation of GO in water by ultrasonication. The gelation of the GO/water dispersion was then achieved by increasing the solid content or adding some water-soluble polymers. The GO/water dispersion was then subjected to critical point drying or freeze-drying to yield graphene aerogel with extreme surface-area-to-volume ratio of graphene sheets. Controlled thermal treatment of the aerogel reduces the GO to graphene and restores the electrical conductivity of the graphene aerogel.

Sponge-like graphene aerogels were prepared from a GO hydrogel precursor [61]. The GO hydrogel was obtained by the enrichment of GO nanosheet at a

Figure 11.13 (a) Digital photos of the aqueous suspension of graphene oxide, (b) the graphene-based hydrogel, (c) the supercritical CO_2 dried (left) and freeze-dried (right) graphene aerogel, and (d) a 7.1 mg graphene aerogel pillar with the diameter of 0.62 cm and the height of 0.83 cm supporting a 100 g counterpoise, more than 14 000 times its own weight. (Source: Adapted with permission from Ref. [62]. Copyright 2011, The Royal Society of Chemistry.)

solid/liquid interface induced by the interaction between hydroxyl groups of anodic aluminum oxide (AAO) and carboxyl groups of GO sheets. Lyophilization of the GO hydrogel produces graphene aerogel keeping the macro- and micro-structures of the parent hydrogel intact. A reduction process with a low temperature annealing yields the final aerogel macroform which shows a good electrochemical energy storage performance.

Graphene aerogels were also achieved through supercritical drying or freeze-drying of hydrogel precursors obtained from the reduction of GO with L-ascorbic acid [62]. The graphene hydrogel precursor was synthesized through the reduction of GO in a uniform mixture of an aqueous suspension of GO (4 mg/ml) and ascorbic acid (Figure 11.13). The mechanical studies demonstrates that the aerogel can hold more than 14000 times its own weight. The graphene aerogel also show light weight (12–96 mg/cm^3), high conductivity (~102 S/m), large Brunauer-Emmett-Teller (BET) surface area (512 m^2/g), ample volume (2.48 cm^3/g) as well as high specific capacitance (128 F/g at a constant current density 50 mA/g).

11.5
Hydrogel and Organogel as the Host for the Incorporation of Graphene

LMMG-derived hydrogels or organogels were used as host materials for the incorporation of graphenes and in the development of nanocomposite materials. The self-standing thermoreversible gels are formed on the basis of various noncovalent interactions such as hydrogen-bonding, π-stacking, van der Waals, and electrostatic interactions [21, 22]. Dispersion of graphene into the gel matrix could occur via π−π stacking interactions between the gelators and graphene leading to the exfoliation of graphene layers, which yields a supramolecular complex. A minute quantity of graphene could significantly reinforce the gel matrix and tune its various physical properties.

Stable supramolecular hydrogels were obtained from Fmoc (*N*-fluorenyl-9-meth oxycarbonyl)-protected synthetic dipeptides (**1**) and (**2**) (Figure 11.14) [63]. Both the gels hold entangled uniform nanofibrillar network organization in favor of their gel formation. The gelator molecules contain aromatic moieties in the side chain (Tyr/Phe) and in the *N*-terminus (fluorenyl group), which may impart π−π stacking interactions with the exfoliated graphene (EG) sheets. One of the peptides (gelator **1**) was employed for the incorporation of RGO into the hydrogel to make a well-dispersed RGO containing stable hybrid hydrogel. RGO forms an apparently homogeneous solution and stabilized within the peptide (**1**)-based hydrogel system. The formation of a more mechanically rigid and "solidlike" hybrid hydrogel was obtained after the incorporation of RGO into the native hydrogel, although the

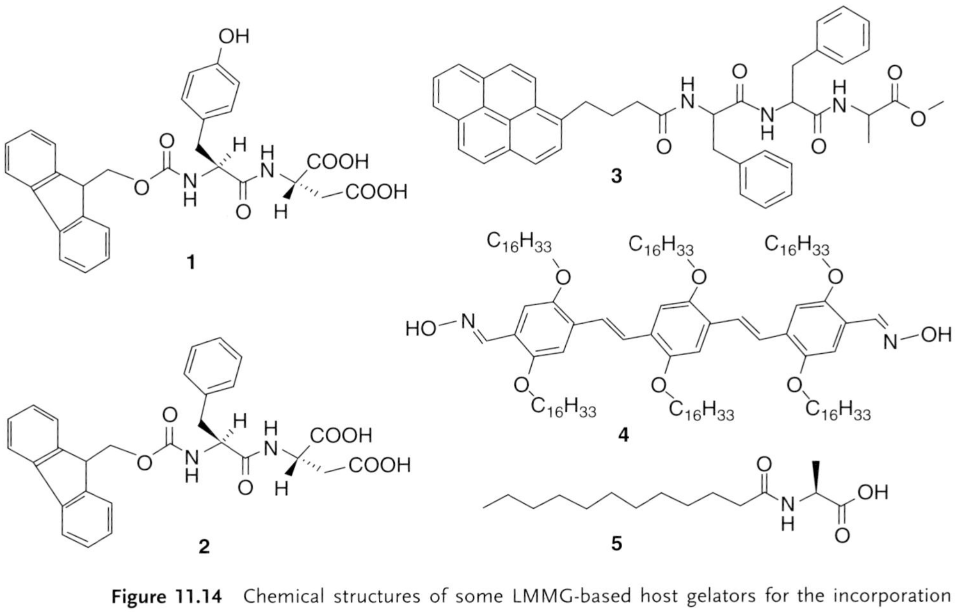

Figure 11.14 Chemical structures of some LMMG-based host gelators for the incorporation of graphene.

morphology of the peptide hydrogel did not change significantly even after the incorporation of small amounts of RGO.

The *N*-terminally pyrene-conjugated oligopeptide (**3**) (Figure 11.14) forms transparent, stable, supramolecular fluorescent organogels in various aliphatic and aromatic organic solvents [64]. Unfunctionalized and nonoxidized graphene has been incorporated into the organogel in *o*-dichlorobenzene (ODCB), which forms a stable hybrid gel-nanocomposite. Graphene is well dispersed into the gel medium through the noncovalent $\pi-\pi$ stacking interactions with the pyrene-conjugated gelator. In the presence of graphene, the minimum gelation concentration of the hybrid organogel was lowered significantly. This suggests that there is a favorable interaction between the graphene and the gelator within the hybrid organogel system. The hybrid gel system showed that the presence of circa three to four layers of EG in ODCB and the presence of both graphene nanosheets and the entangled network of gel nanofibers were observed in proximity. The flow of the hybrid organogel had become more resistant toward the applied angular frequency on the incorporation of graphene into the organogel. The hybrid gel i about seven times more rigid than that of the native gel.

Carbon nanomaterials (CNMs) such as EG, long-chain-functionalized EG, single-walled carbon nanotube (SWNT), and fullerene (C_{60}) have been investigated thoroughly for their interaction with two structurally different gelators based on all-*trans* tri-*p*-phenylenevinylene bis-aldoxime (**4**) and *n*-lauroyl-L-alanine (**5**) in supramolecular organogels [65]. Gelation of the individual gelators occurs in toluene through hydrogen-bonding and van der Waals interactions for **4** and **5** in addition to $\pi-\pi$ stacking specifically in the case of **4** (Figure 11.15). Graphenes and other CNMs contain extended π-conjugated aromatic surface, which is able to interact strongly with **4** through $\pi-\pi$ stacking interactions. These nanocomposites provide significant understanding in terms of the nature of molecular-level interactions of dimensionally different CNMs with structurally different gelators and tunability of their properties in the supramolecular nanocomposites.

Densely wrapped CNMs encapsulated fibrous networks in the resulting composites are observed in all the instances induced by the supramolecular interactions. Owing to the $\pi-\pi$ stacking process between the graphene sheets and the gelator molecules of **4**, the individual graphene sheets were exfoliated from the few-layer graphitic stacks. CNMs promote aggregation of the gelator molecules in solution leading to the hypochromism and quenching of the fluorescence intensity. The gel-CNM composites show significantly increased electrical conductivity compared to that of the native organogel even if a trace quantity of CNM was doped in the gel matrix. Rheological studies of the composites demonstrate the formation of rigid and viscoelastic solidlike assembly because of the reinforced aggregation of the gelators on CNMs. A synergistic behavior is observed in the case of the composite gel of **4** containing a mixture of EG and SWNT when compared with other mixtures of CNMs in all combinations with EG (Figure 11.15). This afforded new nanocomposites with interesting optical, thermal, electrical, and mechanical properties.

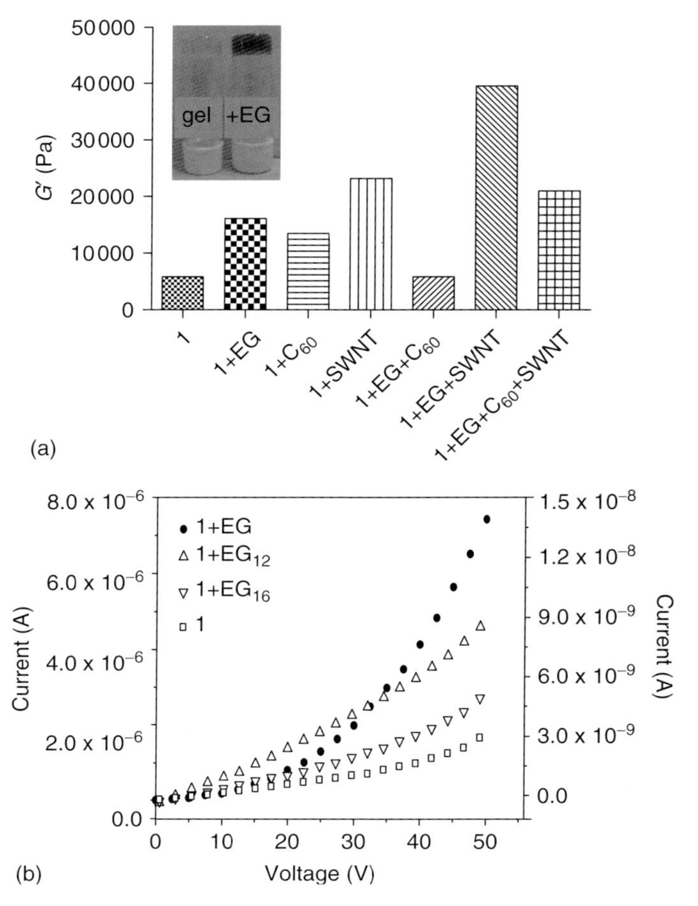

Figure 11.15 (a) Bar diagram showing the storage modulus (G') of the gels of **1** with CNMs and mixtures of CNMs obtained under oscillatory frequency sweep experiment (G' values are at 10 rad s^{-1}, 0.83 wt% CNMs were added individually, [**4**] = 10 mg ml^{-1}), inset showing gel of **4** in toluene and its composite with exfoliated graphene. (b) Electrical conductivity showing $I - V$ characteristics of (b) **4** + EG (left y-axis), **4**, **4** + EG_{12} and **4** + EG_{16} (right y-axis). [**4**] = 10 mg ml^{-1}, [EGs] = 0.99 wt% with respect to **4**. (Source: Adapted with permission from Ref. [65]. Copyright 2011, Wiley-VCH Verlag GmbH & Co. KGaA.)

11.6 Biological Applications Involving Graphene

Interesting properties of graphene offer potential advantages for specific applications in biological sciences, especially in the areas of bioelectronics, biosensors, and drug delivery. In a recent review [66], some of these applications have been discussed briefly. These include biofunctionalization of graphene with DNA, proteins, and other biomolecules; graphene-based FRET (fluorescence resonance energy transfer) biosensors for the detection of DNA and various ions, small molecules and proteins; graphene-based biotechnology for living cell studies; and even drug delivery and cell imaging. Owing to its excellent solubility in aqueous medium, GO finds widespread applications in biological systems. For this purpose, several hosts were selected for the delivery of the GO inside the cellular systems. Also, GO has been used as a carrier for the delivery of bioactive drugs or even DNA. Therefore, achieving a significant binding between the GO and the drugs, preferably through noncovalent interactions, is a prerequisite. Rao *et al.* [67] have shown specific interactions between graphene and the DNA nucleobases and nucleosides. It has been proposed that the van der Waals interaction is the main driving force for the binding between the graphene and the nucleobases. The binding energy of each of the four nucleobases (adenine, guanine, cytosine, and thymine) with the graphene in aqueous solution follows a specific trend depending on various parameters such as types of graphene samples used, solvation and dilution. In another instance, ss-DNA has been adsorbed on the functionalized graphene in aqueous media and studied for the enzymatic cleavage of DNA by DNase [68]. It has been shown that the graphene protects DNA from being cleaved. Owing to the strong binding of ss-DNA with the functionalized graphene through strong molecular interactions, it prevents DNase I from approaching the constrained DNA. Also, the constraint of DNA on the graphene improves the specificity in terms of its response to complementary DNA.

Functionalized GO-based drug delivery and pH-sensitive controlled release systems were established recently [69]. GO–Fe_3O_4 nanohybrid was prepared first and then folic acid (FA), which is a targeting agent toward some tumor cells, was conjugated onto Fe_3O_4 nanoparticles to furnish a multifunctionalized GO. An antitumor drug such as doxorubicin (DOX) hydrochloride was loaded onto the surface of this multifunctionalized GO based on the $\pi-\pi$ stacking interactions. The drug release property depends strongly on the pH values. The DOX loaded GO has been used for the cell uptake studies to estimate the targeted delivery property and toxicity to tumor cells. The multifunctionalized DOX loaded GO exhibit specific transportation of drugs to SK3 cells (human breast cancer cells) and show toxicity to HeLa cells (human cervical cancer cells). This was further evident from the functional nanoscale graphene oxide (NGO) that acts as a nanocarrier for the loading and targeted delivery of the anticancer drugs [70]. The NGO that is functionalized with sulfonic acid groups can form a stable solution in physiological pH. Therefore, covalent binding of FA molecules with the NGO allowed it to specifically target MCF-7 cells (human breast cancer cells) with FA receptors. Thus, controlled loading of two anticancer drugs, DOX and camptothecin (CPT), onto the FA-conjugated NGO (FA–NGO) via $\pi-\pi$ stacking and hydrophobic interactions shows specific targeting. The combined use of two or more drugs displays much better therapeutic efficacy than that of a single drug.

Covalent functionalization of GO with CS through amide linkages was developed to perform as a nanocarrier that imparts a good aqueous solubility and biocompatibility (Scheme 11.1) [71]. Thus, water insoluble anticancer drug *e.g.* CPT can

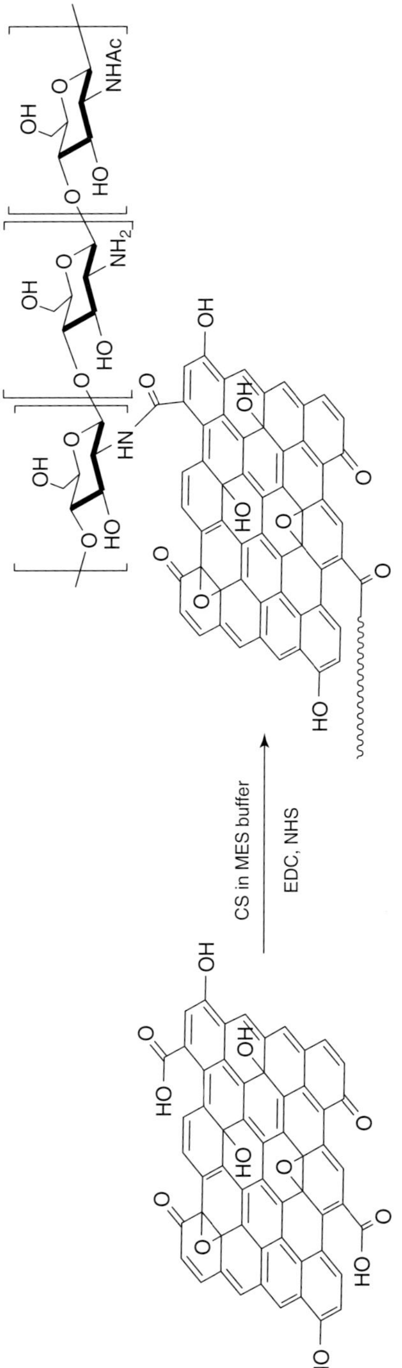

Scheme 11.1 Synthesis of GO–CS [71].

be loaded into the GO–CS nanocarrier through $\pi-\pi$ stacking and hydrophobic interactions. Compared to CPT alone, the GO–CS–CPT complexes exhibit exceptionally high cytotoxicity in HepG2 (human hepatic carcinoma cells) and HeLa cell lines. The GO–CS also forms stable, nanosized complexes with plasmid DNA and the resulting GO–CS/pDNA nanoparticles showed transfection efficiency in HeLa cells. Therefore, the GO–CS based nanocarrier has the ability to load and deliver both anticancer drugs as well as genes.

NGO functionalized polyethylene glycol (PEG) has been prepared for the drug delivery applications [72]. Due to the presence of branched, biocompatible PEG chains, the composite shows high water solubility and serum stability. The composite NGO-PEG has the ability to attach water insoluble aromatic drugs *e.g.* SN38, through van der Waals interactions (Figure 11.16). The drug showed negligible release from the composite in phosphate buffer saline and ~30% release in serum in three days indicating that a strong noncovalent binding of the drug with NGO (Figure 11.17a). The drug loaded NGO-PEG show remarkable killing of the cancer cells including human colon cancer cell line HCT-116 having IC50 values of ~6 nM (Figure 11.17b). However, the NGO-PEG alone was shown to be non-cytotoxic in nature.

PEI-GO has also been developed as a nanocarrier for the delivery of *si*RNA and anticancer drug DOX [73]. The sequential delivery of *si*RNA and DOX by

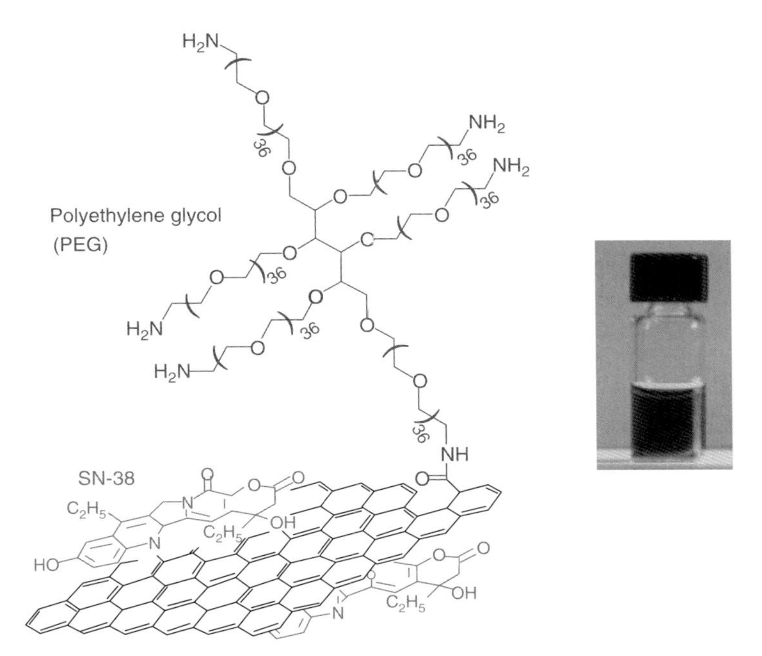

Figure 11.16 SN38 loading on NGO-PEG: schematic drawing of SN38 loaded NGO-PEG and a photo of NGO-PEG-SN38 solution in water. (Source: Adapted with permission from Ref. [72]. Copyright 2008, American Chemical Society.)

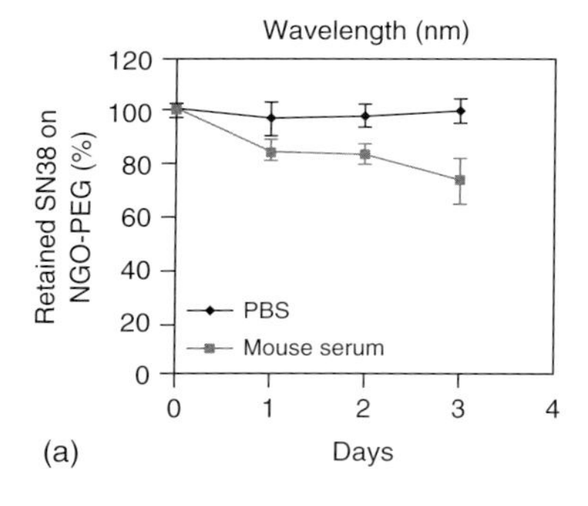

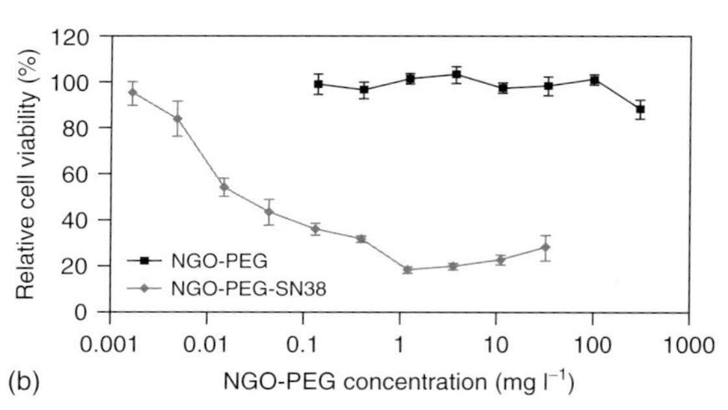

Figure 11.17 (a) Retained SN38 on NGO-PEG over time incubated in PBS and serum, respectively. SN38 loaded on NGO-PEG was stable in PBS and released slowly in serum. Error bars were based on triplet samples. (b) Relative cell viability data of HCT-116 cells after incubation with NGO-PEG with (light gray) and without (black) SN38 loading. Plain NGO-PEG exhibited no obvious toxicity even at very high concentrations. Error bars were based on triplet samples. (Source: Adapted with permission from Ref. [72]. Copyright 2008, American Chemical Society.)

PEI-GO into cancer cells exhibits a synergistic effect that leads to a significantly enhanced chemotherapy efficacy. PEI-GO shows significantly lower cytotoxicity and substantially higher transfection efficacy of *si*RNA than PEI alone.

A graphene–hemin composite conjugated with FA (GFH) has been prepared for selective, quantitative, and fast colorimetric detection of cancer cells [74]. First, GFH was synthesized by covalently grafting poly(allylamine hydrochloride) (PAH) onto graphene through the reaction between epoxy groups of GO and amino groups of PAH (Scheme 11.2). Then, the FA molecules were conjugated with graphene sheets through amidation of the carboxylic acid end groups present on FA and the amine groups on the graphene surface. Finally, hemin was adsorbed on graphene surface to form GFH. GFH shows selective binding with human cervical cancer cells (HeLa) and human breast cancer cells (MCF-7) through targeting FA receptors. The advantage of this process is that the GFH binding could be visualized under bright field microscopy as well as by colorimetric method due to the large size and peroxidase-like activity of GFH. As low as ~1000 cancer cells could be detected through this process. This could be a general method for the cancer cell detection as the folate receptors are over-expressed for different types of cancer cells such as HeLa, MCF-7 as well as NIH-3T3 (mouse embryo fibroblast cell line).

GO based biosensing platform has been accomplished with dye labeled peptides as the probe biomolecules [75]. Based on the FRET mechanism between GO nanosheets and the dye-labeled peptides, the protease activity could be monitored in real-time. The interaction between the hydrophobic basal plane of GO with peptides takes place through the stacking interactions as well as via electrostatic forces. Due to these interactions, the fluorescence of the dye-labeled peptide decreases via strong energy-transfer from the dye molecules to the GO nanosheets (Scheme 11.3). However, addition of protease cause the peptide hydrolysis on GO nanosheet to release the dye-labeled peptide segments which turns on the

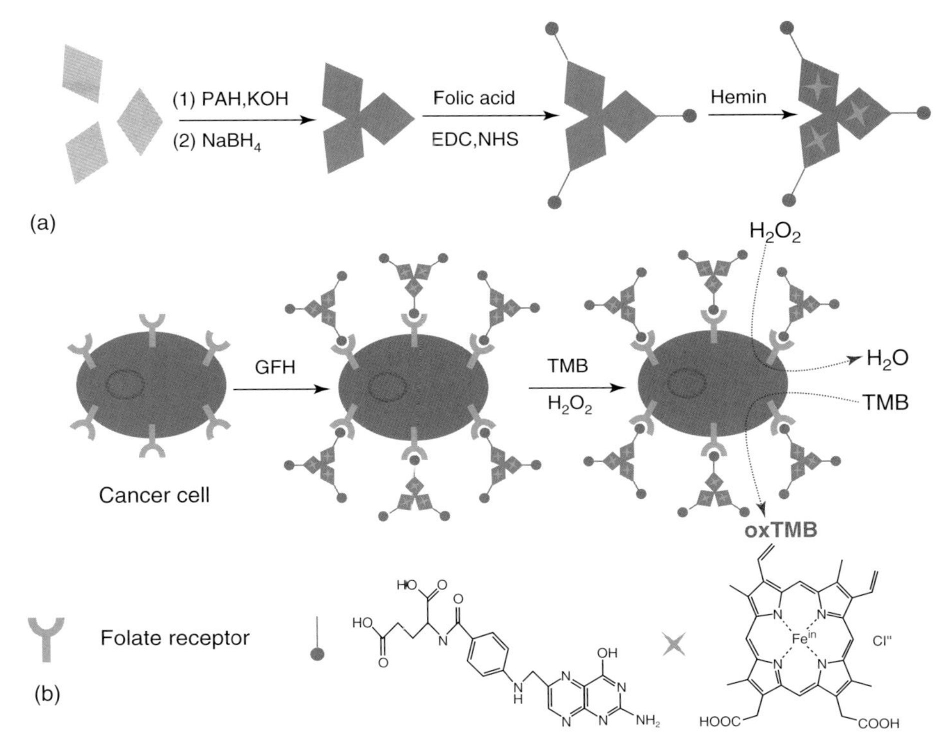

Scheme 11.2 Schematic representation of (a) preparation of GFH and (b) cancer cell detection by using target-directed GFH. (Source: Adapted with permission from Ref. [74]. Copyright 2011, The Royal Society of Chemistry.)

fluorescence. Thus, the protease-catalyzed peptide–GO bioconjugate hydrolysis event can be estimated easily in real time through the measure of fluorescence enhancement.

A biocomposite of EG (GR) and a cationic amphiphile, cholest-5en-3β-oxyethyl pyridinium bromide (PY^+-Chol), has been developed by physically adsorbing the PY^+-Chol on GR in water [76]. This physical adsorption of PY^+-Chol on GR was able to furnish a stable and dark aqueous suspension of GR-PY^+-Chol at room temperature based on the hydrophobic interactions of the cholesterol moiety and the GR (Figure 11.18). The GR-PY^+-Chol suspension could then be used to solubilize tamoxifen citrate (TmC), a currently used breast cancer drug, in water. The resulting TmC-GR-PY^+-Chol remains stable for a long time without any precipitation. TmC-GR-PY^+-Chol remarkably enhances highly selective cell death (apoptosis) of the transformed cancer cells compared to the normal cells. This potency was found to be true for a wide range of transformed cancer cells, namely, A549 (carcinomic human alveolar basal epithelial cells), HeLa, HepG2 (human liver carcinoma cell line), NIH3T3 (mouse embryo fibroblast cell line), MCF-7 (human breast adenocarcinoma cell line), MDA-MB231 (breast cancer cells), and

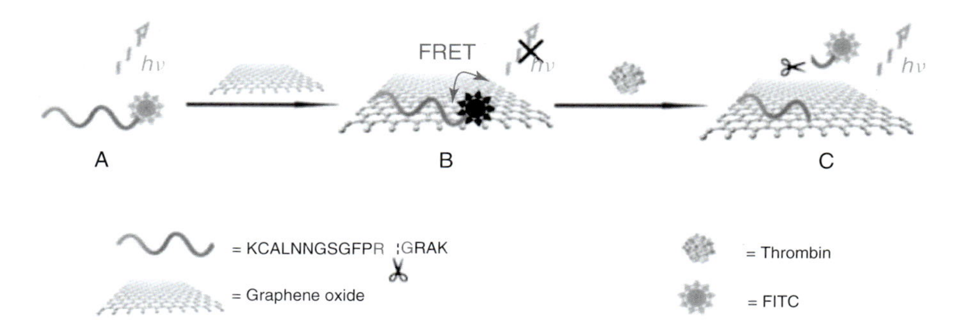

Scheme 11.3 Schematic illustration of the peptide–GO bioconjugate as a sensing probe to monitor the proteolytic activity of thrombin. The peptide was labeled with fluorescein isothiocyanate (FITC). The cleavage point at Arg–Gly bonds is indicated by a dashed line. (Source: Adapted with permission from Ref. [75]. Copyright 2011, The Royal Society of Chemistry.)

HEK293T (transformed human embryonic kidney cells) compared to the normal cell HEK293 (normal human embryonic kidney cells) *in vitro*. Confocal microscopy confirmed the high efficiency of TmC-GR-PY$^+$-Chol in delivering the drug to the cells, compared to the suspensions devoid of GR (Figure 11.18b).

Water-dispersible graphene derivatives such as GO and RGO show antibacterial activity [77]. Through a vacuum filtration protocol the GO and RGO nanosheets were converted into macroscopic, free-standing, robust, and flexible paper (Figure 11.19a,b). The GO nanosheets inhibit the growth of *Escherichia coli* DH5α bacteria while showing minimal cytotoxicity. It also affects the cellular integrity and destroys

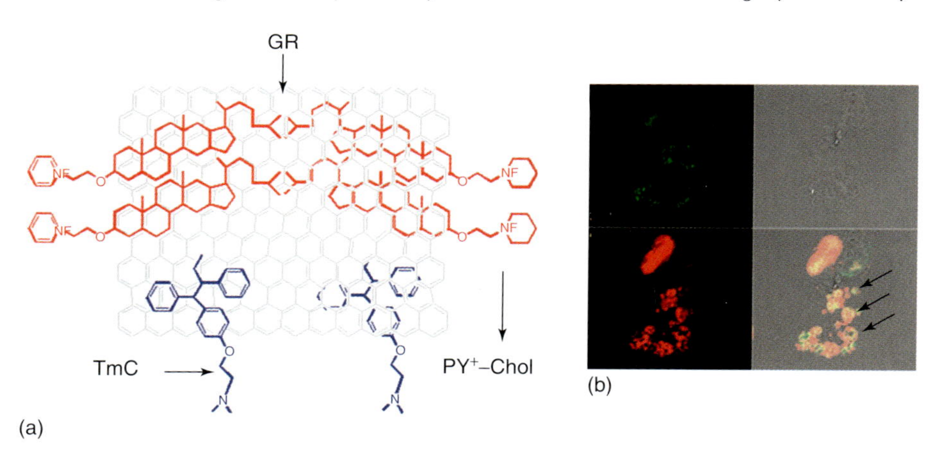

Figure 11.18 (a) Schematic representation showing a molecular level interaction between graphene, cholesterol-based cationic amphiphile (PY$^+$-Chol), and the drug tamoxifen (TmC). (b) Confocal microscopic images of MDA-MB231 cells treated with TmC–GR–PY$^+$-Chol and stained with p53 antibody. Arrow shows the presence of apoptotic cells. (Source: Adapted with permission from Ref. [76]. Copyright 2012, John Wiley and Sons.)

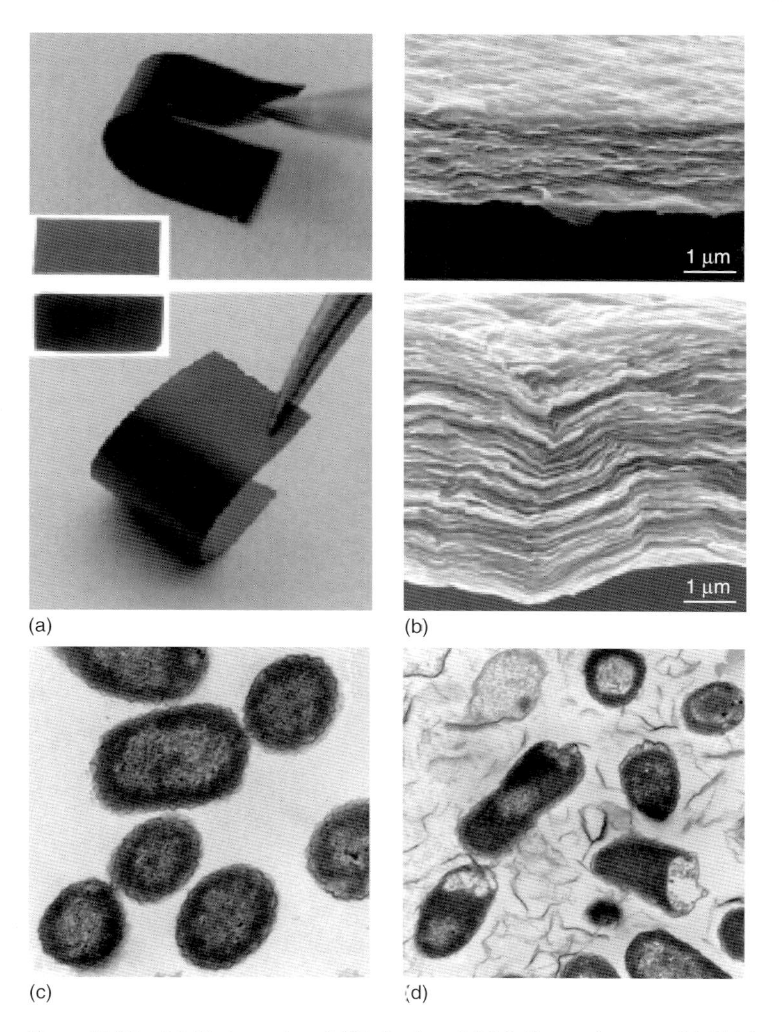

Figure 11.19 (a) Photographs of GO (top) and RGO (bottom) paper. (b) Thickness of GO (top) and RGO (bottom) paper as measured via SEM. TEM images of (c) *E. coli* and (d) *E. coli* exposed to GO nanosheets at 37 °C for 2 h. (Source: Adapted with permission from Ref. [77]. Copyright 2010, American Chemical Society.)

the cell membrane possibly due to oxidative stress or physical disruption processes (Figure 11.19c,d). Also, no evidence of cell growth was observed on the GO paper indicating an efficient antibacterial activity of such papers.

PEG-functionalized nanographene sheets (NGSs) were injected in mice and studied via *in vivo* fluorescence imaging [78]. PEGylated NGS (NGS-PEG) was prepared starting from graphite oxide conjugated with amine terminated six-arm branched PEG (10 kDa) via amide formation (Scheme 11.4). Surprisingly, high tumor accumulation of NGS was shown in several xenograft tumor mouse models due

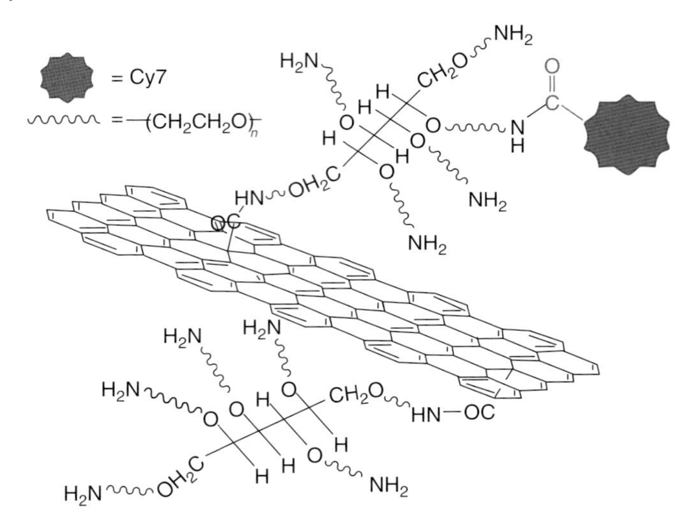

Scheme 11.4 NGS functionalized with PEG. A scheme of an NGS with PEG functionalization and labeling by Cy7. (Source: Adapted with permission from Ref. [78]. Copyright 2010, American Chemical Society.)

to the enhanced permeability and retention (EPR) effect of the cancerous tumors. The NIR optical absorption ability of NGS was utilized for *in vivo* photothermal therapy of cancer. Effective tumor destruction was demonstrated on intravenous NGS injection followed by low-power NIR laser irradiation. The NGS-PEG did not show any sign of toxicity in the injected mice. The *in vivo* behavior of NGS was studied using NIR fluorescent dye (Cy7) labeled NGS-PEG. The NGS-PEG-Cy7 was then injected intravenously in balb/c mice bearing 4T1 murine breast cancer tumors, nude mice bearing KB human epidermoid carcinoma tumors, and U87MG human glioblastoma tumors. Predominant uptake of NGS was found in the tumor after 24 h for all three types of tumor models as confirmed from the strong fluorescent signals (Figure 11.20, the first row). The *in vivo* photothermal therapy was performed on balb/c mice using 4T1 tumor cells after the injection of NGS-PEG followed by NIR irradiation. The irradiated tumors disappeared effectively one day after the laser irradiation. Therefore, these nano-bio-composites of graphene are potentially useful in cancer therapy.

11.7
Conclusions and Future Directions

Graphene is a relatively new 2D carbon allotrope that shows a number of interesting properties, and, therefore, it finds highly novel applications. Supramolecular self-assembly that leads to the gelation of graphenes in either aqueous or organic media and with or without the assistance of any external cross-linkers/host gelators provides new opportunities. GO, a "daughter" analog of graphene, could form

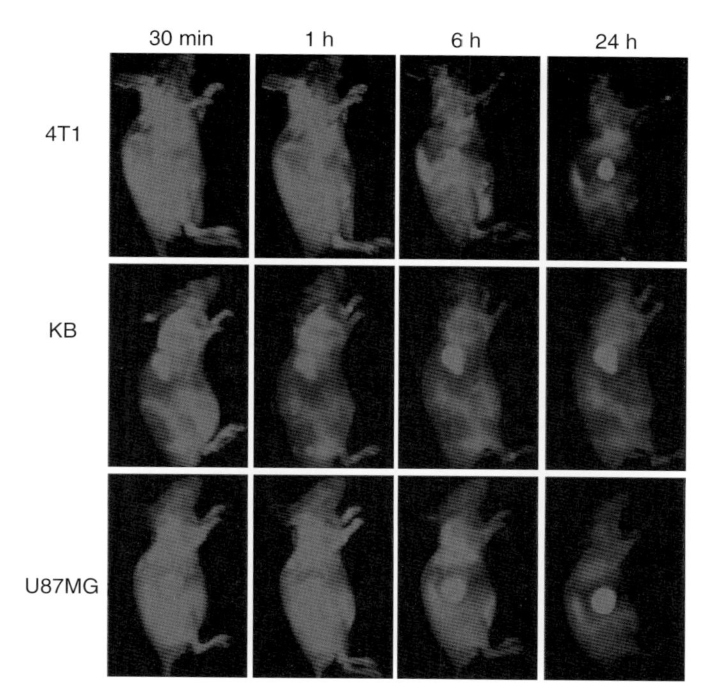

Figure 11.20 Spectrally unmixed *in vivo* fluorescence images of 4T1 tumor bearing Balb/c mice, KB, and U87MG tumor bearing nude mice at different time points postinjection of NGS-PEG-Cy7. Mouse autofluorescence was removed by spectral unmixing in the above images. High tumor uptake of NGSPEG-Cy7 was observed for all of the three tumor models. Hairs on Balb/c mice were removed before fluorescence imaging. (Source: Adapted with permission from Ref. [78]. Copyright 2010, American Chemical Society.)

gel in aqueous media because of its inherent hydrophobicity/hydrophilicity depending on various factors including pH, effect of cross-linker, effect of polymers, cations, biomolecules (Hb and DNA), surfactant, and IL. Various polymer-based nanocomposites assisted by gel formation show superior thermal, mechanical, and electrical properties. A few of these graphene-polymer gels form aerogels by either drying in supercritical CO_2 or via freeze-drying. However, small molecular host gels have also been employed to prepare graphene-based nanocomposites through noncovalent mode of interactions. This shows gelator-induced exfoliation of graphene layers, and simultaneously the integrity of the graphene backbone remains intact.

These gelation-induced nanocomposites are in their initial period of preparation and characterization, and a lot of fine-tuning will be required before their applications in any critical fields such as drug delivery and tissue engineering. Although several examples are reported on the preparation of hydrogel, their biological applications are limited till date such as their toxicity and amount of drug loading. Preparation of nano-biocomposites is in its infancy. Reports based on DNA, peptides, nanoparticles, FA, PEG, PEI, cholesterol, and so on have shown the nature of interaction of each one of them with graphene and graphene analogs toward the formation of appropriate nano-biocomposites. Therefore, graphene-based composites offer both significant potential and challenges, marking it a vibrant area of work for years to come. The improvement and application of these composites will depend on how effectively we can handle these challenges.

References

1. Novoselov, K.S., Geim, A.K., Morozov, S.V., Jiang, D., Zhang, Y., Dubonos, S.V., Grigorieva, I.V., and Firsov, A.A. (2004) *Science*, **306**, 666–669.
2. Kroto, H.W., Heath, J.R., O'Brien, S.C., Curl, R.F., and Smalley, R.E. (1985) *Nature*, **318**, 162–163.
3. Iijima, S. (1991) *Nature*, **354**, 56–58.
4. Geim, A.K. and Novoselov, K.S. (2007) *Nat. Mater.*, **6**, 183–191.
5. Morozov, S.V., Novoselov, K.S., Katsnelson, M.I., Schedin, F., Elias, D.C., Jaszczak, J.A., and Geim, A.K. (2008) *Phys. Rev. Lett.*, **100**, 016602, 4 pp.
6. Chen, J.-H., Jang, C., Xiao, S., Ishigami, M., and Fuhrer, M.S. (2008) *Nat. Nanotechnol.*, **3**, 206–209.
7. Xu, W., Dong, H.M., Li, L.L., Yao, J.Q., Vasilopoulos, P., and Peeters, F.M. (2010) *Phys. Rev. B*, **82**, 125304, 9 pp.
8. Bonaccorso, F., Sun, Z., Hasan, T., and Ferrari, A.C. (2010) *Nat. Photon.*, **4**, 611–622.
9. Ranjbartoreh, A.R., Wang, B., Shen, X., and Wang, G. (2011) *J. Appl. Phys.*, **109**, 014306, 6 pp.
10. Prasad, K.E., Das, B., Maitra, U., Ramamurty, U., and Rao, C.N.R. (2009) *Proc. Natl. Acad. Sci. U.S.A.*, **106**, 13186–13189.
11. Lee, C., Wei, X., Kysar, J.W., and Hone, J. (2008) *Science*, **321**, 385–388.
12. Baladin, A.A., Ghosh, S., Bao, W., Calizo, I., Teweldebrhan, D., Miao, F., and Lau, C.N. (2008) *Nano Lett.*, **8**, 902–907.
13. Du, X., Skachko, I., Barker, A., and Andrei, E.Y. (2008) *Nat. Nanotechnol.*, **3**, 49–495.
14. Li, X.L., Zhang, G.Y., Bai, X.D., Sun, X.M., Wang, X.R., Wang, E., and Dai, H.J. (2008) *Science*, **319**, 1229–1232.
15. Schedin, F., Geim, A.K., Morozov, S.V., Hill, E.W., Blake, P., Katsnelson, M.I., and Novoselov, K.S. (2007) *Nat. Mater.*, **6**, 652–655.
16. Paek, S.-M., Yoo, E.J., and Honma, I. (2009) *Nano Lett.*, **9**, 72–75.
17. Si, Y. and Samulski, E.T. (2008) *Chem. Mater.*, **20**, 6792–6797.
18. Bae, S., Kim, H., Lee, Y., Xu, X., Park, J.-S., Zheng, Y., Balakrishnan, J., Lei, T., Kim, H.R., Song, Y.I., Kim, Y.-J., Kim, K.S., Ozyilmaz, B., Ahn, J.-H., Hong, B.H., and Iijima, S. (2010) *Nat. Nanotechnol.*, **5**, 574–578.
19. Samanta, S.K., Pal, A., Bhattacharya, S., and Rao, C.N.R. (2010) *J. Mater. Chem.*, **20**, 6881–6890.
20. Samanta, S.K., Gomathi, A., Bhattacharya, S., and Rao, C.N.R. (2010) *Langmuir*, **26**, 12230–12236.
21. Terech, P. and Weiss, R.G. (1997) *Chem. Rev.*, **97**, 3133–3159.
22. Bhattacharya, S. and Samanta, S.K. (2009) *Langmuir*, **25**, 8378–8381.
23. Bhattacharya, S., Srivastava, A., and Pal, A. (2006) *Angew. Chem. Int. Ed.*, **45**, 2934–2937.
24. Wang, H., Yang, C., Wang, L., Kong, D., Zhang, Y., and Yang, Z. (2011) *Chem. Commun.*, **47**, 4439–4441.
25. Karinaga, R., Jeong, Y., Shinkai, S., Kaneko, K., and Sakurai, K. (2005) *Langmuir*, **21**, 9398–9401.
26. Suri, A., Campos, R., Rackus, D.G., Spiller, N.J.S., Richardson, C., Palsson, L.-O., and Kataky, R. (2011) *Soft Matter*, **7**, 7071–7077.
27. Grove, T.Z., Osuji, C.O., Forster, J.D., Dufresne, E.R., and Regan, L. (2010) *J. Am. Chem. Soc.*, **132**, 14024–14026.
28. Bird, R., Freemont, T.J., and Saunders, B.R. (2011) *Chem. Commun.*, **47**, 1443–1445.

29. Hummers, W.S. and Offeman, R.E. (1958) *J. Am. Chem. Soc.*, **80**, 1339–1339.
30. Subrahmanyam, K.S., Vivekchand, S.R.C., Govindaraj, A., and Rao, C.N.R. (2008) *J. Mater. Chem.*, **18**, 1517–1523.
31. Chen, D., Li, L., and Guo, L. (2011) *Nanotechnology*, **22**, 325601, 7 pp.
32. Bai, H., Li, C., Wang, X., and Shi, G. (2011) *J. Phys. Chem. C*, **115**, 5545–5551.
33. Si, Y.C. and Samulski, E.T. (2008) *Nano Lett.*, **8**, 1679–1682.
34. Dreyer, D.R., Park, S., Bielawski, C.W., and Ruoff, R.S. (2010) *Chem. Soc. Rev.*, **39**, 228–240.
35. Loh, K.P., Bao, Q., Eda, G., and Chhowalla, M. (2010) *Nat. Chem.*, **2**, 1015–1024.
36. Kim, J., Cote, L.J., Kim, F., Yuan, W., Shull, K.R., and Huang, J. (2010) *J. Am. Chem. Soc.*, **132**, 8180–8186.
37. Bai, H., Li, C., and Shi, G.Q. (2011) *Adv. Mater.*, **23**, 1089–1115.
38. Bai, H., Li, C., Wang, X., and Shi, G. (2010) *Chem. Commun.*, **46**, 2376–2378.
39. Jiang, X., Ma, Y., Li, J., Fan, Q., and Huang, W. (2010) *J. Phys. Chem. C*, **114**, 22462–22465.
40. Guardia, L., Fernandez-Merino, M.J., Paredes, J.I., Solis-Fernandez, P., Villar-Rodil, S., Martinez-Alonso, A., and Tascon, J.M.D. (2011) *Carbon*, **49**, 1653–1662.
41. Ng, S.R., Guo, C.X., and Li, C.M. (2011) *Electroanalysis*, **23**, 442–448.
42. Huang, C., Bai, H., Li, C., and Shi, G. (2011) *Chem. Commun.*, **47**, 4962–4964.
43. Liu, P., Zhang, X., Feng, L., Xiong, H., and Wang, S. (2011) *Am. J. Biomed. Sci.*, **3**, 69–76.
44. Xu, Y., Wu, Q., Sun, Y., Bai, H., and Shi, G. (2010) *ACS Nano*, **4**, 7358–7362.
45. Sridhar, V. and Oh, I.-K. (2010) *J. Ccl-loid Interface Sci.*, **348**, 384–387.
46. Sun, S. and Wu, P. (2011) *J. Mater. Chem.*, **21**, 4095–4097.
47. Alzari, V., Mariani, A., Monticelli, O., Valentini, L., Nuvoli, D., Piccinini, M., Scognamillo, S., Bon, S.B., and Illescas, J. (2010) *J. Polym. Sci., Part A Polym. Chem.*, **48**, 5375–5381.
48. Zu, S.-Z. and Han, B.-H. (2009) *J. Phys. Chem. C*, **113**, 13651–13657.
49. Zhang, L., Zha, D., Du, T., Mei, S., Shi, Z., and Jin, Z. (2011) *Langmuir*, **27**, 8943–8949.
50. Zhang, N., Li, R., Zhang, L., Chen, H., Wang, W., Liu, Y., Wu, T., Wang, X., Wang, W., Li, Y., Zhao, Y., and Gao, J. (2011) *Soft Matter*, **7**, 7231–7239.
51. Zhang, L., Wang, Z., Xu, C., Li, Y., Gao, J., Wang, W., and Liu, Y. (2011) *J. Mater. Chem.*, **21**, 10399–10406.
52. Wu, Q., Sun, Y., Bai, H., and Shi, G. (2011) *Phys. Chem. Chem. Phys.*, **13**, 11193–11198.
53. Zhang, L. and Shi, G. (2011) *J. Phys. Chem. C*, **115**, 17206–17212.
54. Worsley, M.A., Pauzauskie, P.J., Olson, T.Y., Biener, J., Satcher, J.H., and Baumann, T.F. (2010) *J. Am. Chem. Soc.*, **132**, 14067–14069.
55. Kai-Xuan, S., Yu-Xi, X., Chun, L., and Gao-Quan, S. (2011) *New Carbon Mater.*, **26**, 9–15.
56. Yang, X., Qiu, L., Cheng, C., Wu, Y., Ma, Z.-F., and Li, D. (2011) *Angew. Chem. Int. Ed.*, **50**, 7325–7328.
57. Xu, Y., Sheng, K., Li, C., and Shi, G. (2010) *ACS Nano*, **4**, 4324–4330.
58. Lo, C.-W., Zhub, D., and Jiang, H. (2011) *Soft Matter*, **7**, 5604–5609.
59. Hou, C., Zhang, Q., Zhu, M., Li, Y., and Wang, H. (2011) *Carbon*, **49**, 47–53.
60. Wang, J. and Ellsworth, M.W. (2009) *ECS Trans.*, **19**, 241–247.
61. Shao, J.-J., Wu, S.-D., Zhang, S.-B., Lv, W., Su, F.-Y., and Yang, Q.-H. (2011) *Chem. Commun.*, **47**, 5771–5773.
62. Zhang, X., Sui, Z., Xu, B., Yue, S., Luo, Y., Zhanc, W., and Liu, B. (2011) *J. Mater. Chem.*, **21**, 6494–6497.
63. Adhikari, B. and Banerjee, A. (2011) *Soft Matter*, **7**, 9259–9266.
64. Adhikari, B., Nanda, J., and Banerjee, A. (2011) *Chem. Eur. J.*, **17**, 11488–11496.
65. Samanta, S.K., Subrahmanyam, K.S., Bhattacharya, S., and Rao, C.N.R. (2011) *Chem. Eur. J.*, doi: 10.1002/chem.201102572
66. Wang, Y., Li, Z., Wang, J., Li, J., and Lin, Y. (2011) *Trends Biotechnol.*, **29**, 205–212.
67. Varghese, N., Mogera, U., Govindaraj, A., Das, A., Maiti, P.K., Sood, A.K., and

Rao, C.N.R. (2009) *ChemPhysChem*, **10**, 206–210.

68. Tang, Z., Wu, H., Cort, J.R., Buchko, G.W., Zhang, Y., Shao, Y., Aksay, I.A., Liu, J., and Lin, Y. (2010) *Small*, **6**, 1205–1209.
69. Yang, X., Wang, Y., Huang, X., Ma, Y., Huang, Y., Yang, R., Duana, H., and Chen, Y. (2011) *J. Mater. Chem.*, **21**, 3448–3454.
70. Zhang, L., Xia, J., Zhao, Q., Liu, L., and Zhang, Z. (2010) *Small*, **6**, 537–544.
71. Bao, H., Pan, Y., Ping, Y., Sahoo, N.G., Wu, T., Li, L., Li, J., and Gan, L.H. (2011) *Small*, **7**, 1569–1578.
72. Liu, Z., Robinson, J.T., Sun, X., and Dai, H. (2008) *J. Am. Chem. Soc.*, **130**, 10876–10877.
73. Zhang, L., Lu, Z., Zhao, Q., Huang, J., Shen, H., and Zhang, Z. (2011) *Small*, **7**, 460–464.
74. Song, Y., Chen, Y., Feng, L., Rena, J., and Qu, X. (2011) *Chem. Commun.*, **47**, 4436–4438.
75. Zhang, M., Yin, B.-C., Wang, X.-F., and Ye, B.-C. (2011) *Chem. Commun.*, **47**, 2399–2401.
76. Misra, S.K., Kondaiah, P., Bhattacharya, S., and Rao, C.N.R. (2012) *Small*, **8**, 131–143.
77. Hu, W., Peng, C., Luo, W., Lv, M., Li, X., Li, D., Huang, Q., and Fan, C. (2010) *ACS Nano*, **4**, 4317–4323.
78. Yang, K., Zhang, S., Zhang, G., Sun, X., Lee, S.-T., and Liu, Z. (2010) *Nano Lett.*, **10**, 3318–3323.

12
Biomedical Applications of Graphene: Opportunities and Challenges

Manzoor Koyakutty, Abhilash Sasidharan, and Shantikumar Nair

12.1
Introduction

Graphene is a novel 2D allotrope of carbon distinguished by its very unique electronic properties [1] such as high electrical conductivity, attributed to its sp^2-hybridized planar structure, as a result of which there has been an explosion of studies related to potential applications in devices, such as microelectronic or nanoelectronic devices [2], supercapacitors [3], batteries [4], and solar cells [5]. The previous allotropes of carbon namely, 0D carbon or fullerene [6], 1D carbon or carbon nanotube (CNT) [7], and 3D carbon or nanodiamond [8] evolved before graphene; none of which share the extraordinarily attractive electronic properties of graphene. Although conventional carbon-based materials are not of much biomedical use because of their nondegradability and carcinogenic potential [9], the family of "nanocarbons" raises the interesting possibility that they may be cleared from the body and not accumulate in tissues because of their small size. Thus, a wide body of literature had begun to emerge on the potential biomedical applications of the earlier carbon allotropes, namely, fullerenes [10], CNTs [11], and nanodiamonds [12]. Graphene, although sought after for its electronic and electrochemical properties, is nevertheless being investigated [13] for its biomedical applications as well because of its very high surface area (higher than the previous allotropes) and unique 2D structure that may make it a very attractive drug delivery agent and because its special electronic/physical properties could be exploited in biomedical applications, such as photothermal therapy or enhanced radiation therapy.

This chapter briefly reviews the opportunities and challenges involved in the use of graphene as a biomedical material. The opportunities are in drug delivery, biosensors, and other radiation therapies; but the challenges clearly will be the issues of systemic toxicity and long-term ill effects, such as carcinogenesis, and the problem of defining the physicochemical and structural characteristics of graphene that will minimize the toxicity and maximize whatever therapeutic advantages there are to be derived from this material.

Graphene: Synthesis, Properties, and Phenomena, First Edition. Edited by C. N. R. Rao and A. K. Sood.

12.2
Summary of Physical and Chemical Properties of Graphene

Graphene consists of sp^2-bonded carbon atoms ordered in single- or multilayer architecture, produced either by top-down or bottom-up approaches involving chemical vapor deposition, thermal exfoliation, radio frequency exfoliation, laser exfoliation, and so on [14]. Monolayer graphene is the material that has attracted the most interest because of its unique properties [15]. Monolayer graphene can be isolated from graphite by repeated mechanical exfoliation of graphite flakes [16]. Few-layer graphene (FLG) is the flakelike stacks of 2–10 layers of graphene [17]. Some of the important graphene attributes relevant to biomedical applications are described in the following sections.

12.2.1
Surface Chemistry (Biochemistry of Graphene)

The theoretical surface area of monolayer graphene is about 1200 $m^2 g^{-1}$, roughly twice what it would be for CNT or fullerene because both the surfaces of the sp^2-bonded carbon are exposed. In contrast, the surface area of a conventional silica nanoparticle is on the order of 100 $m^2 g^{-1}$. This raises the potential for efficient surface functionalization, which can modify potential toxicity and biological properties, and for highly efficient drug loading, which are potentials that are critical for any nanomedicine.

The as-prepared graphene without any surface modification or functionalization is generally termed as pristine graphene (*p*-G), which is hydrophobic and hence remains nondispersible in physiological media [18]. This property is shared with the other allotropes of carbon [19]. There is an oxidized form of graphene, called graphene oxide (GO), where the graphene edges contain oxygen-containing groups, which is more dispersible and less toxic and also contains a large number of structural defects [20]; and also reduced graphene oxide (rGO), wherein the GO is chemically reduced by the removal of the oxygen-containing side groups [21]. rGO contains more defects than *p*-G because of the inherent high defect content in GO. The basal planes of GO include unmodified graphenic domains that are hydrophobic and capable of $\pi-\pi$ interactions relevant to adsorption of organic biomolecules or drugs, while the sheet edges have a hydrophilic character. The result is that GO is an amphiphilic sheetlike molecule that can act similar to a surfactant [22] and stabilize hydrophobic molecules in solution or collect at interfaces [23].

To make graphene truly dispersible in physiological media, it is necessary to functionalize the graphene surface with suitable biomolecules. This gives rise to functionalized graphene (*f*-G) which, by its very nature, has significantly different biological properties, as is reviewed in the following sections.

12.2.1.1 Interaction of Graphene Surfaces with Biomolecules

PEGylation of nanoparticles with polyethylene glycol is a well-established approach to improve dispersion and prolong circulation in the blood [24]. Graphene too has been PEGylated by noncovalent attachment of PEG molecules to the graphene surfaces by π−π bonds [25]. Direct cytotoxicity of graphene to mammalian cells was reduced by PEGylation [26]. It is, however, yet to be established whether this approach prolongs the circulation time of graphene in the blood *in vivo* as for the case of other nanoparticles.

Carboxylic acid treatment leads to the attachment of COOH side groups, which results in *f*-G that is fully dispersible in the aqueous phase. This *f*-G is efficiently taken up by mammalian cells into the cytoplasm in comparison with *p*-G, which appears to stay localized on the cell membrane surface [18], (Figure 12.1). The mechanism of uptake of graphene by the cells appears to be by endocytosis, given the energy-dependent nature of the process [27], although the particular role of the surface COOH groups in triggering endocytosis is not understood. A more detailed comparison of the *p*-G with the *f*-G in terms of cell function is given in Section 12.3.

Proteins can attach to graphene by physical adsorption or chemical bonding. The bonding can be sufficiently strong to allow for intercalation and exfoliation

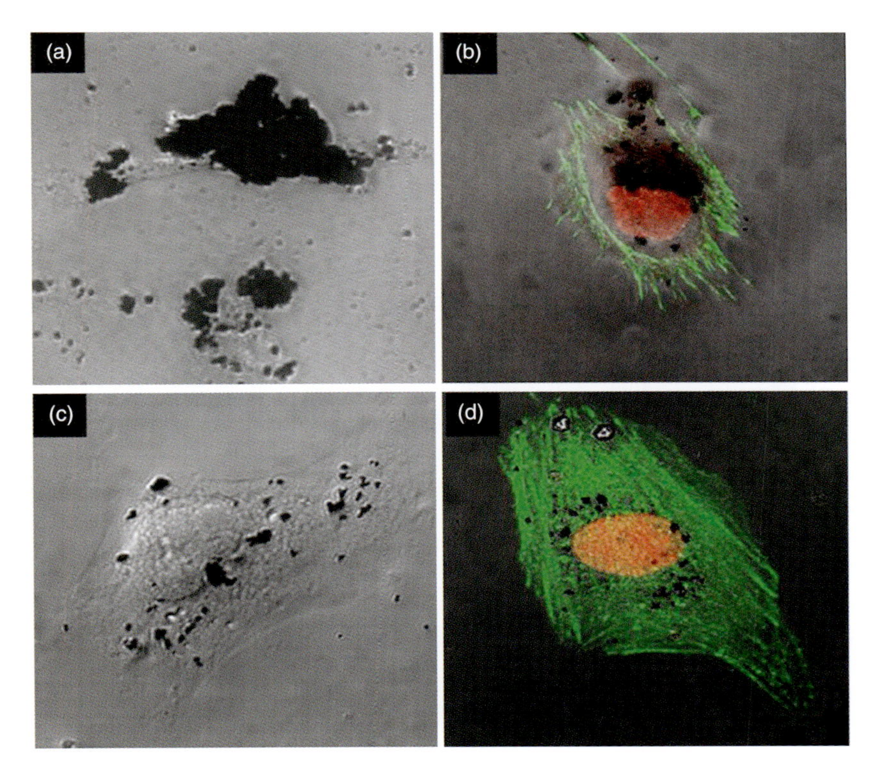

Figure 12.1 Differential interference contrast (DIC) images and fluorescence microscopic images of (a) pristine and (b) functionalized graphene-treated Vero cells (c) DIC image showing the intracellular uptake of COOH-functionalized graphene (f-G) and (d) confocal microscopic image of cytoskeletal F-actin arrangement of cells treated with f-G(Source: From Ref. [18].)

of multilayer graphene sheets [28]. Likewise, peptides [29], aptamers [30], and nucleic acids [31] can be readily functionalized to graphene sheets followed by the attachment of other materials, such as nanoparticles, secondarily through these biomolecules [32]. For example, noncovalent attachment of DNA to graphene followed by attachment of Au nanoparticles to the DNA resulted in a highly water-dispersible graphene solution with the properties of printable inks [33]. In another example, single-stranded DNA (ssDNA) could be covalently attached to the graphene, allowing the graphene to act as a DNA sensor [34]. Graphene has also been dispersed in PEG, pluronic, and deoxycholate (DOC) (sodium deoxycholate). PEG gave the best results in terms of biocompatibility of graphene in aqueous solutions [26].

Further, functionalization with hydrophilic polymers has also been reported to improve the stability of graphene in biological solutions [25]. Other covalent chemistry, such as one using 1,3-dipolar cycloaddition, has also been reported for the modification of graphene surface [35]. For the noncovalent chemistry, polymers such as PEG, poly(ethylenimine) (PEI), dextran, and so on, can be used to functionalize graphene via hydrophobic bindings or $\pi-\pi$ interactions [36, 37]. The surface modification is important from the perspective of reducing the toxic effects as well as loading various aromatic drug molecules, which can be physically adsorbed on the polyaromatic graphene surface by $\pi-\pi$ stacking [36].

12.3
Cellular Uptake, Biodistribution, and Clearance

12.3.1
Influence of Surface Chemistry on Uptake

It was also revealed that despite significant differences in surface chemistry, once water dispersed, *f*-CNTs could enter any type of cells (prokaryotic/eukaryotic and normal/cancer), even under endocytosis-inhibited conditions and thereafter permeate through different intracellular barriers to accumulate in the perinuclear region [38, 39]. In effect, these studies imply that surface functionalization of carbon nanostructures has a significant impact on their nano-biointeractions and possible toxic effects. The effect of surface functionalization on uptake in kidney epithelial cells is also shown in Figure 12.1, which is reproduced from the work of Sasidharan *et al.* [18]. In this study, it was clear that, although *p*-G was not internalized compared to *f*-G, it could have adverse effects because of the accumulation on the plasma membrane. In effect, this study showed that surface functionalization of graphene was critical for pacifying its strong hydrophobic interaction-associated toxic effects. However, being a nonbiodegradable material with greater potential for cellular internalization, the possible long-term adverse effects of functionalized hydrophilic graphene may also be a concern and require further study. Detailed mechanisms of toxicity was also reported using dispersed graphene or GO sheets and in other cell types such as lung epithelial cells [40], fibroblasts [41], and neuronal cells [42]. Single-layer GO sheets were internalized and sequestered in cytoplasmic membrane-bound vacuoles by human lung epithelial cells or fibroblasts.

12.3.2
Uptake of Graphene by Macrophages

Macrophages play an important role in the phagocytosis of foreign bodies as well as in alerting the rest of the immune system against invaders to elicit innate or adaptive immune response. To understand the response of macrophages to hydrophobic versus hydrophilic graphene, Sasidharan *et al.* [27] studied the macrophage uptake of *p*-G and *f*-G by SEM and energy-dispersive X-ray spectrometry (EDX). Compared to the untreated control cells, *p*-G-treated cells showed interaction of graphene flakes at the plasma membrane. In contrast, significant intracellular uptake was seen in *f*-G-treated macrophages (Figure 12.2).

To further investigate the intracellular uptake, these authors used confocal Raman spectral mapping of the *f*-G-treated macrophages. Cell mapping (Figure 12.3) was carried out using characteristic C–H vibration of cells at 2896 cm^{-1} (indicated by blue color), and graphene was mapped using its G-band vibration at 1580 cm^{-1}

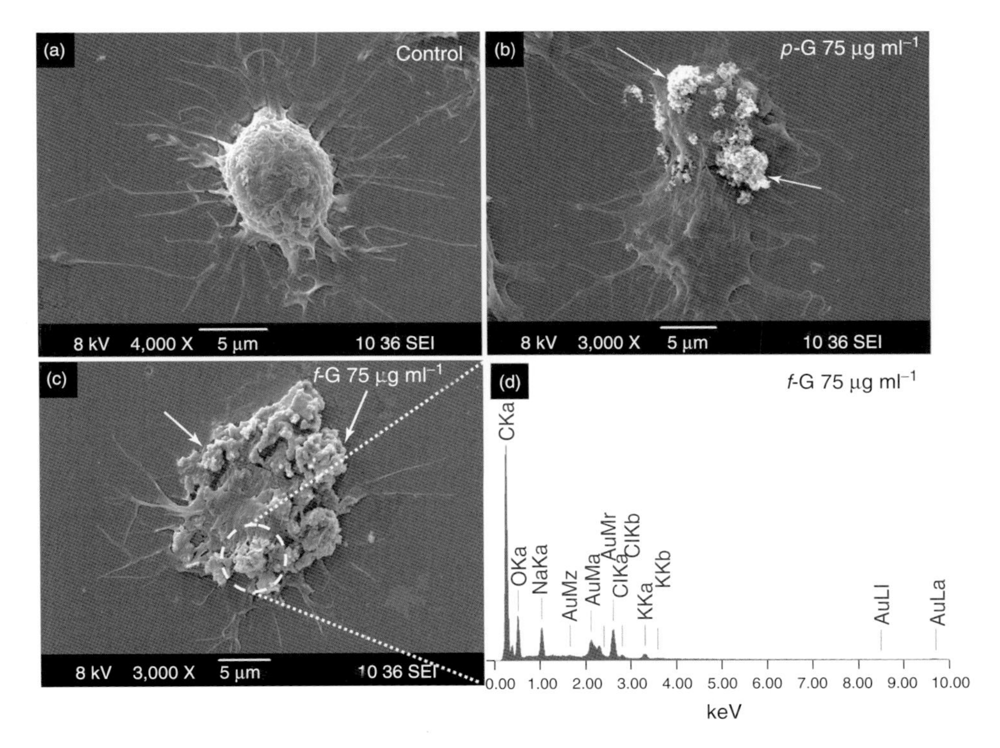

Figure 12.2 SEM images of graphene-treated macrophage cell line RAW 264.7: (a) control, (b) *p*-G 75 μg ml^{-1}, (c) *f*-G 75 μg ml^{-1}, and (d) EDX spectrum of *f*-G-treated cells. (Source: From Ref. [27].)

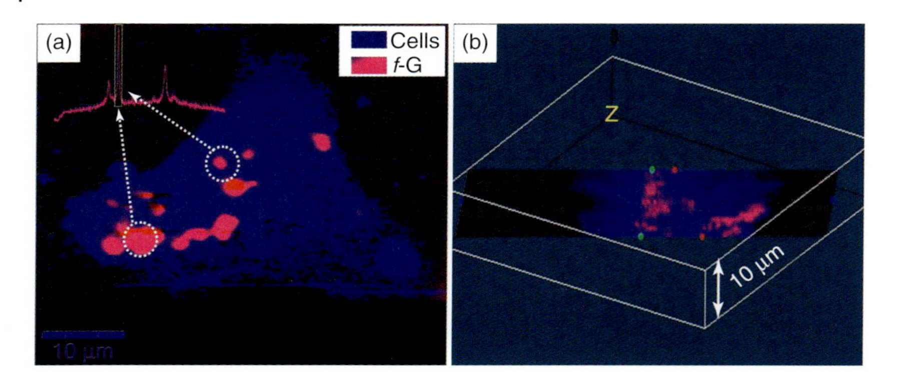

Figure 12.3 Confocal Raman spectral mapping of graphene-treated macrophage cells: (a) 2D Raman image of *f*-G-treated RAW 264.7 cells showing intracellular distribution of *f*-G (red dots) imaged using G-band vibration at 1580 cm^{-1} (inset); (b) cross-sectional view of Z-stacked 3D Raman image showing the presence of graphene at different depths of the cells. Cell is mapped using C–H vibration at 2896 cm^{-1}. (Source: From Ref. [27].)

(indicated by pink color). A cross-sectional image (Figure 12.3b) from the 3D stacked Raman spectral mapping clearly showed the presence of *f*-G at different depths within the cells, thereby confirming its intracellular uptake.

Effect of graphene interactions with the cytoskeleton of macrophage cells was studied using fluorescent confocal microscopy (Figure 12.4). Cytoskeleton plays an important role in the phagocytic defense mechanism of macrophages, and actin filaments are vital for cytokinesis. Under stressed conditions, thiol groups in the actin proteins become highly sensitive to reactive oxygen species (ROS) stress, leading to the inhibition of F-actin and subsequent cytoskeleton dysfunction. Confocal fluorescence images of *p*-G- and *f*-G- (75 $\mu g\ ml^{-1}$) treated macrophages revealed interesting details of graphene interaction with macrophages, where *p*-G remained mostly on the cell membrane with less internalization. The cells were found not attaining its normal stretched morphology, and the filopodial extensions were not clearly visible. In contrast, higher intracellular uptake was seen in *f*-G without any adverse effects on the integrity of cytoskeleton and filopodial extensions.

Although there are numerous reports on the cellular uptake of CNTs and fullerenes, the mechanisms of internalization still remain unclear. Some studies reported an energy-dependent endocytic pathway for CNTs [43], while others showed diffusion of nanotubes through the cell membrane [38, 44]. Similarly, many recent studies reported the internalization of graphene in mammalian cells [18, 45, 46]; however, the cellular uptake pathway remains inconclusive. In the recent work by Sasidharan *et al.* [18], inhibition of uptake at 4 °C compared to 37 °C implied that cellular uptake of *f*-G was mediated by an energy-dependent process. To identify specific possible cellular endocytosis mechanisms, such as clathrin- or caveolae-mediated endocytosis, more detailed experiments need to be conducted for the case of graphene. Because graphene is a molecular sheet, one cannot also rule out new processes that are diffusion limited and thermally activated.

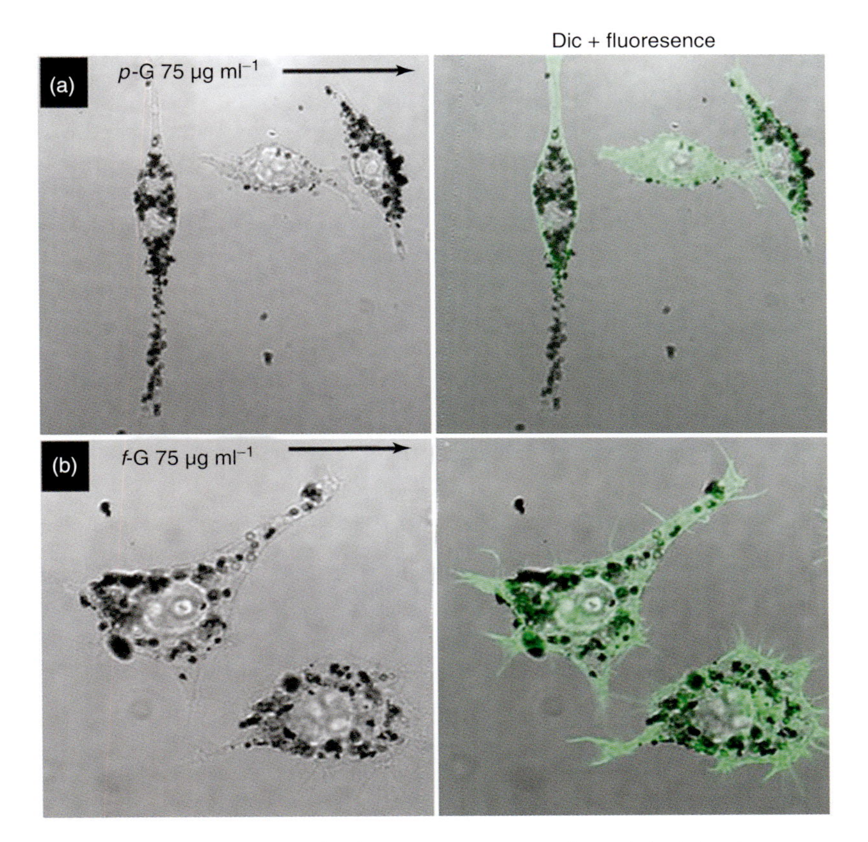

Figure 12.4 Fluorescence confocal microscopic images of cytoskeletal F-actin arrangement in (a) *p*-G- (75 μg ml^{-1}) and (b) *f*-G- (75 μg ml^{-1}) treated RAW 264.7 cells. (Source: From Ref. [27].)

12.4
Toxicity of Graphene

For the case of CNT, toxicity has been attributed to metal ion impurities incorporated into CNT during the processing steps involving metal-catalysis-driven growth. Schipper *et al.* [47] showed that after ultrahigh purification of CNT, toxicity could not be detected in animal samples even after four months. This was a promising result indicating that pure carbon nanostructures may not cause inherent toxic effects. However, it will be difficult to produce ultrahigh pure carbon nanostructures in bulk quantity. But unlike CNTs, graphene is not typically synthesized by catalytic growth, rather it is formed by the exfoliation of highly pure graphite and hence may remain free from residual metal impurities even without additional purification steps. However, in some of the exfoliation processes, toxic solvents, intercalates, or other chemical additives are generally used to separate the layers in the bulk graphite [48] and the same may remain without being fully removed by washing. In addition, preparation of *f*-G or GO uses a variety of reagents such as permanganate, nitrate, sulfate, chromate, peroxide, persulfate, hydrazine, and borohydride and associated cations, typically potassium, sodium, or ammonium, which may leave trace residues within the flakes [49]. Oxidized graphene may also contain lower molecular weight oxidative debris that is noncovalently attached to the primary sheets. These chemical impurities may react with biomolecules or cells and impart toxicity, which may be misinterpreted as that contributed by graphene. Hence, it is important for extensive elemental characterization of graphene used for toxicity studies and biomedical applications.

12.4.1 Macrophage Toxicity

Effect of graphene interactions on the metabolic activity of macrophages was studied by Sasidharan *et al.* [27] after treatment with 0–75 $\mu g\ ml^{-1}$ of *p*-G or *f*-G for 48 h. With increase in concentration of *p*-G sample above 50 $\mu g\ ml^{-1}$, cells started showing reduction in metabolic activity compared to *f*-G-treated cells, which remained viable up to the maximum tested concentration of 75 $\mu g\ ml^{-1}$ (Figure 12.5a). As observed in the confocal images, it appears that strong hydrophobic interactions of *p*-G on the cell membrane eventually lead to cytotoxic effects unlike the case of *f*-G. This result correlated with the earlier data obtained by the same authors on the

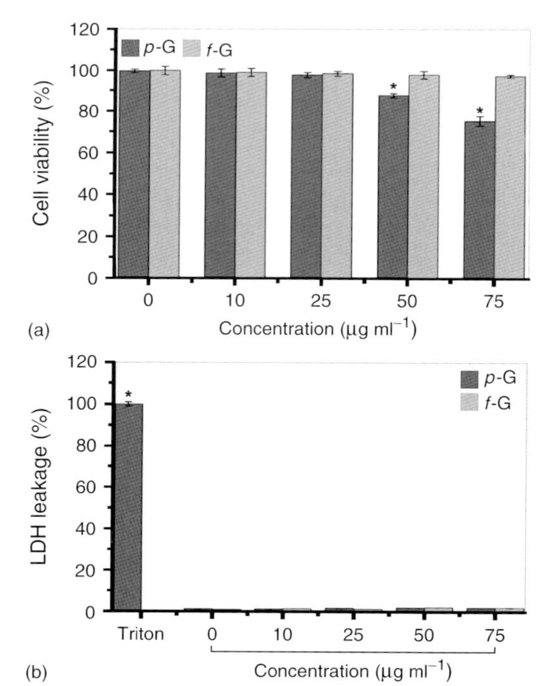

Figure 12.5 Effects of *p*-G and *f*-G on cell viability (a) and plasma membrane integrity (b) of RAW 264.7 cells. (Source: From Ref. [27].)

differential toxic effects of *p*-G and *f*-G on kidney epithelial cell lines [18] and very recent report of Li *et al.* [50] showing toxicity of *p*-G toward macrophage cell line RAW 264.7 by the activation of MAPK and TGF-β signaling. Moreover, this study clearly showed that surface functionalization could pacify the toxic property of *p*-G, whereas even with significant intracellular uptake by the macrophage cells, *f*-G imparted less toxicity than *p*-G under similar *in vitro* test conditions. Considering the strong interactions of both the samples with the plasma membrane of cells, the membrane integrity of *p*-G- and *f*-G-treated cells was also studied using lactate dehydrogenase (LDH) assay. Previous report showed elevated level of LDH leakage from CNT-treated cells [51]. However, no significant LDH leakage was noted from either *p*-G- or *f*-G-treated cells after 48 h of incubation (Figure 12.5b). This implies that although *p*-G-exerted toxicity at higher concentrations and *f*-G was greatly internalized by the cells, the plasma membrane of macrophages remained intact with no leakage of LDH. This implies that unlike the case of CNTs [51], graphene imparted insignificant physical damage to the plasma membrane and the observed toxic effects of *p*-G may be caused by alternative mechanisms.

12.4.2 Hemocompatibility

Hemocompatibility of graphene is an essential requirement for any kind of *in vivo* applications. In Ref. [27], the authors evaluated the hemolytic potential, platelet activation, platelet aggregation, coagulation time factors, proinflammatory cytokine expression, lymphocyte proliferation, and immunotoxic potential of both the graphene systems using human peripheral blood derived from healthy volunteers. This is reviewed in the following sections.

12.4.2.1 Hemolysis

Hemolysis is the loss of membrane integrity of red blood cells (RBCs) leading to the leakage of hemoglobin (Hb) into blood plasma. Hemolysis is one of the basic tests to understand the interaction of nanoparticles with RBCs. Nanoparticles might affect the membrane integrity of RBCs by mechanical damage or ROS [52]. In addition, hemolytic property of nanoparticles can also be influenced by their size, shape, surface charge, and chemical composition [53]. For studying the hemolytic potential of graphene, RBCs from human peripheral blood were treated with 0–75 $\mu g\ ml^{-1}$ of *f*-G and *p*-G for 3 h at room temperature. Figure 12.6a–d shows that, compared to the positive control (Triton X-100), *p*-G- or *f*-G-treated RBCs did not show any hemolysis up to 75 $\mu g\ ml^{-1}$. This is also evident from the optical images (Figure 12.6a, inset) of the blood samples, where triton-treated RBCs showed leakage of Hb into the supernatant, but *p*-G- or *f*-G-treated RBCs showed clear supernatant and the undamaged RBCs settled down to bottom of the tube. This was further confirmed using SEM imaging, where the biconcave morphology of RBCs was found intact even after interaction with *p*-G or *f*-G. These results indicate nonhemolytic property of graphene systems up to tested concentration of 75 $\mu g\ ml^{-1}$.

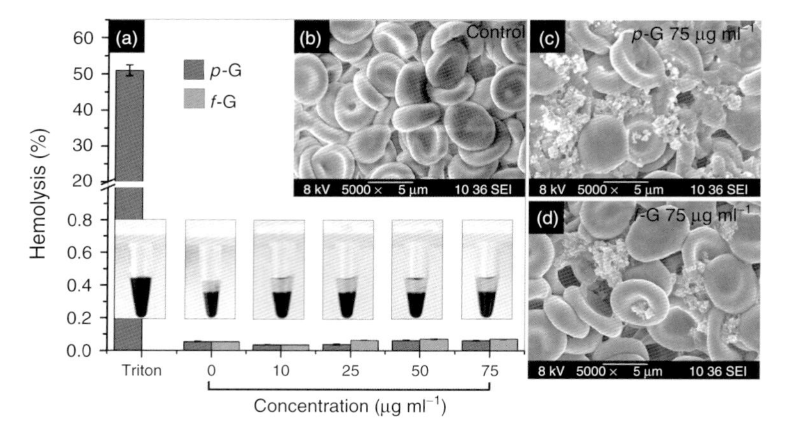

Figure 12.6 Assessment of hemolytic activity. (a) Hemolysis analysis of different concentrations (0–100 μg ml^{-1}) of both the graphene systems treated with whole blood. Inset: optical photograph showing no significant hemoglobin leakage from graphene-treated samples compared to positive control- (Triton X-100) treated sample. SEM image showing the normal morphology of (b) PBS, (c) *p*-G- (75 μg ml^{-1}), and (d) *f*-G- (75 μg ml^{-1}) treated RBCs. (Source: From Ref. [27].)

12.4.2.2 **Effect on Hemostasis: Platelet Activation and Aggregation**

Hemostasis is a multicomponent cascade, which prevents the loss of blood from an injured site and maintains the fluidity of blood for circulation. Circulating platelets and plasma proteins are the main constituents of the hemostatic system. Platelets are ~2 μm sized cells, more fragile than RBCs, originate from megakaryocytes in the bone marrow, and are responsible for thrombus formation, a final step in hemostasis. Numerous studies have reported that carbon nanoparticles can cause platelet activation and aggregation [54–56]. Radomski *et al.* [57] compared *in vitro* and *in vivo* platelet functions in different types of carbon-based materials and reported their platelet aggregation capability in rat models. Very recently, Singh *et al.* [58] reported platelet activation and thrombogenic potential of GO compared to rGO. We have studied the effects of *p*-G and *f*-G on platelet activation and aggregation by flow cytometry and SEM. Platelet-rich plasma (PRP) was incubated with both the graphene systems, and platelet activation was studied by monitoring the expression of granular membrane protein CD62P and resting platelet surface membrane glycoprotein CD42b (Figure 12.7). Flow cytogram showed that compared to controls, phosphate buffer solution (PBS), and ADP, the graphene samples showed maximum ~6.4% of platelet activation, which is relatively insignificant. This was further confirmed by SEM imaging, where the graphene-treated platelets (Figure 12.7e,f) remained discrete and maintained their discoidal shape. In order to confirm these results, the platelet aggregation potential of graphene systems was evaluated. As a part of maintaining hemostasis and preventing excessive blood loss, platelets aggregate along with other blood components to form blood clot. Exaggerated platelet aggregation may cause stroke. PRP was incubated with 0–75 μg ml^{-1} of *p*-G or *f*-G for 30 min, and Figure 12.7g shows that while ADP-treated platelets

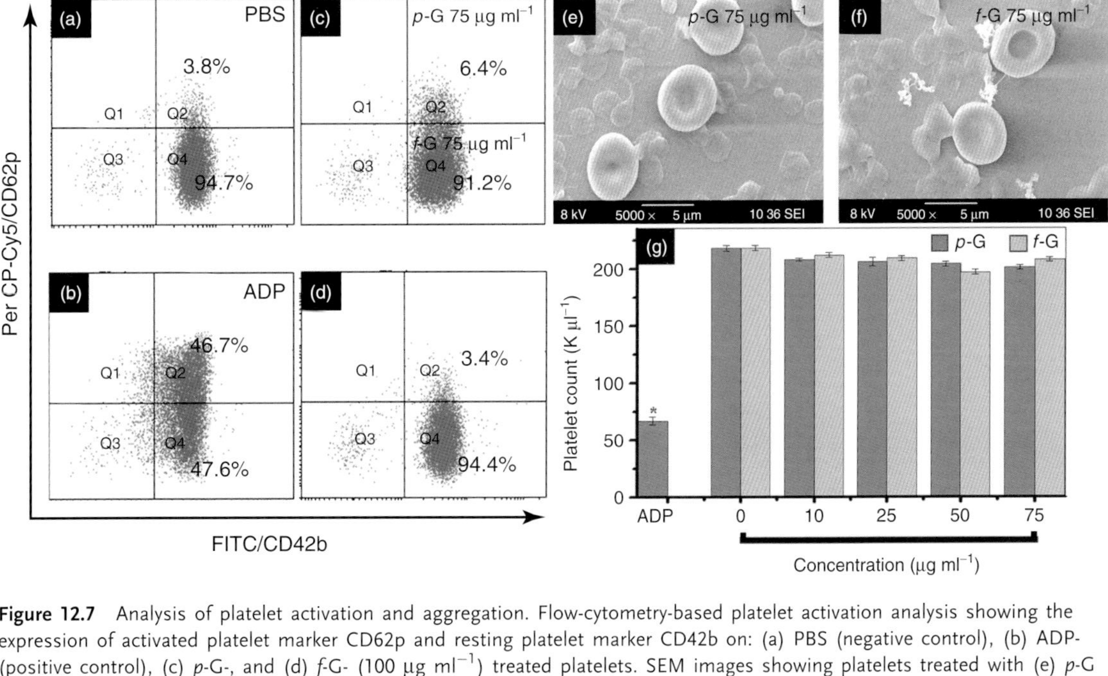

Figure 12.7 Analysis of platelet activation and aggregation. Flow-cytometry-based platelet activation analysis showing the expression of activated platelet marker CD62p and resting platelet marker CD42b on: (a) PBS (negative control), (b) ADP- (positive control), (c) *p*-G-, and (d) *f*-G- (100 μg ml^{-1}) treated platelets. SEM images showing platelets treated with (e) *p*-G and (f) *f*-G samples confirming nonactivated and well-separated platelets (g) Platelet count analysis of whole blood treated with varying concentrations (0-75 μg ml^{-1}) of p–G and f–G, which shows normal platelet counts. (Source: From Ref. [27].)

yielded significant reduction in the count, graphene-treated platelets maintained normal count within the range of 150–450 K μl^{-1}. This suggests that neither *p*-G nor *f*-G caused aggregation of platelet and clotting of blood. This result is interesting because other carbon nanoparticles and CNTs reported to show activation and aggregation of platelets mediated by Ca^{2+} influx [56]. In effect, our study clearly shows that both the graphene systems do not interfere with the platelet functions, at least up to 75 $\mu g\ ml^{-1}$, which is relatively high from the perspective of a drug delivery vehicle.

12.4.2.3 **Effect on Plasma Coagulation**

Plasma coagulation cascade consists of intrinsic and extrinsic pathways, both converging at a common point where factor X gets activated to Xa, which sequentially activates prothrombin to thrombin and subsequently converts fibrinogen to fibrin. Nanoparticle-induced *in vitro* plasma coagulation can serve as a model to study the *in vivo* thrombogenic potential. Several recent studies reported intrusion of CNTs in intrinsic plasma coagulation pathway [59]. We have studied this possibility with graphene by monitoring prothrombin time (PT) and activated partial thromboplastin time (aPTT) coagulation time factors. PT measurement detects abnormalities of the factors involved in the extrinsic pathway, including factors VII, X, V, II, and fibrinogen. However, it was observed that PT values for all the tested concentrations fell well within the normal range of 12–15 s for both the graphene samples (Figure 12.8a). The intrinsic pathway takes place during abnormal physiological conditions such as hyperlipidemic states or bacterial infiltration, which can lead to thrombosis. aPTT registers abnormalities in intrinsic clotting pathway factors, including I, II, V, VIII, IX, X, XI, XII, and proteins such as prekallikrein (PK), high-molecular weight kininogen (HMWK), and fibrinogen. Figure 12.8b reveals that the aPTT ratio variation falls within the normal range of 0.9–1.2 s for all the tested concentrations (0–75 $\mu g\ ml^{-1}$) of *p*-G and *f*-G. Collectively, all the above results on hemostasis analysis shows that both the graphene systems did not interfere with the normal functioning of platelets or influence the coagulation pathways and hence may remain nonthrombogenic.

12.4.3 Inflammatory Response

During an inflammatory response, activated immune cells produce soluble, low-molecular weight glycoproteins and cytokines, which recruit more immune cells to tackle the foreign substances that have entered into the body [60]. There have been reports of SiO_2 and ZnO nanoparticles inducing proinflammatory cytokines [61, 62]. However, the influence of graphene on the immune system has hardly been appraised, although Shvedova *et al.* [63] demonstrated that CNTs interact with the immune system and induce severe inflammatory response *in vivo*. The potential of *p*-G and *f*-G to cause inflammatory response by measuring the production of major proinflammatory cytokines such as IL-8, IL-1β, IL-6, IL-10, tumor necrosis factor (TNF), and IL-12p70 in endotoxin-free graphene-treated peripheral

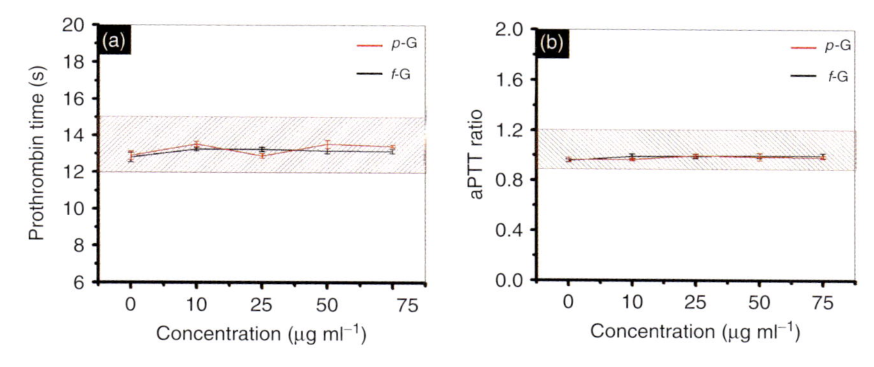

Figure 12.8 (a) Prothrombin time (PT) and (b) activated partial thromboplastin time (aPTT) ratio of *p*-G- and *f*-G- (0–75 μg ml^{-1}) treated blood plasma samples showing no significant variation from normal range, which is depicted as shaded region in both graphs. (Source: From Ref. [27].)

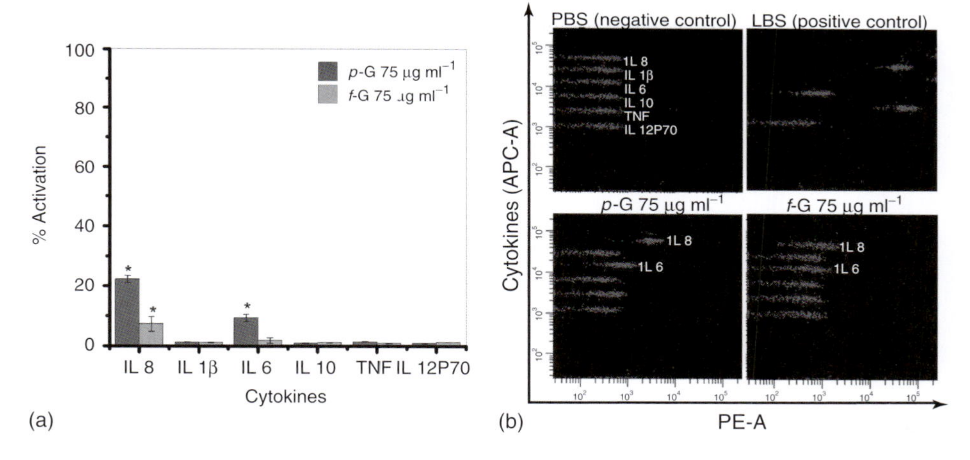

Figure 12.9 (a) Flow cytometric analysis showing the expression of proinflammatory cytokines from *p*-G- and *f*-G-treated PBMCs. (b) Dot plots showing relative expression of various cytokines in *p*-G- and f-G-treated PBMCs. LPS, lipopolysaccharide. (Source: From Ref. [27].).

blood derived mononuclear cells (PBMCs) was also studied by Sasidharan *et al.* [27]. By taking PBS as the negative control and lipopolysaccharides (LPSs) as the positive control (100%), the percentage activation of the above-mentioned set of proteins plotted in Figure 12.9a,b shows that compared to *f*-G, the *p*-G sample induced relatively higher expression of IL-8 and IL-6. All other cytokines showed basal level expression for both the samples. This indicates a possibility of higher inflammatory potential of *p*-G compared to *f*-G, which needs to be confirmed by *in vivo* studies.

12.4.3.1 Immune Cell Stimulation and Suppression

Most of the existing reports clearly suggest that nanomaterials interact with immune cells and may cause their stimulation or suppression, leading to chronic ailments such as allergy, autoimmune diseases, or cancer [64]. On activation of lymphocytes by a foreign material or antigen, clonal proliferation of the cells takes place as a part of the immune response. To study the possibility of graphene to trigger lymphocyte proliferation, four different concentrations (0–75 $\mu g\ ml^{-1}$) of *p*-G and *f*-G were treated with lymphocytes for 72 h and proliferation was quantified using Alamar Blue assay (Figure 12.10a). Phytohemagglutinin-M (PHA-M), a mitogenic stimulator, which causes lymphocyte proliferation, was used as a positive control. Interestingly, it was observed that while PHA-M induced significant proliferation of lymphocytes, *p*-G- or *f*-G-treated cells showed normal proliferative characteristics similar to those of negative control (PBS), as shown in Figure 12.10b. Confocal differential interference contrast (DIC) images of lymphocytes treated with PHA-M showed cell proliferation and agglutination (Figure 12.10c), which was absent in *p*-G- and *f*-G-treated cells (Figure 12.10d,e). This result suggests that unlike single- and multiwalled CNTs, which caused allergic responses in mice [65], graphene may remain largely nonimmunogenic irrespective of its varied surface chemistry.

Immunosuppression by carbon nanoparticles is another major concern because of the possible cytotoxicity to immune cells. Mitchell *et al.* [66] have reported that CNTs cause immunosuppression in mice by TGF-β production. We have studied whether graphene interferes with the normal proliferative capacity of lymphocytes in response to a mitogen. Lymphocytes were treated with different concentrations (0–75 $\mu g\ ml^{-1}$) of both the *p*-G and *f*-G in the presence of a mitogen, PHA-M, and the effects of graphene in providing any inhibitory effects on cell proliferation were investigated. Alamar Blue assay results in Figure 12.11a showed no significant variation between PHA-M alone treated cells versus graphene and PHA-M-treated

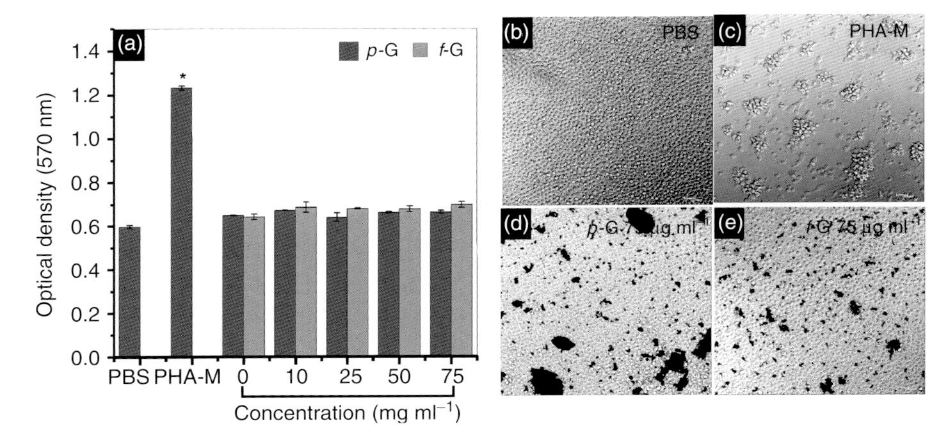

Figure 12.10 Immunostimulation analysis. (a) Proliferation of lymphocytes on exposure to different concentrations of *p*-G and *f*-G for three days. Optical microscope image cells treated with (b) PBS, (c) PPHA-M (positive control), (d) *p*-G, and (e) *f*-G [27].

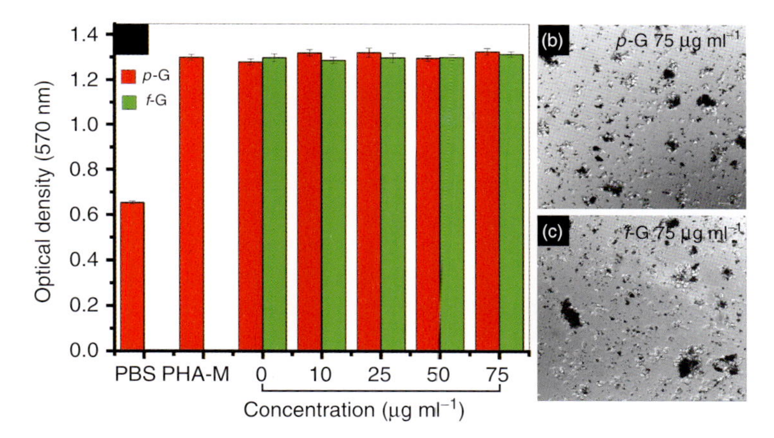

Figure 12.11 Immunosuppression analysis. (a) Cell viability analysis on lymphocytes exposed to mixture of PHA-M + *p*-G or *f*-G. Optical microscopic images of (b) *p*-G- and (c) *f*-G-treated cells. (Source: From Ref. [27].)

lymphocytes, indicating that both the *p*-G and *f*-G did not impede innate lymphocyte proliferation. This was further confirmed by confocal microscopy (Figure 12.11b,c), which showed that both the *p*-G and *f*-G in the presence of PHA-M-induced lymphocyte agglutination, depicting its normal function. In effect, these studies conclude that both the graphene systems neither stimulated nor suppressed the regular functioning of immune cells.

12.4.4
Toxicity Mechanisms

It was found that in addition to nanoscale features such as size-scale, shape (fibrilous nature), or structure (single- or multiwalled), a critical aspect that significantly influenced its nano-biointeractions and toxicity was its dispersion characteristics in physiological medium, which is determined by its surface chemistry. It is well known that pristine CNTs are highly hydrophobic, whereas surface functionalization (carboxylated, aminized, or PEGylated) renders hydrophilicity and dispersibility in aqueous phase, enabling varied interactions with biological systems [67, 68]. For example, pristine CNTs were reported to impart serious adverse toxic effects, including reactive oxygen stress, inflammation, immune response, and cytotoxicity due to their strong hydrophobic interactions, whereas functionalized CNTs (*f*-CNTs) showed much less toxicity even to the stem cell population [69] and induced toxicity at doses above 20 $\mu g\ ml^{-1}$ after 24 h. Zhang *et al.* [42] also reported that FLG increased the intracellular generation of ROS and induced mitochondrial injury in neuronal cells after 4 and 24 h at a dose of 10 $\mu g\ ml^{-1}$, whereas for the same concentrations, CNT was better tolerated by cells.

An important nanotoxicity mechanism manifested in a majority of carbon-based nanoparticles is *oxidative stress*, defined as an imbalance between ROS and the physiological antioxidants that protect oxidation-sensitive biological molecules. Oxidative stress has been reported in a variety of target cells following exposure to graphene [42] and CNTs [63, 70, 71], although in some cases, the effect has been attributed to redox-active metal impurities. A useful biomarker for oxidative stress is the depletion of GSH or the ratio of active reduced glutathione (GSH) to its inactive oxidized form, the dimer GSSG. Glutathione is the major endogenous antioxidant produced by cells. It deactivates free radical species and peroxides by donation of H^+ and e^- by the internal cysteine thiol group in reactions that can be accelerated by glutathione peroxidases. GSH is the most abundant intracellular thiol with concentrations in the range of 2–10 mm; GSH is also the major extracellular antioxidant in lung lining fluid, where it protects lung epithelial cells against oxidants and chemical-induced toxicity [72]. It was found that graphene and other carbon nanostructures can deplete intracellular GSH depletion following direct exposure [73]. Graphene can deactivate antioxidants by direct surface reaction involving bound oxygen intermediates [74]. This route can contribute to oxidative stress pathways and toxicity for carbon nanostructures. This aspect of nanotoxicity by graphene was also emphasized by various other reports. For example, Chang *et al.* [40] show that GO exposure induces oxidative stress in A549 cells even at lower concentrations, and with increasing concentrations, ROS stress increases significantly, indicating that ROS stress mechanism is one of the common features of graphene toxicity.

With recent experiments with neural phaeochromocytoma cells, Zhang *et al.* [42] demonstrated that the shape of carbonaceous nanomaterials also plays an extremely important role in how they interact with cells and potentially other biological systems, such as tissues and organisms. Both graphene and single-walled carbon nanotubes (SWCNT) induced concentration- and shape-dependent cytotoxic effects, and these effects were attributed to the generation of ROS in graphene-treated cells. This indicates that, as of now, ROS is a main mediator of graphene-induced toxicity. Very recently, a detailed study on the toxicity mechanisms in pristine versus *f*-G was conducted by Sasidharan *et al.* [18, 27], which is reviewed in the following section.

12.4.4.1 Intracellular ROS and Apoptosis in Macrophages

ROS production is part of a defense mechanism against foreign substances by macrophages and is a marker for its activation. In addition, ROS production and the resulting oxidative damage have been reported as one of the major toxicity mechanisms of CNTs against phagocytic cells such as monocytes and macrophages [75, 76]. Oxidative stress response of *p*-G versus *f*-G in macrophage cell line was studied by flow cytometry and confocal microscopy after 24 h of exposure to graphene. The flow cytogram and confocal microscopic images (Figure 12.12) revealed that while untreated cells (Figure 12.12a1–a3) showed no green fluorescence related to ROS, *p*-G-treated cells (Figure 12.12b1–b3) showed bright green fluorescence, indicating the formation of intracellular

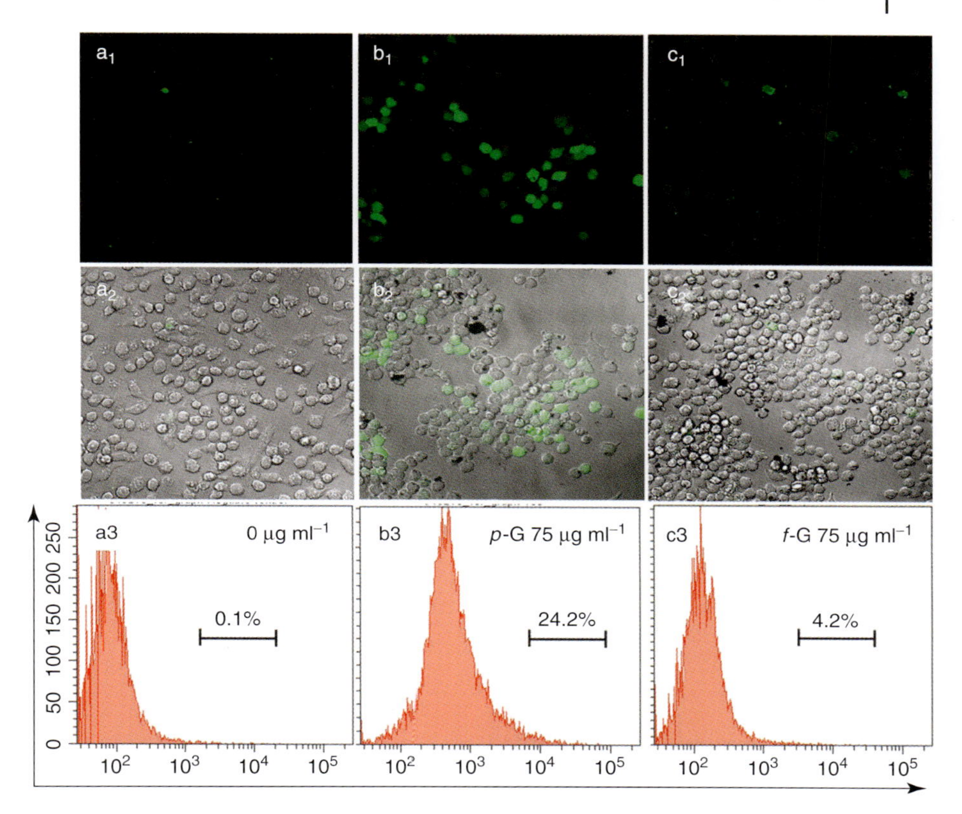

Figure 12.12 Assessment of intracellular ROS. Confocal microscopy and flow cytogram showing the expression of reactive oxygen species (dichlorofluorescein diacetate (DCF)-DA assay) in RAW 264.7 cells treated with: 0 (a_1, a_2, and a_3), 75 μg ml^{-1} *p*-G (b_1, b_2, and b_3), and 75 μg ml^{-1} *f*-G (c_1, c_2, and c_3). (Source: From Ref. [28].)

ROS in ~24.2% of cells. In contrast, *f*-G-treated cells (Figure 12.12c1–c3) virtually exhibited insignificant fluorescence signal (4.2%), suggesting low ROS generation, even though the intracellular uptake of graphene was very high.

To confirm the above-mentioned differential effects by *p*-G and *f*-G, the authors also performed apoptosis/necrosis assay (Figure 12.13). Compared to the control cells, *p*-G-treated cells showed green fluorescence indicating the staining of phosphatidylserine by annexin V, a unique marker for apoptosis. This was found absent in *f*-G-treated cells, which appeared similar to that of untreated cells. This was further confirmed by flow cytometry, where both the control (a_1, a_2, and a_3) and 75 μg ml^{-1} *f*-G-treated (c_1, c_2, and c_3) cells showed no sign of annexin V/PI staining (percentage of cells in Q1, Q2, and Q4 regions), but ~14% of *p*-G-treated cells (b_1, b_2, and b_3) registered apoptosis (Q4). This result is in agreement with the observed effect of *p*-G on metabolic activity (Alamar Blue assay) and induction of

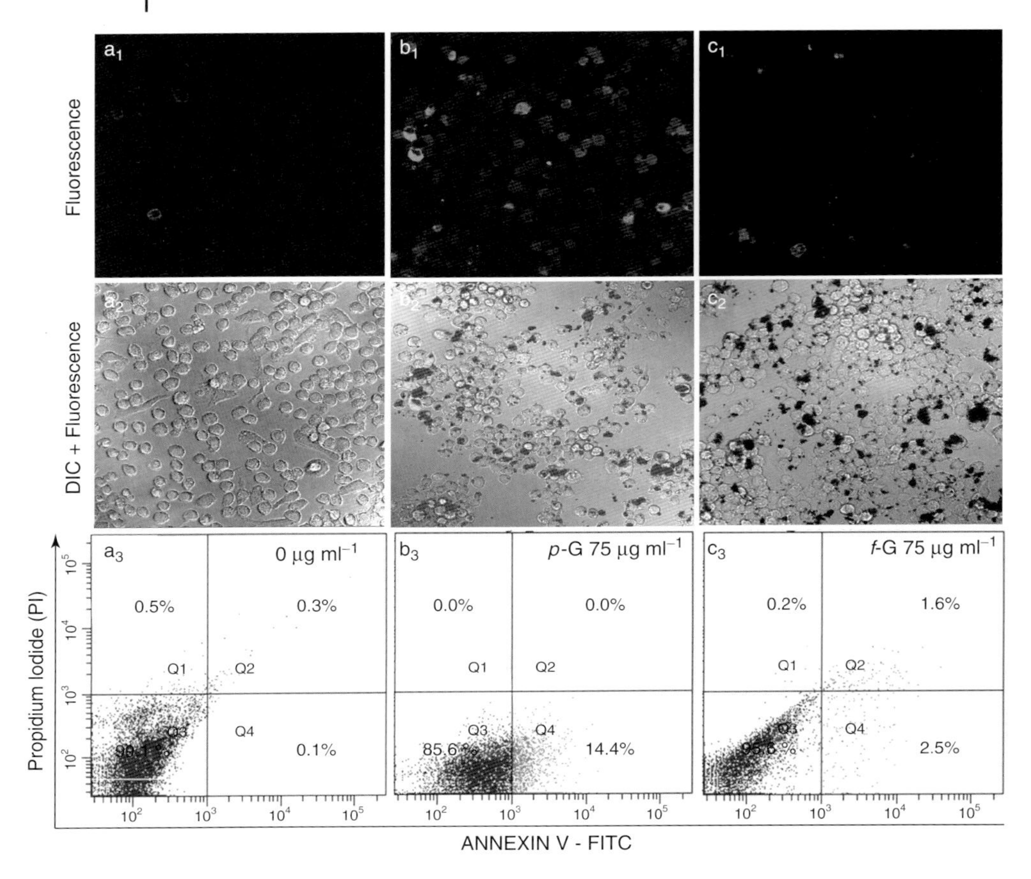

Figure 12.13 Apoptosis assay by annexin V/propidium iodide staining. Confocal images and flow cytogram showing annexin V/PI-stained RAW 264.7 cells treated with: 0 (a_1, a_2, and a_3), 75 µg ml^{-1} *p*-G (b_1, b_2, and b_3), and 75 µg ml^{-1} *f*-G (c_1, c_2, and c_3) samples. FITC, fluorescein isothiocyanate. (Source: From Ref. [27].).

ROS. Collectively, all the above-mentioned results clearly suggest that the surface functionalization has a great impact on pacifying the toxicity of *p*-G.

12.5
Mitigation of Toxicity by Surface Modifications

Although graphene possesses some level of ROS-induced toxicity, it is important to explore the opportunities to pacify the toxic effects by surface modifications or compositing with other biocompatible materials. It appears that graphene causes more toxicity if it is internalized by cells or accumulated on the plasma membrane in large concentrations. This indication which is derived from the reports reveals that if cells are grown on graphene substrates, instead of causing toxicity, cells proliferate and differentiate much better compared to the controls. This is evident from the reports on the behavior of cells grown on graphene films and CNTs [77]. A number of cell function assays such as proliferation, cell morphology, focal adhesion kinase, and gene expression studies by reverse transcriptase-polymerase chain reaction (RT-PCR) experiments reveal that the cells grow much better on graphene films than CNTs. More importantly, the graphene was also found to influence the gene delivery efficacy as the cells grown on the graphene could be transfected 250% greater compared to cells grown on normal cover glass. This study suggests that graphene or carbon nanomaterials can provide biocompatible surface coatings for implants, without inducing notable deleterious effects while enhancing some cellular functions such as gene expression. In the similar work, Fan *et al.* [78] reported better biocompatibility of graphene-reinforced polysaccharide composites using GO–chitosan, which can be used for tissue engineering applications as scaffolds. Graphene–chitosan composite films produced by solution casting shows that with the addition of $\sim$0.1–0.3 wt% of graphene, the elastic modulus increased over 200% while maintaining excellent biocompatibility, tested on L929 cell lines showing adherence and proliferation as good as those of pure chitosan film. In another work on biocompatible nanocomposites of graphene, Park *et al.* [79] have fabricated a strong and biocompatible freestanding paper using chemically reduced graphene and TWEEN 20. This paper exhibited excellent stability in aqueous medium and biocompatibility to mammalian cell lines, indicating possibilities of using this for medical devices such as transplant devices and implants.

It was also found that protein adsorption on graphene plays a major role in pacifying its toxicity [46]. It was believed that the cytotoxicity of GO nanosheets must be arising from direct interactions of cell membrane or proteins. It was shown that this effect could be largely attenuated by treating graphene with 10% bovine serum. This indicates that protein coating may be a potential method to mitigate the toxicity of graphene. In a similar manner, other biocompatible materials were also coated on GO to make it more biocompatible. In one such work, dextran was used to pacify the toxicity of rGO [37]. Dextran is a polysaccharide and one of the most widely used biocompatible biopolymers. It was also shown that dextran can be harnessed as a stabilization agent for graphene. An environmentally friendly approach was presented for the synthesis of biocompatible rGO and rGO-based nanocomposites using dextran as a reducing and stabilization agent for GO.

12.6 *In vivo* Toxicity

Although cell-based *in vitro* toxicity *assays* provide preliminary information about the general toxicity character of nanomaterials, a more realistic evaluation is possible by *in vivo* studies. The *in vivo* behaviors and long-term fate of graphene and its derivatives are not well established so far. To date, very few *in vivo* toxicology studies have been published in peer reviewed journals. Liu's group [80] first studied the toxicological profile of graphene in mice models using ^{125}I-labeled GO-PEG with 10–30 nm lateral dimensions. They have demonstrated that after intravenous administration of 20 mg kg^{-1}, the graphene nanoparticles of size 10–30 nm accumulated mainly in the liver and spleen and there was very negligible accumulation in lungs. After three months, graphene was found to be completely cleared from the animal body, probably from renal as well as hepatic route without causing any toxicity. As the authors have mentioned, although this result is very encouraging, critical issues such as the fate of graphene with relatively larger size >50 nm or different surface chemistry and exact mechanism of clearance are not known. As expected, two other studies have shown the potential *in vivo* toxicity of graphene [81, 82] having larger lateral dimensions. Wang *et al.* [81] studied the concentration-dependent toxicity of intravenously administered 1–2 μM sized monolayer graphene. The results demonstrated that above a dose of 0.4 mg, GO induced severe pathological changes including granuloma formation in the major organs such as liver, spleen, and kidney. In another study, Zhang *et al.* [82] showed the long-term accumulation of GO in lungs. Although, when compared to CNTs, graphene showed extended blood circulation time and relatively less clearance through reticuloendothelial system, significant pathological changes, including inflammation, cell infiltration, pulmonary edema, and granuloma were seen in the lungs at the dosage of 10 mg kg^{-1} of body weight. This suggests that more detailed studies on the long-term effects of graphene in different organs need to be conducted. Effect of alternative surface modification on *in vivo* fate of graphene was also reported very recently. Zhang *et al.* [83] studied dextran-conjugated graphene and found that compared to bare GO, graphene–dextran conjugates show reduced sheet sizes and significantly improved stability in a physiological medium. Cellular experiments revealed that dextran coating on GO offers remarkably reduced cytotoxicity. However, *in vivo*, the GO–dextran (DEX) showed accumulation in the reticuloendothelial system including liver and spleen, and more importantly, shows clearance after a week without causing noticeable toxic effects. This work provides an alternative functionalization method to produce biocompatible graphene conjugates for potential biomedical applications. Later, Duch *et al.* [84] demonstrated that the pulmonary toxicity of GO dramatically reduced in pluronic-dispersed graphene. In this study, the authors did not mention any lateral dimension. Very recently, Donaldson's group [85] administered ~5 μm lateral sized graphene directly into lungs of the mice and studied the possibility of inhalation toxicity. They observed the inflammogenic potential of graphene in lungs as well as the pleural space.

From the above-mentioned preliminary reports, it is almost clear that lateral size and surface chemistry of graphene are major players in determining the *in vivo* toxic effects of graphene. As Yang *et al.* [80] discusses, if the graphene nanoparticles are relatively small, that is, <30 nm and also PEGylated, there is a possibility of safe clearance without major accumulation in various organs. However, the practical challenge is that most of the time, lateral dimensions are not strictly controlled with

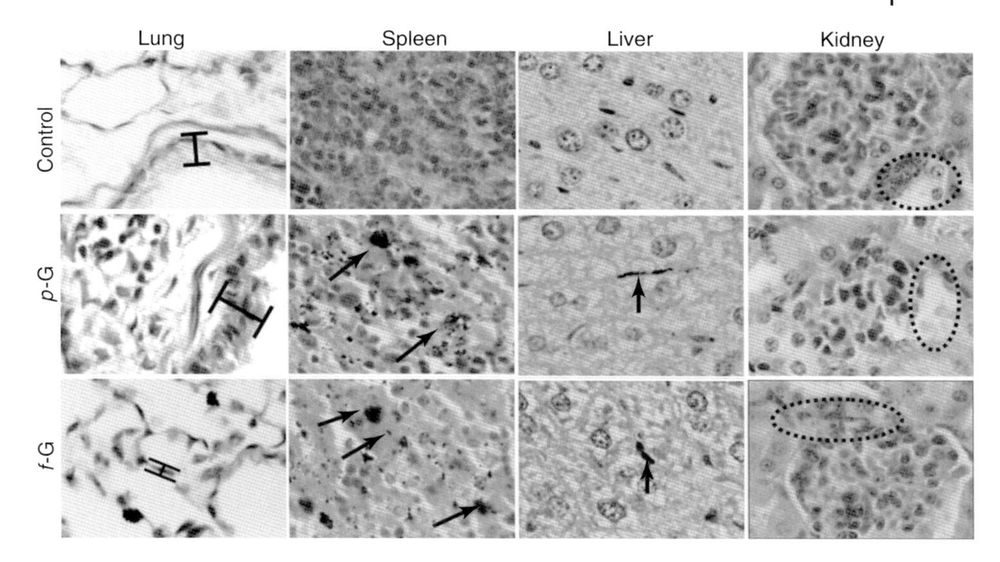

Figure 12.14 Representative H&E images of lung, spleen, liver, and kidney of mice treated with 0.16 mg g^{-1} of pristine and functionalized graphene showing significant cellular and structural damage to all major organs.

this <30 nm regime and will extend over a few hundreds of nanometers. Hence, it is also important to study the *in vivo* effect of relatively larger nanoflakes with different surface chemistries.

In view of the above, very recently, we have studied the acute (one to eight days) and subchronic (three months) toxicities of highly hydrophobic pristine and hydrophilic, carboxyl-functionalized, or PEGylated graphene of size 100–200 nm in mice models after tail vein injection at the dose of ~20 mg kg^{-1}. Biodistribution studies using ^{99}Tc-tagged samples showed extended accumulation in the lung over a period of >96 h, followed by circulation in the body with significant accumulation in all major organs such as liver, spleen, and kidney. Histopathological evaluation and gene expression studies revealed that irrespective of the surface modification, graphene showed significant cellular and structural damage to critical organs such as lungs, liver, spleen, and kidney, which ranged from mild congestion to necrosis and fibrosis even within 24 h, as depicted in Figure 12.14.

Lung sections of untreated control mice and *p*-G- and *f*-G-treated mice are shown in the first column. After 24 h, compared to the control, *p*-G-treated sample shows hyperplasia (marked region) of bronchial epithelium, suggesting that *p*-G has irritated the tissues. The thickness of bronchial epithelium increased from that of one cell layer to three layers, indicating acute inflammation in the lung. Section of *f*-G-treated mice indicates that graphene is getting aggregated not in the bronchial epithelium but inside the alveolar wall causing thickening of the cell wall. Presence of such aggregates in the lung tissue is a matter of concern because the chronic damage may remain permanent.

Spleen is a lymphoid organ involved in the filtration of blood and is the production site of antibodies and activated lymphocytes, making it an important organ in the defense against blood-borne antigens [86]. Any inert particles in blood are actively phagocytized by spleen macrophages. Figure 12.14 shows the normal histology of mouse spleen having clear demarcation between white pulp, marginal zone, and red pulp. However, large accumulation of graphene deposits is seen in *p*-G- and *f*-G-treated animals. In some portions, large aggregates and granular deposits inside the macrophages can be seen. Macrophages appear orange to yellow in color because of the presence of hemosiderin pigment. Presence of such large aggregates in spleen clearly indicates acute immune response by both the types of graphene. This is not a good sign and probably conjugation with PEG may be needed to reduce the endocytosis by macrophages.

Liver is another important and the biggest organ in animal body. Liver is responsible for metabolizing, neutralizing, and eliminating toxic substances in blood. Many intravenously injected nanoparticles or drug molecules are directly deposited in the liver and are metabolized or cleared through reticuloendothelial system. Thus, the interaction of graphene with liver cells and how the liver is able to manage these nonbiodegradable nanoparticles is an important consideration. Figure 12.14 shows the liver histopathology sections of control mice and *p*-G- and *f*-G-treated mice after 24 h. After intravenous injection of graphene in mice, 24 h *p*-G samples show many Kupffer cells containing black granular particles, suggesting the engulfment of graphene nanoparticles. Once graphene reaches the sinusoids through the hepatic artery, resident Kupffer cells might have engulfed these particles. There are no inflammatory changes that can be appreciated at this stage. In contrast, *f*-G samples show clear signature of apoptotic changes (arrow mark) in hepatocytes. It appears that *f*-G is internalized by the hepatocytes and causes cytotoxicity. As we have seen, after intravenous injection, graphene enters the liver through the hepatic artery and gets collected in the sinusoids, where it is engulfed by the Kupffer cells. If graphene particles are larger and cannot pass through fenestrations, they cannot enter into the perisinusoidal space to reach hepatocytes. This may be the case with *p*-G, which is highly hydrophobic and aggregated, whereas the smaller *f*-G particles with hydrophilic character may get suspended very well in blood and pass freely through the fenestrations into the perisinusoidal space and reach the hepatocytes while they are trying to remove high- and low-molecular-weight serum components and nutrients for storage and processing. More detailed investigations will be needed to confirm and identify the actual cause of apoptosis found in G samples.

Interaction of nanoparticles with kidney is another critical component of *in vivo* nanotoxicology as many nanoparticles clear the renal route. Each kidney contains 1–1.4 million functional units called nephrons. Renal corpuscle is the part of the nephron that filters substances from blood into the tubules of nephron. The renal corpuscle is a small mass of capillaries called the glomerulus housed within a bulbous glomerular capsule. Filtrate is produced in the corpuscle when blood plasma is forced under pressure across the filtration membrane of the glomerular capillary wall and through the filtration slits (30–40 nm wide) between the pedicels of podocyte processes. Effectively, this means that nanoparticles of size less than 30 nm can only be cleared through the renal route.

The kidney sections of control mice and *p*-G- and *f*-G-treated mice after 24 h are shown in Figure 12.14. After 24 h, *p*-G sample shows degenerated glomeruli and juxtaglomerular apparatus (circled region). These are the signs of structural damage by way of disappearance of juxtaglomerular apparatus. A number of darkly stained necrotic cells are also seen, possibly because of the blockade of capillaries by graphene leading to necrosis of tissues. There is also dilation of tubules seen in kidney. In case of *f*-G, the damage is more severe as dilated tubules containing eosinophilic deposits represent dead tissue accumulated in tubule, leading to its distension and formation of casts. It can be understood that graphene might have damaged the glomerulus and blocked the circulation, leading to ischemia and necrosis. This necrotic debris is deposited in the tubules. As the particle size is greater than 40 nm, it could not pass through the glomerular filter and hence remained in circulation with repeated encounter with the glomerular filter, leading to significant structural damage. It is possible that this may not happen with graphene with <40 nm size.

Compared to the above organs, interestingly, we have not found any accumulation or damage in the brain and testis. Accordingly, in any of the animal samples that we have studied, we could not see the presence of graphene aggregates or any histopathological changes of inflammation, hyperplasia or hypertrophy in brain. This suggests that relatively larger (200 nm) graphene flakes were not able to penetrate through blood-brain barrier (BBB), unlike the case of more needular CNTs. However, collectively, our results suggest that bilayer graphene having $\sim$200 nm size induces significant toxic effects at the tested concentration of 20 mg kg^{-1} and demands further extensive toxicological studies using smaller particles and lower doses for biomedical applications.

12.7 Potential Application Areas: Opportunities

12.7.1 Drug Delivery

Among many potential biomedical application areas of graphene, the most discussed one is the area of drug delivery. Controlled loading of hydrophobic drugs onto graphene or its functional derivatives by electrostatic interactions, covalent conjugations, or other physisorptions can increase the solubility and stability of drug molecules in physiological medium and aid in improved disease-specific delivery [87]. Adsorption of molecules on carbon surfaces is favored for molecules with low solubility, hydrophobicity, or cationic charge and for molecules with conjugated π-bonds, which impart planarity and facilitate π−π interactions with graphenic carbon surfaces. Most of the attention on small molecule interactions has been related to drug delivery, where graphene flakes and nano-GO are used as carriers for small molecule drugs [87].

GO has been shown to produce a potential platform for the delivery of water-insoluble anticancer drugs such as doxorubicin (Dox) and camptothecin (CPT) [25]. Modification of GO surface with sulfonic acid was used to produce stable suspension of GO in physiological medium and the same is covalently bonded to folic acid (FA) to target folate receptor expressing cancer cells. The FA-conjugated GO was loaded with Dox and CPT by $\pi-\pi$ stacking. This FA-NGO drug (nanoscale graphene oxide) conjugates exhibited specific targeting to folate overexpressing breast cancer cell lines and imparted significantly high toxicity compared to GO-alone or FA-unconjugated counter parts. An important aspect of this chapter is that, this was the first report on loading of multiple drugs to same GO material. Authors proposed that, as the use of drugs in combination is an essential clinical practice, nanodelivery vehicles carrying multiple drug molecules and their targeted delivery have great significance.

In another similar work, Yang *et al.* [88] reported high-efficiency loading and pH-dependent slow release of Dox using GO. It was found that loading of Dox on GO increased almost linearly with the increasing concentration of drug up to 2.35 mg mg^{-1} of GO. The pH-dependent loading and releasing were attributed to the hydrogen-bonding interactions. Fluorescence and electrochemical characterization show that strong $\pi-\pi$ stacking interactions exist between GO and DOX. In another work, Dox was loaded on PEGylated GO, where the PEGylation density is presumed to be low enough that sp^2 domains remain available on the GO surface for drug binding [36]. Dox was reported to be adsorbed by π-stacking and desorbed at reduced pH because of its higher hydrophilicity and high solubility at an intracellular endosomal pH of 5.5. It was expected that other drug molecules with titratable amine groups might also show this higher solubility and partial reversibility at low intracellular pH because of increased positive charge.

In another work, Kakran *et al.* [89] used functionalized GO with biocompatible surfactants such as Pluronic F38, Tween 80, and maltodextrin (MD) for the delivery of hydrophobic anticancer drug, ellagic acid (EA). This was the first report on the loading of EA in graphene-like systems and they have achieved a maximum loading of 1.14 g of drug per gram of GO composites. Among the three, MD-conjugated GO showed maximum drug uptake probably because the larger number of hydroxyl groups and the lower molecular weight of MD molecule led to the greater amount of MD grafted onto GO compared to F38 and T80. The release of the drug from these nanocarriers was studied in water and buffer solutions of pH values 4 and 10 at 37 °C. They have also demonstrated that attachment of EA to the GO did not affect its antioxidant activity. Approximately, 36–38% drug was released within three days at pH 10. The cytotoxicity of EA loaded onto the GO was higher than that of the free drug, as demonstrated in human breast carcinoma cells (MCF7) and human colon sadenocarcinoma cells (HT29).

GO was also studied for sustained antibiotic delivery. Functionalized GO-modified polysebacic anhydride composites were prepared by esterification and loaded with an antibiotic levofloxacin [90]. Release kinetics of drug-loaded GO composites showed much longer drug release duration, and the same could be controlled by changing the molecular weight of the composites. It was shown that ~2% GO-modified polysebacic anhydride exhibited linear release behavior with an extended release time of ~80 days and an effective drug release of >95%. GO was also demonstrated for its potential to deliver photosensitizer (PS) drug molecules used for photodynamic therapy. Huang *et al.* [91] reported FA-conjugated chlorin e6 (Ce6) loaded into GO. PSs are light-sensitive porphyrin-based molecules that can release ROS and induce cell death under light activation. One of the major limitations of PS drugs is their nonspecific accumulation throughout the body because of their lipophilic or hydrophobic nature. Using the same hydrophobic interactions and $\pi-\pi$ stacking, PS could be loaded into GO, between the oxygen functional groups on GO surfaces. Further, folate-conjugated GO was also used for the delivery of Ce6 [92]. Significant reduction in cell viability was reported in human stomach cancer cell line positive for FA receptors, when exposed to FA-GO/Ce6 following irradiation with the appropriate wavelength. In yet another approach, Misra *et al.* [93] demonstrated the efficient delivery of highly hydrophobic anticancer drug tamoxifen using gexfoliated graphene, and showed promising anticancer activity in different cells.

Overall, nano-GO and its functionalized derivatives were found to be effective sorbents for carrying hydrophobic compounds, thanks to the amphiphilicity of GO with its patchwork of oxygenated and unsubstituted graphenic surfaces. Similar to CNTs, the graphene surface having delocalized π-electrons can effectively hold the aromatic anticancer drugs via $\pi-\pi$ stacking. Very high surface area of graphene with all atoms exposed to other molecules allowed for very high drug loading efficiency. In case of PEGylated graphene, the terminals of PEG chains can be used for conjugating with targeting ligands such as antibodies, peptides, or small molecules such as FA. Although found promising for drug delivery applications, there are many challenges and open-ended questions remaining unanswered regarding the possible use of graphene in a clinical setting. The major concern is the toxicity and the fate of the graphene career after the delivery of the drug. We are not sure about the clearance pathways of the drug and its interconnectivity with the lateral size, shape, and surface chemistry. Answers to these questions are possible only by extensive *in vivo* studies and at present not much data is available on these issues.

12.7.2
Gene Delivery

Besides the delivery of drug molecules, another interesting application area of graphene is gene delivery that emanates from its strong interactions with nucleic acids such as DNA and RNA [94]. Graphene shows several unique modes of interaction with nucleic acids, including preferential adsorption of ssDNA over double-stranded (ds) DNA, steric protection of adsorbed nucleotides from nuclease enzymes, and DNA intercalation by small graphene layers [95]. The unique modes of DNA/RNA interaction open up both opportunities such as gene/siRNA delivery as well as challenges such as potential genotoxic effects. In one of the first report on the gene delivery applications of graphene, Feng *et al.* [96] used PEI- (10k and 1.2k molecular weight) composited GO for gene delivery. Cytotoxicity studies reveal that GO-PEI-10k is compatible with treated cells compared to the free PEI of same concentration. The cationic GO–PEI complexes were able to bind with anionic plasmid DNA for intracellular transfection of the enhanced green fluorescence protein, as demonstrated in HeLa cells. Compared to PEI-1.2k, high green fluorescent protein (GFP) transfection was possible with GO-conjugated PEI-1.2k. GO-PEI-10k also showed similar GFP transfection efficiency but lower toxicity compared with PEI-10k. In another similar work by Dong *et al.* [97], a simple nanocarrier of PEI-grafted graphene nanoribbon (GNR) like structure was used to transfect cells. The GNR was formed by a unique method of unzipping multiwalled CNTs, followed by sonication with strong acids to get carboxyl-functionalized GO. The PEI-g-GNR appeared to protect locked nucleic acid; the apoptosis induced by the PEI-g-GNR was negligible under optimal conditions. In another similar work, Chen *et al.* [98] covalently conjugated GO with PEI for siRNA loading. Sequential delivery of siRNA against Bcl-2 and Dox was done by GO–PEI conjugate showing significantly improved cell killing by synergistic effect.

Protecting nucleic acids from DNAse or RNAse of the body fluid is a major challenge for gene and siRNA delivery system. In this regard, Tang *et al.* [99] showed that ssDNA adsorbed on graphene surface was effectively protected from enzymatic cleavage by DNAse I. This was an encouraging result for gene delivery community, considering the major challenge posed by complex cellular and biofluids. Anisotropy, fluorescence, NMR, and CD studies suggest that ssDNA is promptly adsorbed onto the surface of *f*-G forming strong molecular interactions that prevent DNase from approaching the constrained DNA. Although this argument appears quite fragile, constraining a ssDNA probe on graphene showed improvement in the specificity of its response to a target sequence. Considering the low cost of producing graphene on a large scale, these findings will promote the use of graphene in both fundamental research and applied level. In another important work by Lu *et al.* [100], oligonucleotide molecular beacon was found to effectively absorb on GO surface and release in the presence of cDNA, implying that the ssDNA adsorbs preferentially to the ds form. Intracellular desorption of the oligo in the presence of complementary RNA and protection of DNA by GO from enzymatic cleavage were evident in their experiments. It is argued that steric hindrance prevents nucleases from effectively attacking the adsorbed phase DNA.

Another example of strong binding of ssDNA to GO was found in the synthesis of 3D self-assembly of GO and DNA to form multifunctional hydrogels [101]. The dsDNA mixed with GO and heated to 90 °C caused uncoiling to ssDNA and conversion of the mixture to a stable hydrogel. The gelation was attributed to noncovalent ssDNA interactions with GO that bridge the individual sheets. At the level of a single GO sheet, this interaction was related to ssDNA adsorption and its selective adsorption relative to dsDNA. These hydrogels also showed high mechanical strength and environmental stability. In an independent work, the binding affinity of DNA nucleobases and nucleosides with graphene was evaluated by Varghese *et al.* [102]. The interaction energies showed an increase from thymine to cytosine to guanine. Numerical calculations also showed trends in the order of G > A and T > C. These results were similar to those of single-walled CNTs. At high ionic strength and low pH, the adsorption was found to be increased for small oligomers, probably because of the interaction of negative charges on DNA and graphene. Regarding the question about differential interaction of graphene with dsDNA and ssDNA, it is postulated that, in dsDNA, as the bases are protected inside the double helix, the outer charged phosphate groups may show low affinity toward GO surfaces. In ssDNA, the DNA bases can bind to graphene through $\pi-\pi$ stacking and hydrophobic forces. At high ionic strength and low pH where protonation causes reduction in graphene charges, the $\pi-\pi$ interactions and hydrophobicity can overcome repulsive effects of electrostatic interactions.

The strong interaction of GO with DNA was recently used for developing biosensor devices. Sheng *et al.* [103] used a GO platform in an aptamer-based sensing method for the model fungal food contaminant, ochratoxin A. The concept was to selectively bind the toxin with a fluorescently tagged aptamer and use GO to adsorptively quench the unbound aptamer, but not the bound form, which adopts an antiparallel G-quadruplex confirmation that is less prone to adsorption on the planar surface of GO. It was found that achratoxin adsorption limits the sensitivity of the technique. However, polyvinylpyrrolidone (PVP), coating on GO improved the sensitivity. In another integrated sensing system, graphene-mesoporous silica-gold nanoparticle hybrids (GSGHs) were used for DNA detection [104]. It was shown that the new sensing strategy could detect target DNA with a high sensitivity of 10 fM and has a good capability to investigate the single nucleotide polymorphisms. The detection limit for target DNA was nearly five orders of magnitude less than the field effect transistor.

In addition to the above-promising effects of GO interactions with DNA, graphene also caused adverse effects by way of DNA intercalations. It was found that NGO can intercalate between base pairs in dsDNA, in a manner similar to organic compounds such as chemidrugs or genotoxins [95]. It was shown that, in the presence of metallic ions such as Cu^{2+}, GO induced DNA cleavage. It is believed that this property relies on a combination of the planar single-atom thick GO structure, which can be inserted in the molecular spaces between base pairs, and the peripheral carboxyl groups provide binding sites for Cu^{2+}. These two features bring active Cu^{2+} in direct contact with vulnerable target sites in DNA double helix and catalyze the oxidative damage. This indicates potential toxic effects of graphene nanoparticles. But another school of thought is the utility of this toxicity against cancer-like disease, similar to chemodrugs.

12.7.3 Biosensing Using Graphene

Because of the high electrocatalytic activity of graphene, it could be used as an excellent electrode material for oxidase biosensors. There are a number of reports on graphene-based glucose biosensors. Shan *et al.* [105] reported the first graphene-based glucose biosensor with graphene-PEI-functionalized ionic liquid nanocomposite electrode showing wide linear glucose response with excellent reproducibility and stability. In another work, Zhou *et al.* [106] reported glucose biosensor using rGO. This biosensor exhibits substantially increased amperometric signals of wide linear range, very high sensitivity, and low detection limit, <2 mm. The linear range for glucose detection was greater than that for CNTs [107], nanofibers [108], exfoliated graphite platelets [109], and mesoporous carbon [110]. The response at the GOD/rGO/GC electrode to glucose was very fast and stable, which makes them as potential candidates to continuously measure the plasma glucose level for diagnostic purpose. Kang *et al.* [111] employed biocompatible chitosan-composited graphene for glucose sensing. Chitosan was found to facilitate dispersion of graphene and immobilize the enzyme molecules, leading to very high sensitivity (37.93 mA mM^{-1} cm^{-2}) and better stability suitable for measuring glucose. Graphene/metal-nanoparticle-based biosensors have also been reported by Shan *et al.* [105], where chitosan-Au-graphene composites were used for the detection of hydrogen peroxide and oxygen. In a similar manner, Wu *et al.* [110] reported graphene/platinum-nanoparticle/chitosan glucose biosensor with 0.6 mM level detection of glucose. These improved performances were attributed to the large surface area and better electrical conductivity of graphene facilitated by synergistic effect of metallic nanoparticles in graphene mesh.

In another work by Alwarappan *et al.* [112], chemically synthesized graphene nanosheet was used as electrode materials and their electrochemical properties were compared with SWCNTs. It was found that graphene surface possesses greater sp^2 character than the SWCNTs, and it was shown that the conductivity of graphene particles was $\sim$63 mS cm^{-1}, which was $\sim$60 times greater than that with CNTs. Following this, many graphene-based electrochemical sensors for various neurotransmitters (dopamine and serotonin) were studied and compared with those of CNTs. All these studies reported better sensitivity and enhanced signal-to-noise ratio than SWCNTs.

Another interesting area of graphene is electrochemical DNA sensors. Graphene-based systems offer extremely high sensitivity, selectivity, and low operational cost for the detection of DNA sequences, leading to the opportunities for the development of simple, inexpensive, and accurate platform for disease diagnosis. Out of many types of electrochemical DNA sensing, the method of direct oxidation of DNA is the simplest. In this category, Zhou *et al.* [106] demonstrate that signals of the four bases of DNA were separated efficiently using a chemically rGO/GC electrode, indicating that these devices can simultaneously detect freebases and single nucleotide polymorphism without hybridization or additional labeling. Another interesting work by Feng *et al.* [113] describes the use of aptamer-conjugated graphene electrochemical detection of individual cancer cells using clinically validated AS1411 aptamer. AS1411 has high binding affinity and specificity to the nucleolin receptors on the cancer cells and the electrochemical sensor could detect those cells. In a similar cancer cell sensor, Song *et al.* [114] showed selective and quantitative detection of cancer cells using target-directed graphene. FA/conjugated graphene/hemin composite device could detect as low as 1000 cancer cells. Compared to other systems that require antibodies or natural enzymes, this composite is more robust and not susceptible to denaturation or decomposition.

f-G was also used for ultrasensitive electrochemical immunoassay for BRCA1 gene by Cai *et al.* [115]. BRCAl is an antioncogene, which is genetically mutated leading to breast and ovarian cancers. The detection of BRCA1 is therefore important to characterize and screen patients against these two cancer types. In this study, the authors designed a new label and fabricated a novel sandwichtype electrochemical immunoassay for the ultrasensitive detection of BRCA1. Horseradish peroxidase was entrapped in the pores of amino-functionalized mesoporous silica with adjustable pore size from 5 to 30 nm. A secondary antibody was conjugated to porous silica by covalent bond. Ionic liquid was also added to the mixture to increase the electrochemical activity of horse radish peroxidase (HRP) and promote electron transport. The sensitivity of this sandwichtype immunosensor was much higher than that of control devices. Under optimal conditions, this immunoassay device exhibited a wide working range from 0.01 to 15 ng ml^{-1} and detection limit of $\sim$4.86 pg ml^{-1} of BRCA1.

Given the above-promising results from graphene-based biosensors, it can be concluded that biosensing is one of the most promising application areas of graphene. Graphene is not only superior to CNTs but also less expensive for large-scale manufacturing. However, controlled methods of graphene synthesis with sufficient control over the number of layers are a major challenge. In addition, suitable functionalization protocols need to be developed exclusively for enhancing the sensitivity and specificity of the device.

12.7.4 Graphene for Cellular Imaging

Another interesting biomedical application of graphene was the fluorescent labeling of cells using GO loaded with organic fluorophores. Peng *et al.* [45] reported that PEGylated GO conjugated with fluorescence for intracellular imaging. The PEG was used to prevent GO-mediated quenching of fluorescein luminescence. The conjugate thus prepared exhibited excellent fluorescent properties and efficiently taken up by the treated cells, rendering imageability by confocal microscope. In a similar line, Liu *et al.* [116] reported a gelatin *f*-G nanosheets using gelatin as a reducing reagent. Gelatin played a key role in the prevention of aggregation of the graphene nanosheets. The obtained biocompatible gelatin/graphene nanosheets (GNS) composites showed excellent stability in water and various physiological fluids, including cellular growth media and serum. Cellular toxicity test suggested biocompatibility up to 200 mg ml^{-1}. Further, an anticancer drug, Dox, was loaded onto the gelatin/graphene composite sheets by physisorption for cellular imaging and drug delivery. The drug-loaded conjugates exhibited a high toxicity to MCF-7 cells and experienced a gelatin-mediated sustained release *in vitro*, which has the potential advantage of increasing the therapeutic efficacy. In another similar work, Chen *et al.* conjugated fluorescent quantum dots (QDs) with graphene for imaging live cells using optical methods [117]. The challenge of QD fluorescence quenching by graphene was avoided using a bovine serum albumin (BSA) bridge, resulting in a novel highly fluorescent nanoprobe. BSA-capped QDs are firmly grafted onto PEI/poly(sodium 4-styrenesulfonate)-coated reduced graphene oxide using layer-by-layer assembly. The strong luminescence of the graphene-QDs provides a potential for noninvasive optical *in vitro* imaging. The results for the imaging of live cells indicated that the cell-penetrating graphene-QDs could be a promising nanoprobe for intracellular imaging and therapeutic applications. However, toxicity of the QD-graphene system is a major concern and remains to be fully understood.

12.7.5
Graphene for Tissue Engineering

Graphene and its functionalized derivatives could be used to develop materials and surfaces for culturing human cells for tissue engineering applications. Graphene-incorporated scaffolds may offer much better mechanical properties. Unlike the case of CNTs, which also offered excellent mechanical properties with the tissue engineering scaffolds, graphene is relatively free from toxicity issues associated with residual metal impurities. Effectively, the success of graphene-based tissue engineering lies with how compatible graphene is and how good nanocomposites can be prepared by blending with polymers or proteins that support the tissue growth together with the supply of nutrients. Toward the preparation of making graphene/polymer composites, the major challenge was to achieve proper dispersion and maximum interaction between the graphene and polymer matrix. Recently, Liang *et al.* [118] reported the preparation of poly(vinyl alcohol) (PVA) nanocomposites with GO using solution processing method. Efficient load transfer was found between graphene and PVA matrix, and the mechanical properties are found significantly improve with ~76% increase in tensile strength and a ~62% improvement of Young's modulus at 0.7 wt% of GO. In a similar manner, PVA-graphene hydrogels were also reported recently [119]. GO/PVA composite hydrogels were prepared by simple freeze and thaw method. Compared to pure PVA hydrogels, ~132% increase in tensile strength and ~36% improvement of compressive strength were achieved with the addition of 0.8 wt% of GO, which suggests an excellent load transfer between the GO and the PVA. The incorporation of certain amount of GO into composite hydrogels does not affect the biocompatibility of PVA to osteoblast cells. Such an improvement in simple PVA-based graphene composites indicates great promise for the tissue engineering applications. Another material that was extensively studied for tissue engineering was chitosan [120]. Chitosan was extensively studied for drug delivery and tissue engineering applications owing to unique properties such as biocompatibility, antibacterial properties, and easy processability into various geometries and forms such as porous structures. Chitosan can also deliver morphogenic factors and pharmaceutical agents in a controlled manner. Recently, Fan *et al.* [78] reported the preparation of chitosan-graphene composites for tissue engineering applications. Graphene-chitosan films were produced by solution casting and with the addition of very small amount of graphene (0.1–0.3 wt%), the elastic modulus of composite increased to >200%. The cell adhesion results indicated excellent biocompatibility for the composite, as efficient as that of chitosan. This suggests that graphene-chitosan composites are potential candidates as scaffold materials in tissue engineering. The structure-process-property relationship of GO–chitosan composites on the proliferation and growth of osteoblasts was also studied recently by Depan *et al.* [121]. Chitosan-graphene scaffolds were prepared by covalently connecting carboxyl groups of GO and amine groups of chitosan. It was shown that hydrophilic nature, high water retention ability, and high degree of interconnectivity facilitating the movement of growth factors and cells throughout the structure resulted in enhanced cell attachment and proliferation and improved the stability against enzymatic degradation. Direct evidence for graphene-mediated accelerated differentiation of stem cells was also reported recently by Nayak *et al.* [122]. It was shown that graphene does not hamper the proliferation of human mesenchymal stem cells (hMSCs), rather accelerates their specific differentiation into bone cells. The differentiation rate was comparable to the one achieved with common growth factors, demonstrating graphene's potential for stem cell research and tissue engineering. In similar lines, another study on the differentiation of human neural stem cells (hNSCs) into neurons by Park *et al.* [123] showed that graphene substrate could enhance the differentiation of hNSCs into neurons. Furthermore, it was demonstrated that graphene has a unique surface property that can promote the differentiation of hNSCs toward neurons rather than glial cells. Collectively, all the above works indicate potential application of graphene in cell proliferation and differentiation for tissue engineering applications.

12.7.6
Anticancer Therapy: Photothermal Ablation of Cancer

One of the most interesting applications of graphene that gained great attention was its potential to ablate tumor cells and tissues under photoexposure. Being one of the highly absorbing materials of light, graphene can efficiently convert light energy into thermal energy and ablate biomolecules or cellular components. This property was effectively used in tumor cells and tissues *in vitro* and *in vivo*. rGO sheets with high near-infrared (NIR) light absorbance and biocompatibility having ~20 nm in average lateral dimension and functionalized noncovalently by amphiphilic PEGylated reported ~sixfold higher NIR absorption than nonreduced nano-GO [124]. Further, using a targeting peptide-bearing Arg-Gly-Asp (RGD) motif to nano-rGO resulted in selective cellular uptake in cancer cells (U87MG) and highly effective photoablation of cells *in vitro*. In the absence of any NIR irradiation, nano-rGO exhibited little toxicity *in vitro* at concentrations well above the doses needed for photothermal heating. Comparing GO with CNTs for its photothermal properties revealed that GO is better than CNT, despite their lower NIR-absorbing capacity than the latter. The superior photothermal sensitivity of graphene was attributed to their better dispersivity in physiological medium. The mechanisms of graphene-mediated photothermal killing of cancer cells involved oxidative stress and mitochondrial membrane depolarization. This was further substantiated by *in vivo* experiments [125], where ultrahigh accumulation of PEGylated graphene in tumor was observed. Highly efficient tumor ablation was noted after intravenous administration of PEG-graphene at low-power NIR laser irradiation on the tumor. Histology examination of various organs and blood chemistry analysis showed no major adverse effects to the normal functions of mice/organs. Although a lot more efforts are required to further understand the *in vivo* fate and long-term toxicity of graphene, this work demonstrated the success of using this material for efficient *in vivo* photothermal therapy of tumor.

Further efforts to combine the photothermal ablation of graphene with chemotherapy was presented recently by Zhang *et al.* [126] Dox-loaded PEGylated nano-GO was used to combine chemotherapy and photothermal therapy in one system. The ablation of tumor both *in vivo* and *in vitro* using functional GO/PEG-Dox system was demonstrated. Compared with chemotherapy or photothermal therapy alone, the combined treatment demonstrated a synergistic effect, resulting in higher therapeutic efficacy. In a similar work, Tian *et al.* [92] combined photodynamic therapy with photothermal ablation using PS drug-loaded GO. GO-PEG-Ce6 conjugate was prepared and treated with cancer cells for combinatorial photodynamic-thermaltherapy (PDT). Owing to the significantly improved intracellular uptake of the conjugates, remarkably improved cancer cell photodynamic destruction effect was registered compared to free Ce6. More importantly, this study shows that the photothermal effect of graphene can be used to promote the delivery of PS molecules and there by enhance the PDT efficacy against cancer cells.

12.8 Conclusions

In summary, reports suggesting potential biomedical applications of graphene are increasing day by day and in parallel, the concerns regarding potential toxic effects are also rising. In this chapter, we have reviewed literature on many biomedical applications, including drug delivery, gene delivery, tissue engineering, biosensing, and anticancer therapy. The nature of most of these reports remains largely within the arena of "novel concepts" rather than in-depth studies that are needed for assessing the actual translational potential of graphene-based systems. This means that the proposals for biomedical applications of graphene are in their infancy stage and lot more evaluations will be needed for more realistic assessment. However, even at this stage, the available data suggest that the applications requiring only *ex vivo* or *in vitro* use of graphene, such as biosensors, have quick translational potential simply because such devices can be made operational without any direct contact with human body; hence, the regulatory approval may be obtained without much difficulties unlike the case of implantable devices. All other applications requiring *in vivo* administration of graphene demand thorough examination on the short- and long-term toxic effects, including that on carcinogenicity and developmental toxicology.

References

1. Novoselov, K.S., Geim, A.K., Morozov, S.V., Jiang, D., Zhang, Y., Dubonos, S.V. *et al.* (2004) *Science (New York, NY)*, **306** (5696), 666–669.
2. Britnell, L., Gorbachev, R.V., Jalil, R., Belle, B.D., Schedin, F., Mishchenko, A. *et al.* (2012) *Science (New York, NY)*, **335** (6071), 947–950.
3. Stoller, M.D., Park, S., Zhu, Y., An, J., and Ruoff, R.S. (2008) *Nano Lett.*, **8** (10), 3498–3502.
4. Ji, L., Tan, Z., Kuykendall, T., An, E.J., Fu, Y., Battaglia, V. *et al.* (2011) *Energy Environ. Sci.*, **4** (9), 3611.
5. Gomez De Arco, L., Zhang, Y., Schlenker, C.W., Ryu, K., Thompson, M.E., and Zhou, C. (2010) *ACS Nano*, **4** (5), 2865–2873.
6. Kroto, H.W., Heath, J.R., O'Brien, S.C., Curl, R.F., and Smalley, R.E. (1985) *Nature*, **318** (6042), 162–163.
7. Iijima, S. (1991) *Nature*, **354** (6348), 56–58.
8. Mochalin, V.N., Shenderova, O., Ho, D., and Gogotsi, Y. (2012) *Nat. Nanotechnol.*, **7** (1), 11–23.
9. Poland, C.A., Duffin, R., Kinloch, I., Maynard, A., Wallace, W.A.H., Seaton, A. *et al.* (2008) *Nat. Nanotechnol.*, **3** (7), 423–428.
10. Partha, R. and Conyers, J.L. (2009) *Int. J. Nanomed.*, **4**, 261–275.
11. Yang, W.R., Thordarson, P., Gooding, J.J., Ringer, S.P., and Braet, F. (2007) *Nanotechnology*, **18** (41), 412001.
12. Xing, Y. and Dai, L. (2009) *Nanomedicine (London, England)*, **4** (2), 207–218.
13. Feng, L. and Liu, Z. (2011) *Nanomedicine*, **6** (2), 317–324.
14. Pati, S.K., Pati, S.K., Enoki, T., and Rao, C.N.R. (2011) *Graphene and its Fascinating Attributes*, World Scientific Publishing. ISBN# ISBN: 13-978-981-4329-35-4.
15. Geim, A.K. and Novoselov, K.S. (2007) *Nat. Mater.*, **6** (3), 183–191.
16. Subrahmanyam, K., Vivekchand, S., Govindaraj, A., and Rao, C. (2008) *J. Mater. Chem.*, **18** (13), 1517–1523.
17. Rao, C.N.R., Sood, A.K., Subrahmanyam, K.S., and Govindaraj, A. (2009) *Angew. Chem. Int. Ed.*, **48** (42), 7752–7777.
18. Sasidharan, A., Panchakarla, L.S., Chandran, P., Menon, D., Nair, S., Rao, C.N.R. *et al.* (2011) *Nanoscale*, **3** (6), 2461–2464.
19. Prato, M., Kostarelos, K., and Bianco, A. (2008) *Acc. Chem. Res.*, **41** (1), 60–68.
20. Dreyer, D.R., Park, S., Bielawski, C.W., and Ruoff, R.S. (2010) *Chem. Soc. Rev.*, **39** (1), 228–240.
21. Graphene Oxide (2010) *Based Mater.*, 1–13.
22. Park, S., An, J., Jung, I., Piner, R.D., An, S.J., Li, X. *et al.* (2009) *Nano Lett.*, **9** (4), 1593–1597.
23. Lotya, M., King, P.J., Khan, U., De, S., and Coleman, J.N. (2010) *ACS Nano*, **4** (6), 3155–3162.
24. Otsuka, H., Nagasaki, Y., and Kataoka, K. (2003) *Adv. Drug Deliv. Rev.*, **55** (3), 403–419.
25. Liu, Z., Robinson, J.T., Sun, X., and Dai, H. (2008) *J. Am. Chem. Soc.*, **130** (33), 10876–10877.
26. Wojtoniszak, M., Chen, X., Kalenczuk, R.J., Wajda, A., £apczuk, J., Kurzewski, M. *et al.* (2012) *Colloids Surf. B, Biointerfaces*, **89**, 79–85.
27. Sasidharan, A., Panchakarla, L.S., Sadanandan, A.R., Ashokan, A., Chandran, P., Girish, C.M. *et al.* (2012) *Small*. **8** (8), 1251–1263.
28. Laaksonen, P., Kainlauri, M., Laaksonen, T., Shchepetov, A., Jiang, H., Ahopelto, J. *et al.* (2010) *Angew. Chem. Int. Ed. Engl.*, **49** (29), 4946–4949.
29. Bhunia, S.K. and Jana, N.R. (2011) *ACS Appl. Mater. Interfaces*, **3** (9), 3335–3341.
30. Wang, Y., Li, Z., Hu, D., Lin, C., Li, J., and Lin, Y. (2010) *J. Am. Chem. Soc.*
31. Bonanni, A., Ambrosi, A., and Pumera, M. (2012) *Chemistry (Weinheim an der Bergstrasse, Germany)*, **18** (6), 1668–1673.
32. Muszynski, R., Seger, B., and Kamat, P.V. (2008) *J. Phys. Chem. C*, **112** (14), 5263–5266.

33. Liu, F., Choi, J.Y., and Seo, T.S. (2010) *Chem. Commun (Cambridge, England)*, **46** (16), 2844–2846.
34. Husale, B.S., Sahoo, S., Radenovic, A., Traversi, F., Annibale, P., and Kis, A. (2010) *Langmuir*, **26** (23), 18078–18082.
35. Quintana, M., Spyrou, K., Grzelczak, M., Browne, W.R., Rudolf, P., and Prato, M. (2010) *ACS Nano*, **4** (6), 3527–3533.
36. Sun, X., Liu, Z., Welsher, K., Robinson, J., Goodwin, A., Zaric, S. *et al.* (2008) *Nano Res.*, **1** (3), 203–212.
37. Kim, Y., Kim, M., and Min, D. (2011) *Chem. Commun.*, **47** (11), 3195.
38. Kostarelos, K., Lacerda, L., Pastorin, G., Wu, W., Wieckowski, S., Luangsivilay, J. *et al.* (2007) *Nat. Nanotechnol.*, **2** (2), 108–113.
39. Lacerda, L., Raffa, S., Prato, M., Bianco, A., and Kostarelos, K. (2007) *Nano Today*, **2** (6), 38–43.
40. Chang, Y., Yang, S.T., Liu, J.H., Dong, E., Wang, Y., Cao, A. *et al.* (2010) *Toxicol. Lett.*, **200** (3), 201–210.
41. Liao, K.-H., Lin, Y.-S., Macosko, C.W., and Haynes, C.L. (2011) *ACS Appl. Mater. Interfaces*, **3** (7), 2607–2615.
42. Zhang, Y., Ali, S., Dervishi, E., Xu, Y., Li, Z., Casciano, D. *et al.* (2010) *ACS Nano*. **4** (6), 3181–3186.
43. Kam, N.W.S., Liu, Z., and Dai, H. (2006) *Angew. Chem. Int. Ed. Engl.*, **45** (4), 577–581.
44. Kam, N., Liu, Z., and Dai, H. (2006) *Angew. Chem. Int. Ed.*, **45** (4), 577–581.
45. Peng, C., Hu, W., Zhou, Y., Fan, C., and Huang, Q. (2010) *Small*, **6** (15), 1686–1692.
46. Hu, W., Peng, C., Lv, M., Li, X., Zhang, Y., Chen, N. *et al.* (2011) *ACS Nano*, **5** (5), 3693–3700.
47. Schipper, M.L., Nakayama-Ratchford, N., Davis, C.R., Kam, N.W.S., Chu, P., Liu, Z. *et al.* (2008) *Nat. Nanotechnol.*, **3** (4), 216–221.
48. Hernandez, Y., Nicolosi, V., Lotya, M., Blighe, F.M., Sun, Z., De, S. *et al.* (2008) *Nat. Nanotechnol.*, **3** (9), 563–568.
49. Zhu, Y., Murali, S., Cai, W., Li, X., Suk, J., Potts, J. *et al.* (2010) *Adv. Mater.*, **22** (35), 3906–3924.
50. Li, Y., Liu, Y., Fu, Y., Wei, T., Le Guyader, L., Gao, G. *et al.* (2012) *Biomaterials*, **33** (2), 402–411.
51. Hirano, S., Kanno, S., and Furuyama, A. (2008) *Toxicol. Appl. Pharmacol.*, **232** (2), 244–251.
52. Dobrovolskaia, M.A., Aggarwal, P., Hall, J.B., and McNeil, S.E. (2008) *Mol. Pharm.*, **5** (4), 487–495.
53. Yu, T., Malugin, A., and Ghandehari, H. (2011) *ACS Nano*, **5** (7), 5717–5728.
54. Bihari, P., Holzer, M., Praetner, M., Fent, J., Lerchenberger, M., Reichel, C.A. *et al.* (2010) *Toxicology*, **269** (2–3), 148–154.
55. Semberova, J., De Paoli Lacerda, S., Simakova, O., Holada, K., Gelderman, M., and Simak, J. (2009) *Nano Lett.*, **9** (9), 3312–3317.
56. Lacerda, S.H., Semberova, J., Holada, K., Simakova, O., Hudson, S.D., and Simak, J. (2011) *ACS Nano*, **5** (7), 5808–5813.
57. Radomski, A., Jurasz, P., Alonso-Escolano, D., Drews, M., Morandi, M., Malinski, T. *et al.* (2005) *Br. J. Pharmacol.*, **146** (6), 882–893.
58. Singh, S., Singh, M., Nayak, M., Kumari, S., Shrivastava, S., Gracio, J. *et al.* (2011) *ACS Nano*, **5** (6), 4987–4996.
59. Shvedova, A.A., Kisin, E.R., Mercer, R., Murray, A.R., Johnson, V.J., Potapovich, A.I. *et al.* (2005) *Am. J. Physiol., Lung Cell. Mol. Physiol.*, **289** (5), L698–L708.
60. Dobrovolskaia, M.A. and McNeil, S.E. (2007) *Nat. Nanotechnol.*, **2** (8), 469–478.
61. Liu, X. and Sun, J. (2010) *Biomaterials*, **31** (32), 8198–8209.
62. Gojova, A., Guo, B., Kota, R.S., Rutledge, J.C., Kennedy, I.M., and Barakat, A.I. (2007) *Environ. Health Perspect.*, **115** (3), 403–409.
63. Shvedova, A.A., Kisin, E., Murray, A.R., Johnson, V.J., Gorelik, O., Arepalli, S. *et al.* (2008) *Am. J. Physiol., Lung Cell. Mol. Physiol.*, **295** (4), L552–L565.

64. Zolnik, B.S., González-Fernández, A., Sadrieh, N., and Dobrovolskaia, M.A. (2010) *Endocrinology*, **151** (2), 458–465.
65. Nygaard, U.C., Hansen, J.S., Samuelsen, M., Alberg, T., Marioara, C.D., and Løvik, M. (2009) *Toxicol. Sci.*, **109** (1), 113–123.
66. Mitchell, L.A., Lauer, F.T., Burchiel, S.W., and McDonald, J.D. (2009) *Nat. Nanotechnol.*, **4** (7), 451–456.
67. Raffa, V., Ciofani, G., Vittorio, O., Riggio, C., and Cuschieri, A. (2010) *Nanomedicine (London, England)*, **5** (1), 89–97.
68. Lacerda, L., Ali-Boucetta, H., Herrero, M., Pastorin, G., Bianco, A., Prato, M. *et al.* (2008) *Nanomedicine (London, England)*, **3** (2), 149–161.
69. Zhu, L., Chang, D.W., Dai, L., and Hong, Y. (2007) *Nano Lett.*, **7** (12), 3592–3597.
70. Manna, S.K., Sarkar, S., Barr, J., Wise, K., Barrera, E.V., Jejelowo, O. *et al.* (2005) *Nano Lett.*, **5** (9), 1676–1684.
71. Magrez, A., Kasas, S., Salicio, V., Pasquier, N., Seo, J.W., Celio, M. *et al.* (2006) *Nano Lett.*, **6** (6), 1121–1125.
72. Jones, D.P., Brown, L.A., and Sternberg, P. (1995) *Toxicology*, **105** (2–3), 267–274.
73. Pichardo, S., Gutiérrez-Praena, D., Puerto, M., Sánchez, E., Grilo, A., Cameán, A.M. *et al.* (2012) *Toxicol. In Vitro*. **26** (5), 672–677.
74. Liu, X., Sen, S., Liu, J., Kulaots, I., Geohegan, D., Kane, A. *et al.* (2011) *Small*, **7** (19), 2775–2785.
75. Brown, D.M., Donaldson, K., and Stone, V. (2004) *Respir. Res.*, **5**, 29.
76. Brown, D., Kinloch, I., Bangert, U., Windle, A., Walter, D., Walker, G. *et al.* (2007) *Carbon*, **45** (9), 1743–1756.
77. Ryoo, S.R., Kim, Y.K., Kim, M.H., and Min, D.H. (2010) *ACS Nano*, **4** (11), 6587–6598.
78. Fan, H., Wang, L., Zhao, K., Li, N., Shi, Z., Ge, Z. *et al.* (2010) *Biomacromolecules*, **11** (9), 2345–2351.
79. Park, S., Mohanty, N., Suk, J., Nagaraja, A., An, J., Piner, R. *et al.* (2010) *Adv. Mater.*, **22** (15), 1736–1740.
80. Yang, K., Wan, J., Zhang, S., Zhang, Y., Lee, S.T., and Liu, Z. (2011) *ACS Nano*, **5** (1), 516–522.
81. Wang, K., Ruan, J., Song, H., Zhang, J., Wo, Y., Guo, S. *et al.* (2010) *Nanoscale Res. Lett.*, **6** (8).
82. Zhang, X., Yin, J., Peng, C., Hu, W., Zhu, Z., Li, W. *et al.* (2010) *Carbon*.
83. Zhang, S., Yang, K., Feng, L., and Liu, Z. (2011) *Carbon*, **49** (12), 4040–4049.
84. Duch, M.C., Budinger, G.R., Liang, Y.T., Soberanes, S., Urich, D., Chiarella, S.E. *et al.* (2011) *Nano Lett.*, **11** (12), 5201–5207.
85. Schinwald, A., Murphy, F.A., Jones, A., Macnee, W., and Donaldson, K. (2012) *ACS Nano*, **6** (1), 736–746.
86. Kuma, V., Abbas, A.K., Fausto, N., and Mitchell, R. (2007) *Robbins Basic Pathology*, 8th edn, Saunders/Elsevier.
87. Zhang, L., Xia, J., Zhao, Q., Liu, L., and Zhang, Z. (2010) *Small*, **6** (4), 537–544.
88. Yang, X., Wang, Y., Huang, X., Ma, Y., Huang, Y., Yang, R. *et al.* (2011) *J. Mater. Chem.*, **21** (10), 3448.
89. Kakran, M., Sahoo, N.G., Bao, H., Pan, Y., and Li, L. (2011) *Curr. Med. Chem.*, **18** (29), 1–10.
90. Gao, J., Bao, F., Feng, L., Shen, K., Zhu, Q., Wang, D. *et al.* (2011) *RSC Adv.*, **1** (9), 1737.
91. Huang, P., Xu, C., Lin, J., Wang, C., Wang, X., Zhang, C., Zhou, X., Guo, S., and Cui, D. (2011) *Theranostics*, **1**, 240–250.
92. Tian, B., Wang, C., Zhang, S., Feng, L., and Liu, Z. (2011) *ACS Nano*, **5** (9), 7000–7009.
93. Misra, S.K., Kondaiah, P., Bhattacharya, S., and Rao, C.N.R. (2012) *Small*. **8** (1), 131–143.
94. Siwy, Z.S. and Davenport, M. (2010) *Nat. Nanotechnol.*, **5** (10), 697–698.
95. Ren, H., Wang, C., Zhang, J., Zhou, X., Xu, D., Zheng, J. *et al.* (2010) *ACS Nano*, **4** (12), 7169–7174.
96. Feng, L., Zhang, S., and Liu, Z. (2011) *Nanoscale*, **3** (3), 1252.
97. Dong, H., Ding, L., Yan, F., Ji, H., and Ju, H. (2011) *Biomaterials*, **32** (15), 3875–3882.

98. Chen, A.M., Zhang, M., Wei, D., Stueber, D., Taratula, O., Minko, T. *et al.* (2009) *Small*, **5** (23), 2673–2677.
99. Tang, Z., Wu, H., Cort, J.R., Buchko, G.W., Zhang, Y., Shao, Y. *et al.* (2010) *Small*, **6** (11), 1205–1209.
100. Lu, C.-H., Zhu, C.-L., Li, J., Liu, J.-J., Chen, X., and Yang, H.-H. (2010) *Chem. Commun. (Cambridge, England)*, **46** (18), 3116–3118.
101. Xu, Y., Wu, Q., Sun, Y., Bai, H., and Shi, G. (2010) *ACS Nano*, **4** (12), 7358–7362.
102. Varghese, N., Mogera, U., Govindaraj, A., Das, A., Maiti, P.K., Sood, A.K. *et al.* (2009) *ChemPhysChem*, **10** (1), 206–210.
103. Sheng, L., Ren, J., Miao, Y., Wang, J., and Wang, E. (2011) *Biosens Bioelectron.*, **26** (8), 3494–3499.
104. Du, Y., Guo, S., Qin, H., Dong, S., and Wang, E. (2011) *Chem. Commun. (Cambridge, England)*, **48** (6), 799–801.
105. Shan, C., Yang, H., Han, D., Zhang, Q., Ivaska, A., and Niu, L. (2010) *Biosens Bioelectron.*, **25** (5), 1070–1074.
106. Zhou, M., Zhai, Y., and Dong, S. (2009) *Anal. Chem.*, **81** (14), 5603–5613.
107. Liu, G. and Lin, Y. (2006) *Electrochem. Commun.*, **8** (2), 251–256.
108. Wu, L., Zhang, X., and Ju, H. (2007) *Biosens. Bioelectron.*, **23** (4), 479–484.
109. Shao, Y., Wang, J., Wu, H., Liu, J., Aksay, I., and Lin, Y. (2010) *Electroanalysis.* **22** (10) 1027–1036.
110. Wu, H., Wang, J., Kang, X., Wang, C., Wang, D., Liu, J. *et al.* (2009) *Talanta*, **80** (1), 403–406.
111. Kang, X., Wang, J., Wu, H., Aksay, I.A., Liu, J., and Lin, Y. (2009) *Biosens. Bioelectron.*, **25** (4), 901–905.
112. Alwarappan, S., Liu, C., Kumar, A., and Li, C.-Z. (2010) *J. Phys. Chem. C*, **114** (30), 12920–12924.
113. Feng, L., Chen, Y., Ren, J., and Qu, X. (2011) *Biomaterials*, **32** (11), 2930–2937.
114. Song, Y., Qu, K., Zhao, C., Ren, J., and Qu, X. (2010) *Adv. Mater. (Deerfield Beach, Fla)*, **22** (19), 2206–2210.
115. Cai, Y., Li, H., Du, B., Yang, M., Li, Y., Wu, D. *et al.* (2011) *Biomaterials*, **32** (8), 2117–2123.
116. Liu, K., Zhang, J., Cheng, F., Zheng, T., Wang, C., and Zhu, J.J. (2011) *J. Mater. Chem.*, **21** (32), 12034.
117. Chen, M., Liu, J., Hu, B., and Wang, J. (2011) *Analyst*, **136** (20), 4277.
118. Liang, J., Huang, Y., Zhang, L., Wang, Y., Ma, Y., Guo, T. *et al.* (2009) *Adv. Funct. Mater.*, **19** (14), 2297–2302.
119. Bai, H., Li, C., Wang, X., and Shi, G. (2010) *Chem. Commun. (Cambridge, England)*, **46** (14), 2376–2378.
120. Di Martino, A., Sittinger, M., and Risbud, M.V. (2005) *Biomaterials*, **26** (30), 5983–5990.
121. Depan, D., Girase, B., Shah, J.S., and Misra, R.D.K. (2011) *Acta Biomater.*, **7** (9), 3432–3445.
122. Nayak, T.R., Andersen, H., Makam, V.S., Khaw, C., Bae, S., Xu, X. *et al.* (2011) *ACS Nano*, **5** (6), 4670–4678.
123. Park, S., Park, J., Sim, S., Sung, M., Kim, K., Hong, B. *et al.* (2011) *Adv. Mater.*, **23** (36), H263–H267.
124. Robinson, J.T., Tabakman, S.M., Liang, Y., Wang, H., Casalongue, H.S., Vinh, D. *et al.* (2011) *J. Am. Chem. Soc.*, **133** (17), 6825–6831.
125. Yang, K., Zhang, S., Zhang, G., Sun, X., Lee, S.-T.T., and Liu, Z. (2010) *Nano Lett.*, **10** (9), 3318–3323.
126. Zhang, W., Guo, Z., Huang, D., Liu, Z., Guo, X., and Zhong, H. (2011) *Biomaterials*, **32** (33), 8555–8561.

Index

Graphene: Synthesis, Properties, and Phenomena, First Edition. Edited by C. N. R. Rao and A. K. Sood.